Renewable Energy Sources and Emerging Technologies

Third Edition

D.P. KOTHARI

Chairman, Board of Governors
THDC Institute of Hydropower Engineering and Technology
Tehri, Uttarakhand

Former Vice-Chancellor
VIT University, Vellore

Former Director-in-Charge
Indian Institute of Technology Delhi

K.C. SINGAL

Former Chief Engineer
Haryana State Electricity Board

RAKESH RANJAN

Vice-Chancellor
Himgiri Zee University, Dehradun

Former Director–Principal
International Institute of Technology and Business
Sonepat, Haryana

PHI Learning Private Limited

Delhi-110092

2025

In fond memory of **Shri Asoke K. Ghosh** (October 1942 – February 2024), Founder Chairman
and Managing Director of PHI Learning, whose vision endlessly inspires.

The Legacy Continues....

Published by Pushpita Ghosh, PHI Learning Private Limited, Rimjhim House, 111, Patparganj
Industrial Estate, Delhi-110092 and Printed by Star Print-O-Bind, F-31, Okhla Industrial Area
Phase 1, New Delhi-110020

₹695.00

RENEWABLE ENERGY SOURCES AND EMERGING TECHNOLOGIES, Third Edition
D.P. Kothari, K.C. Singal, and Rakesh Ranjan

ISBN-978-93-89347-89-0 (Print Book)
ISBN-978-93-89347-90-6 (e-Book)

The export rights of the book are vested solely with the publisher.

To my wife **Shobha**

— D.P. Kothari

To my wife **Brij Bala**

— K.C. Singal

To my wife **Jyoti**

— Rakesh Ranjan

CONTENTS

3. Solar Radiation and Its Measurement   49–69

4. Solar Thermal Energy Collectors   70–105

5. Solar Thermal Energy Conversion Systems **106–135**

6. Solar Photovoltaic System   **136–164**

10. Geothermal Energy   252–268

11. Electric Power Generation by Ocean Energy   269–321

PREFACE

Encouraging responses and constructive feedback from the readers have motivated us to Come up with the third edition of the book. Since, 2008 and 2011 when the first and second editions were published respectively, many advancements have taken place in the area of energy. While technological development has improved the efficiency of Energy Conversion Systems (ECS) from renewable resources, Energy Conservation is also given due importance. The authors have tried to incorporate such changes in this revised and updated third edition.

In the third edition, chapters have been revised thoroughly and up-to-date data are presented to give real time picture to students and faculty. Supercritical technology for coal-fired thermal generation, National Power Tariff Policy of India are added in Chapter 1. Global Environmental Awareness, International Emission Trading (IET), Paris Conference details with National Electricity Policy (NEP) of India-13th plan are added in Chapter 2. Advancement in Solar Photo-Voltaic System (SPS) with several case studies are added in Chapter 6. Global Small Hydro Power scenario *vis-à-vis* Indian situation with case studies makes the chapter more comprehensive. Chapter on biomass energy has been thoroughly revised with contemporary data. Readers will find the Hybrid Energy System chapter informative with interesting data. Appendices have been modified with the recent data.

The book is intended to serve as an introductory text for the subject of energy and environment for all branches of Engineering and Environmental Sciences.

We are indebted to our colleagues, the students and professors for their invaluable feedback to improve the second edition. We also thank PHI personnel and our families for their help and support during the preparation of the revised manuscript.

At the end, we thank all who directly or indirectly contributed during the revision of the book. We welcome constructive criticism to improve it further.

D.P. Kothari
K.C. Singal
Rakesh Ranjan

PREFACE TO THE SECOND EDITION

The overwhelming response from students and faculty, their feedback and demand to include chapters on Environment and Hybrid Systems motivated us to publish this second revised edition. Since the appearance of the first edition in 2008, some advances have taken place in the area of energy. The same have now been incorporated in this edition.

The second edition also includes two new chapters. A chapter on Hybrid Systems has been added to introduce contemporary practices in Renewable Technologies and fill the gap of various universities syllabus on related subject. In the chapter on Environment, an attempt has been made to cover the contents of the mandatory subject 'Environmental Sciences' for all undergraduate courses. Besides, the Appendices are provided to equip students with the recent topics such as, Smart Grid and Grid Systems in India, Remote Village Electrification (RVE) with renewable energy sources, and Indian Electricity Act 2003 which support exploration of Renewable Energy to ensure life supporting environment and energy security.

The book is intended to serve as an introductory text for the subject of energy and environment of all branches of Engineering and Environmental Sciences.

As renewable energy is a growing field, data for national and international achievements are updated. We have endeavoured to equip each chapter with MNRE annual reports.

We are indebted to our colleagues, the students and professors for their invaluable feedback to improve the first edition for their encouragement and various useful suggestions.

We also thank PHI personnel and our families who supported us during this period and given all possible help so that this book could see the light of the day.

We welcome any constructive criticism of the book and will be grateful for any appraisal by the readers.

D.P. Kothari
K.C. Singal
Rakesh Ranjan

PREFACE TO THE FIRST EDITION

Energy is a vital input for economic growth in agriculture and industry. Fossil fuels are depleting fast due to over-exploitation, besides increasing the environmental protection costs. Search for renewable energy sources and their technology development is of paramount importance to have a balanced and buoyant environment for better quality of life. Energy supply from renewable sources is therefore an essential part of every country's strategy, especially when there is a serious threat of environment degradation and challenge for maintaining sustainability of fossil fuels.

India is perhaps the only country having a full-fledged Ministry of New and Renewable Energy (MNRE) sources. It shows our commitment and importance attached to the development of the renewable energy sources. Creating awareness in young generation about renewable sources of energy is therefore our dictum to write this book. Renewable energy is an interdisciplinary subject and so requires a special effort to develop the diverse and competent manpower as well. For technical development, MNRE has established solar energy centres, and a centre for wind energy technology, etc. Infrastructure established by BHEL (PV solar, thermal and wind), IIT Delhi (biomass characterization), IIT Roorke (micro-hydel projects), IIT Bombay (testing of gasifiers) and Tata Energy Reserach Institute (TERI) are also some of the steps towards the development of renewable energy sources in India.

The development of renewable energy sources and their technologies is a subject of well-proven technical and economical importance worldwide. This book will be useful for the higher-level courses in all undergraduate energy programmes and multidisciplinary postgraduate courses in Science and Engineering. Further, since many practising energy professionals would not have had a general training in renewable energy, this book will be of immense benefit to them. Hence, this book can be used for understanding the basic principles and applications of renewable energy sources by students as well as practising professionals. Essential and useful references are cited in the book for further perusal of studies in this area.

Important basic principles are revised at the end of every chapter in the form of review questions, and numerical problems are included wherever required in the chapters to correlate theory and typical practical values. Economical considerations to harness energy from renewable sources are discussed and serious efforts have been made to highlight the present state of technologies with a view to emphasizing the importance of developing renewable energy sources as cost-effective power generation alternatives.

Inputs from various universities, ministries, government and non-government organizations were sought and authors wish to thank all of them for their cooperation. Mr. K.C. Singal, one of the authors, appreciates the supportive role played by his two sons Shri Ravi Kant IAS and Shri Ashwini Kumar IAS in updating the technical inputs during the preparation of the manuscript. Gratitude is also due to all our colleagues at the Centre for Energy Studies (CES), IIT Delhi, VIT, Vellore, and Institute of Technology and Management, Gurgaon who have helped us directly or indirectly in completing the book. All the three authors thank their respective families for their patience and encouragement shown in completion of this book.

The first author would like to thank Hon. Chancellor Shri G. Viswanathan for his constant encouragement for completing this project.

We would thankfully welcome constructive suggestions and comments for further improvement of the book.

D.P. Kothari
K.C. Singal
Rakesh Ranjan

ENERGY RESOURCES AND THEIR UTILIZATION

1.1 A PERSPECTIVE

Life on the planet earth is the manifestation of energy. The origin of fire, heat and light is energy. It is required to grow food grains which enable humans and animals to survive and work. Energy causes the great universal movement of the earth on its axis and around the sun.

The term 'energy' can be described as 'capacity to do work'. In early days, human beings used their own strength in carrying loads and collecting their food, and later started depending on natural energy sources like the power of falling water used for grinding corn and wind energy for sailing boats. In industry, initially the energy source was fire that used to be obtained by burning wood. Subsequently, wood became a source of charcoal that was used to extract metals from ores.

The fossil fuels were exploited as surface deposits of asphalt, peat and coal, oil from surface seepage, and gas venting from underground reservoirs. The widespread use of petroleum began during the 20th century, particularly for cars and buses, aeroplanes and industries. The use of energy got enhanced with the invention of electricity and development of electric energy generating stations, consuming either fossil fuels or potential energy of water. The Second World War ended in 1945 with the invention and use of nuclear energy.

1.2 CONSERVATION AND FORMS OF ENERGY

The phrase 'conservation of energy' was coined and made popular by German physicists Helmholtz and Joule. They demonstrated that energy could not be annihilated but only be transformed. The following is a review of different forms of energy and their conversion from one form to another.

Kinetic energy: The energy of an object in motion is called 'kinetic energy'. If the mass of an object is m and the object is moving with a velocity v, its kinetic energy in joules is expressed as: $KE = (1/2)mv^2$, where m is in kg and v in m/s.

Potential energy: The energy which a body possesses as a result of its position in the earth's gravitational field is called 'potential energy' and is expressed in joules as: $PE = mgh$, where the mass m is in kg, g is the acceleration due to gravity in m/s^2, and h is the height in metre.

Heat energy: Heat is an intrinsic energy of all the combustible substances. It is the kinetic energy of molecules. Heat energy, for example, can cause gases to expand, drive engines and raise the temperature of water.

Chemical energy: Chemical energy is tied up in fossil fuels such as coal, oil and natural gas. Fossil fuels are used to generate electricity, power vehicles and railway engines. Chemical energy in the food helps us to sustain our life.

Radiant energy: Solar radiation is the manifestation of radiant energy that is received on the earth. Radio waves, X-rays, infrared and ultraviolet electromagnetic radiations contain radiant energy.

Electrical energy: Electrical energy arises out of the arrangement of movement of electrons to produce heat, magnetic field and electromagnetic radiations. It is a highly versatile form of energy, and can be easily converted to other forms for utilization.

Nuclear energy: Matter can be changed into energy when larger atoms are split into smaller ones (atomic fission) or when smaller ones combine to form larger atoms (atomic fusion). Albert Einstein established that the amount of energy produced, when matter disappears, is governed by the equation $E = mc^2$, where E represents the energy generated, m is the loss in mass of the system and c is the velocity of light (3×10^8 m/s). This equation explains the energy released when U^{235} nucleus undergoes fission in a nuclear reactor. It is also the basis when a deuterium and tritium (^2_1H and ^3_1H) fuse in a thermonuclear reaction to release huge amount of energy.

Energy and Mass: Relation between energy and mass is given by

$$E = mc^2$$

It depicts that mass can be converted into energy or mass and energy are equivalent.

Till now, this has been a hypothesis, but recently France's National Centre for Scientific Research have set down the calculations for estimating the mass of protons and neutrons, the particles at the nucleus of atoms. Particle physics explains, protons, and neutrons comprise smaller particles known as quarks which in turn are bound by gluons.

The mass of gluons is zero, and the mass of quarks is only 5%. Then, where is the balance 95%? The answer is; it comes from the energy, the movement, and interactions of quarks and gluons. Thus, energy and msss are equivalent.

Green power: Electrical power generated from renewable energy sources is called "Green Power". Green Power is environmental friendly and non-polluting. Though, environmental assessment is necessary to safeguards utilization of Land area.

Renewable sources are sun (solar), wind, small hydropower and biomass. Green power development in India has been largely driven by private sector investments and has flourished on imports. At present India is advancing to 'Make in India' programme and is expected that indigenous technology and local industries may caters the countries demand at competitive cost. India's wind industry has developed and benefitted because most of its components are manufactured in India.

Renewable energy certificates (green certificates or green tags) needs to be encouraged with incentives to support consumers and corporate sector.

1.3 ELECTRIC ENERGY FROM CONVENTIONAL SOURCES

Thermal plants (coal, oil, gas) nuclear and hydropower stations are the major conventional methods of generating electrical energy. Rise in the cost of fossil fuels has created an urgency to conserve these fuels, and engineers across the world are looking for alternative renewable sources of energy. A few such sources of energy being experimented are: solar, wind, and biomass. The conventional source of energy is also called finite energy and the difference between the finite energy source and the renewable energy source is shown in Figure 1.1.

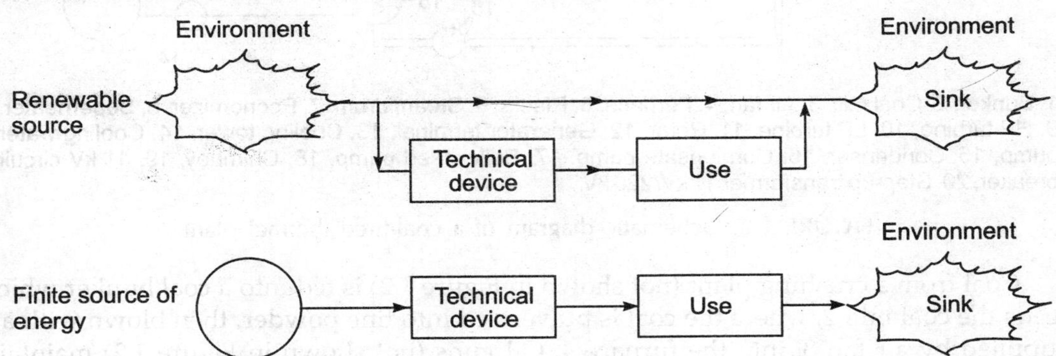

FIGURE 1.1 Generation of power from environmental sources of energy.

From Figure 1.1, it may be observed that the environmental source of energy is tapped by technical devices to generate power without disturbing the balance of environments. Different energy sources, conventional and renewables, in the context of our electric energy requirements are briefly dealt with here.

1.3.1 Thermal Plants (Coal Fuelled)

In thermal plants, the chemical energy of coal is first transformed into mechanical energy and then into electrical energy. The thermodynamic cycle performed in a steam power unit comprises the supply of coal to a furnace to generate steam in the boiler, then expansion of steam in a turbine which acts as the prime mover and drives the generator installed on the same shaft. The spent steam from the turbine is cooled in a condenser. The condensate so produced is re-circulated into the boiler drum by high-pressure pumps and the steam/water cycle is repeated. A schematic installation of a coal-fired thermal plant is shown in Figure 1.2.

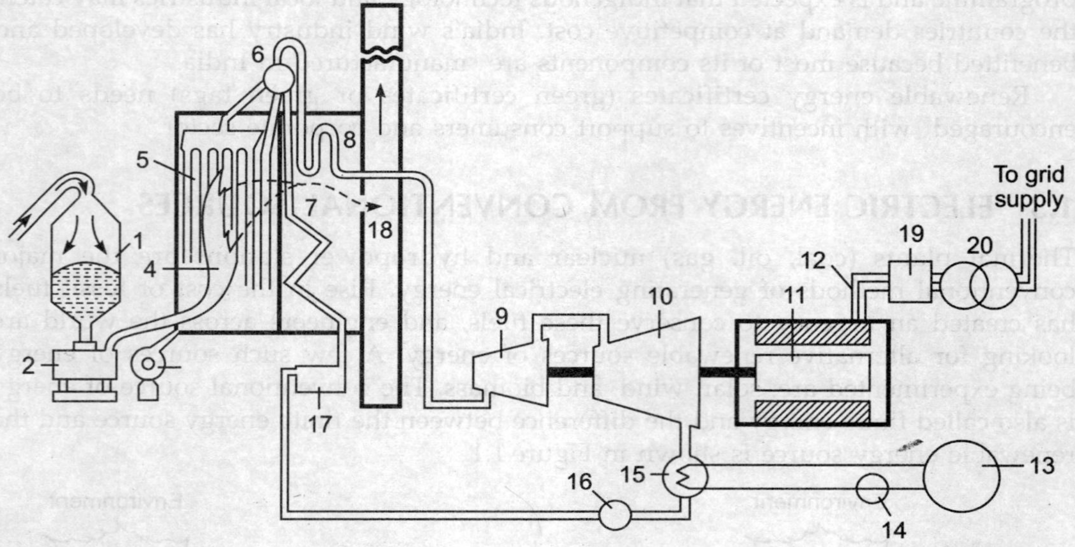

1. Bunker, 2. Coal mill, 3. Air fan, 4. Furnace, 5. Risers, 6. Steam drum, 7. Economizer, 8. Superheater, 9. HP turbine, 10. LP turbine, 11. Rotor, 12. Generator terminal, 13. Cooling tower, 14. Cooling water pump, 15. Condenser, 16. Condensate pump, 17. Boiler feed pump, 18. Chimney, 19. 11 kV circuit breaker, 20. Step-up transformer 11 kV/220 kV.

FIGURE 1.2 Schematic diagram of a coal-fired thermal plant.

Coal from a crushing plant (not shown in Figure 1.2) is fed into a coal bunker which feeds the coal mill 2, where the coal is pulverized into fine powder, then blown with air supplied by air fan 3, into the furnace 4. Oil guns (not shown in Figure 1.2) maintain the flame and the coal burns to produce heat in the furnace. Demineralized water, flowing in furnace wall tubes called risers 5, is turned into steam. The water is fed into boiler drum 6 by the boiler feed pump 16, through economizer 7. The high pressure steam collected in the boiler drum, is allowed to pass over superheater 8, installed in flue gas path to increase steam temperature. Superheated high pressure steam flows into the High Pressure (HP) turbine 9 and then into the Low Pressure (LP) turbine 10. Both turbines revolve thus converting steam energy into mechanical energy, and rotate rotor 11 of the generator thus producing electrical energy. The spent steam moves from LP turbine to condenser 15 where it condenses. Cooling is carried out by water circulation through cooling tower and pump 14. Condensate is the demineralized water and re-circulated back into boiler drum 6. Demineralization improves the efficiency of the thermal plant and to avoid scaling of boiler tubes and drum, demineralized water is used for boiler filling and steam production. Demineralized water carries no hardness, contains silica less than 0.01 ppm, and all ionised salts are also removed.

The generated electric energy is tapped from the stator terminals at 11 kV, stepped up to 220 kV by transformer 20 as shown in Figure 1.2 and finally fed into the grid.

The combustion of coal generates flue gases and ash. Flue gases pass through electrostatic precipitators which trap the particulate matter that drops down on the ground. The ash gets collected at the bottom of the furnace, which is removed. Flue gases then escape to the atmosphere through the chimney.

The overall efficiency of a thermal plant is low (about 35%) due to high heat losses in combustion gases, and a large quantity of heat is rejected to the condenser. The operation of a steam power station is based on the modified Rankine cycle, modified to cover superheating of steam, feed water heating and steam reheating to reduce heat losses. The approach is to increase the thermal efficiency by raising the temperature (540°C) and pressure (155 kg/cm^2) of steam entering HP turbine.

Another feature of thermal plant is economy in fuel consumption and reduced level of pollution. Minimum oil and coal consumption required for a 210 MW set is 3.35 ml/kWh and 764 g/kWh respectively. The larger units more than 250 MW require less fuel per kWh and have low heat and ash pollution.

The super thermal power stations at Singrauli (UP), Korba (Chhattisgarh) and Talcher (Orissa) have installed 500 MW units. The only constraint with larger units is the high intensity disturbance to the power system, when such a large machine trips.

1.3.2 Integrated Gasification Combined Cycle (IGCC) Power Generation

Considering the limited reserve of crude oil and natural gas, the coal provides an economical solution for enhanced generation capacity. However, the environmental pollution and poor efficiency of combustion are the two major challenges to the power generation system. Thermal plants use the Pulverized Coal Combustion (PCC) technology that pollutes the environment and has low efficiency up to 35%.

The Integrated Gasification Combined Cycle (IGCC) is an emerging technology that addresses the efficiency and environmental needs of the 21st century. In the IGCC system, a boiler is replaced with a gasifier, where the coal is gasified in a controlled supply of air/oxygen. The product is fuel gas (the mixture of $CO + H_2$) which is passed through a clean-up stage where the particulate matter and polluting compounds (sulphur and nitrogen) are removed as detailed in Figure 1.3.

The fuel gas is then burnt in the combustion chamber of the gas turbine generator for power generation. The heat absorbed by water from the gas cooler and the heat exhaust is used to generate steam which is fed into the steam turbine for power generation. This combined cycle concept is based on thermodynamic cycles—the Brayton gas turbine cycle and the Rankine steam cycle. The Brayton cycle operates at very high temperature and discharges exhaust gases with high heat content. The Rankine cycle operates with steam and is ideally suited for recovering waste heat from the gas turbine exhaust. When operating together, the combined cycle recovers more energy from the fuel than either of the cycles working alone.

The total cycle efficiency in the IGCC system is higher (38–45%) compared to 35% for the conventional steam cycle, and more importantly, any grade of coal can be gasified with low emission level of CO_2 making it an environmentally advantageous programme.

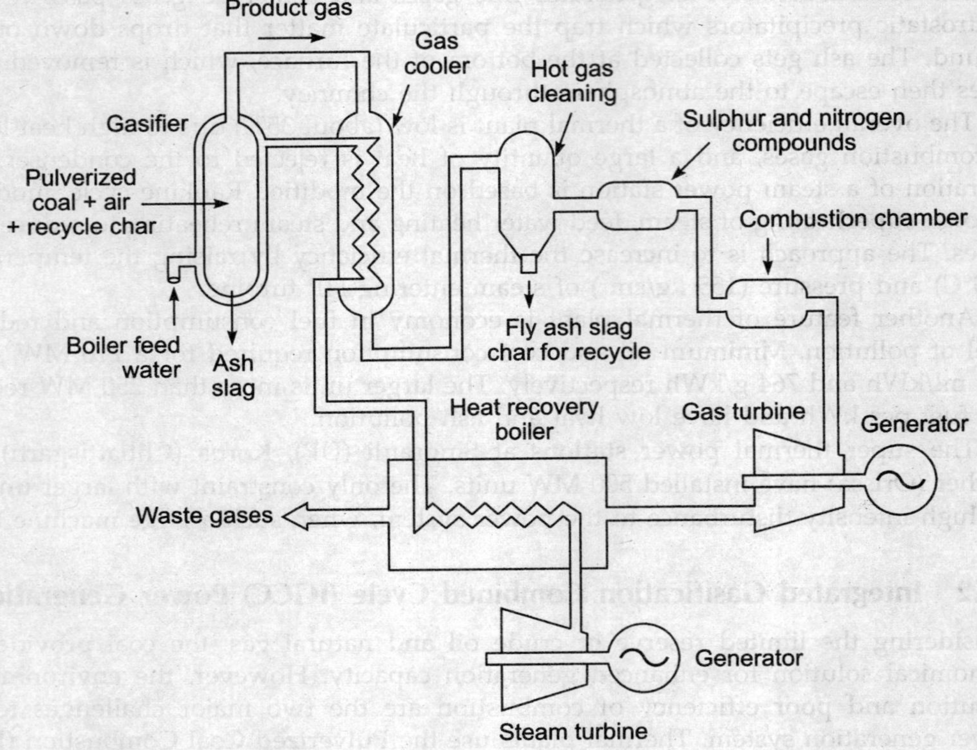

FIGURE 1.3 Process schematic for IGCC power plant.

1.3.3 Super-critical Technology for Coal Fired Thermal Generator

Super-critical technology enhance the efficiency of coal fired thermal generator by bringing down GHG emissions. This breakthrough achieved with advancement in metallurgy, availability of better material, adoption of higher parameters of pressure and temperature beyond critical points. Super-critical technology is now used in India.

Initial super critical units in the country were based on steam parameters of 246 kg/cm^2; 535/565°C at turbine inlet, but recently higher steam temperatures of 565/593°C resulted in higher efficiency as detailed in Table 1.1.

TABLE 1.1 Efficiency gains for various super-critical parameters

Unit size (MW)	Pressure (kg/cm^2)	Temperature (°C)	Gross Design Efficiency (%)
660/800	246	535/565	39.56
660/800	246	565/593	40.24
660/800	246	600/600	40.56

By adopting higher parameters, efficiency gain of about 2% is reported over conventional sub-critical units. Efficiency gain of 2% may appears small but it reduces emissions of CO_2, NO_x and SO_x by over 20% and about 4% of coal saving.

The first super-critical unit of 660 MW was commissioned in December 2010 at Mudra, Gujrat. This initiative has led to UN Framework on Climate Change to certify the installation with "Clean Development Mechanism".

1.3.4 Gas-Turbine Plant

With improved availability of natural gas (methane), short gestation gas turbine plants may become an alternative solution to meet power shortage. Gas combustion products build up high temperature and pressure and are mixed with hot air to operate a gas turbine. An additional advantage of the gas turbine plant is that the exhaust of the gas turbine contains high heat content which is sufficient to raise the steam to operate a turbogenerator. This is called the Combined Cycle Gas-Turbine (CCGT) plant, shown schematically in Figure 1.4.

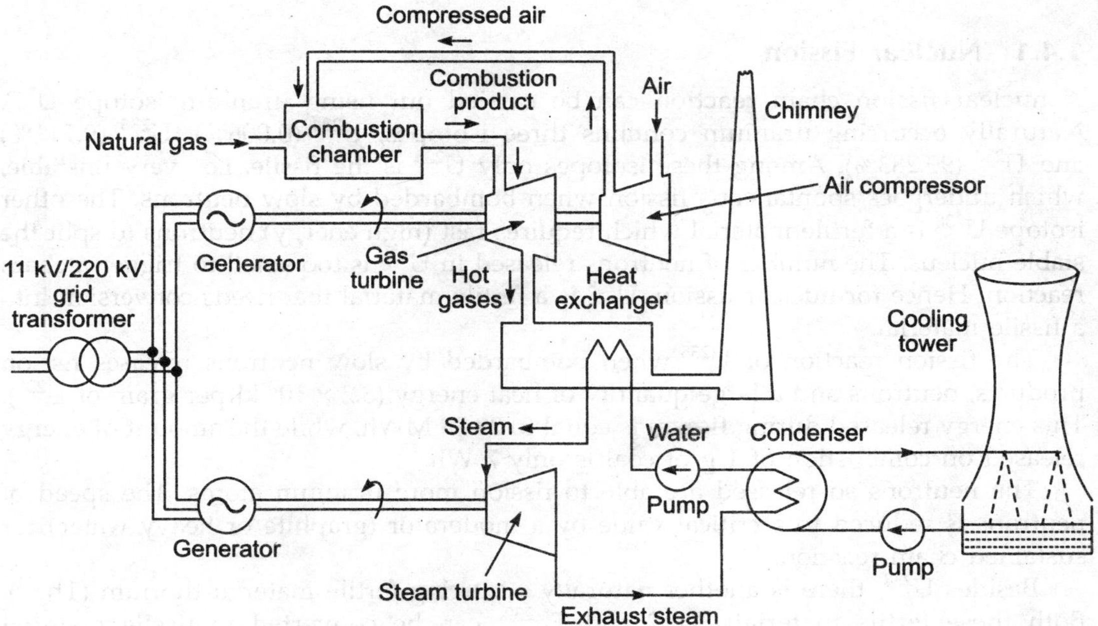

FIGURE 1.4 Combined cycle gas-turbine plant.

The CCGT plant can be started from cold condition; it takes about three minutes for gas turbine and 20 minutes for the steam turbine. It has overall system efficiency of about 45% against 35% with a steam turbine alone. Gas power plants are eco-friendly as they do not increase air pollution.

Gas power projects are suitable to meet the peak load demand of an area. Seven CCGT projects are in operation since 2003 in India (Table 1.2).

TABLE 1.2 CCGT projects in India

Project	Capacity (MW)
Dadri (UP)	817
Auraiya (UP)	652
Anta (Rajasthan)	413
Faridabad (Haryana)	430
Kawas (Gujarat)	645
Gandhar (Gujarat)	648
Kayamkulam (Kerala)	350

1.4 NUCLEAR POWER

With the fast depletion of oil and natural gas, an alternative source of large-scale electric energy generation is the nuclear energy. To generate nuclear energy, two nuclear reactions, fission and fusion, can be used.

1.4.1 Nuclear Fission

A nuclear fission chain reaction can be carried out using uranium isotope U^{235}. Naturally occurring uranium contains three isotopes, U^{234} (0.006%), U^{235} (0.711%) and U^{238} (99.283%). Among these isotopes only U^{235} is the fissile, i.e., very unstable, which undergoes spontaneous fission when bombarded by slow neutrons. The other isotope U^{238} is a fertile material which requires fast (high energy) neutrons to split the stable nucleus. The number of neutrons released in U^{238} is too small to trigger a chain reaction. Hence for nuclear fission, U^{238} is a fertile material that needs conversion into a fissile material.

The fission reaction of U^{235} when bombarded by slow neutrons releases fission products, neutrons and a large quantity of heat energy (8.2×10^7 kJ per gram of U^{235}). This energy released during fission is equal to 22.78 MWh, while the amount of energy released on combustion of 1 g of coal is only 7 Wh.

The neutrons so released are able to fission more uranium atoms. The speed of neutrons is reduced to a critical value by a moderator (graphite or heavy water) for sustained chain reaction.

Besides U^{238}, there is another naturally occurring fertile material thorium (Th^{232}). Both these fertile materials, U^{238} and Th^{232}, can be converted to fissile material plutonium 239 (Pu^{239}) and uranium 233 (U^{233}) respectively, by neutron bombardment. These reactions are shown in Figures 1.5(a) and (b).

The neutrons generated by the fission reaction serve two activities. First they convert a fertile material into fissile material and then sustain the fission reaction for the fissile material so formed. These two reactions are called breeder reactions as they produce more fissile material than they consume in the process. The nuclear reactor in which such a reaction occurs is called a breeder reactor. (It is as if fuel is breeding.)

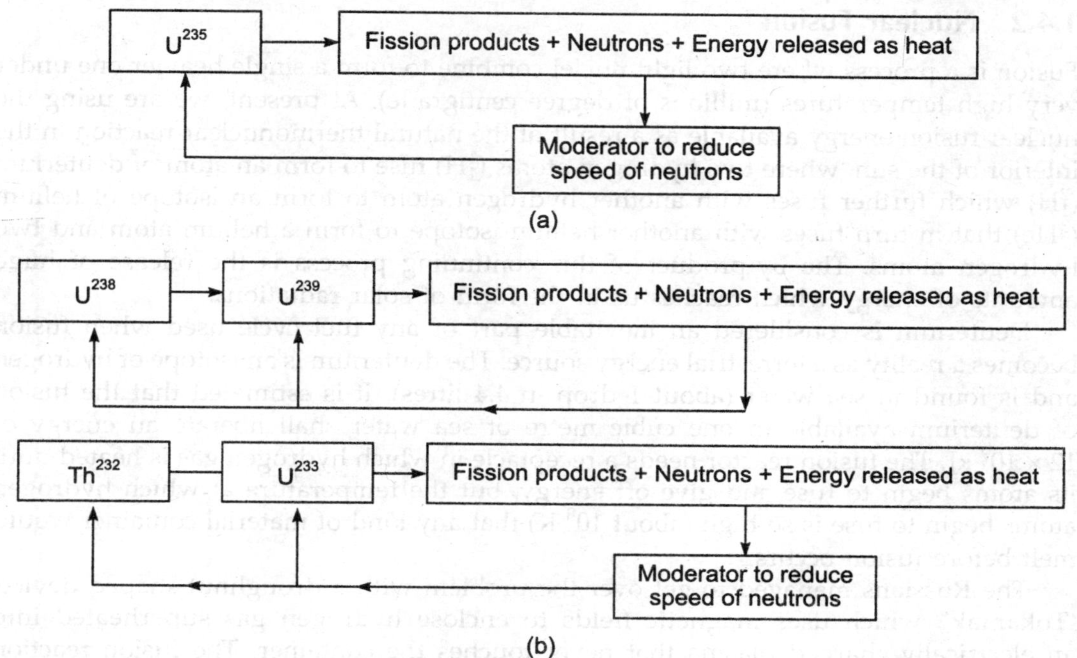

FIGURE 1.5   (a) Fission reaction of U^{235}, and (b) breeder reactions for U^{238} and Th^{232}.

In case of U^{238} to Pu^{239} breeders, neutrons are not slowed down by a moderator, so they are known as Fast Breeder Reactors (FBRs). According to estimates the natural supply of fissile element will last for decades and that of fertile elements for hundred years provided that fast breeder reactors (FBRs) are used.

The FBR technology is being intensely developed to extend the availability of nuclear fuel for several centuries. In India, a Fast Breeder Test Reactor (FBTR) of 40 MW (Thermal) and 13.2 MW (Electrical) was commissioned at Kalpakkam (Chennai) in 1985. Another 500 MWe prototype FBR is under construction.

For breeding operation in FBR, the conversion ratio (fissile material produced/fissile material consumed) is greater than unity. It is made possible by fast moving neutrons, and the moderator is not required. For cooling the reactor, liquid sodium is used as the coolant and the reactor uses the mixed plutonium–uranium carbide as the fuel. With FBR technology in use, it is expected that the cost of electrical energy will reduce near to that of conventional thermal plants. The development of FBR technology carries an advantage for India as there are abundant thorium reserves of over 500,000 tonnes in the monazite beach sands of Kerala.

Spent fuel from uranium fuelled reactors is reprocessed to obtain Pu^{239}. The plutonium is used in fast breeder reactors to breed a uranium isotope U^{233} from thorium, which can then be used in an Advanced Heavy Water Reactor (AHWR) to generate electric power.

1.4.2 Nuclear Fusion

Fusion is a process where two light nuclei combine to form a single heavier one under very high temperatures (millions of degree centigrade). At present, we are using the nuclear fusion energy available as a result of the natural thermonuclear reaction in the interior of the sun, where two hydrogen atoms ($^1_1$H) fuse to form an atom of deuterium ($^2_1$H) which further fuses with another hydrogen atom to form an isotope of helium ($^3_2$He) that in turn fuses with another helium isotope to form a helium atom and two hydrogen atoms. The by-product of this continuing process is the release of huge amounts of energy which reaches us in the form of solar radiations.

Deuterium is considered an inevitable part of any fuel cycle used when fusion becomes a reality as a terrestrial energy source. The deuterium is an isotope of hydrogen and is found in sea water (about 1 drop in 4.4 litres). It is estimated that the fusion of deuterium available in one cubic metre of sea water shall liberate an energy of 12×10^9 kJ. The fusion reactor needs a receptacle in which hydrogen gas is heated until its atoms begin to fuse and give off energy, but the temperature at which hydrogen atoms begin to fuse is so high (about 10^8 K) that any kind of material container would melt before fusion occurs.

The Russians managed to get over the problem with a droughnut-shaped device, 'Tokamak', which uses magnetic fields to enclose hydrogen gas superheated into an electrically charged plasma that never touches the container. The fusion reaction between the hydrogen isotopes, deuterium and tritium follows the pattern given in Figure 1.6.

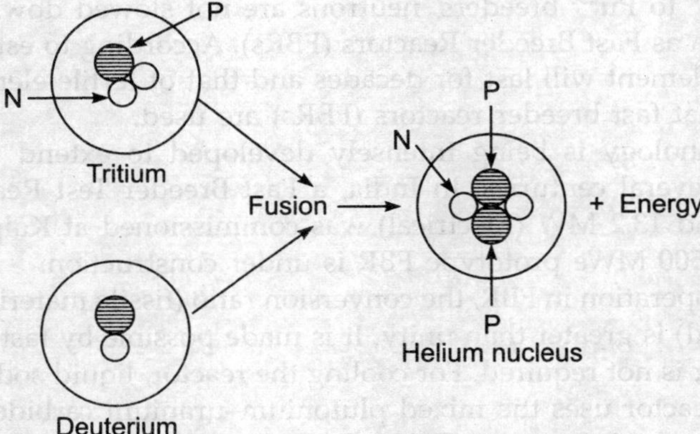

FIGURE 1.6 Generation of energy during fusion reaction.

Further research is in progress to develop a fusion reactor to conduct a controlled thermonuclear reaction, by a group of countries including the European Atomic Energy Community, China, Japan, South Korea, Russia and the USA. The group is holding a $ 5 billion experimental project at France for an International Thermonuclear Experimental Reactor (ITER) based on nuclear fusion technologies. The Government of India has decided to participate in this project. This is the world's latest effort to secure cheap nuclear energy. The objective of ITER is to demonstrate the feasibility of producing electricity from a fusion reaction at an extremely high temperature inside a giant electromagnetic ring.

Deuterium, the major fuel to operate the reactor, will be extracted from sea water which is inexhaustible. If this project succeeds it will result in a potential new source, for cheap supply of energy, replacing oil and gas. The advantages of energy production by nuclear fusion are:

- Unlimited energy can be produced without any risk of atmospheric pollution and catastrophic climate change.
- The nuclear device shall be safe against accidents at nuclear power plants in operation.
- Disposal of radioactive waste will not involve the long time scales that are required in case of conventional fission nuclear power.

1.5 ENERGY RESERVES OF INDIA

1.5.1 Coal

Coal is the prime source of energy in India and accounts for 55% of commercial energy requirement. According to a recent Geological Survey of India report, India has total coal reserves of 519,020 million tonnes. Of this, 73% is of non-coking quality used for power generation and the balance 27% is the coking variety suitable for metallurgical processes.

Coal production which was 211.73 million tonnes in 1990–91 rose to 360 million tonnes in 2002–03. India is now the second largest coal producer in the world, ranking behind China.

India's "Energy Security" is based on coal, as coal is the primary source of fuel. The present per capita primary energy consumption in India is 479 kg oe/year (kg oil equivalent per year), which is quite low. 637 kg oe/year (in 2014) which shows that there is still a large margin for even higher energy consumption in the coming years. With limited petroleum and natural gas reserves, eco-conservation constraints on hydro-electric projects and geo-political approach for nuclear power, coal will continue to occupy the centre-stage of India's energy scenario.

1.5.2 Oil

Crude oil is extracted from oil production wells, drilled either on-shore or off-shore. It is then refined in refineries to obtain petrol, diesel, kerosene, furnace oil, lubricating oil, paraffins, asphalt, tar, etc. In addition, fuel gases like butane, propane and methane, LPG, LNG are also the important products.

India is the world's third largest energy consumer, but produces only 25% of the 141 million tonnes of petroleum products it consumes. Some important information regarding oil reserves is presented as follows:

Reserves and production

Balance recoverable reserves of crude oil in the country have been declining, falling from 806 million tonnes in 1991 to 732 million tonnes in 2001–02. Crude oil production in 2021 stood at 2.51 million tonnes.

Oil and gas discovery

During the year 2019–2020, ONGC made 12 discoveries (7 onland, 5 offshore). 10 discoveries were made in 2020-2021 [3 onland, 7 offshore]. Bengal Basin is the 8th sedimentary basin of India from which hydrocarbon is commercially produced.

Refineries

The oil refining capacity of 18 refineries in the country has reached around 250 million tonnes per annum which meets the present requirement. However, the demand is growing by 5% annually. It was planned to increase the refining capacity by 26.33 MT by 2007. Further capacity addition envisaged during the 11th Five-Year Plan (2007–2012) is 24 MT by setting up new refineries and 32 MT by expansion of existing refineries.

The MoPNG (Ministry of Petroleum and Natural Gas) monitors the entire chain of activities in the oil industry, i.e., exploration and production of crude oil and natural gas, refining, distribution, exports and imports of crude oil and petroleum products.

EIL in year 2020 managed 89 refineries, 44 oil and gas processing plants, over 40 offshore process, 50 pipeline and 10 petroleum companies.

1.5.3 Natural Gas

Natural gas is found either free in drilled wells or in association with crude oil. Free gas when present occupies the upper part of the reservoir. The extraction rate of associated gas depends on the rate of crude oil production.

Recoverable natural gas reserves during 2021 were 48.76 trillion cubic feet (BCM) while Reliance discovered gas in Krishna Godavari basin 150 km off the Andhra Coast and Gulf of Cambay, Cairn Energy in Barmer district of Rajasthan and Gulf of Khambhat (Gujarat), and ONGC in the Arabian Sea and the Bay of Bengal. In March 2004, the gas reserves were 1340 BCM and at present consumes 1,462 cubic feet of natural gas per capita every year. India's natural gas consumption has increased by 4.5%.

1.5.4 National Grid for Gas Distribution

Gas Authority of India Limited (GAIL) is responsible for transportation and marketing of natural gas. GAIL was set up in 1984, and operates 4000 km of pipeline, including the 2702 km long Hazira– Bijapur–Jagdishpur (HBJ) pipeline extending from the Western Coast to North India and over 1300 km of pipeline in other different states.

1.5.5 Gas Conservation

The pressure of associated gas is quite low and thus needs boosting before transporting through pipelines. It had been customary in oil industry to burn or flare the low pressure gas as it cannot be easily transported for use. In the wake of energy shortages, ONGC took measures to conserve the gas. It was flaring over 21% of its natural gas output in 1991–92 and this figure was reduced to 5% by March 2002. However, a small degree of flaring has to continue for technical reasons.

1.6 HYDROELECTRIC POWER POTENTIAL

Water flowing in rivers from high mountains possesses potential energy, which is utilized to generate electrical power. India is endowed with an enormous hydropower potential which is assessed by Central Electricity Authority as 148,700 MW. In February 2010, Minister of state of Energy reiterated the commitment to raise the hydrocapacity from current 26% to 40% of national power production. Hydropower is environmental friendly and meets the peak power requirement which stabilizes the power system. It involves no fuel cost and can provide energy security by saving costly fossil fuel.

India has a big potential (19,749 MW) for a large number of micro (up to 100 kW), mini (101–1000 kW) and small (1 MW to 25 MW) hydel plants in hilly states of Himachal, Jammu and Kashmir, Uttarakhand, Haryana, Sikkim, West Bengal and Arunachal Pradesh. These projects can be set up on rivers, canals or at small dams where generated power can be supplied to nearby villages situated far away from the grid power. The installed capacity of such projects was 4380 MW at the end of 31 March 2017.

There are areas where hydropower is not available and peak load is controlled by the pumped storage system. A pump storage plant consists of two reservoirs, one at the upper level and the other at a lower site near the powerhouse. These are equipped with reversible turbine-generator sets which are also used as motor-pump sets. The upper reservoir can store sufficient water to operate the installed generating machines for more than six hours. This plant operates as a hydropower station during peak load hours (05:30 p.m. to 10:30 p.m.) when demand is maximum and the generation cost is the highest. During the minimum load period (12 midnight to 05:30 a.m.) water from the lower reservoir is pumped back into the upper one for the next day's operation of the power station during the peak load hours. During the minimum load period, generators change the mode of operation to synchronous motors while turbines change to as pumps. Grid power supply feeds the motor and pump action of the power station. At present 56 pumped storage projects are operating in the country with installed capacity of 94,000 MW. A pump storage plant keeps the frequency of the power system within limits (48.5–51.5 Hz) and saves it from power collapse. Another advantage is the improvement in the supply voltage by operating in synchronous condenser mode, when it generates reactive power for VAR compensation of the power network.

A power grid dominated by thermal power generation, hydropower and pump storage schemes smooths the peaks and covers the troughs of the daily load demand curve.

1.7 INDIA'S POWER SCENE

India's largest coal reserves are a major asset to the country, accounting for more than 70% production of electricity. The installed capacity of the power generated in the country in 31 November 2017 stood at 330,860 MW which included 218,959 MW thermal coal-based plants. In spite of the various environmental problems with the use of coal for power generation, it will continue to be the cheapest, reliable and long-term source of electric power. Coal consumption in the power sector would rise from the present level of 285 million metric tonnes (MMT) per year to 675 million metric tonnes in 2013. India's coal production has decreased by 3% in 2021.

Strategies to reduce the environmental impact of coal-based generation on atmosphere, water and land would therefore need urgent consideration. It is more relevant, as funding agencies like World Bank will pose searching questions regarding Energy and Environment Management System before releasing financial assistance to new power projects. A long-term strategy is to develop Clean Coal Technologies (CCTs) as detailed below:

- To increase the efficiency of coal combustion at existing thermal plants through renovation and retrofit with Fluidized Bed Combustion (FBC) technology.
- New power plants need to utilise Circulating Fluidized Bed Combustion (CFBC). In this, high pressure air is blown through finely ground coal and the particles become entrained in the air and form a fluidized bed. The bed behaves like a fluid in which constituent particles collide with one another. It can burn low grade coal and even lignite.
- Introduce power generation with Integrated Gasification Combined Cycle (IGCC). It makes use of two cycles: the heat from gasification of coal is first used to run a gas turbine and generate electricity. The waste heat from this cycle is used to run the steam turbine and produce more electricity. The IGCC plants up to 250 MW have been developed abroad with efficiency levels of 45–55% and SO_2 removal rate up to 99%.

1.7.1 Gas-based Generating Plants

Natural gas is becoming a popular fuel for power generation. The installed capacity of gas turbine plants is 14734 MW which is nearly 11% of the total capacity. For rising requirement of electric power, the gas-based power stations will have high priority on the power scene due to manifold advantages such as:

- Low capital investment
- Low gestation period
- High efficiency
- Eco-friendly, as burning of natural gas does not produce toxic gas
- Low operating and maintenance cost.

Gas turbines were first used for meeting the peak load demand of an area. In the present scene of power shortage, natural gas is a preferred fuel due to recent gas discoveries in India and also the import of gas is cheaper than the import of oil to generate energy.

At present 65 gas-based power plants are operating successfully in different parts of the country.

Coal transportation by railways for the future thermal plants poses congestion problems causing shift to gas-based plants. Recently the approved gas fuelled power generation projects in Haryana are: Faridabad 1065 MW and Jhajjar 1050 MW. The Gas Authority of India Ltd. (GAIL) would lay the necessary gas transmission network for these projects.

1.7.2 Nuclear Power Programme

The Indian nuclear power programme is based on a three-stage strategy. In the first stage, the PHWRs will use natural uranium as fuel, and heavy water as both moderator and coolant. Their performance is good with an average capacity factor of 80%. The second stage is the construction of fast breeder reactors that would use plutonium reprocessed from the spent fuel of the PHWRs. The initiative for this stage was taken when the 13 MWe fast breeder test reactor at Kalpakkam went critical in 1985. In October 2004, construction started for a 500 MWe Prototype Fast Breeder Reactor (PFBR) which was scheduled to be commissioned by 2010. The construction of the reactor suffered multiple delays and is expected to be operational by October 2022. It marked the start of the second stage of the country's nuclear electricity programme, to be followed by a series of breeder reactors. This PFBR technology will be the basis for the generation of 500,000 MWe, which is likely to provide energy security to the country.

In the third stage, reactors shall be built to use India's abundant reserves of thorium. The objective is to build pure thorium-uranium-233 based reactors for electricity generation. The Advanced Heavy Water Reactor (AHWR) will form the first phase of the third stage. The AHWR will use naturally available thorium as fuel and convert it into Uranium-233 which will then undergo fission to generate electricity. The AHWR will be in a self-sustaining mode with U-233 as fuel. When U-233 is consumed for electricity generation, the same amount of U-233 will be produced in the reactor, though it will require a certain amount of plutonium as a kind of driver fuel. The AHWR has an innovative concept which is to be implemented in the public domain, its design is being reviewed by the Atomic Energy Regulatory Board from the viewpoint of safety considerations.

India's R&D has made a beginning with the final phase of the third stage with the development of Compact High Temperature Reactor (CHTR), which will generate fission energy at 1000°C. The important aspect is the temperature at which the energy is available. In the PHWR, energy is obtained at 300°C, and in the FBR at 500°C. For other energy conversion applications, a temperature of 1000°C is required for which CHTR is being developed.

Nuclear Power Plants Capacity in (i) operation (ii) under construction and (iii) planned projects as stood in September 2020 are as under:

1. Nuclear power plants in operation

Power station	Total capacity (MW)
Kaiga (Karnataka)	880
Kakrapar (Gujarat)	1140
Kalpakkam (Tamil Nadu)	440
Narora (Uttar Pradesh)	440
Rawatbhata, Kota (Rajasthan)	1180
Tarapur (Maharashtra)	1400
Kudankulam (Tamil Nadu)	2000
Total	7480

2. Under construction

Kalpakkam (Tamil Nadu)	500
Kakrapar (Gujarat)	700
Rawatbhata (Rajasthan)	1400
Kudankulam (Tamil Nadu)	4000
Gorakhpur (Haryana)	1400
Total	8000

3. Planned projects

Gorakhpur (Uttar Pradesh)	2800
Chutka (Madhya Pradesh)	1400
Mahi, Banswara (Rajasthan)	2800
Kaiga (Karnataka)	1400
Kavali (Andhra Pradesh)	6000
Chutka (MP)	1400
Tarapur (MH)	300
Jaitapur (Maharashtra)	9900
Kovada (Andhra Pradesh)	6600
Total	32,600

Currently 23 nuclear power reactors are in operation producing around 7480 MW. 10 reactors are under construction and expected to generate an additional 8000 MW, while another 33 are planned. Today nuclear energy constitutes around 3.11% of the total power produced in the country and the vision is to install 63 GWe by 2032.

National Energy Mix: India aims to achieve a sustainable nuclear fuel cycle using its vast thorium reserves by a mix of indigenous Pressurized Heavy Water Reactors (PHWRs), Fast-Breeder Reactors (FBRs) and Light Water Reactors (LWRs) with foreign technical co-operation. By the year 2032 large expansion based on FBRs and later thorium-based reactors are envisioned. The 500 MW Prototype FBR (PFBR), first commercial nuclear reactor designed to generate more fuel than it burns, stationed in Kalpakkam that drives India closer to harness its vast thorium reserves.

With fast construction time frame, high plant load factor, long life span and low fuel costs nuclear energy is competitive even though it is capital intensive. In India, nuclear energy is cheaper outside the coal belt and a viable option in terms of Long Range Marginal Cost (LRMC). If we take Tarapur 1 and 2 it has been found to have delivered for four decades electricity at ₹1.00/kWh.

India has over 400 reactor for commercial nuclear operation without any major safety-related accident and survived tsunami and earthquake also. Nuclear energy is thus viable and economical option to meet India's electric energy demand. In addition it offers primary and secondary economic benefits of the nuclear projects for the local, state and national level.

1.7.3 Core Drivers for Development of Renewable Energy in India

(a) **Energy security:** At present around 70% of India's power generation capacity is based on coal. Net coal import is nearly 3% in 2021. In addition India has to import 28% oil to meet country's total energy needs.

(b) **Electricity shortage:** Despite increase in installed capacity by more than 110 times in last 62 years, India is still short of power. The peak power deficit in the year 2020–21 was 0.4% and union power minister has said that peak power deficit in the country was almost wiped out.

(c) **Energy access:** India faces a challenge to ensure availability to reliable energy to all its citizens. Almost 70% of rural households depend on solid fuel for their cooking needs and only 55% of all rural households have access to electricity. Lack of rural lighting is leading to large scale use of Kerosene. However, Government is trying to improve Energy access to all and in some states use of Kerosene is banned.

(d) **Climate change:** India is committed to reduce emission intensity of its GDP by 20–25 per cent from 2005 levels by 2020. In the recently concluded, Paris climate change talk in December 2015 ended with an agreement. The green- house gas emissions, primarily emitted because we needed energy for development solutions. With the increased share of renewable energy is expected to contribute toward achieving the committed goal.

1.8 RENEWABLE ENERGY SOURCES

Renewable energy sources occur in nature which are regenerative or inexhaustible like solar energy, wind energy, hydropower, geothermal, biomass, tidal and wave energy. Most of these alternative sources are the manifestation of solar energy as shown in Figure 1.7.

India is implementing one of the world's largest programmes in renewable energy. The country ranks second in the world in biogas utilization and fifth in wind power and photovoltaic production. Renewable sources contribute to about 5% of the total power generating capacity in the country. The major renewable energy sources and devices in use as on December 2009 are shown in Table 1.2 indicating the potential and installed capacity.

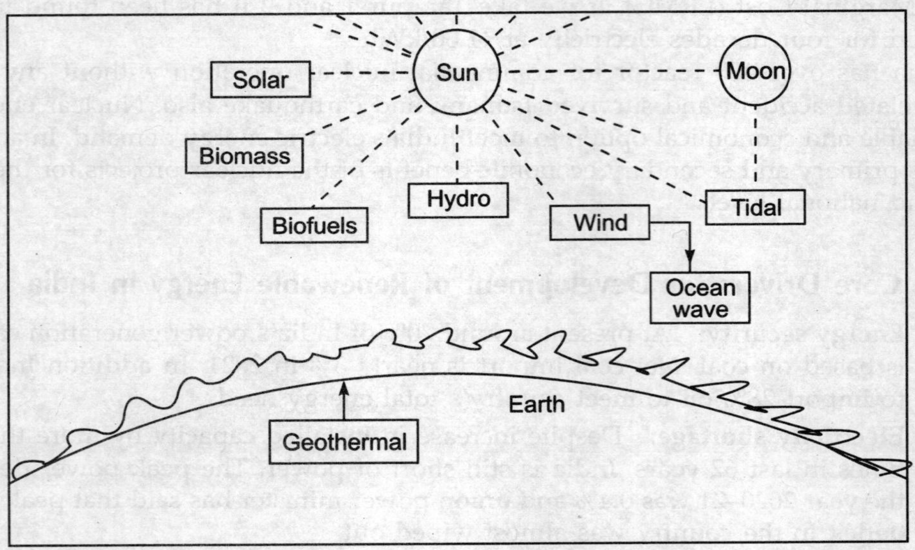

FIGURE 1.7 Renewable sources of energy.

TABLE 1.2 Renewable energy potential (31 October 2017) and achievement (as on 31 July 2019) in India

Sector	Estimated potential	Installed capacity (GW)
1. Grid-interactive power		
(i) Wind Power	102772 MW	38.7
(ii) Biomass Power and Gasification +	17536 MW+	
(iii) Bagasse Cogeneration	5000 MW	0.2
(iv) Waste to Power	2554 MW	0.2
(v) Small Hydro Tower	19749 MW	4.7
(vi) Solar Power (SVP)	748990 MW	38.8
2. Off-Grid/Captive Power (Capacities in MW) on 31 December 2016		
(i) Waste to energy		0.19
(ii) Biomass (non-bagasse cogeneration)		652
(iii) Biomass Gasifier		0.17
(a) Rural		18

(Contd...)

TABLE 1.2 Renewable energy potential (31 October 2017) and achievement (as on 31 July 2019) in India (*Contd...*)

Sector	Estimated potential	Installed capacity (GW)
(b) Industrial		168
(c) Aero-Generators/Hybrid Systems		2.97
(iv) SPV systems		0.94
(v) Water Mills/Micro Hydel		2690/72
(vi) Biogas Based Energy System as on December 2024		5
3. Other Renewable Energy Systems		
(i) Family Biogas Plants (number in lakhs)		50.28
(ii) Solar Water Heating—collector area as on December 2014 (million m^2)		8.63
(iii) National Biogas and Manure Management Programme (Plants in NO$_s$)	123,393,00	49,09,639

Small hydropower is under renewable source. Large hydropower is also renewable in nature, but has been utilized all over the world for many decades and hence not included in the term 'alternate or renewable'. Municipal and industrial waste is also a useful source of energy, but are different forms of biomass.

The Ministry of New and Renewable Energy (MNRE) have made efforts during the past few decades to develop and utilize various renewable energy resources in the country. Consequently, wind electric generators, solar water heaters, solar lanterns, street lights, biogas plants, biomass gasifiers and small hydro-electric generators have become commercially available. Wind farms, solar arrays, hydro and biomass power generation are all environmentally benign unlike fossil fuel and nuclear plants.

At present renewable sources contribute 38% of power generation while India has set a target of generating 50% of electricity from renewable sources by 2030. It is planned to cover electrification of all those remote villages which are not approachable by grid power supply.

1.9 ENERGY PARAMETERS

To improve efficiency, it is necessary to adopt measures for maximizing production with minimum energy consumption. To explain it we define GDP and GNP.

Gross Domestic Product (GDP): The total value of goods produced and services provided in a country in one year.

Gross National Product (GNP): The Gross Domestic Product plus the total net income from abroad.

Energy Intensity: It is defined as the ratio of energy consumption and GDP.

$$\text{Energy intensity} = \frac{\text{Energy consumption}}{\text{GDP}}$$

Thus, energy intensity is a measure of efficient utilization of energy in developing national economy. Developed countries have reduced 'energy intensity', resulting in less energy consumption and achieving higher production. Lower energy intensity indicates that a lower cost is incurred in converting energy resources is to national GDP. India's energy intensity is 3.0 times higher than that of Japan and double than that of USA.

There is a scope for reducing our energy requirement by reducing the energy intensity by adopting energy conservation and improved efficiency.

Energy elasticity: It is defined as the growth in energy requirement per GDP

$$\text{Energy elasticity} = \frac{\text{Growth in energy requirement}}{\text{GDP}}$$

The lower the value of elasticity, the higher is the overall efficiency. The energy should contribute in increasing the GDP. The value of elasticity for the developed countries ranges from 0.8 to 1.0 whereas for India it is around 1.2. We can save about 25% energy in energy intensive industries like Aluminium,Iron,Steel, Textile and Chemical industries through improved capacity utilization and cogeneration.

1.10   COGENERATION

Large quantity of heat generated in thermal power stations is lost in condensers and cooling towers. It is economical to save fuel by the simultaneous generation of electricity and steam or hot water for process heat. It is called cogeneration, and with this mode the efficiency of fuel utilisation can be increased up to 60%.

Cogeneration of steam and electricity is highly energy efficient and is especially suitable for sugar mills, textile, paper, fertilizer and crude oil refining industries. In India, almost all sugar mills operate in cogeneration mode and generate more electricity than what they require. The additional power is purchased by state governments with facilities of wheeling power on existing lines. Cumulative achievement up to 31-12-2009 is 1308 MW.

Cogeneration of heat and electricity can be dealt with in two ways:

(i) Topping cycle
(ii) Bottoming cycle

In the topping cycle mode, fuel is burnt to generate electric power and the discharged heat from the turbine is supplied as process heat. In bottoming cycle, fuel is consumed to produce process heat, and waste heat is then utilised to generate power.

1.11   RATIONAL USE OF ENERGY

Technological and social progress in developed countries has increased energy consumption beyond limits. The richest 10% of the world's population consume a very large quantity of energy compared to that used by the remainder of the 90% population, thus leading to an ecologically imbalance. The level of industrial production and consumption of energy are linked with irreversible environmental risks and damages. An action plan for 'economical and rational use of energy' needs to be enforced to check climatic degradation.

In Europe, "German Federal Government." implemented an environmental policy, and imposed an ecological tax on mineral oil products and electricity consumption to induce people for economical use of energy. It resulted in 11% rise in their GDP between 1990–2000 with 5% reduction in energy consumption. A new concept was coined: 'decoupling of economic growth and energy consumption'. This has also created awareness to use green energy, i.e., solar, wind, and fuel cell.

1.12 ENERGY EFFICIENCY AND CONSERVATION

Energy conservation can be said to be the cheapest form of new source of energy. It actually minimizes wastage of energy without affecting productivity and human comfort. Energy efficiency and energy conservation involve all sectors of economy. Motors and drive systems in industry and agriculture sectors consume major chunks of energy. It is advisable to use energy-efficient and correct-capacity (not oversized) motors. Considerable reduction in energy consumption is also possible by reducing high lighting levels in domestic, commercial and industrial installations.

A 14 W LED may be used instead of a 60 W incandescent lamp or a 40 W fluorescent tube to obtain the same lumen output. Reflector systems and luminaries should be installed to project light towards operational areas for better illuminance.

Compared to increasing the capacity by installing new power plants, energy efficiency and energy conservation measures can provide a quick way out of the crisis.

To cope increased energy demand by installing energy production plant is expensive and long term option, hence better way is to conserve (save) energy. Developed countries have reduced their energy consumption by strict energy conservation measures. Japanese primary energy consumption per unit GDP is the lowest in the world, achieved by applying energy conservation measures.

Government of India realized the vast potential of energy saving and advantages of energy efficiency and enacted the 'Energy Conservation Act 2001'. The Act provides the legal framework, institutional arrangement and a regulatory mechanism at the state and Central Govt. levels to achieve an energy conservation target in the country.

Further, central government also established a 'Bureau of Energy Efficiency (BEE)'applicable with effect from 1st March 2002. It is responsible to form energy consumption codes. BEE acts as a facilitator for providing self regulatory system to save energy and bring commercial concept in the organization.

Various Factors of Energy Conservation

Energy conservation qualifies reduction in energy consumption by cutting down losses and wastage using energy efficient equipment and gadgets from generation to utilization. Two important aspects of energy conservation are:

1. *Economic gain:* Energy conservation provides us economic benefit as cost of production per item is reduced ensuring competitiveness. It implies application of new technology and replacement of old machines resulting new job opportunities.
2. *Environmental gain:* Energy conversion in generation and utilization adversely affect environment as energy loss reflects as heat to the surroundings, resulting in global warming. Energy conservation minimizes environmental damage.

1.13 NEW TECHNOLOGIES

New and emerging technologies like hydrogen energy, fuel cells and biofuels hold major promises for meeting the future energy needs of the country.

1.13.1 Hydrogen Energy Systems

Hydrogen, high in energy content, is a clean and efficient energy carrier with a potential to replace liquid fossil fuels. Globally, 95% of hydrogen is produced from hydrocarbons. About 4% is produced through electrolysis of water using electricity and by thermal decomposition of water through solar energy or nuclear power.

Hydrogen can be converted effectively back to electricity either directly in IC engines or through fuel cells. Hydrogen is also a good fuel for aircraft and automobiles that could encourage its large-scale commercial production, storage and distribution.

1.13.2 Fuel Cells

A fuel cell is an electrochemical device that converts fuel energy into electricity and heat without combustion. A fuel cell is similar to a battery having electrodes, positive and negative terminals, and an electrolyte. To operate a fuel cell, hydrogen is supplied to the negative electrode and oxygen (or air) to the positive. Hydrogen and oxygen react to produce water and electricity. A fuel cell continues to work as long as fuel is supplied. Despite their benefits, fuel cells are not in wide use due to their high cost ($ 2000/kW). The state-of-the art fuel cells have been developed, and being tested, costing $ 1200/kW, and which cost is comparable to that achievable with a conventional coal-fired power plant.

1.13.3 Biofuels

The organic material of plants is called biomass which may be converted by anaerobic digestion into methane, and through fermentation process into alcohol. Examples are oil extraction from oil seeds, and transesterification of oil with alcohol which produces biofuel (ethanol). Production of biofuels is encouraged due to 5% compulsory blend of ethanol in petrol in nine states (Andhra Pradesh, Goa, Gujarat, Haryana,

Karnataka, Maharashtra, Punjab, Tamil Nadu and Uttar Pradesh) and four union territories (Chandigarh, Dadar Nagar Haveli, Daman and Diu and Pondicherry). Its advantages are: environmental friendly, blends of renewable sources of energy and savings in foreign exchange spent in importing crude oil. Ethanol requirement for 5% blending is about 320 million litres per year which accounts for only 25% present production in the country.

1.14 DISTRIBUTED ENERGY SYSTEMS AND DISPERSED GENERATION

1.14.1 Distributed Energy Systems

Conventional energy sources constitute fossil fuels such as coal, crude oil products (petrol, diesel, and natural gas) and nuclear fission fuel. These energy resources do not get replenished after their consumption and are likely to be exhausted. Fossil fuels are available in bulk quantities at their source. They pose environmental problem at the extraction, transport and generation stages.

Non-conventional energy or renewable energy resources that are renewed by nature periodically. These are, solar, wind, small hydro, biomass, bio diesel, geothermal, ocean tide and wave energy etc. These sources are not affected by rate of consumption. However, these sources are not available in concentrated form at one location, so they are called distributed energy systems and are classified as:

Solar energy distributed system

Solar energy is a flow and not a stock, solar energy reaching per square metre of the Earth's atmosphere is 1.36 kW, i.e., 16.32 kWh in 12 hours. This energy is attenuated by the earth's atmosphere, thus, actual energy reaching the earth's surface varies with atmospheric conditions, time of the day, month and also latitude of the place. Nearly 35 per cent of solar energy received at the earth's atmosphere is reflected back into the space, 18 per cent is absorbed by the atmosphere and drives the winds, balance 47 per cent reaches the earth. Energy received per square metre in India varies in the range 4.0 kWh to 7.5 kWh per day under normal clear sky conditions. Solar energy is widely distributed, can be tapped at the location of its consumption.

Distributed system of wind energy

Wind energy is the kinetic energy associated with the movement of large mass of air resulting from the uneven heating of the atmosphere by the Sun. As this energy is not sufficient to be tapped at every location, so, wind energy maps are prepared using monthly and annual mean wind speeds known for selected locations. These maps show energy available in kWh in one square metre area of wind stream at 10 metre height from the ground level. Sites of high wind locations are in Gujarat, Tamil Nadu, coastal belts of Bay of Bengal and Arabian sea islands, Rajasthan, Madhya Pradesh and Karnataka. In India, to maximise energy availability during a year rated speed is in the range of 20–25 kmph which would correspond to 9–17 kmph annual average. Wind energy is available at specific locations in a distributed form.

Distributed system of small hydro

Small, mini and micro hydel power generation result from the dispersal of their sites. Most of the potential is in Himalayan States as river-based projects and in other states as an irrigation canal. Favourable parameters of dispersed *small hydro system* are:

1. Cost-effective, i.e., ideal locations
2. Short gestation period
3. No deforestation, so environmentally benign
4. Moderate potential 'Greenhouse Gases' abatement
5. Improves quality of life in remote low income areas
6. Requires participation of house hold and local communities.

Small hydro energy is available at favourable locations in a distributed form.

Distributed system of biomass

Biomass includes both terrestrial and aquatic matter which can be grouped into new plant growth, plant residue and wastes.

New plant growth includes wood, short rotation trees, herbaceous plants, arid area plantation, algae and aquatic plants.

Plant residue cover crop material such as straws rice husk, cotton stalks, maise cobs, coconut shells, bagasse, etc. There are secondary level products such as cow dung, animal droppings, forest residues like bark, wood shavings, and saw dust etc.

Waste comprises disposable material like municipal garbage, night soil, sewage solids and industrial refuse.

Total biomass is not centrally available at one location but widely distributed in entire country, i.e., in forests, plants in arid areas, water hyacinth in water bodies, crop material in villages, municipal waste in towns and cities.

1.14.2 Dispersed Generation

Developed and developing countries of the world are focussing on environmental damage observed due to conventional energy powers. After deliberations, consensus has emerged in favour of renewable energy sources to correct damage done by climate change, a threat to life over the globe.

There are two options; *soft path* (benign) and *hard path* (harmful) with energy concept. It is experimentally proved that renewable energy sources are the best options with no adverse environmental impacts, while conventional energy (fossil fuels) options have triggered global warming. Now, the study is, to generate energy with minimum environmental damage.

Hard path represents centralised and planned expansion of electrical energy generation to meet growing demand. It leads to *super thermal* coal based power stations, large hydro power generation and nuclear power installations, susceptible to large scale breakdowns, possible sabotage, line losses leading to uncertain power supply in large area.

Soft path energy concept postulates electrical energy production at site of energy source and nearest to the consumption centre which is more efficient, economical and reliable. A soft energy path constitutes small dispersed generation systems to form larger component of energy production and utilisation.

Parameters delaying the soft path are:

- Having disperse energy generating facilities nearest to its source.
- Individuals or a group or defense installations in remote area control their own source of energy for efficient operation.
- Utilising natural source of energy called renewable sources.
- Minimum line losses and highest load factor.
- Few renewable energy sources are dealt here with:

Disperse small hydroelectric power generation

Hydro power project harness energy from falling water in rivers, rivulets, storage dams or canals. It covers small, mini or kiosk type micro hydro plants. It is a good source of disperse generation especially in hilly inaccessible areas.

- Keylong town district head quarter of Lahaul spiti in Himachal Pradesh was electrified in 1964 by 2 × 50 kW hydro electric units. This area remains snow bound for 6 months from November to June next year, ambient temperature plumbing down to (–8°C).
- Shansha micro hydro-electric project (2 × 50 kW) on Chenab river tributary operating in Lahaul valley with head of 23 metre and water quantity 350 litre/second.
- Asia's lowest head micro hydel project in Kakroi (Sonepat), Haryana (3 × 100 kW) operating with head of 1.6 metre on Western Yamuna canal with a discharge of 31.6 cumec.
- Three small hydroelectric power stations capacity (2 × 8 mW) each are operating on Western Yamuna canal between Hathnilkund and Dadupure (Yamuna Nagar), Haryana having water head of 12.8 metre with a water discharge of 73.33 m^3/s each.

Disperse solar power generation

There are two basic techniques for converting solar radiation into useful electric power.

- (i) Conversion of solar radiation into heat and then to electricity in a thermodynamic process which is called solar thermal route.
- (ii) Conversion of radiative energy of the sun directly to electricity, called solar photovoltaic route.

Solar thermal application covers solar water heating, solar cooking in rural and urban areas, solar drying, and purification of water. Solar energy is also used for space heating and cooling.

Solar photovoltaic (SPV) technology converts sunlight into DC electricity without any moving parts and utilised for lighting, water pumping computers and telecommunications etc. Stand alone SPV power plants in rural areas provide power for electrification. SPV roof top power plants are used for diesel saving in remote areas and tail end of grid in rural areas.

Studies have shown that entire gamut of solar energy utilisation is ecologically benign source of energy. Solar power generation is highly disperse generation system, a *clean* energy option with no environmental dangers.

Disperse wind energy generation

Wind energy is harnessed through wind turbine. Shaft power from wind turbine can be utilised for electric energy generation, direct pumping and direct mechanical work. A wind turbine system involves a tower mounted multi blade rotor, facing the wind, rotating around a horizontal axis and turning an electric generator.

Wind Energy Generators (WEGs) begin generating at a minimum or cut-in speed and are designed to break or cut out when seed exceeds specified limits to prevent damage to the moving parts and control systems. With varying wind speed, the pitch angle at the rotor is also changed to get maximum generation from available wind.

Areas having potential wind power density greater than 200 W/m^2, ensuring land availability wind farms @ 12 ha/MW are established for grid interactive wind power. These areas are remote and dispersed having favourable wind regimes on islands, coastal areas and mountain regions where large arrays of wind turbines are set up to extract renewable energy.

Wind energy generators located scattered over country side, connected with a battery system or made hybrid with another source of energy they will be little offensive. Scarce water resource is not required in wind energy generating system, contrary to geothermal, solar central receiver, nuclear and thermal power generation where cooling water is a necessity. Thus, dispersed wind energy systems are more environmentally benign than any other source of energy.

Biomass energy dispersed generation system

Biomass as a source of energy has tremendous potential. Dispersed biomass energy utilisation systems are of three types:

(i) Dry biomass is burnt directly to produce heat by house holds in rural areas.
(ii) With biological conversion of biomass, high quality fuels are produced; biogas from bacterial fermentation.
(iii) Thermo chemical conversion of biomass used to generate heat/electricity based on 'pyrolysis' and 'gasification' of wood.

First type of dispersed biomass energy is used by 70% households in rural areas. Firewood is obtained from crop waste and forests. Use of fuel wood directly in homes for cooking causes air pollution causing health hazard to the eyes of women.

Biological conversion process involves bacterial breakdown by micro-organisms. This technique is used to produce biogas (methane) from animal and human waste with anaerobic digestion. Technology is quite simple as waste (cow dung) is fed in a digestor placed below the ground. Microbes present with low temperature heat provided from the Sun, decompose the waste to produce biogas (CH_4 + CO_2 + CO). Methane is a major component with CO, being combustible gases used directly for heating and cooking purpose providing relief to rural women.

Community biogas option is on a larger scale is economical for marginal farmers with a few and no cattle. Biogas produced in community digestor is also suitable for motive power, i.e., one cuft of biogas can produce 0.052 H.P./hour. It is used to operate generator to produce electricity.

In chemical conversion of biomass, charcoal making is the 'Pyrolysis' where vapour and gases are not collected.

Gasification of biomass means thermal decomposition under controlled air supply. Thus, solid carbonaceous fuels are converted into combustible gas mixture (CO and H_2) called producer gas or synthesis gas. This gas can be directly burnt to generator process heat or can operate gas turbines to generate electric power.

India has several programmes to promote biomass-based dispersed electricity generation system namely, biomass gasifiers, biomass combustion and congeneration.

Recently two biomass gasifiers of 10 kW capacity each were commissioned in Kandhal Test Project in Cuttack district of Orissa. It covers 150 families of tribals, for domestic lighting, television programme, children's study at night.

64 projects in states (Assam, Chattisgarh, Gujarat, Jharkhand, M.P., Maharashtra, Odisha, Tamil Nadu, Uttrakhand, and West Bengal, Bihar, Arunachal Pradesh, Karnataka, Nagaland, Manipur Tripura, Mizoram, Jammu) were commissioned.

Biomass energy dispersed systems can supply energy on a large scale, even to the poorest and is ecologically acceptable.

1.15 NATIONAL POWER TARIFF POLICY

In an effort to provide quality power supply to all homes across the country, Govt. of India approved "National Tariff Policy for Electricity" in January2016, which is aimed to tight regulations for fixing power rates and promoting clean energy.

Under this policy electricity regulators have been asked adhere to the policy while framing regulations under Section 61 of the Act. Before the act was passed, the state regulators hardly abide regulations and twist it according to their requirement. The policy is based broadly in four areas:

1. 24 × 7 electricity for all.
2. Higher efficiency in electricity generation to ensure affordable tariffs.
3. Protection of environment from an adverse impact of utilizing energy resources.
4. Ease of doing business in generation, distributes of electrical energy to attract investment.

Henceforth, micro grids will have a provision to feed electricity as and when the grid connection reaches remote locations. In states where there is surplus power, plants are allowed to sell the excess power to the power exchanges. Transmission projects are envisaged to be developed through a reverse bidding process.

1.16 GREEN POWER—AN URGENT NEED OF INDIA

Green power is the electrical power obtained from renewable sources like solar radiation, wind energy, small hydro power and biomass. Solar radiation can be used as (i) solar thermal energy (ii) photovoltaic power producing electricity using solar cells. Wind is air in motion and wind turbines extract wind energy as electrical power. Falling water is a source of energy and can generate electrical power using net head (meter) and flow in cubic meter per second.

Biomass is solid carbonaceous material derived from plants and animals. It can generate electrical energy by two processes. (i) Combustion and incineration (ii) Bio-chemical conversion.

India has formulated its ambitious plan to increase its renewable energy capacity to 450 Giga Watts (GW) by 2030. Status as on 31–12–2016 and proposed capacity is:

Position on 31 December 2016

Solar	13,652 MW	100,000 MW
Wind	32,562 MW	60,000 MW
Biomass	8,181 MW	10,000 MW
Small hydro	4,389 MW	5000 MW

India is currently dependent on ground-mounted solar projects and Central Electricity Regulatory Commission suggest 2.2 ha of land for 1 MW solar PV. To save land, solar energy can be harnessed from roof tops. Few success stories shall be discussed to project how renewable energy can provide a clean environment and raise standards of living for entire communities.

- According to Ministry of New Renewable Energy (MNRE) reports Chandigarh is generating about 6.5 million units of solar electricity annually. All plants are under net metering mechanism, so that the solar energy generated is first used in the local building and then excess energy is fed to the grid.

- 100 MW Solar thermal Power Project at Pokhran, Jaisalmer (Rajasthan).It is the largest investment done in solar thermal technology in India. When fully operated it will generate about 250 million units of clean solar electricity per year. The project is located in village Dhursar of Pokhran in Jaisalmer district of Rajasthan. It is spread over an area of 340 hectares of barren land.

The project provided an impetus to community development by providing employment opportunities, contributing to rural economy and reducing the rate of migration. Project shall provide permanent employment to more than 200 skilled and unskilled workers for operation and maintenance.

REVIEW QUESTIONS

1. Briefly discuss the different forms of energy.
2. Discuss renewable and conventional forms of energy. Highlight their merits and demerits.
3. Explain with the help of diagrams, the operation of
   (a) a coal-fired thermal power plant
   (b) an IGCC power plant.
4. How is nuclear fission different from nuclear fusion? Discuss the method of energy generation in both the cases.
5. Briefly discuss the power scenario in India.
6. What are the renewable energy resource? Discuss their importance in India's power requirement context.
7. Define the terms:
   (a) Energy intensity
   (b) Energy–GDP elasticity
   (c) Cogeneration
   (d) Demineralized water
8. Write short notes on 'Energy Conservation' and 'Energy Efficiency'.
9. What are 'Biofuels' and 'Fuel cells'?
10. Discuss and differentiate between 'decentralized' and 'dispersed generations'.
11. Explain how the cogeneration of heat the electricity can be dealt with in 'Topping cycle' and 'Bottoming Cycle'.
12. Explain Nuclear Power Program of stage 1, 2 and 3.
13. Discuss in detail, how nuclear electric power generation is competitive with coal based power generation?
14. Explain how super critical unit of thermal generation is superior to conventional thermal power generating station?

2 ENVIRONMENTAL ASPECTS OF ELECTRIC ENERGY GENERATION

2.1 INTRODUCTION

Energy conversion and environment are interrelated. With the increase in electric power generation, environmental degradation has become a serious problem. To meet the bulk electric energy demand in industrial and agricultural sectors, India has to move forward and build many large thermal, hydro and nuclear power projects. All of these projects have environmental ramifications. We all live in an environment, which constitutes air, water, land and other biological organisms present in the biosphere.

Air, water and the surrounding environment are all polluted by emissions from energy conversion plants and industries. Clean air in the atmosphere, natural pure water and good growth of trees are the basic requirements for human survival. Nature has created self-cleaning processes like photosynthesis, water cycle, carbon and nitrogen cycles, winds and four important seasons in a year.

However, the large-scale fossil fuel combustion causes atmospheric pollution, effluent discharge in water, particulate matter and fly ash—that all adversely affect the environment and it then becomes beyond the nature's capacity to clean and create ecological balance. It causes irreversible damages to water bodies, i.e., lakes and rivers; produces acid rain that damages agriculture and forests and creates ozone layer holes and global warming. Emphasis is now being laid on alleviation of the situation for sustainable development, with appropriate technology.

2.2 ATMOSPHERIC POLLUTION

Due considerations have been given to treat the pollution caused by thermal plants that burn coal. India's energy security is largely based on fossil fuel generating plants, supported by hydro and nuclear plants which are also responsible for environmental hazards.

The major pollutants which are released from coal-based generating plants are: SO_2, nitrogen oxides (NO_x), CO and CO_2, hydrocarbons, fly ash and suspended particulates. Indian coal carries 0.6% to 1% sulphur and its ash content varies from 30% to 50%. Various pollutants are dealt below with their possible impact and related issues.

2.2.1   Oxides of Sulphur (SO_2)

Coal containing sulphur, on burning in the combustion chamber, produces SO_2 which is released through chimney. It causes respiratory ailments in concentrations of 20 mg/m^3 and constitutes danger to life in amounts of 400 mg/m^3. In atmosphere, SO_2 is further oxidized to H_2SO_4 and falls down on the earth as acid rain. It is injurious to plants and causes damage to buildings and marble structures (e.g. the marble monument like Taj Mahal).

Sulphur emissions can be removed from the coal by gasification or floatation processes.

Use of chemical reaction is recommended to remove sulphur oxides from flue gas. Installing limestone scrubbers in the power plants also reduces the sulphur emission.

2.2.2   Oxides of Nitrogen (NO_x)

Oxides of nitrogen that pollute the air include NO, NO_2 and N_2O. Of these, nitrogen oxide (NO_2) is a major pollutant. It is highly injurious; if inhaled in concentration of 150–200 ppm NO_2 can damage respiratory tissues and may cause even pneumonia. Emission of NO_x can be reduced by:

- Installing advanced technology burners in the boiler to ensure complete combustion and reduction of these oxides.
- Providing tall stacks for wider dispersion of air pollutants that can lower pollution level in the ambient air (100 $\mu g/m^3$).

2.2.3   Oxides of Carbon (CO, CO_2)

CO is a toxic gas and affects human metabolism. If released to the atmosphere, it gets converted to CO_2. The concentration of CO_2 reduces in the air through the natural process of photosynthesis to generate oxygen and organic matter. High concentration of CO_2 is also a major cause of global warming.

To control CO_2 emissions, the combustion efficiency of boilers should be improved which also results in reducing coal consumption.

2.3   HYDROCARBONS

In the boiler combustion chamber, during the process of oxidation, a few specific hydrocarbons are formed. These compounds contribute to photochemical reaction, which causes damage to atmospheric ozone layer.

2.4 PARTICULATES (FLY ASH)

Particulates comprise fine particles of carbon, ash and other inert material with size greater than 1 μm. It gets emitted from chimney in the form of fly ash. Particulates suspended in air with pollution level 300 μg/m^3 cause poor visibility, lungs inflammation and bronchitis.

2.4.1 Control of Particulates

Emission of particulates is controlled by installing 'Electrostatic Precipitators' (ESPs) in the path of flue gases before discharging them through stack. The ESPs remove particles from the flue gases and thus clean gases before such gases are released to atmosphere.

An electrostatic precipitator consists of sets of two electrodes, one is the collecting electrode in the form of parallel plates connected to positive polarity of 25 kV dc supply and the other is the emitting electrode of thin wires connected to negative potential. It creates 'corona discharge' that emits electrons and ionizes gas molecules, thereby producing negative ions. The ions are driven towards the collecting electrodes, get neutralized and deposited on positive grounded plates. Accumulated dust is dislodged by rapping the electrode. The dust then falls into a hopper for disposal.

Precipitators have high efficiency of 99%. The fly ash which escapes ESP is smaller than 2 microns in size.

2.4.2 Upgrading ESPs by Pulse Energization

Emission reduction is further optimized by introducing pulse energization in place of the conventional dc system in the existing ESPs.

The pulse energization technique constitutes a multi-pulse unit that provides high voltage pulses. High frequency and high voltage pulses are triggered for fraction of a millisecond and then blocked for a few milliseconds. High voltage pulses provide good distribution of corona current, cause effective charging of dust and better use of collecting area.

This technique is being used at Badarpur Thermal Power Station by M/s. FLACT, Sweden, for high resistivity ash where the efficiency has gone up 2.9 times. The advantages of this system are:

- High peak voltage can be supplied at low current level.
- Spark voltage of pulse energization increases.
- Dust is charged by high electrical field without back corona.
- Corona discharging occurs uniformly on discharge electrode even with dust deposits.

2.5 THERMAL POLLUTION

Steam from the low-pressure turbine flows into a condenser and gets converted into water at the lowest possible temperature to attain the highest possible thermodynamic efficiency. Even with the Integrated Gasification Combined Cycle (IGCC) technology, the improved steam cycle efficiency obtained is only 40%. Unused 60% heat in steam at the cycle end is dissipated to the atmosphere.

2.6 HYDROELECTRIC PROJECTS

Large hydroelectric projects generate electricity and maintain ecological balance by channelling surplus water to irrigation deficit areas. More food and foliage, check on soil erosion, recharging of underground acquifers and industrial built-up are the major positive outcomes of these projects.

The negative side of these projects is the ecological imbalance caused during construction and its aftermath. The environmental impact caused due to hydropower construction is shown in Figure 2.1.

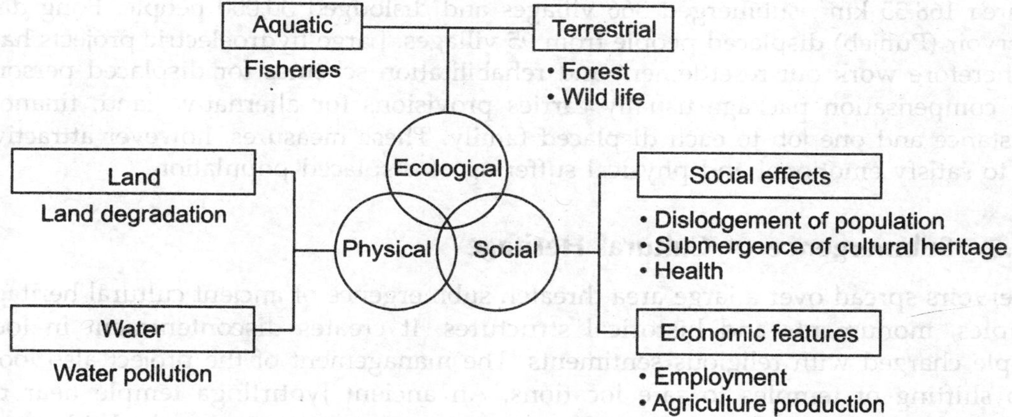

FIGURE 2.1 Environmental impact of hydropower construction.

2.6.1 Terrestrial Effects

The construction activity of a hydropower project begins with the clearance of forest for infrastructural facilities like roads, communication systems, and housing colonies on the project site. Destruction of forest and damage to flora and fauna do take place due to submergence of vast land in artificial lakes. Depletion of forest areas creates a highly negative effect on biological diversity.

It has thus become mandatory to compensate the loss of forest cover by planting more trees on denuded land.

2.6.2 Wild Life

Reduction in vegetation cover at the construction site disturbs wild life habitat. Blasting operations, movement of heavy machines, noise and dust produced during construction, drive away wild animals. The risk of fire endangers jungle conservation.

2.6.3 Aquatic Life

Construction activities like localized sedimentation and impoundment of intake area directly affect the aquatic habitats. Excavated soil materials produce sediment load on the river bed. Construction work and movement of heavy vehicles on water course, contamination of water by accidental spilling of oil, all lower the quality of water for fish breeding. Project personnel carrying fishing activities with explosives also cause damage to aquatic life.

2.6.4 Social Problems

Large hydroelectric projects are associated with dam construction and formation of big lakes. Such projects uproot the local population who is forced to shift to a new environment. Bhakra multipurpose project (Himachal Pradesh) created a lake of area 168.35 km^2, submerged 366 villages and dislodged 30,000 people. Pong dam reservoir (Punjab) displaced people from 95 villages. Large hydroelectric projects have to therefore work out resettlement and rehabilitation schemes for displaced persons. The compensation package usually carries provisions for alternative land, financial assistance and one job to each displaced family. These measures, however attractive, fail to satisfy emotional and physical suffering of displaced population.

2.6.5 Submergence of Cultural Heritage

Reservoirs spread over a large area threaten submergence of ancient cultural heritage, temples, monuments and historical structures. It creates discontentment in local people charged with religious sentiments. The management of the project also looks into shifting of temples to safe locations. An ancient Jyotirlinga temple near the Omkareshwar dam (MP) was saved by constructing a special road and a bridge, as a project component.

2.6.6 Health Concern

Large hydropower projects employ a substantially big element of workforce of engineers, managers, and artisans. Villagers around supply unskilled labour, and they, too, require preventive and curative health facilities. Hospitals with medical staff, trained health workers, medical equipment and medicines become a mandatory requirement in the project area itself. Local people are also benefited by such clinical facilities. They are made aware about proper drainage system, clean drinking water, proper sanitation and child care. It prevents outbreak of diseases among the workers and the villagers. These are a few positive aspects of hydropower projects.

2.6.7 Economic Aspect

Effort is made to employ local people as unskilled and semi-skilled workforce. The displaced, compensated people are given the first priority to work. A consistent, fair and equitable employment policy is adopted. Local farmers are benefited by new road networks and can thus diversify the sale of their farm products.

2.6.8 Physical Effects

Land degradation

The construction of a project entails mass failure, landslides and slope failures on steep gradients. Soil erosion and heavy precipitation accelerate these processes. Quarrying, earth excavation and tunnel-muck dumping become notable sites of land degradation. Progressively, land is restored by adopting bioengineering techniques.

Water pollution

Project construction progresses with the basic tenet of minimum water pollution. Garbage dumping, sewage disposal and septic tanks are built away from the water courses to obviate contamination by seepage or direct runoff.

2.7 OPERATIONAL PHASE OF HYDROPOWER PROJECTS

Sedimentation of reservoirs

A hydropower project generates pollution-free electricity, supplies water to grow food and foliage in arid areas and builds groundwater aquifers. However, certain apprehensions are also argued such as that flow obstruction by a dam causes dips and flanks of the storage basin. Sediment observations carried out on large dams in UP, AP, Bihar, MP and Maharashtra have shown that the average annual percentage loss of water storage lies between 0.3 and 0.9. However, the survey also shows that the economic life of a dam reservoir is about 100 years.

Reservoir induced seismicity

Environmental critics predict disastrous shocks in every dam storage, irrespective of the fact whether it is located in a seismic zone or not. To address one such apprehension the behaviour of 'Nurek' Dam in Ukraine, with height of 305 metre, located in the high seismic zone of 6.5–7.0 magnitude on Richter scale, was studied but the computation of readings showed no adverse effect on the dam structure. It may therefore be seen that the issue raised by environmentalists that impounded reservoirs would cause earthquake is not substantiated by any adverse findings.

2.8 NUCLEAR POWER GENERATION AND ENVIRONMENT

India is transforming itself from a developing country to a developed country with multifold increase in electric power demand. A nuclear power plant is a quick and bulk-power generator that requires minimum land area compared to the hydroelectric power plant; its contribution to atmospheric pollution is also quite low compared to the coal-based plant. However, there is a concern in public about radioactivity release and its effects. It is a misconceived conception, as safety in design and operation of nuclear power plants ensures clean and cost-effective power generation.

2.8.1 Natural Radiation

It is a known fact that we all live in a sea of natural radiation. It comes from cosmic rays, and terrestrial deposits of uranium and thorium. Man-made radiation comes through X-rays during medical check-up, from fall-outs of nuclear weapons tests, and use of mobile communication sets, etc.

2.8.2 Radioactive Pollution

The areas of possible radioactive releases by Nuclear Power Plants (NPPs) are:

- Radiations from radio nuclides during uranium mining
- Processing of uranium ore as fuel for nuclear reactors
- Operation of nuclear reactors for power generation
- Accidental radiological hazards
- Contamination from nuclear waste.

Mining

Uranium ore is obtained from mines at Jaduguda in Chotanagpur. During extraction of ore, the safety of workers is ensured by controlling radon and airborne radioactive dust by proper ventilation.

Processing of ore

Processing of ore is carried out in three stages: crushing, grinding, and leaching. Purified uranium is precipitated as yellow cake. Uranium forms only 0.5% of the ore and the remaining bulk is rejected as waste called tailings. The waste contains radio nuclides like radon–222 and emits radiations. Tailings are neutralized with lime in ponds, whereas solid tailings are retained in ponds.

A typical value of radiation constant in pond measured with a dosemeter is reported to be 0.75 micro gray an hour [gray (Gy) is the SI unit of absorption dose of ionizing radiation, corresponding to one joule per kg of absorbing medium]. Radiation constant decreases with the increase in distance which is 0.2 gray an hour on the embankment of the pond. No danger of radiation pollution is possible in nearby populated areas.

2.9   OPERATIONAL SAFETY IN NUCLEAR POWER PLANTS

Operational safety in nuclear plants concerns protecting individuals, society and the environment from radiation hazards as defined by the International Atomic Energy Agency (IAEA). Safety considerations for Pressurized Heavy Water Reactors (PHWRs) are:

- Control of reactor power during shutdown and maintenance
- Providing adequate cooling
- Containment of radioactivity.

In reactors, nuclear chain reaction takes place and energy is released. During normal operation, heat is removed by the circulatory primary coolant, which in turn is cooled in the steam generators. Cooling is required during shutdown as heat continues to be generated, though at small levels. An ultimate heat sink is provided catering to the safety function of decay heat removal. Water for cooling is kept available all the time in an earthquake resistant hold-up structure.

The PHWRs have special provisions for maintaining sub-criticality under long-term shutdown. After shutdown, the sub-criticality margin increases because of xenon build-up and then decreases due to xenon decay. Adequate provisions are made for maintaining the required minimum sub-criticality margin after shutdown.

Containment of radioactivity

The third important safety function is the containment of radioactivity and to prevent its release into the public domain. Various barriers preventing the release of radioactivity are shown in Figure 2.2.

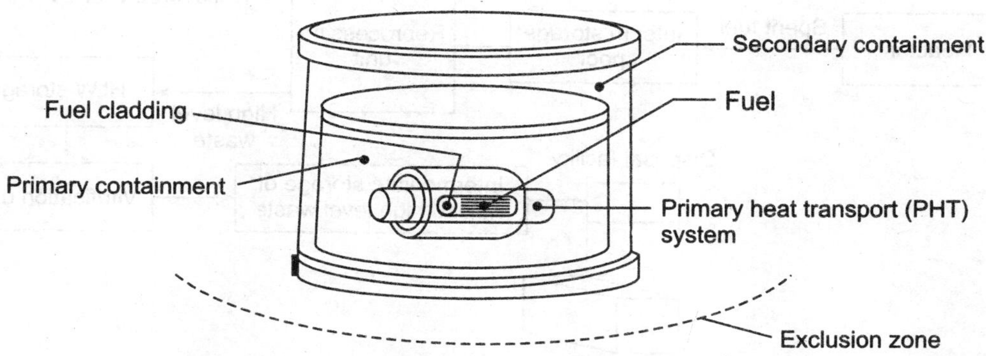

FIGURE 2.2   Barriers to prevent release of radioactivity.

The fuel pellet retains the fission products within its matrix. The cladding that surrounds the fuel serves as the next barrier to the release of radioactivity. Any release of radioactivity on account of cladding failure remains confined within the Primary Heat Transport (PHT) system boundary. The final barrier is the double containment where the primary containment is surrounded by the secondary containment.

Accidental safety

Following a severe accident, hydrogen may get generated within the containment by a metal–water reaction. Hydrogen concentration is safe up to 4%, and any further increase in concentration may lead to its burning and above 14% detonation may occur. The PHWR containment volume is good enough to keep hydrogen concentration just below 4%. The containment system carries a number of engineered safety features which confine radioactive release within acceptable levels.

Multiple electric power supplies

Safety equipment need power supply to operate. The design adopted for nuclear power plants covers the exigencies of equipment failure and provides four classes (Class I to IV) of back-up supply. Class IV supply is from the grid, Class III from a diesel generator set, Class II from a dc motor-alternator set and Class I from a 220 V dc system.

2.10 DISPOSAL OF NUCLEAR WASTE

The disposal of radioactive waste is a serious issue. Major wastes generated in a nuclear fuel cycle are 'low-active wastes' which are handled like other wastes and dispersed. 'Medium-level wastes' do not create any problems and can be disposed with the techniques of dilution and decay. Difficulty arises with high-level radioactive wastes which require special technology to handle for a long-term solution to their disposal. A flow chart showing the handling, the storage and the final disposal of wastes is shown in Figure 2.3.

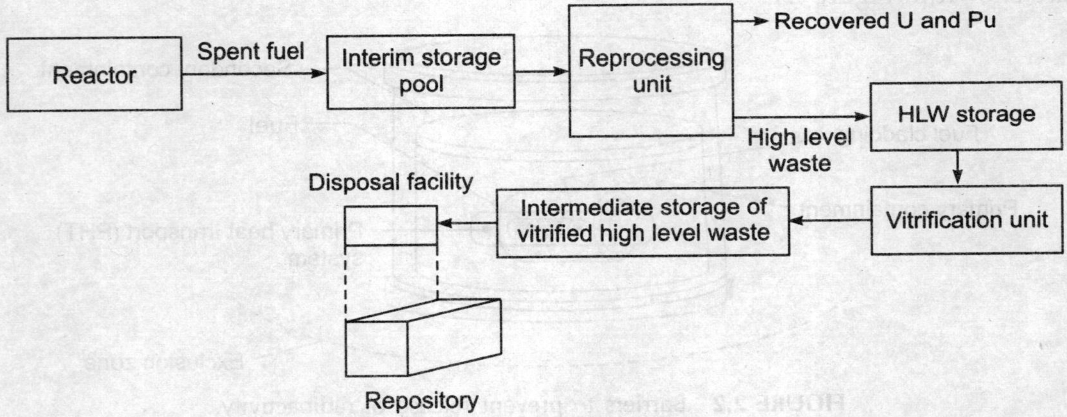

FIGURE 2.3 A nuclear waste disposal scheme.

Spent fuel discharged from the reactor is kept in an interim storage pool located in reactor premises. It is then sent to a reprocessing unit where wastes are separated and stored before shipping to a vitrification unit for immobilization in a glass matrix. The vitrified wastes are filled in special casks for final disposal in a repository (not yet opened in India).

At the world level, US is conducting studies in the Yucca mountain of Nevada desert for suitability of a first permanent high level nuclear waste repository.

The safety provisions made in nuclear power plants with strict surveillance enforcement, release of radioactivity to the environment and impact in public domain should be within acceptable limits during operation and accident conditions. Factual projections of nuclear power plants aptly confirm that electric power generated from nuclear plants is clean and cost effective.

2.11 GLOBAL ENVIRONMENTAL AWARENESS

Normal concentration of CO2 in earth's atmosphere is 0.3% which is necessary to maintain average surface temperature of about 15°C which necessary to sustain life. In absence of CO2 earth's surface temperature would have been –18°C, which is not suitable for survival of life. Due to developmental activities such as power generation by coal based thermal stations, fossil fuel burning in transport and manufacturing industry, Green House Gases (GHG), i.e. CO_2, CH_4 NO_x SO_2, hydro-fluorocarbons increased substantially. This has caused 'global warming'. Creating heat waves, floods, draughts and rising sea level. CO2 emissions from developed countries accounts for 80–85% of total greenhouse gas emissions of the world.

A first international effort to mitigate global warming was taken in 1992 by the formation of United Nations Framework Convention on Climate Change (UNFCCC). It was created to explore the possibilities of mitigating the problem of global warming so as to save the planet for coming catastrophe.

2.11.1 Kyoto Protocol

A agreement was signed by 84 countries in Kyoto, Japan in 1997, popularly Known as Kyoto Protocol. The Kyoto protocol recognizes that developed countries are responsible for high concentration of GHGs in earth's atmosphere due to intense 150 years of industrial activities. The protocol calls up 38 developed countries to cut their GHG emissions between the year 2008 and 2012 to levels 5 (below their 1990 levels). Though, USA which generates 37% of global emissions, highest in the world refused to follow Kyoto Protocol.

The Kyoto protocol is a legally binding agreement on international level. It came into force on 17 February 2005 when Russia accorded its approval. About 183 countries including India have ratified this protocol. It also recognized the necessity of some developing countries and permitted them a higher than 1990 level of GHG emissions facilitating them to accelerate industrial activities.

Kyoto Protocol Mechanism

The agreement lays down three 'market based mechanisms' to developed countries to meet their target to reduce their green house gas through their own measures. For the countries pollution more of than assigned limits, the mitigation mechanisms are:

1. Clean Development Mechanism (CDM)
2. Joint Implementation (JI)
3. International Emission Trading (IET)

Clean Development Mechanism (CDM): The CDM provides an opportunity to industrialized countries to earn credits by investing in emission reducing projects in developing countries to offset their excess emissions than the assigned amount of GHGs. A CDM activity may be a rural electrification project using solar plant, or adopting more efficient boilers, reduction in fossil fuel driven traffic to reduce the greenhouse gas emissions. Such projects can earn saleable Certified Emission Reduction (CER) credits. Each CER credit is equivalent to the reduction of 1 ton of CO_2 emission and is considered to meet GHG emission target.

Example: 1000 solar pumps of 3 kWh if operate 5 hours, i.e. at 19% plant load factor will save $1000 \times 3 \times 5 \times 300$ kg GH (considering 300 sunshine days)

$$= 4500 \times 10^3 \text{ kg } CO_2 \text{ or}$$
$$= 4500 \text{ Carbon credit.}$$

However, every nation has flexibility in deciding the means to attain its to emission targets.

India established its own National CDM Authority to conduct awareness and build capacity for preparing such project. By 2021, India registered 3334 projects, this is the highest for any country in the world. Most of these projects are in sectors of power generation, fuel switching, industrial processes, forestry and municipal solid waste management.

Joint Implementation (JI): Joint Implementation (JI) is another approach for countries (Annex I of Kyoto Protocol) to earn credits by investing emission reduction projects in another countries (of same Annex I) with Kyoto target.

The use of **JI** mechanism envisages emission reduction to the host country to earn Emission Reduction Units (ERU). These ERUs are subtracted from the host countries allowed emissions and are added to the total allowed emissions of the investor country.

International Emission Trading (IET): International Emission Trading allows buying and selling of emission credits,i.e. emission reduction units (ERU) among the countries which are the members in Kyoto document.

The treaty is a step to solve the environmental challenges of global warming posed by emission of green house gases. Dominant warming occurred in continental land mass between 40° and 70° N latitude.

This measures envisages to achieve the objective by adopting the following:

- Maximum use of renewable energy
- Identify and Restrict industries causing pollution
- Improve energy efficiency
- Manage transport Industry to reduce emission

Paris Conference (Nov. 29 to Dec. 12–2015)

The 21st Conference of Parties to the United Nations Framework Convention on Climate Change was held in Paris, France from November 29 and ended on December 12, 2015. Purpose was to conduct a negotiation process to a productive end. i.e, a consensus among 196 nations (195 Nations + the European Union) on how to tackle climate change?

India's Prime Minister Narendra Modi spoke on the opening day of the Paris conference about the need of developed country to vacate carbon space and on the need to keep alive the idea of 'differentiation' of developed and developing countries. He also spoke about climate justice and sustainable life style. As he put it, "The life styles of a few must not circumvent opportunities for the many who are on the first steps of the development ladder."

Kyoto Protocol and Paris Agreement

Paris agreement was 'historic' in many ways:

- It erases the "historical responsibility" of developed countries for causing climate change. Post-Paris there are no longer requirement to cut emissions drastically and vacate the carbon space.
- It removes all liabilities of developed countries for the loss and damage suffered by the poor of the world due to climate change.
- It makes carbon market voluntary, developed or developing countries will meet their emission reduction targets. Now, everyone can trade in carbon credits with willing country to trade.
- In this agreement, forests in the developing countries can be used to generate carbon credits to offset emissions from cars, factories and power plants.
- It globalizes creative carbon accounting, ensuring that the developed world does little domestically and trades with poorest nations to meet targets.

This agreement, therefore is monumental in a sense that from now on, the burden of mitigating, as well as paying for the impacts of climate change, has decisively shifted to developing countries.

The agreement has recognized that the increase in global average temperature has to remain well below 2°C and the world would pursue effort to limit such increase to 1.5°C. This is an ambitious target. 1.5°C reduces the risks of worst impacts of climate change significantly.

Future expectation: Paris agreement is not conducive for India. By 2030, if nations do not cut emissions, 60–75% of the 2°C carbon budget will be exhausted and nothing will be left for 1.5°C carbon budget. In 2030, India's Human Development Index will be less than 0.7. By 2030, India would need to double its forest expansion rate and further need cumulative carbon sink of 2.5–3 billion tonnes of carbon dioxide equivalent.

2.11.2 Copenhagen Climate Change Summit

Climate change conference was held at Bella Centre in Copenhagen during December 6–18, 2009. Terms of Kyoto Protocol expiration date decided was 31 December 2012 (as first commitment period) and 31 December 2020 (second commitment period). To keep the process on the line, there was a need for a new Climate Protocol. The conference in Copenhagen in the Framework Convention on climate change was attended by 193 countries from all over the world.

Nations chalked out strategies to achieve greater energy efficiency and a shift to renewable energy sources. Indian delegation expressed 'At present India's per capital emission is 1 tonne to 1.2 tonnes with 8.9% yearly GDP growth.' It was informed that India's energy intensity of production was falling with improvement in energy efficiency. Our emission intensity declined by 17.6% between 1990–2005 and futher 20–25 per cent reduction took place from 2005 to 2020. India's eleventh plan includes increasing energy efficiency up to 30% by 2026–27. Development of science and technology related to mitigation and adoptation to climate change was highlighted.

The summit introduced 'Copenhagen Accord' with a new kind of dynamics in global climate policy.

2.12 IMPACT OF RENEWABLE ENERGY GENERATION ON ENVIRONMENT

The environment is scarcely polluted by plants generating electricity from solar radiant energy, wind, biomass, geothermal and Ocean Thermal Energy Conversion (OTEC) processes. It is considered expedient to examine the extent to which renewable energy sources are environmentally benign. Some of the environmental issues associated with renewable sources are discussed below:

2.12.1 Solar Energy

Solar energy is available in abundance and considered the easiest and the cleanest means of tapping renewable energy. For direct conversion of solar radiation into usable form, the routes are: solar thermal, solar photovoltaic, and solar architecture. However, the main problem associated with tapping solar energy is the requirement to install large solar collectors. The other related problems are:

- High cost in populated areas because of a large area of land required. No reclaimation of such land is possible till the plant is decommissioned.
- Solar thermal systems use heat transfer fluids like glycol nitrates and sulphates. For high temperature applications, CFCs and aromatic alcohols are required. Solar thermal systems may pose a health hazard in public domain due to their careless disposal.
- Solar photovoltaic modules pose disposal problems due to the presence of arsenic and cadmium.
- Solar power generators need battery banks with inverters for storage capacity to provide power during nights and on cloudy days, with a back-up diesel generator. The total system contains several pollutants.
- Hazards to eyesight from solar reflectors.
- Solar thermal collectors and roof top photovoltaic (PV) systems have now become an integral part of high-rise buildings in metropolis. Large-scale use in densely populated cities limits the exposure of people to daylight due to changes in albedo.

2.12.2 Wind Energy

A wind turbine generator produces electricity with virtually no adverse effect on global environment. It produces no water and air pollution. In practice, a few environmental considerations associated with wind energy generation may be noted as follows:

- Wind power development requires a large land area to keep distance between turbines and turbine rows. Wind farm development in a forest area needs cutting of trees leading to degradation of environment.
- Visual intrusion of wind turbines on the existing landscape gives negative public response.
- Wind turbines degrade environment by noise pollution.
- Large wind turbines do interfere with television signals through reflection.
- Wind generators are hazards for birds, especially those in a migration route.

2.12.3 Biomass Energy

Biomass material obtained from agricultural, agro-industrial (bagasse) and forestry operations, serves 70% of India's population for energy needs. It is renewable, widely available and carbon-neutral. It supports soil fertilization, checks water runoff and stops desertification. On the other side of environmental analysis, several adverse impacts are:

- Combustion of biomass produces air pollution.
- Large-scale production of biomass and its harvesting accelerates soil erosion and nutrient loss.
- Energy-crop plantation on a large scale is water consuming with increased use of pesticides and fertilizers. It causes water pollution and flooding.
- Domestic use of biomass in rural areas creates air pollution, a health-hazard for women and children.

2.12.4 Geothermal Energy

Geothermal energy is the heat from the earth's interior, obtained as trapped hot water or steam. Geothermal resources differ with location, but environmentally pollute air and water. The chemical contents in geothermal fluids widely differ with site and rock structure of each reservoir. Steam and water from geothermal fields, contain non-condensable gases CO_2, CH_4, NH_3 and H_2S besides several toxic chemicals in suspension and collidal form. Environmental concerns of geothermal energy are:

- Gases escape into the atmosphere. Gases containing H_2S are oxidized to SO_2 and H_2SO_4 and drop down as acid rain.
- Chemicals like sulphates, chlorides and carbonates of lead, boron and arsenic, pollute soil and water.
- Discharge of waste hot water infects rivers, adversely affecting drinking water, farming and fisheries.

- Noise pollution, caused by exhausts, blow downs and centrifugal separation, is a health hazard. It is controlled by installation of silencers.
- Large-scale withdrawal of underground fluids from geothermal fields may trigger ground subsidence, causing damage to surface structures. The problem is mitigated by re-injection of spent fluids back into the reservoir.
- Existing geothermal electric plants emit an average of 122 kgs of CO_2 per MWh of electricity. It is a small fraction of the emission intensity of conventional thermal plants.

2.12.5 Ocean Thermal Energy Conversion (OTEC)

The Ocean Thermal Energy Conversion (OTEC) plants convert the thermal energy of ocean water, acquired from solar radiations, into electrical energy. Being environmentally benign compared to conventional power plants, it may still carry threats of adverse effects on quality of ocean water, prominently affecting the marine ecosystem.

- An OTEC plant displaces nearly 4 cumec water per MW generation. Massive flow disturbs thermal balance, changes salinity gradient and turbidity. It creates adverse impacts on marine environment.
- Mining of warm and cold water near the surface, develops convection of sinking cold water. It triggers thermal effects, i.e., variation in temperature by 4°C forcing mortality among coral and fishes.
- Ammonia is used as working fluid in closed cycle OTEC system; its leakage may cause great damage to the ocean ecosystem.

2.13 GHG EMISSIONS FROM VARIOUS ENERGY SOURCES

Greenhouse gases emitted by various energy sources are tabulated in Table 2.1.

TABLE 2.1 Life cycle emissions from various energy sources

Energy source	CO_2 (g/kWh)	SO_2 (g/kWh)	NO_x (g/kWh)
Coal	955	11.8	4.3
Oil	818	14.2	4.0
Diesel	772	1.6	12.3
Natural Gas (CCGT)	430	–	0.5
Large Hydro	11.6	0.024	0.006
Small Hydro	9	0.03	0.07
Wind	9	0.09	0.06
Solar Photovoltaic	167	0.34	0.30
Solar Thermal Electric	38	0.27	0.13
Biomass Energy Crop	27	0.16	2.5
Geothermal	9	0.02	0.28

It is evident from Table 2.1 that the renewable energy sources make little contribution to anthropogenic CO_2 emissions.

Ecological cost

Ecological true energy costs are higher than presently paid by the consumer. Ecological cost covers all expenses incurred to correct the environmental damage occurred during production and disposal of waste from exhaustible energy sources. This is also termed 'Green Accounting'.

2.14 COST OF ELECTRICITY PRODUCTION FROM DIFFERENT ENERGY SOURCES

Market penetration of any technology depends upon its comparative economic and financial advantages. The cost of energy produced from conventional sources like thermal (coal, oil, gas), large hydro and nuclear plants is cheaper than that obtained from renewable energy sources. The comparative cost details are indicated below:

Thermal	:	₹ 4.75/kWh + Ecological cost
Large Hydro	:	₹ 2.75/kWh
Nuclear	:	₹ 3.75/kWh + Ecological cost
Geothermal	:	₹ 4.5/kWh + Ecological cost
Small Hydro	:	₹ 4.0/kWh
Wind	:	₹ 4.0/kWh + Ecological cost
Biomass and bagasse	:	₹ 4.25/kWh
Solar thermal	:	₹ 10.49/kWh
Solar Photovoltaic	:	₹ 10.95/kWh

The above cost figures are representative for the year 2015.

Comparative study of electricity production cost from renewable sources and fossil fuel are quite matching except solar thermal and solar PV. However, Kyoto Protocol covers this aspect in article 2, and called upon the signatories to take remedial steps for these market imperfections.

With growing population and desire for better standard of living, the cost of electricity should be affordable ensuring concurrently a non-polluting environment.

2.15 ELECTROMAGNETIC RADIATION FROM EXTRA HIGH VOLTAGE (EHV) OVERHEAD LINES

The electromagnetic field in proximity to conductors of 400 kV and 765 kV power transmission lines, has an unfavourable effect on human organism. According to biologists, life is sensitive to electrical processes. Electrochemical and all biochemical processes are subject to possible variations in cells brought nearer to electromagnetic radiations. Magnetic and electric radiations induce a specific effect (not clearly understood as yet) on all living organisms.

British Medical Journal studied more than 29,000 children with cancer, including 9700 with leukaemia, born between 1962 and 1995. They found that youngsters, living within 200 metre of power lines, have 70% more chances to develop leukaemia compared with who lived beyond 600 metre (according to Scottish daily *The Scotsman*). Accordingly, a ban exists in the USA and Sweden for constructing new houses near EHV lines.

For information, it may be noted that directly under an overhead line of 400 kV, the electrostatic field strength is 11 kV/m and the magnetic flux density (based on current) may have a value of 40 µT. However, no significant effects were observed on humans at voltage stress ranging from 7.5 kV/m to 27 kV/m for a period of 30 minutes at 50 Hz.

2.15.1 Visual and Audible Impacts of EHV Lines and Substations

Environmental concerns are caused by factors given below:

- Land is acquired to install towers.
- Lines converging to an EHV substation spoil ambience and the landscape.
- Radio interference does occur which can be countered by providing a Faraday cage.
- Phenomenon of corona (a violet glow on the conductor surface with a hissing noise) is audible under the line. Towers are checked for tightness of joints with no sharp edges and use of an earth screen shield keeps the audible noise within acceptable limits.
- Engineers and technicians inside a power plant tolerate coal dust, high pitch noise, vibrations on turbine floor, steam blow-out and high temperatures near the furnace. They observe the necessary precautions and undergo regular medical examination.

2.16 ENERGY OPTIONS FOR INDIAN ECONOMY

With a population of around 139 crores, India cannot afford to follow the high energy consumption pattern of the West which resulted in an indiscriminate exploitation of fossil fuels and contributed to high pollution levels. To streamline the energy consumption power development, National Electricity Policy (NEP) stands approved by Govt. of India in January 2016. The policy guidelines are:

- To add 116,900 MW generation capacity during 13th plan (2017–2022). The existing installed capacity is 2,78,733 MW as on 30-9-2015.
- To increase the per capita consumption of power up to 1010 kWh/year by 2014–15, which increases to 1208 kWh in 2019–20.
- To cater for suitable margins and returns for investors in electricity sector.
- To encourage captive power plants to supply their surplus power to the grid and to rural areas locally as distributed generation. It would accelerate development of renewable energy sources like solar thermal and PV, wind, biomass and small hydro. Adapt bio-fuels in energy mix to save fossils.
- To concentrate on tapping unharnessed estimated hydropower of 148,700 MW at average load factor. The installed capacity in 30-9-15 was only 42,283 MW.

- Nuclear energy to contribute 40,000 MW of power with technology development for thorium-based reactors. Thorium fuel is abundantly available in the country.

Nuclear power has now emerged as a promising way of reducing carbon emissions. This will also slow down global warming.

2.17 INDIA'S AMBITIOUS PLAN OF RENEWABLE POWER BY 2022

India in October 2015 formalized its ambitious plan to increase its renewable energy capacity to 175 GW. Centre plans to install 17500 MW each year between 2020 and 2022, which is 17.5 times higher than the maximum it has ever installed in one year (1000 MW). China holds the distinctions of adding the maximum capacity in a year which is 12000 MW.

The development of power lines for exchange of power across the states has to be created to support the growing renewable energy capacity. The Discoms (Power Distribution Companies) of various states expressed that the share of renewable energy in the national grid should be between 20 to 30% of the total capacity. The estimated investment for the development of green corridor transmission lines is around ₹49000 crores.

Share of green power in different sectors to be developed by 2022 and capacity achieved till August 2015 is given in Table 2.1.

TABLE 2.1 Capacity of Green Power

Sector	Capacity till Aug. 2015 in MW	Proposed capacity in MW up to 2022
Solar	4229	100,000
Wind	24088	60,000
Biomass	4546	10,000
Small Hydro	4146	5,000

Development of solar power is land intensive and acquiring land is a difficult task. Estimate of Central Electricity Regulatory Commission suggests 2.2 ha of land is required for 1 MW of ground mounted installation of solar PV.

REVIEW QUESTIONS

1. Discuss the atmospheric pollution threat posed to environment from thermal power plants.
2. What do you mean by 'particulates'? How can particulates be controlled.
3. Discuss the technical and social problems associated with hydropower plants.
4. Briefly discuss the operational safety considerations of nuclear power plants.
5. Discuss various methods of nuclear waste disposal.
6. What is 'Kyoto Protocol'? Briefly discuss the importance of Kyoto Protocol for environmental safety.

7. How does environment get affected by the use of the following sources of energy?
   (a) Solar energy sources
   (b) Biomass energy sources
   (c) Wind energy

8. Discuss the effect of electromagnetic radiation on humans.

9. Discuss and elaborate 'Geothermal' and 'Ocean Thermal' energy options for India.

10. Write your views on energy options for India, keeping the socio-economic environmental considerations in mind.

11. How the environment is affected by operation of geothermal power plant?

12. Explain major points of agreement at 21st United Nation Framework Convention on Climate Change was held in Paris (Dec. 2015).

13. Explain
    (i) Global warming
    (ii) GHG and their effects
    (iii) Clean development mechanism (CDM),
    (iv) International Emission Trading.

14. Discuss India's Ambitious Plan for development of renewable power by 2022.

SOLAR RADIATION AND ITS MEASUREMENT

3.1   A PERSPECTIVE

The sun is a hydrodynamic spherical body of extremely hot ionized gases (plasma), generating energy by the process of thermonuclear fusion. The temperature of the interior of the sun is estimated at 8×10^6 K to 40×10^6 K, where energy is released by fusion of hydrogen to helium.

Energy radiated from the sun is electromagnetic waves reaching the planet earth in three spectral regions, ultraviolet 6.4% ($\lambda < 0.38$ µm), visible 48% (0.38 µm $< \lambda < 0.78$ µm) and infrared 45.6% ($\lambda > 0.78$ µm) of total energy. Due to the large distance between the sun and the earth (1.495×10^8 km) the beam radiation received from the sun on the earth is almost parallel.

3.2   SOLAR CONSTANT

The sun, being at a very large distance from the earth, solar rays subtend an angle of only 32 minutes on earth, as shown in Figure 3.1. Energy flux received from the sun before entering the earth's atmosphere, is a constant quantity.

The solar constant, I_{sc}, is the energy from the sun received on a unit area perpendicular to the solar rays at the mean distance from the sun outside the atmosphere. Based on the experimental measurements, the standard value of the solar constant is 1367 W/m^2 or 1.958 langley per minute (1 langley/min is the unit, equivalent to 1 cal/cm^2/min). In terms of other units, I_{sc} = 432 Btu/ft^2/h or 4.921 MJ/m^2/h.

3.3   SPECTRAL DISTRIBUTION OF EXTRATERRESTRIAL RADIATION

Extraterrestrial radiation is the measure of solar radiation that would be received in the absence of atmosphere. A typical spectral distribution of extraterrestrial radiation is shown in Figure 3.2. The curve rises sharply with the wavelength and reaches the maximum value of 2074 W/m²/µm at a wavelength of 0.48 µm. It then decreases asymptotically to zero, showing that 99% of the sun's radiation is obtained up to a wavelength of 4 µm.

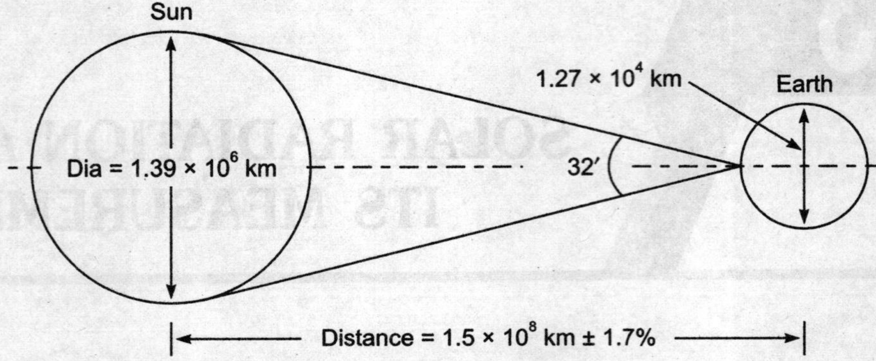

FIGURE 3.1 Sun–Earth geometry.

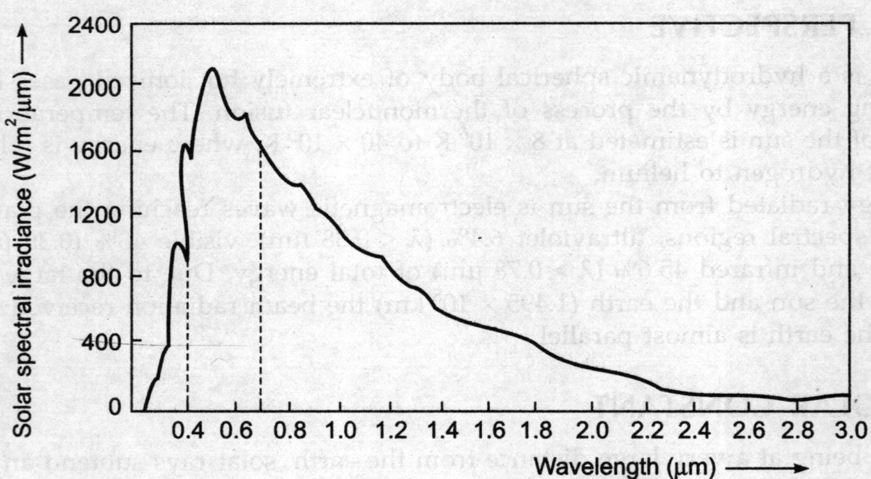

FIGURE 3.2 Spectral distribution of extraterrestrial radiation.

The distance between the sun and the earth varies due to the elliptical motion of the earth. Accordingly, the extraterrestrial flux also varies, which can be calculated (on any day) by the equation

$$I_\mathrm{n} = I_\mathrm{sc}\left(1 + 0.033\cos\frac{360n}{365}\right) \tag{3.1}$$

where n is the day of the year counted from the first day of January.

Solar radiation reaching the earth is essentially equivalent to blackbody radiation. Using the Stefan–Boltzmann law, the equivalent blackbody temperature is 5779 K for a solar constant of 1367 W/m².

3.4 TERRESTRIAL SOLAR RADIATION

For utilisation of solar energy, a study is required to be carried out of radiations received on the earth's surface. Solar radiations pass through the earth's atmosphere and are subjected to scattering and atmospheric absorption. A part of scattered radiation is reflected back into space.

Short wave ultraviolet rays are absorbed by ozone and long wave infrared rays are absorbed by CO_2 and water vapours. Scattering is due to air molecules, dust particles and water droplets that cause attenuation of radiation as detailed in Figure 3.3. Minimum attenuation takes place in a clear sky when the earth's surface receives maximum radiation.

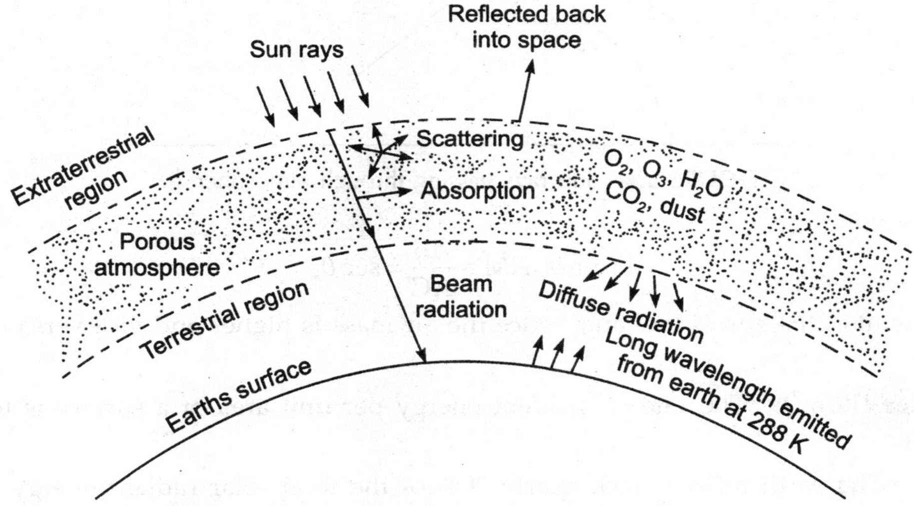

FIGURE 3.3 Solar radiation atmospheric mechanisms.

The terms pertaining to solar radiation are now defined as below:

Beam radiation (I_b): Solar radiation received on the earth's surface without change in direction, is called *beam* or *direct radiation*.

Diffuse radiation (I_d): The radiation received on a terrestrial surface (scattered by aerosols and dust) from all parts of the sky dome, is known as *diffuse radiation*.

Total radiation (I_T): The sum of beam and diffuse radiations ($I_b + I_d$) is referred to as total radiation. When measured at a location on the earth's surface, it is called *solar insolation* at the place. When measured on a horizontal surface, it is called *global radiation* (I_g).

Sun at zenith: It is the position of the sun directly overhead.

Air mass (AM): It is the ratio of the path length of beam radiation through the atmosphere, to the path length if the sun were at zenith. At sea level AM = 1, when the sun is at zenith or directly overhead; AM = 2 when the angle subtended by zenith and line of sight of the sun is 60°; AM = 0 just above the earth's atmosphere. At zenith angle θ_z, the air mass is calculated as (see Figure 3.4):

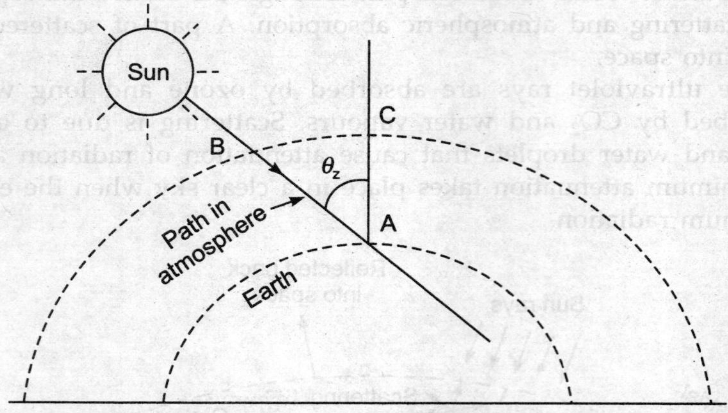

FIGURE 3.4 Sun rays passing through atmosphere.

$$\text{Air mass, AM} = \frac{AB}{AC} = \sec \theta_z \qquad (3.2)$$

During winter, the sun is low and hence the air mass is higher and vice versa during summer.

Irradiance (W/m²): The rate of incident energy per unit area of a surface is termed *irradiance*.

Albedo: The earth reflects back nearly 30% of the total solar radiant energy to the space by reflection from clouds, by scattering and by reflection at the earth's surface. This is called the *albedo* of the earth's atmosphere system.

3.5 SOLAR RADIATION GEOMETRY

Solar radiation varies in intensity at different locations on the earth, which revolves elliptically around the sun. For the calculation of solar radiation, the position of a point P on the earth's surface with regard to sun's rays can be located, if the latitude ϕ, the hour angle ω for the point and the sun's declination δ are known. These basic angles for a location P on the northern hemisphere are shown in Figure 3.5 and defined as follows:

Latitude (ϕ): The latitude ϕ of a place is the angle subtended by the radial line joining the place to the centre of the earth, with the projection of the line on the equatorial plane. Conventionally, the latitude for northern hemisphere is measured positive.

Declination (δ): Declination δ is the angle subtended by a line joining the centres of the earth and the sun with its projection on the earth's equatorial plane. Declination occurs as the axis of the earth is inclined to the plane of its orbit at an angle 66½°, as shown in Figure 3.6.

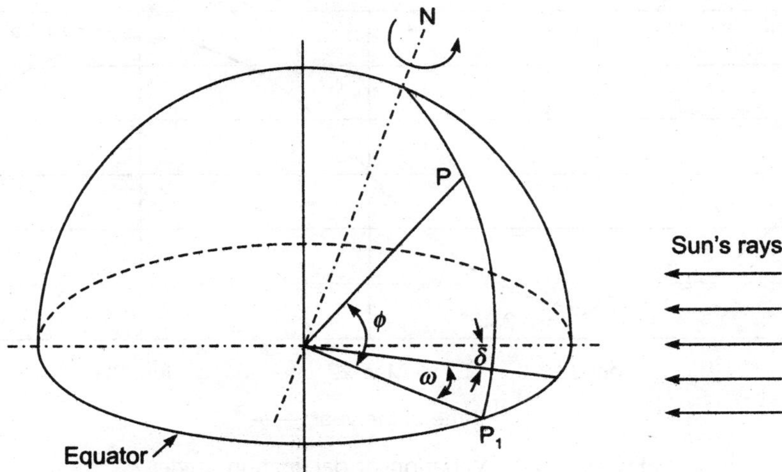

FIGURE 3.5 Latitude ϕ, hour angle ω and sun's declination δ.

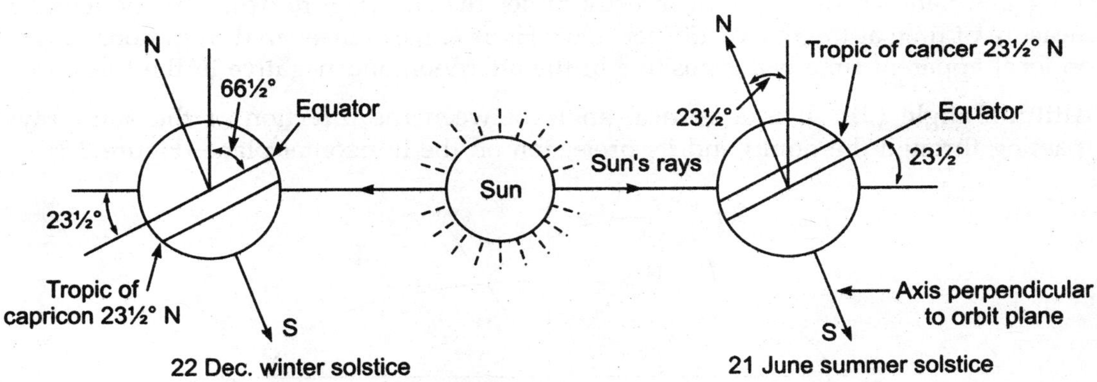

FIGURE 3.6 Tropics and northern hemisphere.

The declination angle changes from a maximum value of +23.45° on June 21 to a minimum of −23.45° on December 22. The declination is zero on two equinox days, i.e., March 22 and September 22. The angle of declination may be calculated as suggested by Cooper (1969)

$$\delta \text{ (in degrees)} = 23.45 \sin\left[\frac{360}{365}(284 + n)\right] \tag{3.3}$$

where n is the total number of days counted from first January till the date of calculation.

[For example for June 21, 2004, $n = 31 + 29 + 31 + 30 + 31 + 21 = 173$]

The variation of declination angle δ with the nth day of the year is shown in Figure 3.7.

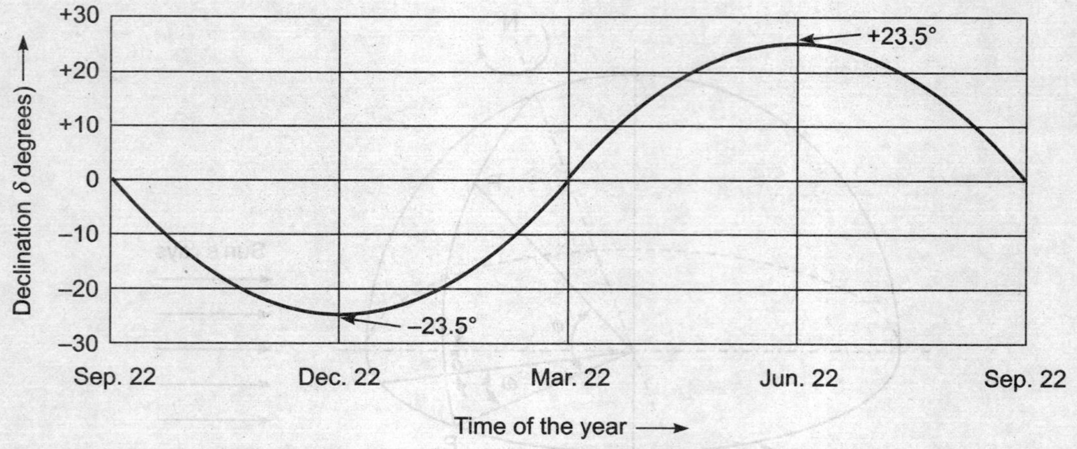

FIGURE 3.7 Variation of declination angle.

Hour angle (ω): Hour angle ω is the angle through which the earth must rotate to bring the meridian of the point directly under the sun (Figure 3.5). It is the angular measure of time at the rate of 15° per hour. Hour angle is measured from noon, based on local apparent time being positive in the afternoon and negative in the forenoon.

Altitude angle (α): It is a vertical angle between the direction of the sun's rays (passing through the point) and its projection on the horizontal plane (Figure 3.8).

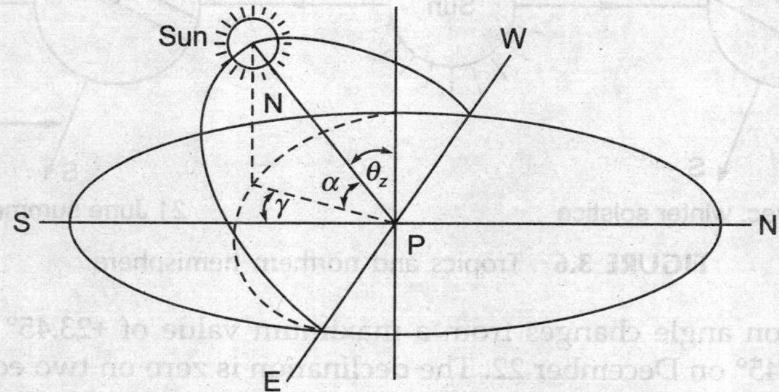

FIGURE 3.8 Sun's zenith, altitude and azimuth angles (northern hemisphere).

Zenith angle (θ_z): It is the vertical angle between the sun's rays and the line perpendicular to the horizontal plane through the point. It is the complimentary angle of the sun's altitude angle. Thus,

$$\theta_z + \alpha = \frac{\pi}{2}$$

Surface azimuth angle (γ): It is an angle subtended in the horizontal plane of the normal to the surface on the horizontal plane (Figure 3.8). By convention, the angle is taken positive if the normal is west of south and negative when east of south in northern hemisphere, and vice versa for southern hemisphere.

Tilted surface: The basic angles for a location P on a tilted surface are shown in Figure 3.9.

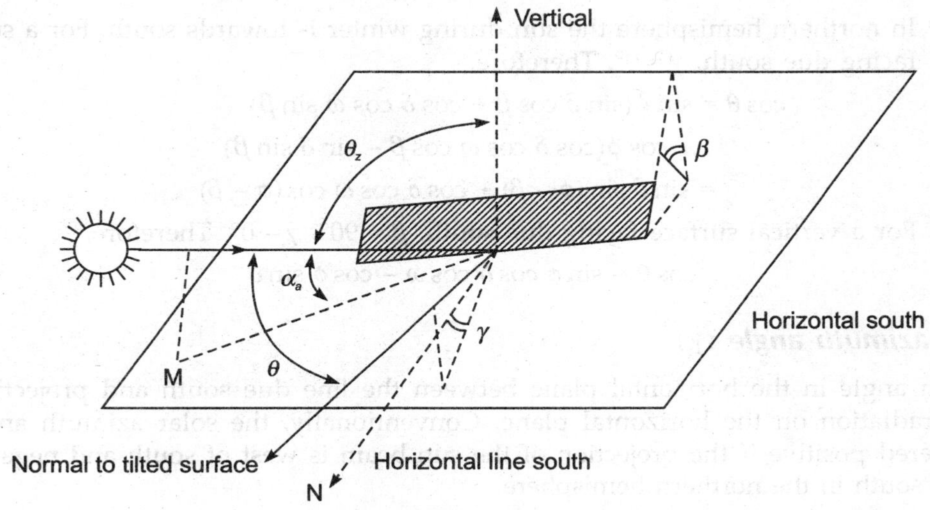

FIGURE 3.9   Diagram showing the angle of incidence q, the zenith angle θ_z, the solar altitude angle α_a, the slope β, and the surface azimuth angle γ for a tilted surface.

Slope (β): It is an angle made by the plane surface with the horizontal surface. The angle is taken as positive for a surface sloping towards south, and negative for a surface sloping north (Figure 3.9).

3.6   COMPUTATION OF COS θ FOR ANY LOCATION HAVING ANY ORIENTATION

To compute the beam energy falling on a surface having any orientation, the incident beam flux I_b is multiplied by cos θ, where θ is the angle between the incident beam and the normal to the tilted surface (Figure 3.9). The angle θ depends on the position of the sun in the sky.

A general equation showing the relation of angles is

$$\cos\theta = \sin\phi(\sin\delta\cos\beta + \cos\delta\cos\gamma\cos\omega\sin\beta)$$
$$+ \cos\phi(\cos\delta\cos\omega\cos\beta - \sin\delta\cos\gamma\sin\beta)$$
$$+ \cos\delta\sin\gamma\sin\omega\sin\beta \tag{3.4}$$

Use of Eq. (3.4) can be demonstrated as:

(i) For a vertical surface, $\beta = 90°$. Therefore,

$$\cos \theta = \sin \phi \cos \delta \cos \gamma \cos \omega - \cos \phi \sin \delta \cos \gamma + \cos \delta \sin \gamma \sin \omega \qquad (3.5)$$

(ii) For a horizontal surface, $\beta = 0°$. Therefore,

$$\cos \theta = \sin \phi \sin \delta + \cos \phi \cos \delta \cos \omega \qquad (3.6)$$

In this case, the angle θ is the zenith angle θ_z (shown in Figure 3.9).

(iii) In northern hemisphere the sun during winter is towards south. For a surface facing due south, $\gamma = 0°$. Therefore,

$$\cos \theta = \sin \phi (\sin \delta \cos \beta + \cos \delta \cos \omega \sin \beta)$$
$$+ \cos \phi (\cos \delta \cos \omega \cos \beta - \sin \delta \sin \beta)$$
$$= \sin \delta \sin (\phi - \beta) + \cos \delta \cos \omega \cos (\phi - \beta) \qquad (3.7)$$

(iv) For a vertical surface facing due south, $\beta = 90°$, $\gamma = 0°$. Therefore,

$$\cos \theta = \sin \phi \cos \delta \cos \omega - \cos \phi \sin \delta \qquad (3.8)$$

Solar azimuth angle (γ_s)

It is an angle in the horizontal plane between the line due south and projection of beam radiation on the horizontal plane. Conventionally, the solar azimuth angle is considered positive if the projection of the sun beam is west of south and negative if east of south in the northern hemisphere.

3.7 SUNRISE, SUNSET AND DAY LENGTH

The times of sunrise and sunset and the duration of the day-length depend upon the latitude of the location and the month in the year. At sunrise and sunset, the sunlight is parallel to the ground surface with a zenith angle of 90°. The hour angle pertaining to sunrise or sunset (ω_s) is obtained from Eq. (3.6) as

$$\cos \omega_s = -\tan \phi \tan \delta$$

or

$$\omega_s = \cos^{-1} (-\tan \phi \tan \delta) \qquad (3.9)$$

The value of hour angle corresponding to sunrise is positive, and negative corresponding to sunset. The total angles between sunrise and sunset is given by

$$2\omega_s = 2 \cos^{-1} (-\tan \phi \tan \delta) \qquad (3.10)$$

Since 15° of hour angle corresponds to one hour, the corresponding day-length (T_d) in hours is given by

$$T_d = \frac{2}{15} \cos^{-1} (-\tan \phi \tan \delta) \qquad (3.11)$$

Local apparent time (LAT)

The time used for calculating the hour angle ω is the 'local apparent time' which is not the same as the 'local clock time'. It can be obtained from the local time observed on a clock by applying two corrections. The first correction arises due to the difference between the longitude of a location and the meridian on which the standard time is determined. This correction has a magnitude of 4 minute for each degree difference in longitude. The other correction is known as the 'equation of time correction' which is required due to the fact that the earth's orbit and the rate of rotation are subject to certain fluctuations. This correction is applied by results of experimental observations as plotted in Figure 3.10.

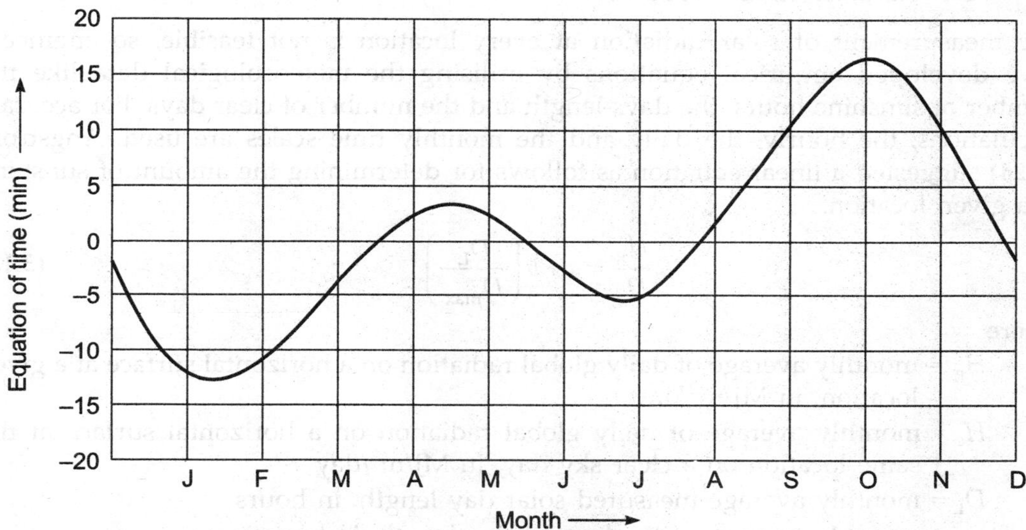

FIGURE 3.10 Graph for the 'equation of time correction'.

Therefore,

Local apparent time (LAT) = Standard time ± 4 (Standard time longitude
– Longitude of location) + (Time correction) (3.12)

The positive sign in the first correction is for the western hemisphere while the negative sign is applicable for the eastern hemisphere.

EXAMPLE 3.1

Determine the local apparent time corresponding to 13:30 IST on July 1, at Delhi (28°35′ N, 77°12′ E). The 'equation of time correction' on July 1 from Figure 3.10 is –4 minutes. In India, the standard time is based on 82°30′ E.

Solution:

Local apparent time = 13.50 h – 4 [(82.50) – (77.2)] min + (–4 min)

= 13.50 h – 4 (82.50 – 77.2) min – 4 min

= 13.50 h – 21.20 min – 4 min

$$= 13.50 \text{ h} - 25.20 \text{ min}$$
$$= 13.50 \text{ h} - 0.42 \text{ h}$$
$$= 13.08 \text{ h} = 13 \text{ h } 4 \text{ min } 48 \text{ s}$$

3.8 EMPIRICAL EQUATION FOR ESTIMATING THE AVAILABILITY OF SOLAR RADIATION

The measurement of solar radiation at every location is not feasible, so engineers have developed empirical equations by utilising the meteorological data like the number of sunshine hours, the days-length and the number of clear days. For accurate calculations, the hourly, the daily and the monthly time scales are used. Angstrom (1924) suggested a linear equation as follows for determining the amount of sunshine at a given location.

$$\frac{H_g}{H_c} = a + b\left(\frac{D_L}{D_{max}}\right) \tag{3.13}$$

where

H_g = monthly average of daily global radiation on a horizontal surface at a given location, in MJ/m^2/day

H_c = monthly average of daily global radiation on a horizontal surface at the same location on a clear sky day, in MJ/m^2/day

D_L = monthly average measured solar day length, in hours

D_{max} = monthly average of the longest day-length, in hours

a, b = constants for the location.

It is difficult to define a clear sky day, so it was proposed that H_c in Eq. (3.13) should be replaced by H_o. Here, H_o is the monthly average of daily extra terrestrial radiation that would fall on a horizontal surface at the given location. Thus,

$$\frac{H_g}{H_o} = a + b\left(\frac{D_L}{D_{max}}\right) \tag{3.14}$$

The values of a and b obtained for 20 Indian cities by conducting a city-wise regression analysis are given in Table 3.1.

The value of H_o can be obtained by the following empirical equation

$$H_o = \frac{24}{\pi} I_{sc}\left(1 + 0.033 \cos \frac{360n}{365}\right)(\omega_s \sin \phi \sin \delta + \cos \phi \cos \delta \sin \omega_s) \tag{3.15}$$

where

I_{sc} = solar constant per hour = 1367 W/m^2 in SI units

ω_s = sunset hour angle

n = day of the year.

TABLE 3.1 Constants a and b in Eq. (3.14) for 20 Indian cities

Location	a	b	Mean error per cent
Ahmedabad	0.28	0.48	3.0
Bangalore	0.18	0.64	3.9
Bhavnagar	0.28	0.47	2.8
Bhopal	0.27	0.50	–
Kolkata	0.28	0.42	1.3
Goa	0.30	0.48	2.1
Jodhpur	0.33	0.46	2.0
Kodaikanal	0.32	0.55	2.9
Chennai	0.30	0.44	3.5
Mangalore	0.27	0.43	4.2
Minicoy	0.26	0.39	1.4
Nagpur	0.27	0.50	1.6
New Delhi	0.25	0.57	3.0
Pune	0.31	0.43	1.9
Roorkee	0.25	0.56	–
Shillong	0.22	0.57	3.0
Srinagar	0.35	0.40	4.7
Trivandrum	0.37	0.39	2.5
Vidisha	0.27	0.50	–
Vishakhapatnam	0.28	0.47	1.2

EXAMPLE 3.2

At Nagpur, the following observations were made:

Theoretical maximum possible sunshine hours = 9.5 h
Average measured length of a day during April = 9.0 h
Solar radiation for a clear day, H_o = 2100 kJ/m²/day
Constants: $a = 0.27$, $b = 0.50$.

Calculate the average daily global radiation.

Solution:

$$H_g = H_o \left[a + b \left(\frac{D_L}{D_{max}} \right) \right]$$

$$= 2100 \left[0.27 + 0.50 \left(\frac{9.0}{9.5} \right) \right]$$

$$= 1554 \text{ kJ/m}^2/\text{day}$$

Monthly average daily diffuse radiation

As a result of study of field data conducted by Liu and Jordan, they arrived at a result that the daily diffuse to global radiation ratio could be correlated with the daily global to extraterrestrial radiation ratio. It was expressed by a cubic equation

$$\frac{H_d}{H_g} = 1.390 - 4.027\,K_T + 5.531\,K_T^2 - 3.108\,K_T^3 \tag{3.16}$$

where

H_d = monthly average for daily diffuse radiation on a horizontal surface, in kJ/m²/day

$$K_T = \frac{H_g}{H_o} = \text{monthly average clearness index.} \tag{3.17}$$

It was indicated by Kreith that Eq. (3.16) was obtained with a value of 1394 W/m² for the solar constant.

When the Indian data was analyzed, two linear equations were finalized.

$$\frac{H_d}{H_g} = 1.411 - 1.696\,K_T \tag{3.18}$$

$$\frac{H_d}{H_g} = 1.354 - 1.570\,K_T \tag{3.19}$$

The above equations provide similar results, and are valid for

$$0.3 < \left(\frac{H_d}{H_g}\right) < 0.7$$

Solar radiation on an inclined surface

The total solar radiation incident on a surface has three components.

    (i) Beam solar radiation
    (ii) Diffuse solar radiation
    (iii) Reflected solar radiation from ground and surroundings.

To obtain maximum solar energy, flat plate collectors always face the sun using a sun tracking equipment. It, therefore, infers that the solar radiation collecting appliances are tilted at an angle to the horizontal. However, the measuring instruments generally measure the values of solar radiation falling on a horizontal surface. Thus, mathematical analysis is necessary to convert the values measured on horizontal surfaces to the corresponding values obtainable on the inclined surfaces.

Beam radiation

Generally, the inclined surface faces south to obtain maximum solar radiation even during winter, i.e., $\gamma = 0°$. Therefore,

$$\cos \theta = \sin \delta \sin (\phi - \beta) + \cos \delta \cos \omega \cos (\phi - \beta)$$

While for a horizontal surface ($\theta = \theta_z$), and therefore,

$$\cos \theta_z = \sin \phi \sin \delta + \cos \phi \cos \delta \cos \omega$$

The ratio of beam radiation falling on an inclined surface to that falling on a horizontal surface is termed *tilt factor for beam radiation*. It is represented by the notation R_b. Thus,

$$R_b = \frac{\cos \theta}{\cos \theta_z} = \frac{\sin \delta \sin (\theta - \beta) + \cos \delta \cos \omega \cos (\phi - \beta)}{\sin \phi \sin \delta + \cos \phi \cos \delta \cos \omega} \tag{3.20}$$

Other equations for R_b can be derived complying to conditions, when the inclined surface is oriented in different directions with $\gamma \neq 0°$.

Diffuse radiation

The ratio of diffuse radiation falling on a tilted surface to that falling on a horizontal surface is known as *tilt factor for diffuse radiation*, symbolized by R_d. Considering the sky as an isotropic source of diffuse radiation, R_d for an inclined surface with a slope β may be calculated from

$$R_d = \frac{1 + \cos \beta}{2} \tag{3.21}$$

where $(1 + \cos \beta)/2$ is the *radiation shape factor* for an inclined surface with reference to the sky.

Reflected radiation

Since $(1 + \cos \beta)/2$ is the *radiation shape factor* for an inclined surface with reference to the sky, so $(1 - \cos \beta)/2$ is the radiation shape factor for the surface with respect to surroundings. Accepting that the beam and diffuse radiation after reflection from the ground is diffuse and isotropic, and the reflectivity is ρ, the *tilt factor for reflected radiation* is expressed as:

$$R_r = \frac{\rho(1 - \cos \beta)}{2} \tag{3.22}$$

Total radiation

The total radiation flux falling on an inclined surface at any instant is expressed as:

$$I_T = I_b R_b + I_d R_d + (I_b + I_d) R_r \tag{3.23}$$

Dividing Eq. (3.23) by I_g, we get the ratio of solar flux reaching on an inclined surface at any instant to that on a horizontal surface. That is,

$$\frac{I_T}{I_g} = \left(1 - \frac{I_d}{I_g}\right) R_b + \frac{I_d}{I_g} R_d + R_r \qquad (\because \ I_g = I_b + I_d) \tag{3.24}$$

For evaluating R_r, the diffuse reflectivity ρ can be taken as 0.2 for the surface of concrete or grass and 0.7 for a surface with snow cover.

The monthly average of daily radiation reaching a tilted surface is required in dealing with liquid flat-plate collectors and in other applications. Liu and Jordan have suggested that the ratio of the daily radiation falling on an inclined surface (H_T) to the daily global radiation on an horizontal surface (H_g) can be represented by an equation similar to Eq. (3.24). Thus,

$$\frac{H_T}{H_g} = \left(1 - \frac{H_d}{H_g}\right) R_b + \frac{H_d}{H_g} R_d + R_r \tag{3.25}$$

For a surface facing south ($\gamma = 0°$), Liu and Jordan proposed:

$$R_b = \frac{\omega_{si} \sin\delta \sin(\phi - \beta) + \cos\delta \sin\omega_{si} \cos(\phi - \beta)}{\omega_{sh} \sin\phi \sin\delta + \cos\phi \cos\delta \sin\omega_{sh}} \tag{3.26}$$

$$R_d = \frac{1 + \cos\beta}{2} \tag{3.27}$$

$$R_r = \rho\left(\frac{1 - \cos\beta}{2}\right) \tag{3.28}$$

where ω_{si} and ω_{sh} in Eq. (3.26) are sunrise or sunset hour angles (in radians) for an inclined surface and a horizontal surface respectively.

EXAMPLE 3.3

Find the angle subtended by beam radiation with the normal to a flat-plate collector at 9 a.m. for the day on November 3, 2003. The collector is in Delhi (28° 35′ N, 77° 12′ E), inclined at an angle of 36° with the horizontal and is facing due south.

Solution:

Given $\gamma = 0°$ and $n = 307$ for November 3, 2003
From Eq. (3.3),

$$\delta = 23.45 \sin\left[\frac{360}{365}(284 + 307)\right]$$

$$= 23.45 \sin 582.9°$$

$$= 23.45 (-0.681) = -15.96°$$

At 9.00 a.m. (local apparent time) $\omega = 45°$. From Eq. (3.7),

$$\cos\theta = \sin\delta \sin(\phi - \beta) + \cos\delta \cos\omega \cos(\phi - \beta)$$

or $\quad \cos\theta = \sin(-15.96°) \sin(28.58° - 36°) + \cos(-15.96°) \cos45° \cos(28.58° - 36°)$

$$= (-0.275)(-0.129) + 0.961 \times 0.707 \times 0.99 = 0.709$$

∴ $\quad\quad \theta = 44.85°$

EXAMPLE 3.4

Compute the monthly average hourly solar flux received on a flat-plate collector facing due south ($\gamma = 0°$) having a slope of 12°. The collector is located at a place 15° 00′ N on 20th day of October. The data given are:

Time 11:12 h (local apparent time)

$H_g = 2408$ kJ/m²/h

$H_d = 1073$ kJ/m²/h

Ground reflectivity, $\rho = 0.25$, $\omega = 7.5°$

Solution:

Given $\gamma = 0°$ and $n = 293$ for 20th October.

From Cooper's equation, given in Eq. (3.3)

$$\delta = 23.45 \sin\left[\frac{360}{365}(284 + 293)\right]$$

$$= -11.40°$$

Substituting the given data in Eq. (3.20), we have

$$R_b = \frac{\sin(-11.4°)\sin(15° - 12°) + \cos(-11.40°)\cos 7.5° \cos(15° - 12°)}{\sin 15° \sin(-11.4°) + \cos 15° \cos(-11.40°)\cos 7.5°}$$

$$= 1.08$$

From Eqs. (3.21) and (3.22),

$$R_d = \left(\frac{1 + \cos 12°}{2}\right) = 0.989$$

$$R_r = 0.2\left(\frac{1 - \cos 12°}{2}\right) = 0.0022$$

Equation (3.25) is valid for evaluating the average daily radiation reaching on an inclined surface if the value of ω is taken at the middle of the hour. Similarly, the monthly average hourly value H_T can be calculated by using a representative day of the month. The modified form of Eq. (3.25) becomes

$$\frac{H_T}{H_g} = \left(1 - \frac{H_d}{H_g}\right)R_b + \frac{H_d}{H_g}R_d + R_r$$

Thus,

$$\frac{H_T}{H_g} = \left(1 - \frac{1073}{2408}\right)1.08 + \frac{1073}{2408}0.989 + 0.0022$$

$$= 1.04165$$

Therefore,

$$H_T = 2508 \text{ kJ/m}^2/\text{h}$$

3.9 SOLAR RADIATION MEASUREMENTS

The solar radiation data bank is required for many purposes, e.g. solar energy appliances, hydrology and weather forecast. A few instruments used to measure solar radiation are discussed below:

Pyranometer

The pyranometer measures global or diffuse radiation on a horizontal surface. It covers total hemispherical solar radiation with a view angle of 2 steradians.

The pyranometer designed by the Eppley laboratories, USA, operates on the principle of thermopile. It consists of a black surface which heats up when exposed to solar radiation. Its temperature rises until the rate of heat gain from solar radiation equals the heat loss by conduction, convection and radiation. On the black surface the hot junctions of a thermopile are attached, while the cold junctions are placed in a position such that they do not receive the radiation. An electrical output voltage (0 to 10 mV range) generated by the temperature difference between the black and the white surfaces indicates the intensity of solar radiation. The output can be obtained on a strip chart or on a digital printout over a period of time. This is a measure of global radiation.

The pyranometer can also measure diffuse sky radiation by providing a shading ring or disc to shade the direct sun rays. The shading ring is provided with an arrangement such that its plane is parallel to the plane of the sun's path across the sky. Consequently, it shades the thermopile element at all times from direct sunshine and the pyranometer measures only the diffuse radiation obtained from the sky. A continuous record can be obtained either on an electronic chart or on an integrated digital printout system. As the shading ring blocks a certain amount of diffuse sky radiation besides direct radiation, a correction factor is applied to the measured value.

Data acquisition system for measurement of solar radiation

This system does not require an instrument operator to measure the radiation data. With a personal computer (PC), the system uses an analog-to-digital conversion (ADC) card, which serves as a vital interface between the sensor and the PC to obtain analog data from the sensor. The data so received is processed in the PC with an appropriate software.

The radiation falling on the pyranometer generates thermo-electric emf which is fed into one of the channels of the ADC card provided with the PC. The numerical value of the instantaneous voltage in the digital form is stored in a Programmable Peripheral Interface (PPI). A printout of the solar flux can be obtained by processing the data. The block diagram of such a radiation measuring system is shown in Figure 3.11.

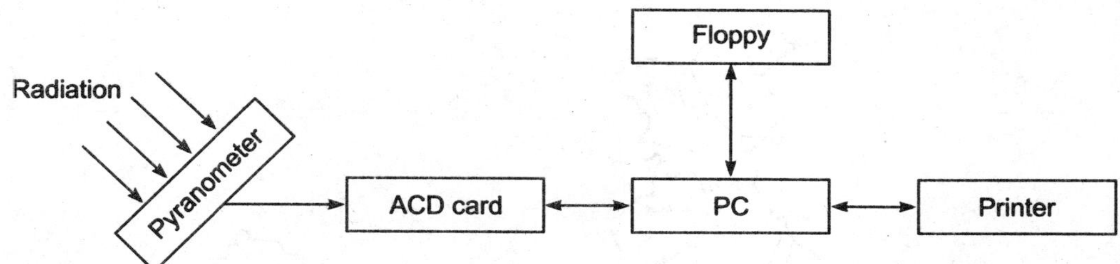

FIGURE 3.11 Block diagram of a radiation measuring system.

Pyrheliometer

A pyrheliometer is an instrument which measures beam radiation on a surface normal to the sun's rays. The sensor is a thermopile and its disc is located at the base of a tube whose axis is aligned in the direction of the sun's rays. Thus, diffuse radiation is blocked from the sensor surface. The pyrheliometer designed by Eppley Laboratories, USA, consists of bismuth silver thermopile, with 15 temperature-compensated junctions connected in series. It is mounted at the end of a cylindrical tube, with a series of diaphragms (the aperture is limited to an angle of 5.42°). The instrument is mounted on a motor-driven heliostat which is adjusted every week to cover changes in the sun's declination. The output of the pyreheliometer can either be recorded on a strip chart recorder or integrated over a suitable time period. The pyrheliometer readings give data for atmospheric turbidity and provide a clearness index.

Sunshine recorder

The duration in hours of bright sunshine in a day is measured by a sunshine recorder. It consists of a glass sphere installed in a section of spherical metal bowl, having grooves for holding a recorder card strip. The glass sphere is adjusted to focus sun rays to a point on the card strip. On a bright sunshine day, the focused image burns a trace on the card. Through the day the sun moves across the sky, the image moves along the strip. The length of the image is a direct measure of the duration of bright sunshine.

3.10 SOLAR RADIATION DATA FOR INDIA

India lies within the latitudes of 7° N and 37° N, with annual intensity of solar radiation between 400 and 700 cal/cm^2/day. Most parts of India receive 4–7 kWh/m^2/day of solar radiation with 250–300 sunny days in a year. The annual average daily global solar radiation in India (in kWh/m^2/day) is shown in Figure 3.12. A similar map can also be drawn for average daily diffuse radiation.

The highest annual radiation energy is received in the western Rajasthan while the north-eastern region receives the lowest annual radiation.

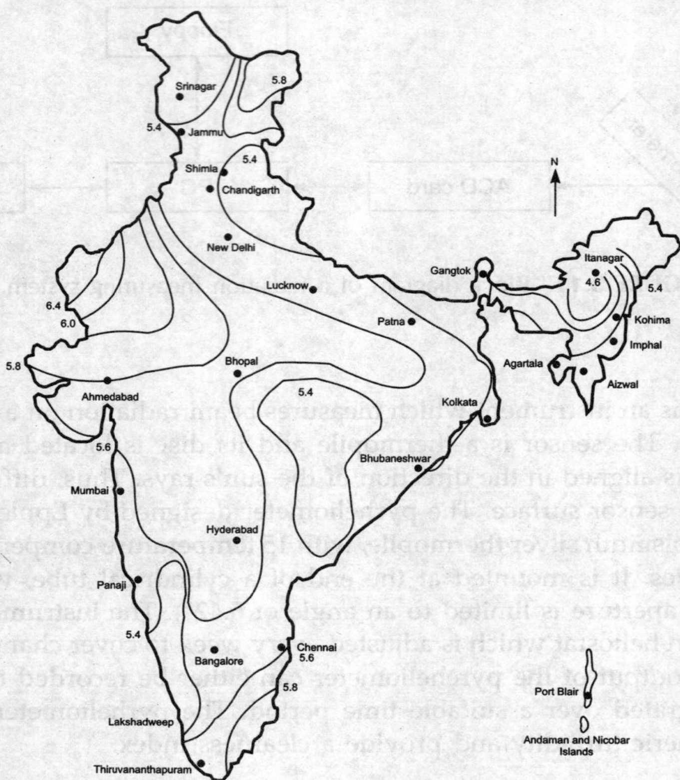

FIGURE 3.12 Solar radiation map.

Annual solar radiation pattern

India is divided into five regions as shown in Figure 3.13, with changing solar radiation pattern between January and December. It gives the annual average of global solar energy received on a horizontal plane. The daily record of global radiation data is useful for industry as India lies in the sunny regions of the world. Other countries having a rich solar flux belt are Saudi Arabia, Central Australia and South Africa. Solar energy can be used through two routes. One is the thermal route for water heating, cooking, drying, water purification and power generation. The other is photovoltaic route that converts solar radiation into electricity which can be used for pumping water, communications and power supply in unelectrified areas.

The daily solar insolation values over selected cities in India with seasonal variations are shown in Table 3.2. The peak values are measured from March to May, when the western Rajasthan and Gujarat receive over 600 cal/cm^2/day (25,100 kJ/m^2/day). During monsoon and winter months the daily solar radiation decreases to 400 cal/cm^2/min (16,700 kJ/m^2/min).

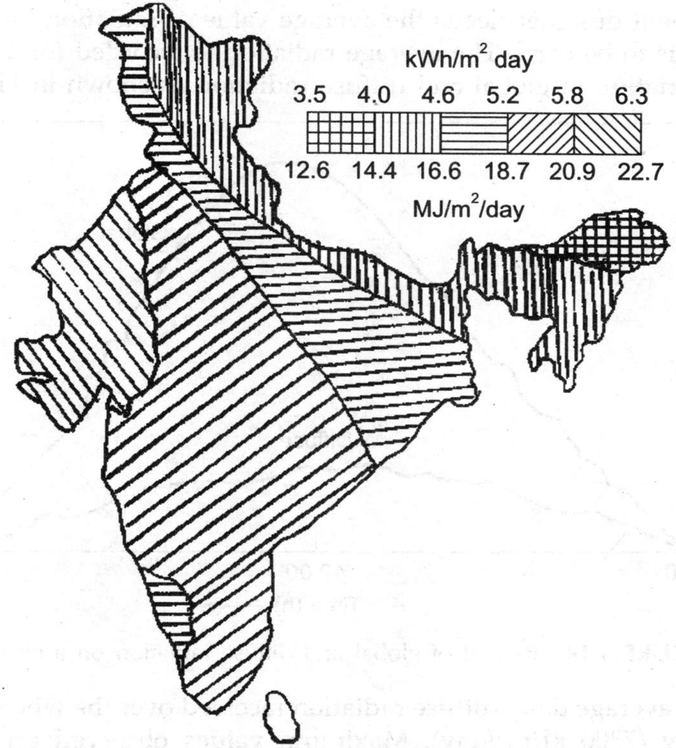

FIGURE 3.13 Annual average of global solar energy.

TABLE 3.2 Annual average solar insolation values (cal/cm²/day) with seasonal variation for selected Indian cities

Name of city	Winter period (Dec. to Feb.)	Summer period (March to May)	Monsoon period (June to Sept.)
New Delhi	364	576	479
Jodhpur	414	611	459
Ahmedabad	429	611	466
Bhavnagar	442	618	452
Goa	492	588	416
Mangalore	467	534	354
Trivandrum	493	520	455
Chennai	451	580	476
Visakhapatnam	466	565	425
Pune	459	602	435
Nagpur	436	574	421
Kolkata	374	516	391
Shillong	366	465	333
Port Blair	421	438	303

A solar equipment designer needs the average value of radiation for a location where the equipment is to be used. The average radiation is recorded for a month, tabulated with hourly variation in global and diffuse radiation as shown in Figure 3.14.

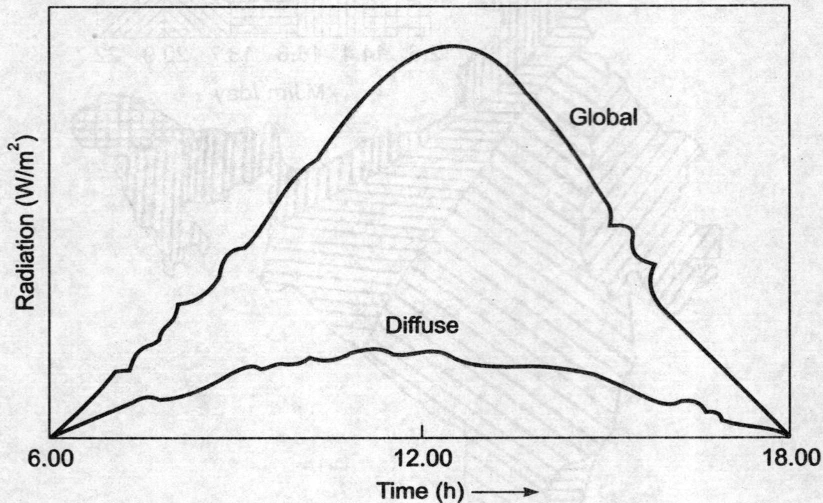

FIGURE 3.14   Record of global and diffuse radiation on a clear day.

The annual average daily diffuse radiation received over the whole country is about 175 cal/cm²/day (7300 kJ/m²/day). Maximum values observed are 300 cal/cm²/day during July in Gujarat, while the minimum values between 75 and 100 cal/cm²/day during December are observed over many locations.

EXAMPLE 3.5

Evaluate the monthly average clearness index for 16 March 2001, at a surface located at latitude 30° N. The monthly average daily terrestrial radiation on a horizontal surface is 28.1 MJ/m²/day.

Solution

For 16th March 2001, $n = 75$
From Eq. (3.3),

$$\delta = 23.45 \sin \left[\frac{360}{365} (284 + 75) \right]$$

$$= -2.4°$$

From Eq. (3.9),

$$\omega_s = \cos^{-1}(-\tan \phi \, \tan \delta)$$

$$= \cos^{-1}[-\tan 30° \tan (-2.4°)] = 88.61°$$

Monthly average of daily extra terrestrial radiation H_o is expressed in joule/m²/day by Eq. (3.15). Therefore,

$$H_o = \frac{24 \times 3600 \times 1367}{\pi} (1.01 \times 0.832)$$

$$= 31.57 \text{ MJ/m}^2$$

Then, the monthly average clearness index is given as

$$K_T = \frac{H_g}{H_o} = \frac{28.1}{31.57} = 0.89$$

REVIEW QUESTIONS

1. Define the following terms and differentiate between their meanings.
   (a) Beam radiation and diffuse radiation
   (b) Surface azimuth angle and solar azimuth angle
   (c) Local clock time and local apparent time

2. Find the day-length in hours at New Delhi (28° 35′ N, 77° 12′ E) on July 1 for a south facing surface tilted at 10°.

3. For the data of Question no. 2, find the local apparent time corresponding to 14:30 IST, with correction of time as −4 minutes, if IST is based on 82° 30′ E.

4. Find the hour angle at the sunrise and the sunset for a horizontal surface with zenith angle of 90°, $\phi = 20°$ and $\delta = -16°$.

5. Find the hour angle at the sunrise and the sunset on March 22 for a surface inclined at an angle 20° facing south at New Delhi (28° 35′ N, 77° 12′ E).

6. Find the angle subtended by beam radiation with the normal to a flat-plate collector at 9 a.m. for the day on 30th October, 2003. The collector is placed at Mumbai (19° 07′ N, 72° 51′ E), inclined at an angle 36° and is facing south.

7. Compute the monthly average hourly solar flux received on 15th October on a flat-plate collector facing south having slope of 15°. The collector is located at Chennai (13°.00 N). The data given is:

   Time : 1–12 h (local apparent time)
   I_g : 2408 kJ/m$^2 \cdot$h
   I_d : 1073 kJ/m$^2 \cdot$h
   The ground reflectivity is 0.2.

8. Find the monthly average daily extra terrestrial radiation for 16 March, 2001 at a surface located at latitude 30° N. The monthly average daily terrestrial radiation on a horizontal surface is 28.1 MJ/m^2.

9. Discuss the following terms:
   (a) Diffuse radiation
   (b) Reflected radiation
   (c) Total radiation

10. Discuss the various types of solar radiation measurement instruments.

SOLAR THERMAL ENERGY COLLECTORS

4.1 INTRODUCTION

A solar thermal energy collector is an equipment in which solar energy is collected by absorbing radiation in an absorber and then transferring to a fluid. In general, there are two types of collectors:

- *Flat-plate solar collector:* It has no optical concentrator. Here, the collector area and the absorber area are numerically the same, the efficiency is low, and temperatures of the working fluid can be raised only up to 100°C.

- *Concentrating-type solar collector:* Here the area receiving the solar radiation is several times greater than the absorber area and the efficiency is high. Mirrors and lenses are used to concentrate the sun's rays on the absorber, and the fluid temperature can be raised up to 500°C. For better performance, the collector is mounted on a tracking equipment to face the sun always with its changing position.

In this chapter, both the above types of solar collectors are discussed in detail.

4.2 FLAT-PLATE COLLECTOR

A schematic cross-section of a flat-plate collector is shown in Figure 4.1. It consists of five major parts as mentioned below:

(i) A *metallic flat absorber plate* of high thermal conductivity made of copper, steel, or aluminium, and having black surface. The thickness of the metal sheet ranges from 0.5 mm to 1 mm.

(ii) *Tubes* or *channels* are soldered to the absorber plate. Water flowing through these tubes takes away the heat from the absorber plate. The diameter of tubes is around 1.25 cm, while that of the header pipe which leads water in and out of the collector and distributes it to absorber tubes, is 2.5 cm.

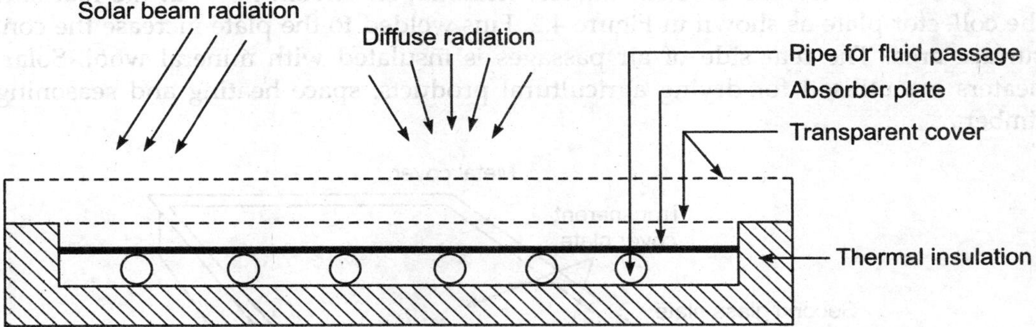

FIGURE 4.1 Schematic cross section of a flat-plate collector.

(iii) A *transparent toughened glass sheet* of 5 mm thickness is provided as the cover plate. It reduces convection losses through a stagnant air layer between the absorber plate and the glass. Radiation losses are also reduced as the spectral transmissivity of glass is such that it is transparent to short wave radiation and nearly opaque to long wave thermal radiation emitted by interior collector walls and absorbing plate.

(iv) Fibre glass insulation of thickness 2.5 cm to 8 cm is provided at the bottom and on the sides in order to minimize heat loss.

(v) A container encloses the whole assembly in a box made of metallic sheet or fibre glass.

In Figure 4.1, since the heat transfer fluid is liquid, so, this type of flat-plate collector is also known as liquid flat-plate collector.

The commercially available collectors have a face area of 2 m². The whole assembly is fixed on a supporting structure that is installed in a tilted position at a suitable angle facing south in the northern hemisphere.

For the whole year, the optimum tilt angle of collectors is equal to the latitude of its location. During winter, the tilt angle is kept 10–15° more than the latitude of the location while in summer it should be 10–15° less than the latitude.

4.3   EFFECT OF DESIGN PARAMETERS ON PERFORMANCE

There are many parameters that affect the performance of a flat-plate collector. However, four important parameters are discussed below:

4.3.1   Heat Transport System

Heat from the absorber plate is removed by continuous flow of a heat transport medium. When water is used, it flows through metal tubes that are welded to the absorber plate for effective heat transfer. Cold water enters the bottom header, flows upwards and gets warmed by the absorber. The hot water then flows out through the top header.

When air is used as the heat transfer fluid, an air stream flows at the rear side of the collector plate as shown in Figure 4.2. Fins welded to the plate increase the contact surface area. The rear side of air passages is insulated with mineral wool. Solar air heaters are utilised for drying agricultural products, space heating and seasoning of timber.

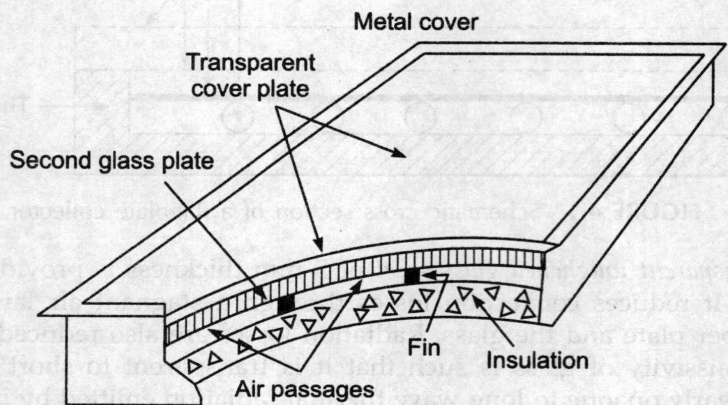

FIGURE 4.2 Solar collector with air as the heat transfer fluid.

4.3.2 Selective Surfaces

Absorber plate surfaces which provide high absorptivity for incoming solar radiation and low emissivity for outgoing radiation are termed *selective surfaces*. Solar radiation lies in short wavelength band up to 4 μm while the absorber plate emits long wave radiation with a maximum at 8.3 μm. Thus, a selective surface needs to have a high absorptivity for wavelengths shorter than 4 μm and a low emissivity for wavelengths longer than 4 μm. No natural surface is available which possesses selective radiation characteristics.

A selective surface is composed of a thin black metallic oxide coated on a bright metal base. Black coating is sufficiently thick to be a good solar radiation absorber. Bright metal base possesses low emissivity for thermal radiation. A successful selective surface can be developed with a black chrome ($Cr–Cr_2O_3$) coating. It is a metal dielectric Cr_2O_3 layer over a Cr particle/Cr_2O_3 composite prepared by electroplating on a steel base. An effective selective surface has solar absorptivity of about 0.95 and thermal emissivity close to 0.1. A selective surface of black chrome is durable with no degradation in performance even in humid atmosphere and operating temperature of 300°C. Selective surfaces are important for low concentration solar equipment operating at high temperatures. For high concentration devices the major requirement is high absorptance rather than low emittance.

4.3.3   Number of Covers

To minimize convection and radiation loss, a solar collector is provided with a transparent glass sheet over the absorber plate. Solar radiation incident on glass sheet passes through the glass cover. Glass sheet also absorbs heat radiation emitted by the hot absorber plate. Thus, the glass sheet cover reduces the heat loss coefficient to 10 W/m$^2 \cdot$K. Experiments show that with two glass covers, the heat loss coefficient further reduces to 4 W/m$^2 \cdot$K.

4.3.4   Spacing

The spacing between the absorber plate and the cover or between two covers also influences the performance of a flat-plate collector. The operating performance varies with the spacing as well as with tilt and service conditions and hence there is no way to specify the exact optimum spacing. However, researchers have suggested a spacing of 4 cm to 8 cm for improved performance. It is also observed that a large spacing reduces the collector area requirements.

4.4   LAWS OF THERMAL RADIATION

Solar energy reaches on the earth by radiation which is important for operation of solar collectors. Solar radiation is electromagnetic energy propagating through space at the speed of light. The sun emits radiation like a 'blackbody' whose surface temperature is 6000 K. Emission of energy with regard to wavelength is not uniform but depends on temperature. 'Planck's law' gives the relation of spectral emissive power with wavelength distribution of radiation and temperature as:

$$E_{\lambda b} = \frac{C_1}{\lambda^5} \frac{1}{(e^{C_2/\lambda T} - 1)} \tag{4.1}$$

where C_1 and C_2 are the Planck's first and second radiation constants respectively, λ is the wavelength and T is the temperature in kelvin. The numerical values of C_1 and C_2 are

$$C_1 = 3.7405 \times 10^{-16} \text{ W} \cdot \text{m}^2$$

$$C_2 = 0.01439 \text{ m} \cdot \text{K}$$

It is possible to calculate the wavelength pertaining to maximum intensity of blackbody radiation, as shown in Figure 4.3 which gives spectral radiation distribution of blackbody from a source at 6000 K, 1000 K and 400 K.

The highest temperature 6000 K represents nearly the surface temperature of the sun (5762 K). The other two temperatures, i.e., 1000 K represents the high temperature solar heated surface while 400 K depicts the low temperature solar heated surface.

Energy emitted by a blackbody at temperature T over the wavelengths is expressed by

$$E_b = \int_0^\infty E_{\lambda b} d\lambda = \sigma T^4 \tag{4.2}$$

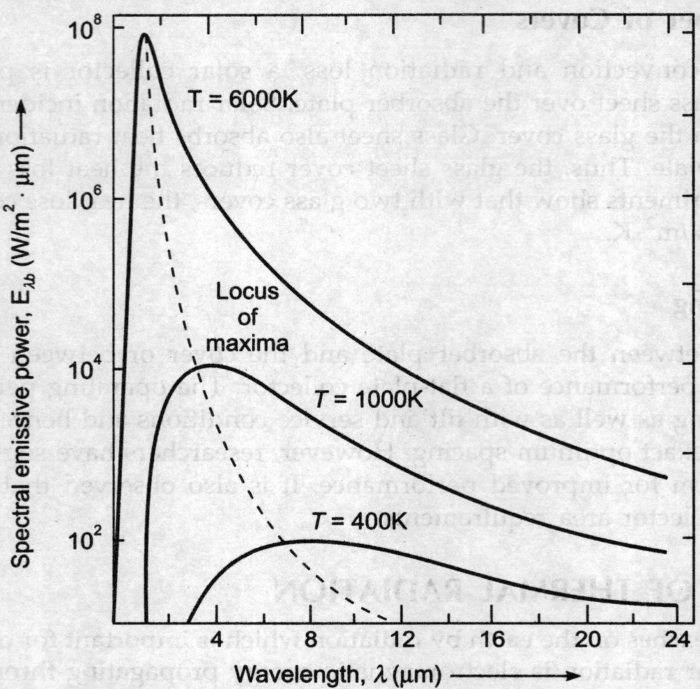

FIGURE 4.3 Thermal radiation graph against wavelength from a source at different temperatures.

where $\sigma = 5.6697 \times 10^{-8}$ W/m²·K⁴ and is called the Stefan–Boltzmann constant. The wavelength corresponding to maximum intensity of blackbody radiation at temperature T is expressed by Wien's Displacement law as

$$\lambda_{max}T = 2897.8 \ \mu m \cdot K \tag{4.3}$$

It shows that an increase in temperature shifts the maximum blackbody radiation intensity towards the shorter wavelength.

The dotted line in Figure 4.3 indicates the displacement of wavelength for maximum intensity as given by Eq. (4.3). The radiation emitted by a real body is a fraction of the blackbody radiation, i.e.,

$$E = \varepsilon \sigma T^4 \tag{4.4}$$

where ε represents the emissivity of a real body surface and is always less than 1.

4.5 RADIATION HEAT TRANSFER BETWEEN REAL BODIES

Radiation exchange between two surfaces takes place from a hot to a cold body. The rate of exchange of heat energy between two closely spaced parallel bodies, one at a temperature T_1 and with emissivity ε_1 and the other at a temperature T_2 and with emissivity ε_2, is given by

$$Q_{rad} = \frac{\sigma A}{[(1/\varepsilon_1) + (1/\varepsilon_2)] - 1}(T_1^4 - T_2^4) \tag{4.5}$$

To evaluate the performance of a solar collector, it is necessary to calculate the radiation exchange between the collector and the sky. The net radiation to a body of surface area A with emittance ε and temperature T from the sky is calculated from

$$Q = \varepsilon A \sigma (T_{sky}^4 - T^4) \tag{4.6}$$

where T_{sky} is called the sky temperature and it is the temperature of the equivalent blackbody.

To estimate T_{sky}, for clear skies, Swinbank (1963) proposed a relation of sky temperature to local air ambient temperature T_{air} (kelvin) as given by the equation

$$T_{sky} = 0.0552\, T_{air}^{1.5} \tag{4.7}$$

Whiller (1967) proposed another simple relation,

$$T_{sky} = T_{air} - 12 \tag{4.8}$$

or the relation

$$T_{sky} = T_{air} - 6 \tag{4.9}$$

4.6 RADIATION OPTICS

Thermal radiation from a high temperature body to a lower temperature body causes transfer of heat through electromagnetic waves up to 0.1 μm–100 μm. The larger part of the terrestrial solar energy lies between 0.3 μm and 3 μm. Thermal radiation is in the infrared range and travels at the speed of light. When radiation strikes a body, a part is reflected, another is absorbed, and the remainder is transmitted through if the body is transparent. The law of conservation of energy dictates that the total sum of radiation components must be equal to incident radiation, i.e.,

$$I_\alpha + I_\rho + I_\tau = I \tag{4.10}$$

and

$$\alpha + \rho + \tau = 1 \tag{4.11}$$

where α, ρ and τ are absorptivity, reflectivity and transmissivity of the light-impinged body. I_α, I_ρ and I_τ are radiation components that are absorbed, reflected and transmitted respectively. The values of α, ρ and τ are always positive within the limits of 0 and 1.

For an opaque surface, $\tau = 0$, so

$$\alpha + \rho = 1$$

For a white surface which reflects all radiation, $\rho = \tau = 0$ and so $\alpha = 1$.

For a blackbody, α and τ are zero and $\rho = 1$ making it a body that absorbs all the energy incident on it.

4.7 TRANSMISSIVITY OF THE COVER SYSTEM

Transmissivity considering reflection only, when a light beam strikes a glass surface there are two losses—one is reflection loss from the top surface and the other is absorption loss as the beam passes through the glass material. First we find transmittance as if there is reflection loss only and then we find transmittance as if there is absorption loss only.

When a beam of light having intensity I_1 travelling in a transparent medium 1 strikes another transparent medium 2, a part of it is reflected and the major part is refracted (Figure 4.4).

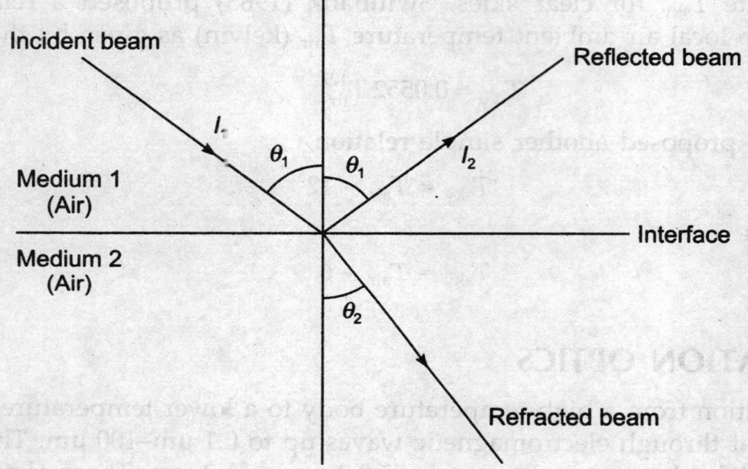

FIGURE 4.4   Reflection and refraction at the interface of two transparent media.

According to the Snell's law of refraction,

$$\frac{\sin \theta_1}{\sin \theta_2} = \frac{n_2}{n_1} \tag{4.12}$$

where θ_1 = angle of incidence, θ_2 = angle of refraction and n_1, n_2 = refractive indices of the two media.

Reflectivity is expressed by $\rho = I_2/I_1$, where I_2 is the reflected beam intensity and I_1 is the incident beam radiation. Also,

$$\rho = \frac{1}{2}(\rho_1 + \rho_2) \tag{4.13}$$

where ρ_1 and ρ_2 are the reflectivities for the two components of polarization — one parallel to the plane of incidence and the other perpendicular to this plane—as given below:

$$\rho_1 = \frac{\sin^2(\theta_2 - \theta_1)}{\sin^2(\theta_2 + \theta_1)}$$

$$\rho_2 = \frac{\tan^2(\theta_2 - \theta_1)}{\tan^2(\theta_2 + \theta_1)}$$

For radiation at normal incidence, $\theta_1 = 0$ and for this case

$$\rho = \rho_1 = \rho_2 = \frac{(n_2 - n_1)^2}{(n_2 + n_1)^2} \tag{4.14}$$

Transmissivity τ can be expressed similar to that for ρ, i.e.,

$$\tau = \frac{1}{2}(\tau_1 + \tau_2) \tag{4.15}$$

where τ_1 and τ_2 are the polarization components of transmissivity.

The cover material used in solar appliances requires transmission of radiation through a slab or sheet having two interfaces per cover where reflection-refraction takes place. The cover interfaces with air on both sides. Multiple reflections and refractions will occur as shown in Figure 4.5.

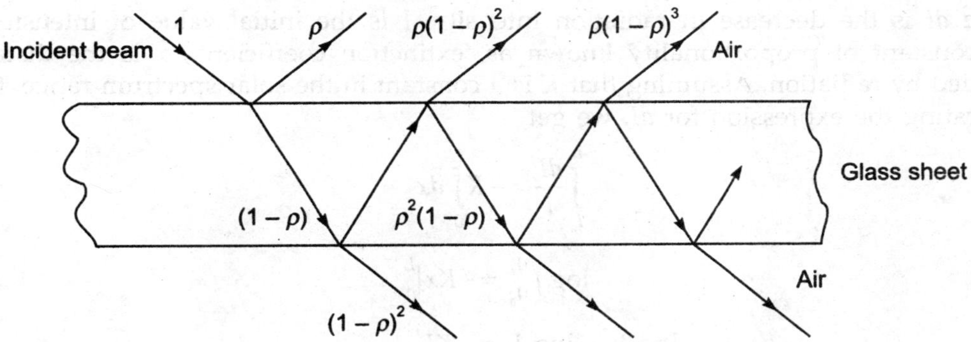

FIGURE 4.5 Ray diagram for single cover multiple reflections and refractions.

For each component of polarization, the incident beam depletes at the second surface. The amount of incidence beam reaching below the interface after reflection is only $(1 - \rho)$, i.e., in case of a unit incident beam after reflection, only $(1 - \rho)$ reaches the second interface. From this, $(1 - \rho)^2$ passes through the interface and $\rho(1 - \rho)$ is reflected back to the first interface, and the process is repeated. Summing up all the terms, the transmittance for a single cover is

$$\tau = (1 - \rho_1)^2 + (1 - \rho_1)^2 \, \rho_1^2 + (1 - \rho_1)^2 \, \rho_1^4 + \cdots$$

$$= (1 - \rho_1)^2 (1 + \rho_1^2 + \rho_1^4 + \cdots)$$

$$= (1 - \rho_1)^2 \frac{1}{1 - \rho_1^2}$$

or

$$\tau_1 = \frac{1 - \rho_1}{1 + \rho_1}$$

Similarly
$$\tau_2 = \frac{1-\rho_2}{1+\rho_2}$$

For a system of N covers and of the same material, therefore, we can write

$$\tau_1 = \frac{1-\rho_1}{1+(2N-1)\rho_2}$$

$$\tau_2 = \frac{1-\rho_2}{1+(2N-1)\rho_1} \tag{4.16}$$

4.7.1 Transmittance Considering Absorption Only

Transmissivity, based on absorption, in a transparent material sheet, can be explained by the Bouger's law, i.e.,

$$dI = -KI\,dx$$

where dI is the decrease in radiation intensity, I is the initial value of intensity, K is a constant of proportionality known as 'extinction coefficient', x is the distance travelled by radiation. Assuming that K is a constant in the solar spectrum range, then integrating the expression for dI, we get

$$\int_{I_0}^{I_L} \frac{dI}{I} = -K\int_0^L dx$$

or
$$\log I\Big|_{I_0}^{I_L} = -Kx\Big|_0^L$$

or
$$\log I_L - \log I_0 = -KL$$

or
$$I_L = I_0 e^{-KL}$$

or
$$\tau_\alpha = \frac{I_L}{I_0} = e^{-KL} \tag{4.17}$$

where τ_α is the transmittance considering only absorption and L is distance travelled by radiation through the medium.

The extinction coefficient K is a physical property of the cover material. For clear white glass, the value of K is 0.04/cm, while for poor quality glass with greenish colour at its edges the value of K is 0.25/cm. A low value of K is preferred.

When the beam is incident at an angle θ_1, the path length through the cover would be $(L/\cos\theta_2)$, where θ_2 is the angle of refraction. Thus, Eq. (4.17) is modified as

$$\tau_\alpha = e^{-KL/\cos\theta_2}$$

The transmissivity of the system allowing for both absorption and reflection is given by

$$\tau = \tau_\alpha \tau_\rho$$

EXAMPLE 4.1

Estimate τ_α, τ_ρ and τ for a glass cover system with the given data:

   Angle of incidence = 10°
   Number of covers = 4
   Thickness of each cover = 3 mm
   Refractive index of glass relative to air = 1.52
   Extinction coefficient of glass = 15 m^{-1}

Solution

$\theta_1 = 10°$, using Snell's law,

$$\frac{n_2}{n_1} = \frac{\sin \theta_1}{\sin \theta_2} \qquad (\theta_1 = 10°)$$

n_2/n_1 = refractive index of glass relative to air = 1.52 (given)

So

$$\theta_2 = \sin^{-1}\left(\frac{\sin 10°}{1.52}\right) = 6.55°$$

$$\rho_1 = \frac{\sin^2(6.55° - 10°)}{\sin^2(6.55° + 10°)} = 0.044$$

$$\rho_2 = \frac{\tan^2(6.55° - 10°)}{\tan^2(6.55° + 10°)} = 0.041$$

$$\tau_{\alpha_1} = \frac{1 - 0.044}{1 + 7 \times 0.041} = 0.733$$

$$\tau_{\alpha_2} = \frac{1 - 0.041}{1 + 7 \times 0.044} = 0.742$$

$$\tau_\alpha = \frac{1}{2}(0.733 + 0.742) = 0.737$$

$$\tau_\rho = e^{-KL/\cos\theta_2}$$

Given in the problem

$$K = 15 \text{ m}^{-1}$$
$$L = 4 \times 3 \times 10^{-3}$$

$\therefore$

$$\tau_\rho = 0.836$$

and

$$\tau = \tau_\alpha \tau_\rho = 0.737 \times 0.836$$
$$= 0.616$$

4.7.2 Transmissivity–Absorptivity Product

For solar collector analysis, it is required to calculate the transmissivity–absorptivity product ($\tau\alpha$). Here, τ is the transmissivity of glass cover and α is the absorptivity of absorber plate. It is defined as the ratio of solar flux absorbed by the absorber plate to the solar flux incident on the cover system.

Solar radiation after passing through the cover system falls on the absorber plate, where some radiation is reflected back to the cover system. Out of the reflected part, a portion is transmitted through the cover system and a part gets reflected back to the absorber plate. This activity of absorption and reflection is shown in Figure 4.6 which goes on indefinitely. However, the quantities involved in the process gradually get reduced.

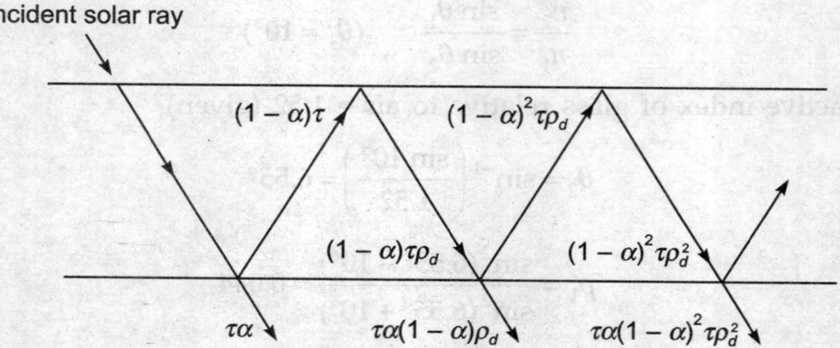

FIGURE 4.6 Absorption and reflection at an absorber plate.

Reflection from the absorber plate is more diffuse and let ρ_d be the reflectivity of glass cover for diffuse radiation. The fraction $(1 - \alpha)\tau$ that reaches the cover plate is diffuse radiation, $(1 - \alpha)\rho_d\tau$ is reflected back to the absorber plate.

Thus, the net radiation absorbed is the summation of

$$(\tau\alpha)_{\text{net}} = \tau\alpha + \tau\alpha(1 - \alpha)\rho_d + \tau\alpha(1 - \alpha)^2\rho_d^2 + \cdots$$

or

$$(\tau\alpha)_{\text{net}} = \frac{\tau\alpha}{1 - (1 - \alpha)\rho_d} \tag{4.18}$$

For an incident angle of 60°, the value of ρ_d is about 0.16, 0.24 and 0.29 for one, two and three glass covers respectively.

4.8 PERFORMANCE ANALYSIS OF A LIQUID FLAT-PLATE COLLECTOR

The performance of solar collector can be improved by enhancing the useful energy gain from incident solar radiation with minimum losses. Thermal losses have three components, namely the conductive loss, the convective loss and the radiative loss.

Conductive loss is reduced by providing insulation on the rear and sides of the absorber plate. Convective loss is minimized by keeping an air gap of about 2 cm between the cover and the plate. Radiative losses from the absorber plate are lowered by applying a spectrally selective absorber coating.

During normal steady-state operation, useful heat delivered by a solar collector is equal to the heat gained by the liquid flowing through the tubes welded on to the underside of the absorber plate minus the losses. The energy balance of the absorber can thus be represented by a mathematical equation, i.e.,

$$Q_u = A_p S - Q_L \qquad (4.19)$$

where

Q_u = useful heat delivered by the collector (watts)

S = solar heat energy absorbed by the absorber plate (W/m^2)

A_p = area of the absorber plate (m^2)

Q_L = rate of heat loss by convection and reradiation from the top, by conduction and convection from the bottom and sides (watts).

From Eq. (3.23), the solar flux falling on an inclined surface is expressed by

$$I_T = I_b R_b + I_d R_d + (I_b + I_d) R_r$$

The flux absorbed is obtained if the above equation is multiplied by the transmissivity–absorptivity product $(\tau\alpha)$. Therefore,

$$S = I_b R_b (\tau\alpha)_b + [I_d R_d + (I_b + I_d) R_r] (\tau\alpha)_d \qquad (4.20)$$

where $(\tau\alpha)_b$ is the transmissivity–absorptivity product for the beam radiation falling on the collector and $(\tau\alpha)_d$ is the transmissivity–absorptivity product for diffuse radiation impinging the collector.

Now, it is necessary to define two terms—*instantaneous collector efficiency* and *stagnation temperature* which are required to indicate the performances of the collector and also for comparing the designs of different collectors.

The instantaneous collector efficiency is defined as the ratio of *useful heat gain* to *radiation* falling on the collector. It is expressed by

$$\eta_i = \frac{Q_u}{A_p I_T} \qquad (4.21)$$

Depending on the given data, the collector aperture area A_a or the collector gross area A_c is used in place of A_p in the above equation. The collector aperature area is the net opening in the top cover through which solar radiation passes into the collector. It is nearly 15% greater than the absorber plate area. The collector gross area is the top cover area including the frame and A_a is about 20% higher than A_p.

In case the flow of liquid through the collector is stopped, the useful heat gain and the efficiency become zero. At this stage, the absorber plate attains a temperature so that $A_p S = Q_L$. This is the maximum temperature that the absorber plate can attain, and is called the *stagnation temperature*. This data helps in selecting an appropriate material for manufacturing the collector.

Since water heating through solar energy occurs comparatively at a slow pace, the time base chosen is an hour. Accordingly, Q_u, useful heat gain in one hour becomes kJ/h and I_T the energy falling on the collector face in one hour becomes kJ/m$^2 \cdot$h.

4.9 TOTAL LOSS COEFFICIENT AND HEAT LOSSES

For mathematical analysis, the heat lost from the collector using the total loss coefficient is given by the equation

$$Q_1 = U_T A_p (T_p - T_a) \qquad (4.22)$$

where

U_T = total loss coefficient

T_p = average temperature of the absorber plate

T_a = ambient temperature of surrounding air

A_p = area of the absorber plate.

The collector loses heat from the top, the bottom and the sides. Thus,

$$Q_1 = Q_t + Q_b + Q_s$$

where

Q_t = rate of heat loss from the top

Q_b = rate of heat loss from the bottom

Q_s = rate of heat loss from the sides.

Each heat loss component can be expressed in terms of the top loss coefficient U_t, the bottom loss coefficient U_b and the side loss coefficient U_s, given by the equations

$$Q_t = U_t A_p (T_p - T_a) \qquad (4.23)$$

$$Q_b = U_b A_p (T_p - T_a) \qquad (4.24)$$

$$Q_s = U_s A_p (T_p - T_a) \qquad (4.25)$$

From the above equations, the total loss coefficient is given by the equation:

$$U_T = U_t + U_b + U_s \qquad (4.26)$$

The total loss coefficient is a relevant parameter as it is a measure of all the losses. Its value ranges from 2 W/m$^2 \cdot$K to 10 W/m$^2 \cdot$K.

4.9.1 Top Loss Coefficient (U_t)

The top loss coefficient U_t can be determined by considering convection and re-radiation losses from the absorber plate, in the upward direction. Four assumptions are made for the determination of U_t: (i) Transparent covers and the absorber plate create a system of infinite parallel surfaces, (ii) the flow of heat is steady and one-dimensional, (iii) there is negligible temperature drop across the thickness of covers, and (iv) for long wavelengths, the transparent glass cover acts to be opaque.

A schematic diagram for a two-cover system is shown in Figure 4.7. When steady state occurs, heat transfer due to convection and radiation (i) between the absorber plate and the first cover, (ii) between the first cover and the second cover, and (iii) between the second cover and the surroundings, should be equal.

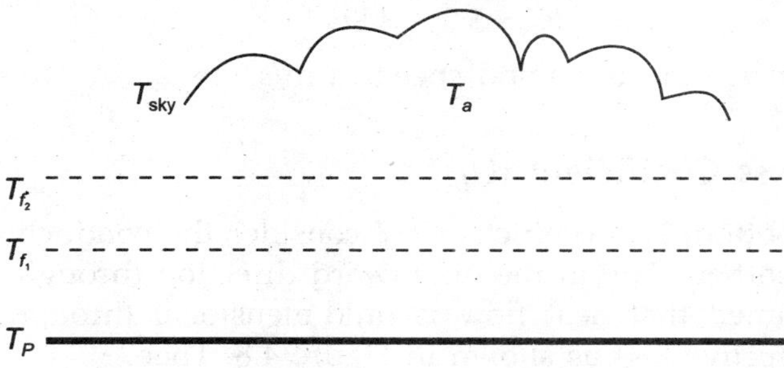

FIGURE 4.7 Schematic diagram of a two-cover system.

Thus,

$$\frac{Q_t}{A_p} = h_{p-f_1}(T_p - T_{f1}) + \frac{\sigma(T_p^4 - T_{f1}^4)}{[(1/\varepsilon_p) + (1/\varepsilon_f)] - 1} \tag{4.27}$$

$$= h_{f_1-f_2}(T_{f1} - T_{f2}) + \frac{\sigma(T_{f1}^4 - T_{f2}^4)}{[(1/\varepsilon_f) + (1/\varepsilon_f)] - 1} \tag{4.28}$$

$$= h_w(T_{f2} - T_a) + \sigma\varepsilon_f(T_{f2}^4 - T_{\text{sky}}^4) \tag{4.29}$$

where

h_{p-f1} = convective heat transfer coefficient between the absorber plate and the first glass cover (W/m$^2 \cdot$ K)

h_{f1-f2} = convective heat transfer coefficient between the first and the second glass covers (W/m$^2 \cdot$ K)

h_w = convective heat transfer coefficient between the second glass cover and the surrounding ambient air (W/m$^2 \cdot$ K).

T_{f1}, T_{f2} = temperatures of glass cover 1 and 2 respectively

T_{sky} = effective temperature of sky with which radiative heat exchange takes place (K)

ε_p = emissivity of absorber plate

ε_f = Emissivity of glass covers.

There are three Eqs. (4.27), (4.28) and (4.29) which are nonlinear, need to be solved for finding the unknowns Q_t, T_{f1} and T_{f2}. Before arriving at the solution it is essential to have some correlation for finding the convective heat transfer coefficients h_{p-f1}, h_{f1-f2} and h_w. Effective sky temperature can be calculated from Eqs. (4.7) and (4.9) already explained earlier.

Convective heat transfer coefficient at top cover

The convection heat transfer coefficient h_w at the top cover is evaluated by an empirical correlation put forward by an engineer McAdams as

$$h_w = 5.7 + 3.8v \tag{4.30}$$

where h_w is in W/m$^2 \cdot$K and v is wind speed in m/s.

4.9.2 Bottom Loss Coefficient (U_b)

To determine the bottom loss coefficient U_b, consider the conduction and convection losses from the absorber plate in the downward direction through the bottom of the collector. It is assumed that heat flow is unidimensional through conduction only, neglecting the convective loss as shown in Figure 4.8. Then,

$$U_b = \frac{K_i}{L_b} \tag{4.31}$$

where

$\quad K_i$ = thermal conductivity of insulation (W/m$\cdot$K)

$\quad L_b$ = thickness of insulation (m).

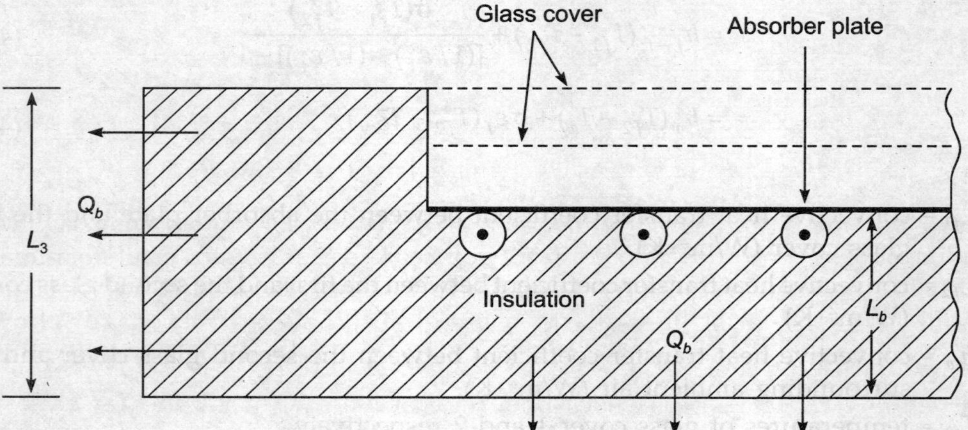

FIGURE 4.8 Side and bottom losses from a flat-plate collector.

4.9.3 Side Loss Coefficient (U_s)

The side loss coefficient is calculated by considering the dimensions of the absorber plate as $L_1 \times L_2$ with height L_3 (Figure 4.8).

The sideways area across which heat flows is

$$A_s = 2(L_1 + L_2)L_3 \tag{4.32}$$

The average temperature drop across side insulation is $(T_p - T_a)/2$. Heat loss through sides, in steady state, if the thickness of insulation is L_s, is given by

$$Q_s = 2L_3 (L_1 + L_2) K_i \frac{T_p - T_a}{2L_s} \tag{4.33}$$

Thus, from Eq. (4.25),

$$U_s = \frac{(L_1 + L_2) L_3 K_i}{L_1 L_2 L_s} \tag{4.34}$$

4.10 SOLAR CONCENTRATING COLLECTORS

While dealing with flat-plate collectors with heat transport medium as water or air, the area of glass cover and that of absorber plate are practically the same. Thus, solar radiation intensity is uniformly distributed over the glass cover and the absorber, keeping the temperature rise of the solar device up to 100°C. If solar radiation falling over a large surface is concentrated to a smaller area of the absorber plate or receiver, the temperature can be enhanced up to 500°C. Concentration is achieved by an optical system either from the reflecting mirrors or from the refracting lenses. These concentrators are used in medium temperature or high temperature energy conversion cycles.

An optical system of mirrors or lenses projects the radiation on to an absorber of smaller area. This process compensates the reflection or absorption losses in mirrors or lenses and losses on account of geometrical imperfections in the optical system. A term called 'optical efficiency' takes care of all such losses. For higher collection efficiency, concentrating collectors are supported by a tracking arrangement that tracks the sun all the time, so that beam radiation is on to the absorber surface. As collectors provide a high degree of concentration, a continuous adjustment of collector orientation is required.

Some new terms that will be encountered in the text hereinafter are defined now for greater clarity. These are:

(i) 'Concentrator' is for the optical subsystem that projects solar radiation on to the absorber. The term 'receiver' shall be used to represent the sub-system that includes the absorber, its cover and accessories.
(ii) 'Aperture' (W) is the opening of the concentrator through which solar radiation passes.
(iii) 'Acceptance angle' ($2\theta_a$) is the angle across which beam radiation may deviate from the normal to the aperture plane and then reach the absorber.
(iv) 'Concentration ratio' (CR) is the ratio of the effective area of the aperture to the surface area of the absorber. The value of CR may change from unity (for flat-plate collectors) to a thousand (for parabolic dish collectors). The CR is used to classify collectors by their operating temperature range.

4.11 TYPES OF CONCENTRATING COLLECTORS

Plane receiver with plane collectors

It is a simple concentrating collector, having up to four adjustable reflectors all around, with a single collector as shown in Figure 4.9. The CR varies from 1 to 4 and the non-imaging operating temperature can go up to 140°C.

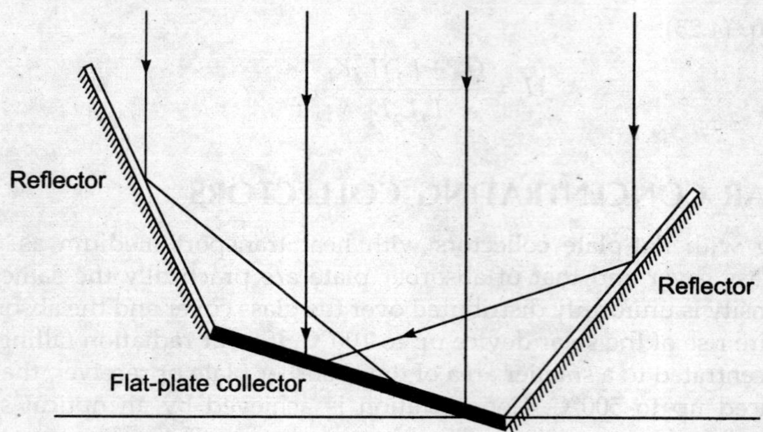

FIGURE 4.9   Plane receiver with plane reflectors.

Compound parabolic collector with plane receiver

Reflectors are curved segments that are parts of two parabolas (Figure 4.10). The CR varies from 3 to 10. For a CR of 10, the acceptance angle is 11.5° and tracking adjustment is required after a few days to ensure collection of 8 hours a day.

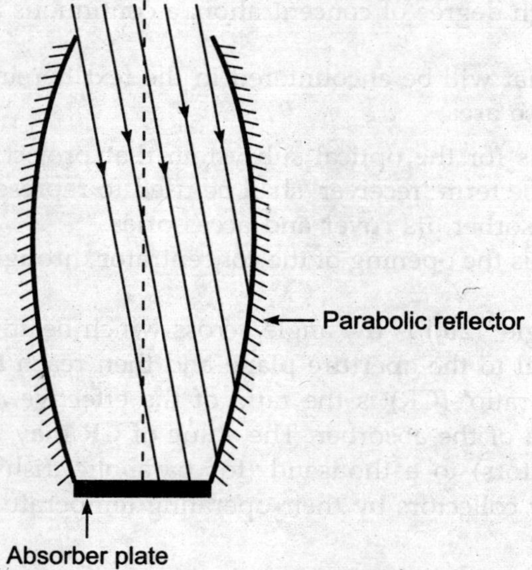

FIGURE 4.10   Compound parabolic collector with a plane receiver.

Cylindrical parabolic collector

The reflector is in the form of trough with a parabolic cross section in which the image is formed on the focus of the parabola along a line as shown in Figure 4.11. The basic parts are: (i) an absorber tube with a selective coating located at the focal axis through which the liquid to be heated flows, (ii) a parabolic concentrator, and (iii) a concentric transparent cover.

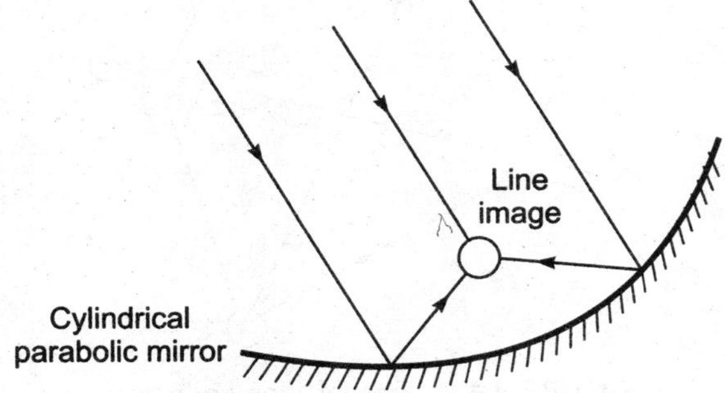

FIGURE 4.11 Cross section of a cylindrical parabolic collector.

The aperture area ranges from 1 m^2 to 6 m^2, where the length is more than the aperture width. The CR range is from 10 to 30.

Collector with a fixed circular concentrator and a moving receiver

The fixed circular concentrator consists of an array of long, narrow, flat mirror strips fixed over a cylindrical surface as shown in Figure 4.12. The mirror strips create a narrow line image that follows a circular path as the sun moves across the sky. The CR varies from 10 to 100.

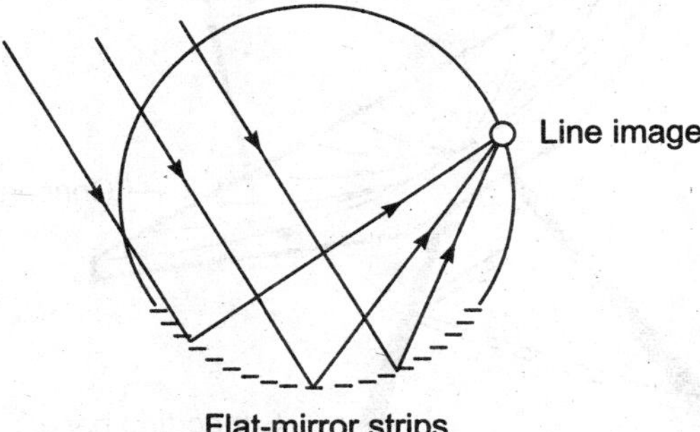

FIGURE 4.12 Cross section of a collector with a fixed circular concentrator and a moving receiver.

Fresnel lens collector

Fresnel lens refraction type focusing collector is made of an acrylic plastic sheet, flat on one side, with fine longitudinal grooves on the other as shown in Figure 4.13. The angles of grooves are designed to bring radiation to a line focus. The CR ranges between 10 and 80 with temperature varying between 150°C and 400°C.

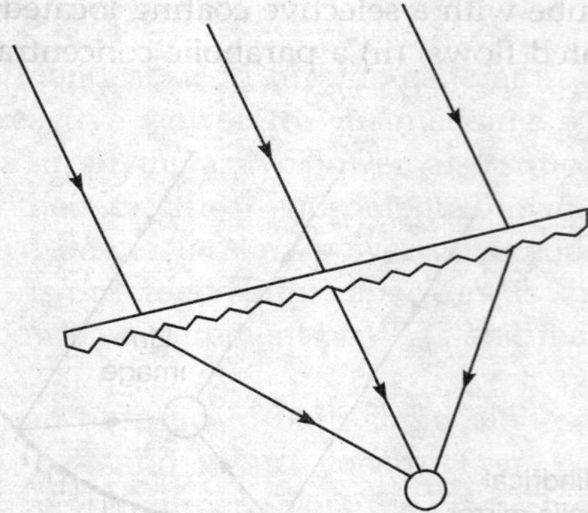

FIGURE 4.13 Fresnel lens collector.

Paraboloid dish collector

To achieve high CRs and temperature, it is required to build a point-focusing collector. A paraboloid dish collector is of point-focusing type as the receiver is placed at the focus of the paraboloid reflector (Figure 4.14).

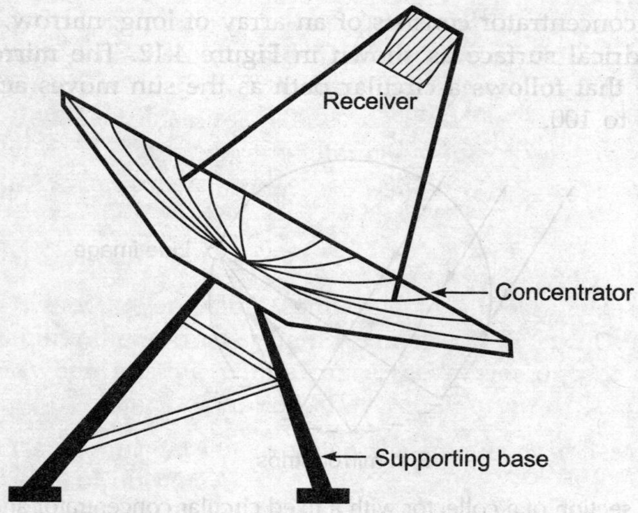

FIGURE 4.14 Paraboloid dish collector.

As a typical case, a dish of 6 m in diameter is constructed from 200 curved mirror segments forming a paraboloidal surface. The absorber has a cavity shape made of zirconium–copper alloy, with a selective coating of black chrome. The CR ranges from 100 to a few thousands with maximum temperature up to 2000°C. For this, two-axis tracking is required so that the sun may remain in line with the focus and vertex of the paraboloid.

Central receiver with heliostat

To collect large amounts of heat energy at one point, the 'Central Receiver Concept' is followed. Solar radiation is reflected from a field of heliostats (an array of mirrors) to a centrally located receiver on a tower (Figure 4.15).

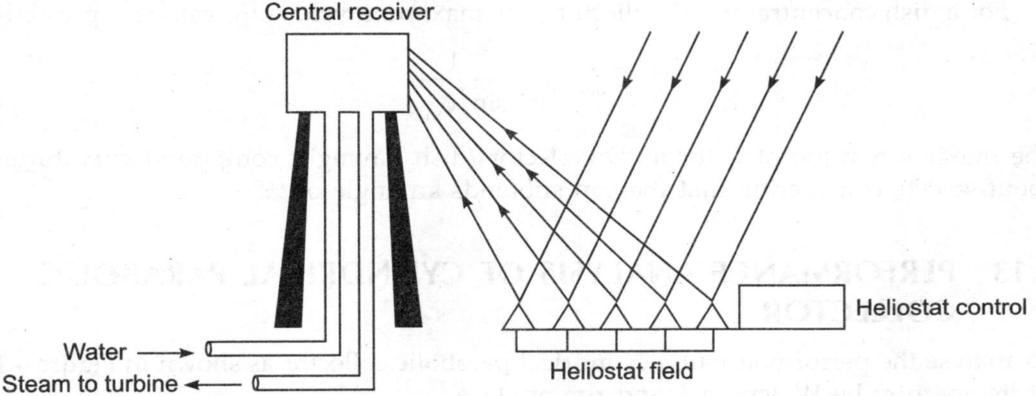

FIGURE 4.15   Central receiver tower with a field of heliostats.

Heliostats follow the sun to harness maximum solar heat. Water flowing through the receiver absorbs heat to produce steam which operates a Rankine cycle turbo generator to generate electrical energy.

With a central receiver optical system, a large number of small mirrors are installed, each steerable to have an image at the absorber on the central receiver. A curvature is provided to the mirrors so as to focus the sunlight in addition to directing it to the tower.

4.12   THERMODYNAMIC LIMITS TO CONCENTRATION

The function of a solar concentrator is to enhance the flux density of solar radiation. A solar concentrator is shown in Figure 4.16 where radiation is incident on an aperture area A_a, which is then concentrated on a smaller absorber plate area A_{ap}.

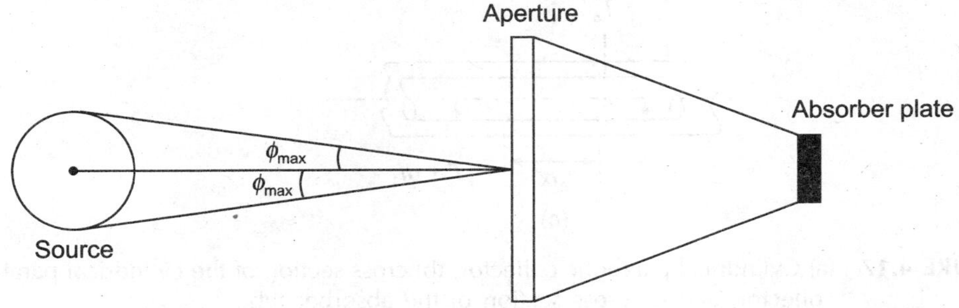

FIGURE 4.16   Schematic of source, aperture and absorber.

If the acceptance angle is $2\phi_{max}$, the concentration ratio (maximum) C_{max} is given by

$$C_{max,\,2D} = \frac{1}{\sin \phi_{max}} \tag{4.35}$$

For a linear 2D collector, the maximum value of C is 212.

For a dish concentrator (3D collector), the maximum value of C can be expressed as

$$C_{max,\,3D} = \frac{1}{\sin^2 \phi_{max}}$$

The maximum value of C for a 3D collector (dish having a compound curvature) is about 40,000, considering that the sun subtends an angle of ½°.

4.13 PERFORMANCE ANALYSIS OF CYLINDRICAL PARABOLIC COLLECTOR

To analyse the performance of a cylindrical parabolic collector as shown in Figure 4.17, let its aperture be W, length L and rim angle ϕ_{rim}.

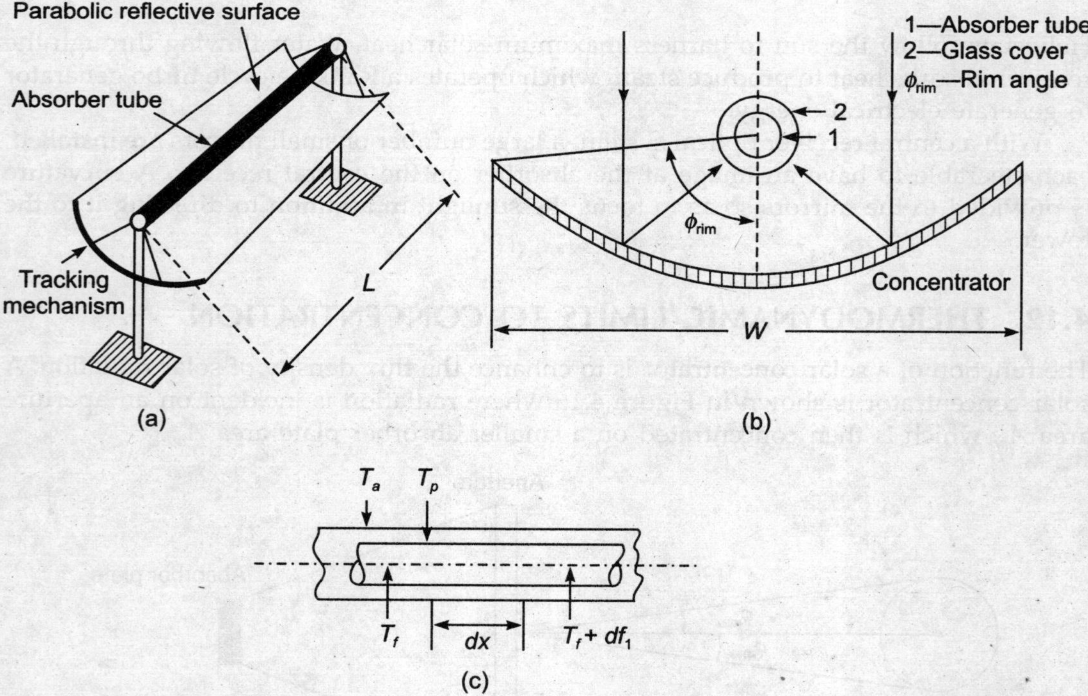

FIGURE 4.17 (a) Cylindrical parabolic collector, (b) cross section of the cylindrical parabolic collector, and (c) cross section of the absorber tube.

The inner diameter of the absorber tube is D_i and the outer diameter D_o. It has either a concentric glass cover or a flat glass/plastic sheet covering the whole aperture area that protects the reflecting surface from weather effects.

The following assumptions are made for analysis:

(a) Radiation flux is the same all along the length of the absorber tube.
(b) The temperature drop across the thickness of the absorber tube and that across the glass cover is negligible.

The 'concentration ratio' of the collector is expressed by

$$C = \frac{\text{effective aperture area}}{\text{absorber tube area}} = \frac{W - D_o}{\pi D_o} \tag{4.36}$$

For energy balance, if we consider an elemental length dx of the absorber tube at a distance x from the inlet, the following equation can then be written for the steady-state condition. Beam radiation normally incident on the aperture is $I_b R_b$.

$$dQ_u = [I_b R_b (W - D_o)\, \rho r\, (\tau\alpha)_b + I_b R_b D_o\, (\tau\alpha)_b - U_1 \pi D_o (T_p - T_a)] dx \tag{4.37}$$

where

dQ_u = useful heat gain rate for a length dx

$\quad I_b$ = beam radiation flux

$\quad R_b$ = beam radiation tilt factor

$\quad \rho$ = specular reflectivity of the concentrator surface

$\quad r$ = intercept factor, the fraction of the specularly reflected radiation intercepted by the absorber tube

$(\tau\alpha)_b$ = transmissivity–absorptivity product for beam radiation

$\quad U_1$ = overall heat loss coefficient

$\quad T_p$ = local temperature of absorber tube

$\quad T_a$ = ambient temperature.

The first term in the above equation represents the incident beam radiation on absorber tube after reflection. The second term indicates the absorbed incident beam radiation directly falling on the absorber tube. The second term can be ignored when the top of the tube is insulated. The third term is the loss by convection and re-radiation.

Now, the absorbed radiation flux S may be given as

$$S = I_b R_b \rho r (\tau\alpha)_b + I_b R_b (\tau\alpha)_b \left(\frac{D_o}{W - D_o} \right) \tag{4.38}$$

Equation (4.37) can be represented as

$$dQ_u = \left[S - \frac{U_1}{C}(T_p - T_a) \right] (W - D_o) dx \tag{4.39}$$

The useful heat gain rate can also be represented as

$$dQ_u = h_f \pi D_i (T_p - T_f)\, dx \tag{4.40}$$

$$= \dot{m}c_p \, dT_f \tag{4.41}$$

where

D_i = inner diameter of the absorber tube

h_f = heat transfer coefficient on the inner surface of the tube

T_f = local fluid temperature

$\dot{m}$ = mass flow rate of the fluid being heated in the collector

T_{fi} = inlet temperature of the fluid

T_{fo} = outlet temperature of the fluid.

Combining Eqs. (4.40) and (4.41) with the elimination of T_p, the relation for the useful heat gain becomes

$$dQ_u = F'\left[S - \frac{U_1}{C}(T_f - T_a)\right](W - D_o)dx \tag{4.42}$$

Here, F' is the collector efficiency factor. Its value is

$$F' = \frac{1}{U_e[(1/V_e) + (D_o/D_i h_f)]} \tag{4.43}$$

Solving Eqs. (4.41) and (4.42), the following differential equation is obtained.

$$\frac{dT_f}{dx} = F'\pi \frac{D_o U_1}{\dot{m}c_p}\left[\frac{CS}{U_1} - (T_f - T_a)\right] \tag{4.44}$$

Integrating and applying the inlet condition at $x = 0$, $T_f = T_{fi}$, we get the temperature distribution as

$$\frac{[(CS/U_1) + T_a] - T_f}{[(CS/U_1) + T_a] - T_{fi}} = -\exp\left(-\frac{F'\pi D_o U_1 x}{\dot{m}c_p}\right) \tag{4.45}$$

The outlet temperature of the fluid can be had by putting $T_f = T_{fo}$ and $x = L$ in Eq. (4.45). With this substitution and then subtracting both sides from unity, we get

$$\frac{T_{fo} - T_{fi}}{[(CS/U_1) + T_a] - T_{fi}} = 1 - \exp\left(-\frac{F'\pi D_o U_1 L}{\dot{m}c_p}\right) \tag{4.46}$$

The useful heat gain rate is

$$G_u = \dot{m}c_p(T_{fo} - T_{fi})$$

or

$$Q_u = \dot{m}c_p\left(\frac{CS}{U_1} + T_a - T_{fi}\right)\left[1 - \exp\left(-\frac{F'\pi D_o U_1 L}{\dot{m}c_p}\right)\right]$$

$$= F_R(W - D_o)L\left[S - \frac{U_1}{C}(T_{fi} - T_a)\right] \tag{4.47}$$

where F_R is the heat removal factor and it is given as

$$F_R = \frac{\dot{m}c_p}{\pi D_o L U_R}\left[1 - \exp\left(-\frac{F'\pi D_o U_1 L}{\dot{m}c_p}\right)\right] \tag{4.48}$$

The *instantaneous collection efficiency* considering beam radiation only, η_{ib}, in percentage (neglecting ground reflected radiation) is given by

$$\eta_{ib} = \frac{Q_u}{(I_b R_b)WL} \times 100 \tag{4.49}$$

In general, the *instantaneous collection efficiency*, η_i, can be expressed as

$$\eta_i = \frac{Q_u}{(I_b R_b + I_d R_d)WL} \times 100 \tag{4.50}$$

The heat loss coefficient U_1 can be calculated from

$$U_1 = \left(\frac{1}{h_{\text{wind}}} + \frac{1}{h_r}\right)^{-1} \tag{4.51}$$

where

h_{wind} = film coefficient due to wind

= 5.7 + 3.8v W/m^2°C

with v as the wind velocity in m/s.

h_r = radiation coefficient.

The radiation coefficient h_r can be calculated as

$$h_r(T_r - T_a) = \sigma\varepsilon_r(T_r^4 - T_a^4)$$

where

σ = Stefan–Boltzmann constant = 5.67 $\times$ 10^{-8} W/m$^2 \cdot$°C

ε_r = emissivity of the surface

T_r = temperature of the radiant surface.

Hence,

$$h_r = \sigma\varepsilon_r(T_r + T_a)(T_r^2 + T_a^2)$$

Assuming that $T_r \approx T_a$ and $\overline{T} = (T_r + T_a)/2$, we have

$$h_r = 4\sigma\varepsilon_r \overline{T}^3 \tag{4.52}$$

EXAMPLE 4.2

For a parabolic collector of length 2 m, the angle of acceptance is 15°. Find the concentration ratio of the collector.

Solution

Concentration ratio,
$$CR = \frac{1}{\sin \phi_{max}}$$

$$\phi_{max} = \frac{\text{angle of acceptance}}{2} = 7.5°$$

So,
$$CR = \frac{1}{\sin 7.5°} = 7.66$$

EXAMPLE 4.3

For a cylindrical parabolic concentrator of 2.5 m width and 9 m length, the outside diameter of the absorber tube is 6.5 cm. Find the concentration ratio of the collector.

Solution

$$\text{Concentration ratio} = \frac{W - D_o}{\pi D_o}$$

$$= \frac{2.5 - 6.5 \times 10^{-2}}{\pi \times 6.5 \times 10^{-2}}$$

$$= 11.93$$

EXAMPLE 4.4

Calculate the heat removal factor, the useful heat gain, the exit fluid temperature and the collection efficiency for a cylindrical parabolic concentrator having 2.5 m width and 9 m length, the outside diameter of the absorber tube being 6.5 cm. The temperature of the fluid to be heated at the inlet is 16°C with a flow rate of 450 kg/h. The incident beam radiation is 700 W/m^2. The ambient temperature is 28°C. The optical properties are as given below:

$\rho = 0.85,$ $(\tau\alpha)_b = 0.78,$ $\tau = 0.93$

$c_p = 1.256$ kJ/kg·°C

Collector efficiency factor, $F' = 0.85$

Heat loss coefficient, $U_1 = 7.0$ W/m^2·°C

Solution

From the given data, $I_b R_b = 700$ W/m^2
Absorbed radiation flux,

$$S = I_b R_b \, \rho\tau(\tau\alpha)_b + I_b R_b \, (\tau\alpha)_b \frac{D_o}{W - D_o}$$

$$S = 700 \times 0.85 \times 0.93 \times 0.78 + 700 \times 0.78 \left(\frac{0.65}{2.5 - 0.65}\right)$$

$$= 431.61 + 0.02 = 431.63 \ \text{W/m}^2$$

Heat removal factor, $\quad F_R = \dfrac{\dot{m}c_p}{\pi D_o L U_1}\left[1 - \exp\left(-\dfrac{F'\pi D_o U_1 L}{\dot{m}c_p}\right)\right]$

$$\dot{m} = \frac{450}{3600} = 0.125 \ \text{kg/s}$$

$$\frac{\dot{m}c_p}{\pi D_o U_1 L} = \frac{0.125 \times 1.256 \times 10^3}{3.14 \times 0.015 \times 7.0 \times 9} = 12.21$$

$\therefore \qquad\qquad F_R = 12.21\left[1 - \exp\left(-\dfrac{0.85}{12.21}\right)\right]$

$$= 12.21[1 - \exp(0.00696)]$$

$$= 12.21(1 - 0.9327)$$

$$= 0.821$$

Concentration ratio, $\qquad C = \dfrac{W - D}{\pi D} = \dfrac{2.5 - 0.065}{3.14 \times 0.065}$

$$= 11.93$$

Useful heat gain Eq. (4.47) is

$$Q_u = F_R(W - D_o) L\left[S - \frac{U_1}{C}(T_{fi} - T_a)\right]$$

$$= 0.821\,(2.5 - 0.65)\,9\left[431.6 - \frac{7}{11.93}(150 - 28)\right]$$

$$= 0.821 \times 2.435 \times 9 \times 360.015$$

$$= 6477.5 \ \text{W}$$

Also, $\qquad\qquad Q_u = \dot{m}c_p(T_{fo} - T_{fi})$

or $\qquad\qquad T_{fo} = \dfrac{Q_u}{\dot{m}c_p} + T_{fi}$

$$= \left(\frac{6477.5}{0.125 \times 1.256 \times 10^3}\right) + 150$$

$$= 191.2°\text{C}$$

$$\eta_{ib} = \frac{Q_u}{(I_b R_b) WL} \times 100 = \frac{6477.5}{700 \times 2.5 \times 9} \times 100\%$$

$$= 41\%$$

4.14 COMPOUND PARABOLIC CONCENTRATOR (CPC)

A two-dimensional CPC is shown in Figure 4.18. It has two segments BE and CF which are parts of parabolas 2 and 1 respectively. BC is the aperture of width W while EF is the absorber surface of width b. Both parabolas are positioned in such a way that the focus of parabola 1 lies at E while that of parabola 2 at F. Also, the tangents drawn at points B and C to the parabolas are parallel to the axis of CPC.

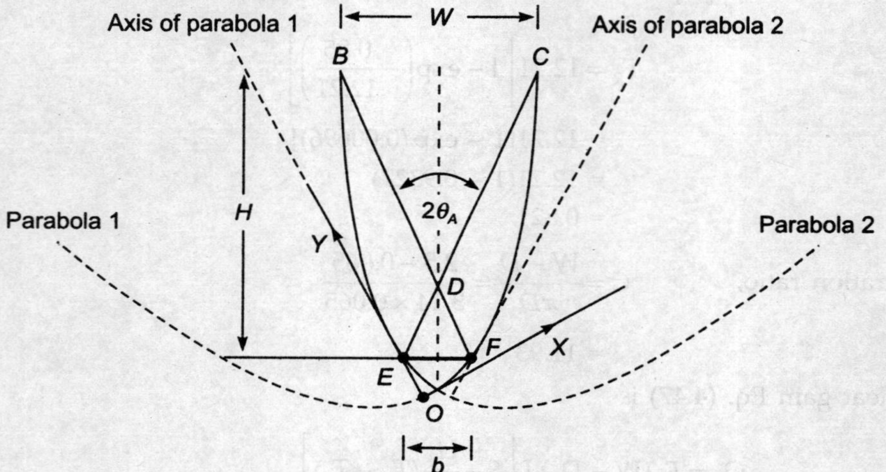

FIGURE 4.18   Geometry of a CPC collector.

Acceptance angle $\angle BDC = 2\theta_A$

Concentration ratio,
$$C = \frac{W}{b} = \frac{1}{\sin \theta_A}$$

Equation of parabola 1 having vertex at O, i.e., the origin of X–Y co-ordinate is given by

$$y = \frac{x}{2b(1 + \sin \theta_A)} \tag{4.53}$$

And focal length,

$$OE = \frac{b}{2(1 + \sin \theta_A)}$$

The coordinates of F are

$$x = b \cos \theta_A$$

$$y = \frac{b}{2(1-\sin\theta_A)}$$

The coordinates of C are

$$x = (b + W)\cos\theta_A$$

$$y = \frac{b}{2}(1 - \sin\theta_A)\left(1 + \frac{1}{\sin\theta_A}\right)^2$$

The ratio of height to aperture can be expressed as

$$\frac{H}{W} = \frac{1}{2}\left(1 + \frac{1}{\sin\theta_A}\right)\cos\theta_A$$

$$= \frac{1}{2}(1 + C)\left(1 - \frac{1}{C^2}\right)^{1/2} \tag{4.54}$$

The surface area of the concentrator can be calculated by integrating along the parabolic arc. However, for a concentration ratio of more than 3, a simple equation provides a nearly correct value as

$$\frac{\text{concentrator area}}{\text{aperture area}} = \frac{A_{conc}}{A_a} = 1 + C$$

EXAMPLE 4.5

A CPC, 1.5 m long has an acceptance angle of 20°. The surface of the absorber is flat with a width of 15 cm. Evaluate the concentration ratio, the aperture height and the surface area of the concentrator.

Solution

$$C = \frac{1}{\sin 10°} = 5.76$$

Aperture, $W = 5.76 \times 15 = 86.4$ cm

$$\frac{H}{W} = \frac{1}{2}\left(1 + \frac{1}{\sin 10°}\right)\cos 10° = 3.328$$

$$\therefore \qquad H = 3.328 \times 86.4 = 287.53$$

$$\frac{A_{conc}}{WL} = 1 + 5.76 = 6.76$$

$$\therefore \qquad A_{conc} = 6.76 \times 0.867 \times 1.5$$

$$= 8.79 \text{ m}^2$$

4.15 TRACKING CPC AND SOLAR SWING

A two-dimensional CPC is normally installed with its length parallel to the horizontal East–West direction, and the aperture plane is kept slopped towards South. For concentration ratio of 10 and acceptance angle of 11.5°, the tracking frequency is decided so as to ensure collection for 7 to 8 hours every day. A lower concentration ratio, say 5, has a higher acceptance angle of 23.1° and needs tracking for longer intervals.

Assume that AB is a vertical stick having its shadow BC in the horizontal plane as shown in Figure 4.19; CDBE is a rectangle in a horizontal plane where CE and DB are the East–West lines.

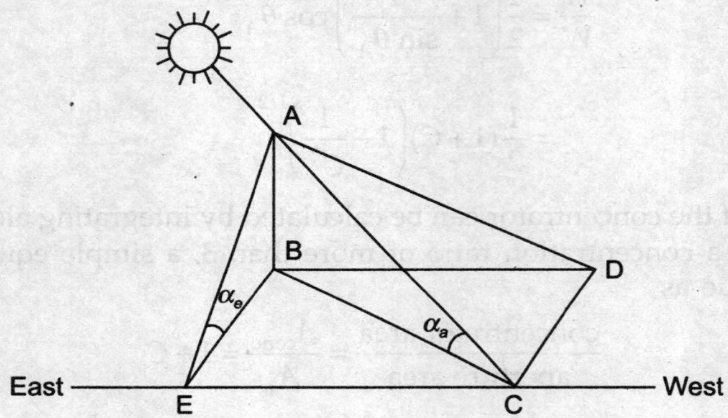

FIGURE 4.19   Geometry for solar swing.

Now, the angle ACB denoted by α_a is called the 'solar angle'. The projection of this angle in a vertical north-south plane is the angle AEB, called the 'solar elevation angle' α_e. The change in angle α_e over a given time period is the 'solar swing'.

$$\tan \alpha_e = \frac{AB}{EB} = \frac{AB}{AC} \bigg/ \frac{CD}{AC} = \frac{\sin \alpha_a}{\cos \angle ACD}$$

Putting the values for $\sin \alpha_a$ and $\cos \angle ACD$ from Eqs. (3.6) and (3.8) respectively, we get

$$\tan \alpha_e = \frac{\sin \phi \sin \delta + \cos \phi \cos \delta \cos \omega}{\sin \phi \cos \delta \cos \omega - \cos \phi \sin \delta} \tag{4.55}$$

At noon the value of $\omega = 0$. Then,

$$\tan \alpha_e = \frac{\cos (\phi - \delta)}{\sin (\phi - \delta)} = \cot (\phi - \delta)$$

Thus,

$$(\alpha_e)_{\omega=0} = \frac{\pi}{2} - (\phi - \delta) \tag{4.56}$$

For correct tracking, it is necessary to know the daily 'solar swings'. The solar swing angle for a period corresponding to an hour angle is the variation in α_e from the time of matching with the angle $\pm \omega_t$ to solar noon. Accordingly,

$$\text{Solar swing} = \left[\left\{ \frac{\pi}{2} - (\phi - \delta) \right\} - \tan^{-1} \left(\frac{\sin\phi \sin\delta + \cos\phi \cos\delta \cos\omega_t}{\sin\phi \cos\delta \cos\omega_t - \cos\phi \sin\delta} \right) \right] \tag{4.57}$$

Equation (4.57) is further simplified by Rabel (1976) in an easy form. According to Rabel (1976), the solar elevation angle α_e is measured with reference to the equatorial plane instead of the horizontal plane. Then, α'_e and α_e are related as

$$\alpha'_e = \alpha_e - \left(\frac{\pi}{2} - \phi \right) \tag{4.58}$$

So,

$$\tan\alpha'_e = -\cot(\alpha_e + \phi)$$

$$= -\left(\frac{\cos\phi - \tan\alpha_e \sin\phi}{\tan\alpha_e \cos\phi + \sin\phi} \right)$$

Substituting the expression for $\tan\alpha_e$ from Eq. (4.55),

$$\tan\alpha'_e = \frac{\tan\delta}{\cos\omega} \tag{4.59}$$

Therefore, the magnitude of the solar swing angle $= |(\alpha'_e)_{\omega=0} - (\alpha'_e)_{\omega=\omega t}|$

$$= \left| \delta - \tan^{-1} \left(\frac{\tan\delta}{\cos\omega_t} \right) \right| \tag{4.60}$$

Though Eqs. (4.57) and (4.60) provide the same values for solar swing angle, the Rabel's (1976) Eq. (4.59) is simpler. From the above discussion, two important observations are made:

(i) Solar swing angle is independent of the latitude.
(ii) Solar swing angle is maximum on solstices days, i.e., June 21 and December 21.

The value of solar swing angle is zero on equinox days, i.e., March 21 and September 21.

EXAMPLE 4.6

A compound parabolic collector installed in Mumbai (19.12°N) collects solar radiation for 8 hours on December 21 with no tracking adjustments. Calculate the minimum acceptance angle needed for the collector.

Solution

For Mumbai, $\qquad\qquad\qquad \delta = -23.45°$ and $\quad \omega_t = 60°$

Substituting these values in Eq. (4.60),

$$\text{Solar swing angle} = \left| -23.45° - \tan^{-1}\left\{ \frac{\tan(-23.45°)}{\cos 60°} \right\} \right|$$

$$= \left| -23.45° - (-40.95°) \right|$$

$$= 17.5°$$

The minimum acceptance angle for the collector is equal to the solar swing angle,

$$2\theta_a = 17.5°.$$

4.16 PERFORMANCE ANALYSIS OF CPC

A compound parabolic collector with an aperture W, acceptance angle θ_a and length L, is shown in Figure 4.20. The surface of the absorber (width b) collects heat and transfers it to a fluid flowing through n tubes, each having inner diameter D_i and outer diameter D_o. The inlet and outlet temperatures of the fluid are T_{fi} and T_{fo} respectively. The mass flow rate of the fluid being heated is denoted by m. The aperture of CPC is covered with a transparent sheet and slopes towards south to ensure that the beam radiation incident on it is within the acceptance angle of the collector.

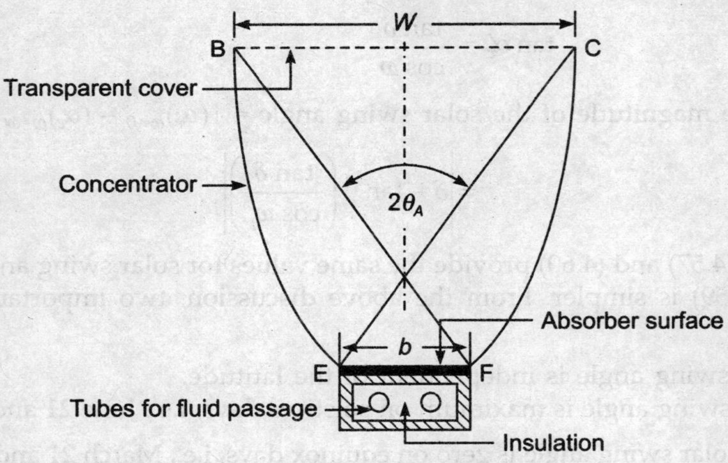

FIGURE 4.20 Compound parabolic collector.

As the acceptance angle of CPC is large, so it receives both beam and diffuse radiation. The beam radiation flux received on the aperture plane is $I_b R_b$, and the diffuse radiation flux received on the aperture plane is $I_d R_d$, and the diffuse radiation flux is expressed by I_d/C. Therefore, the total operative flux entering the aperture is

$$I_b R_b + \frac{I_d}{C}$$

The flux absorbed by the absorber surface is

$$S = \left(I_b R_b + \frac{I_d}{C} \right) \tau \rho_e \alpha \tag{4.61}$$

where

τ = transmissivity of the cover

ρ_e = effective reflectivity of the concentrator for all types of radiation

α = absorptivity of the absorber surface.

To arrive at the useful heat gain rate, consider an energy balance on a small slice dx of the absorber surface at a distance x from the entry point. The energy balance provides the equation

$$dQ_u = \left[S - \frac{U_1}{C} (T_p - T_a) \right] W \, dx \tag{4.62}$$

where T_p is the local temperature of the absorber tube and T_a is the ambient temperature as expressed in Eq. (4.39). Following the same method as used for the cylindrical parabolic collector, we get

$$Q_u = F_R W L \left[S - \frac{U_1}{C} (T_{fi} - T_a) \right] \tag{4.63}$$

$$F_R = \frac{\dot{m} c_p}{b U_1 L} \left[1 - \exp\left(-\frac{F' b U_1 L}{\dot{m} c_p} \right) \right] \tag{4.64}$$

where F' is the collector efficiency factor and its value is given by

$$\frac{1}{F'} = U_1 \left\{ \frac{1}{U_1} + \frac{b}{n \pi D_i h_f} \right\} \tag{4.65}$$

Here n is the number of tubes through which the fluid flows to collect heat from the absorber surface.

The instantaneous collector efficiency can be calculated using Eq. (4.49). The value of overall loss coefficient U_l is difficult to estimate. However, the approximate values are tabulated in Table 4.1.

TABLE 4.1 Overall loss coefficient in a CPC

U_l (W/m²·K)	T_{pm} (°C)	C	ε_p
5	120	3	0.1
15	200	6	0.9

4.17 SOLAR THERMAL ENERGY STORAGE

Solar energy is available only during the sunshine hours. Consumer energy demands follow their own time pattern and the solar energy does not fully match the demand. As a result, energy storage is a must to meet the consumer requirement.

There are three important methods for storing solar thermal energy. These are discussed in subsections below.

4.17.1 Sensible Heat Storage

Heating a liquid or a solid which does not change phase comes under this category. The quantity of heat stored is proportional to the temperature rise of the material. If T_1 and T_2 represent the lower and higher temperature, V the volume and ρ the density of the storage material, and c_p the specific heat, the energy stored Q is given by

$$Q = V \rho \int_{T_1}^{T_2} c_p \, dT$$

For a sensible heat storage system, energy is stored by heating a liquid or a solid. Materials that are used in such a system include liquids like water, inorganic molten salts and solids like rock, gravel and refractories. The choice of the material used depends on the temperature level of its utilization. Water is used for temperature below 100°C whereas refractory bricks can be used for temperature up to 1000°C.

Liquids

The ability to store heat depends upon the product ρc_p and water has the highest value. Largely the solar water heating and space heating systems utilise hot water storage tanks. An optimum tank size for a flat-plate collector system is about 100 litres of storage per square metre of collector area.

A molten inorganic salt may also be used for high temperature applications of 300°C. A mixture of 40% $NaNO_2$, 7% $NaNO_3$ and 53% KNO_3 (by weight), is marketed under the trade name of Hitec. Its melting temperature is 145°C and can be utilised up to a temperature of 400°C. Liquid sodium is also in use as a storage fluid for a solar thermal power plant (0.5 MW) in Spain.

Solids

For sensible heat storage, rocks or gravel packed in an insulated vessel are used with solar heaters and it provides a large and inexpensive heat transfer surface. A typical size of rock piece varies from 1 to 5 cm. This system operates efficiently as the heat transfer coefficient between the air and the solid is high. As a thumb rule, 300 kg to 500 kg of rock per square metre of collector area is sufficient for space heating applications.

Refractory materials like magnesium oxide bricks, silicon oxide and aluminium oxide, are used in storage devices to operate up to 600°C.

4.17.2 Latent Heat Storage (Phase Change Heat Storage)

In this system, heat is stored in a material when it melts, and heat is extracted from the material when it freezes. Heat can also be stored when a liquid changes to gaseous state, but as the volume change is large, such a system is not economical. A few such materials which melt on heating have been experimented for their suitability for solar energy applications. These are organic materials like paraffin wax and fatty acids; hydrated salts such as calcium chloride hexo hydrate ($CaCl_2 \cdot 6H_2O$) and sodium sulphate deca hydrate ($Na_2SO_4 \cdot 10H_2O$); and inorganic materials like ice (H_2O), sodium nitrate ($NaNO_3$) and sodium hydroxide ($NaOH$).

Phase change materials such as sodium sulphate decahydrate (Glauber's salt) melt at 32°C, with a heat of fusion of 241 kJ per kg.

$$Na_2SO_4 \cdot 10H_2O \underset{\text{Energy release}}{\overset{\text{Energy storage}}{\rightleftharpoons}} \underset{\text{Solid}}{Na_2SO_4} + \underset{\text{Solution}}{10H_2O}$$

Paraffin wax possesses a high heat of fusion (209 kJ/kg), and is known to freeze without supercooling. The inorganic material ice is quite suitable if energy is to be stored/extracted at 0°C. Sodium nitrate having a melting point of 310°C is suitable for high temperature applications.

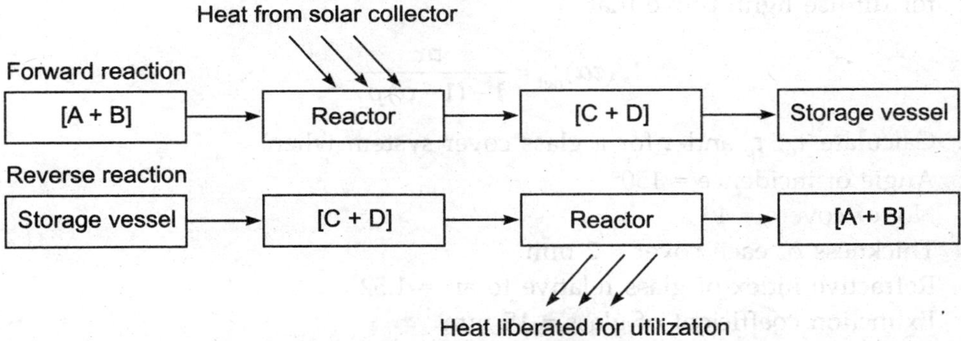

FIGURE 4.21   Schematic representation of thermochemical storage reaction.

4.17.3 Thermochemical Storage

With a thermochemical storage system, solar heat energy can start an endothermic chemical reaction and new products of reactions remain intact. To extract energy, a reverse exothermic reaction is allowed to take place. Actually, the thermochemical thermal energy is the binding energy of reversible chemical reactions.

A schematic representation of thermochemical storage reaction is shown in Figure 4.21. Chemicals A and B react with solar heat and through forward reaction are converted into products C and D. The new products are stored at ambient temperature. When energy is required, the reverse reaction is allowed to take place at a lower temperature where products C and D react to form A and B. During the reaction, heat is released and utilized. Details of some such reactions are shown in Table 4.2.

TABLE 4.2 Chemical energy storage reactions

Reaction	Temperature of forward reaction (°C)	Temperature of reverse reaction (°C)	Energy stored per unit volume of storage material (kJ/m^3)
$CH_4 + H_2O \rightleftharpoons CO + 3H_2$	780	610	209.4×10^3
$SO_3 \rightleftharpoons SO_3 + \frac{1}{2}O_2$	1025	590	460.6×10^3
$NH_4OH_4 \rightleftharpoons NH_3 + H_2O$	498	435	2143.7×10^3
$Mg(OH)_2 \rightleftharpoons MgO + H_2O$	199	335	3098.3×10^3

Like the latent heat storage system, chemical storage has the advantage of releasing heat at constant temperature.

REVIEW QUESTIONS

1. Discuss the parameters governing the performance of flat-plate collectors.

2. In a glazed flat-plate collector, only a fraction α of radiation is absorbed and the rest is reflected. In a multiple reflection cover system with ρ_d as the reflectance for diffuse light, prove that

$$(\tau\alpha)_{net} = \frac{\tau\alpha}{1 - (1 - \alpha)\rho_d}$$

3. Calculate τ_α, τ_ρ and τ for a glass cover system when:
   Angle of incidence = 150°
   No. of covers = 4
   Thickness of each cover = 2 mm
   Refractive index of glass relative to air = 1.52
   Extinction coefficient of glass = 15 mm^{-1}

4. For a multiple reflection glass cover system if

$$\frac{(\tau\alpha)_{net}}{\tau\alpha} = 0.81, \ \alpha = 0.9$$

   calculate the reflectance ρ_d of the cover system.

5. For a cylindrical parabolic concentrator the angle of acceptance is 20°, the width of the absorber is 10 cm, and the absorber tube outer diameter is 6 cm. Calculate the concentration ratio.

6. For a compound parabolic concentrator, if the angle of acceptance is 20° with flat absorber width 10 cm, evaluate the concentration ratio, the aperture height and the surface area of the concentrator.

7. Calculate the heat removal factor of a cylindrical parabolic concentrator of 5 m width and 10 m length, with the absorber tube outer diameter of 6 cm. The other known parameters are

Incident beam radiation = 650 W/m^2

$\rho = 0.85, \quad \tau = 0.9, \quad \tau\alpha = 0.72$

Flow rate = 400 kg/h

8. Calculate the solar swing angle on May 1 from 0800 hrs to 1200 hrs in Pune (18.53°N).

9. A compound parabolic collector installed in Pune (18.53°N) operates without tracking and collects solar radiation for 8 hours on May 1. Calculate the minimum acceptance angle needed for the collector.

10. Write short notes on various solar thermal energy storage systems.

SOLAR THERMAL ENERGY CONVERSION SYSTEMS

5.1 INTRODUCTION

Solar energy is available as a radiant flux; its intensity is greatest when the earth is closest to the sun and is least when the earth is farthest from the sun. India, a tropical country, lies within the latitude of 7°N and 37°N, with annual average intensity of solar radiation between 500 and 600 W/m²/day. In arid and semiarid regions the insolation is more, i.e., about 750 W/m²/day.

This chapter deals with systems, which convert solar energy to thermal energy. The most promising among them is the solar water heating system for domestic and commercial use. Solar industrial water heating systems are used in textile, food processing, dairy, chemical and other industries.

The use of concentrating-type solar collector produces high quality thermal energy, used in thermodynamic cycles to obtain work in solar thermal plants. Solar air-heaters, solar dryers, solar kilns, and space heating with solar passive architecture save fossil fuel energy. Solar cooling and refrigeration is quite attractive as cooling demand is more when the sunlight is strongest. Economic evaluation suggests that solar thermal devices are financially viable as the payback period is within their lifetime.

5.2 SOLAR WATER HEATING

Solar water heating is one of the most common applications of solar energy. A simple solar water heater with natural circulation is shown in Figure 5.1(a). The system consists of a flat-plate solar collector, normally single glazed, and a storage tank kept at a height. It is installed on a roof with the collector facing the sun and connected to a continuous water supply.

The collector comprises copper tubes welded to a copper sheet (both coated with a highly absorbing black coating) with a toughened glass sheet on top and insulting material on the rear. Water flows through the tubes, absorbs solar heat and is stored in a tank. Water circulation is entirely based on the density difference between the solar-heated water in the collector and the cold water in the storage tank. Hot water for use is taken out from the top of the tank. An auxiliary heating system is provided for use on cloudy and rainy days.

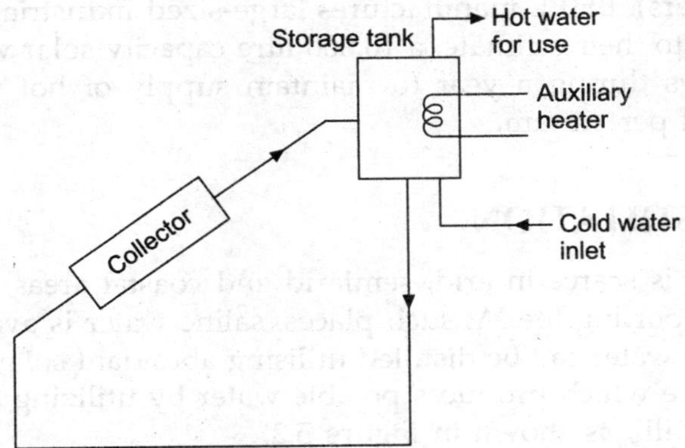

FIGURE 5.1(a) Natural circulation solar water heater.

In India, about 1 million square metres of collector area for water heating had been installed till the end of 2004. Most domestic systems are of capacity ranging from 100–500 litres of hot water per day. A typical solar water heating system can save up to 1500 units of electricity every year, for every 100 litres per day of solar water heating. To ensure quality and performance, flat-plate collectors are manufactured as per Indian Standard (IS-12933:1992).

When a large quantity of hot water is required, natural circulation is not feasible; forced circulation with a water pump is used as shown in Figure 5.1(b). Water is pumped through a collector array where it is heated and flows back into the storage

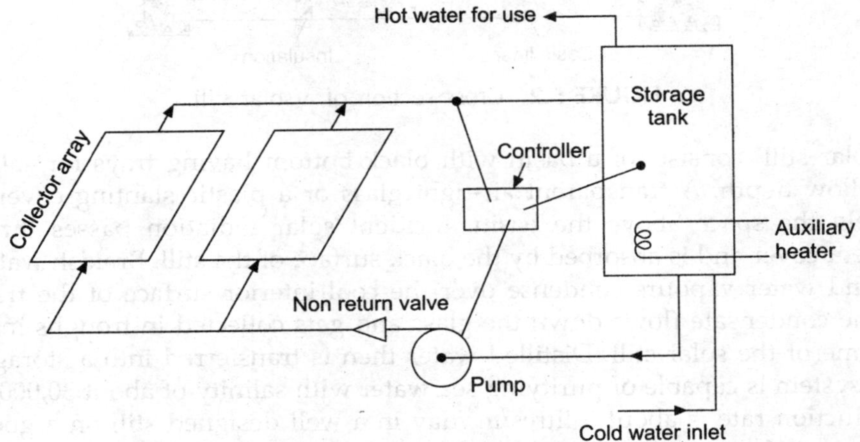

FIGURE 5.1(b) Forced-circulation solar water heater.

tank. Whenever hot water is withdrawn for use, cold water takes its place. The pump motor is actuated by a differential thermostat when the difference in water temperature at the collector array outlet and that at the storage tank exceeds 7°C. Solar water heaters of this type are suitable for industries, hospitals, hostels and offices. A solar water heater is quite economical as it pays back its cost in 3–4 years and lasts for a long time (15–20 years). BHEL manufactures large-sized industrial solar water heating systems. According to their estimate, a 10,000 litre capacity solar water heating system utilised for 300 days during a year to maintain supply of hot water at 60°C saves 30,000 litre of diesel per annum.

5.3 SOLAR DISTILLATION

Safe drinking water is scarce in arid, semiarid and coastal areas, though an essential requirement for supporting life. At such places, saline water is available underground or in the ocean. This water can be distilled utilising abundant solar insolation available in that area. A device which produces potable water by utilizing solar heat energy, is called 'solar water still', as shown in Figure 5.2.

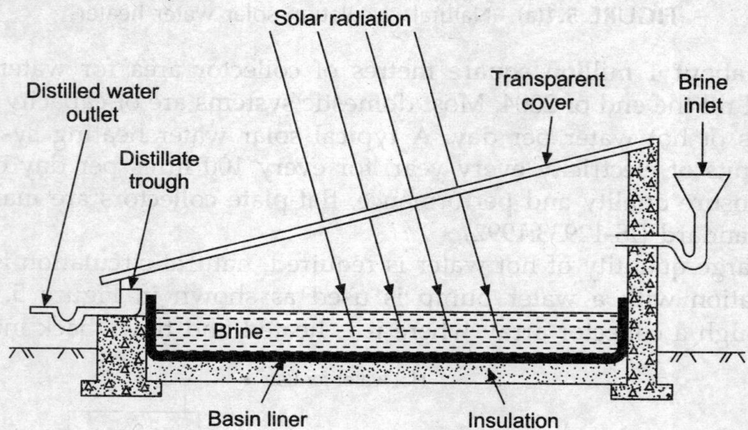

FIGURE 5.2 Cross section of a solar still.

A 'solar still' consists of a basin with black bottom having trays for saline water with shallow depth. A transparent air-tight glass or a plastic slanting cover encloses completely the space above the basin. Incident solar radiation passes through the transparent cover and is absorbed by the black surface of the still. Brakish water is then heated and water vapours condense over the cool interior surface of the transparent cover. The condensate flows down the glass and gets collected in troughs installed as outer frame of the solar still. Distilled water then is transferred into a storage tank.

This system is capable of purifying sea water with salinity of about 30,000 mg/litre. The production rate is about 3 litres/m^2/day in a well designed still on a good sunny day. The cost of water comes to about 50 paise per litre.

The performance of solar still is expressed as the quantity of water produced by each unit of basin area per day. However, the production rate depends on the intensity of solar radiation, the ambient air temperature, wind speed, and the sky condition. Desalination output increases with the rise in ambient temperature and is independent of the salt content in raw feed water. Design parameters that affect production of drinking water, include orientation of still, inclination of glass cover and insulation of the base.

In India the Central Salt and Marine Chemicals Research Institute (CSMCRI), Bhavnagar (Gujarat) has done good work in promoting solar stills in potable water scarce areas. The present status of major installations is shown in Table 5.1.

TABLE 5.1 Large solar stills in India

Location	Capacity (m³/day)	Evaporating area (m²)	Remarks
Salt works, Bhavnagar	1000	350	Drinking water for workers in salt works sea water
Awania village, near Bhavnagar	5000	1866.6	Drinking water to village. Saline water with TDS: 4500 ppm, Fluoride: 10 ppm
Narayana Sarovar, Dist. Kutch, Gujarat	3000	1244.4	Saline water with TDS:15000 ppm
Bhaleri Dist. Churu, Rajasthan	8000	3110	Saline water with TDS: 3800 ppm, Nitrates: 340 ppm, Fluoride: 5 ppm
Bitra Island, Lakshadweep	2000	750	Drinking water for islanders

5.4 LIQUID BATH SOLAR WAX MELTER

Candle and Match industries utilize wax, which needs melting for further processing. Use of solar heat energy is a viable option and cuts down the use of oil and fire wood. For designing a liquid bath solar wax melter the desired properties are:

Melting point = 62.8°C

Specific heat (i) Solid, 25°C = 2.09 kJ/kg·°C

(ii) Liquid, 65°C = 2.3 kJ/kg·°C

Density   (i) Solid, 25°C = 816 kg/m³

(ii) Liquid, 65°C = 781 kg/m³

Field experiments show that 1 m² of flat-plate collector area is required to melt 5 kg of wax.

5.4.1 Solar Wax Melter

It consists of a flat-plate collector connected to a water storage tank built around a wax chamber, as shown in Figure 5.3. Hot water from the flat-plate collector circulates in the storage tank that transfers heat to the solid wax. When temperature reaches the melting point, phase change occurs and the liquid wax is collected.

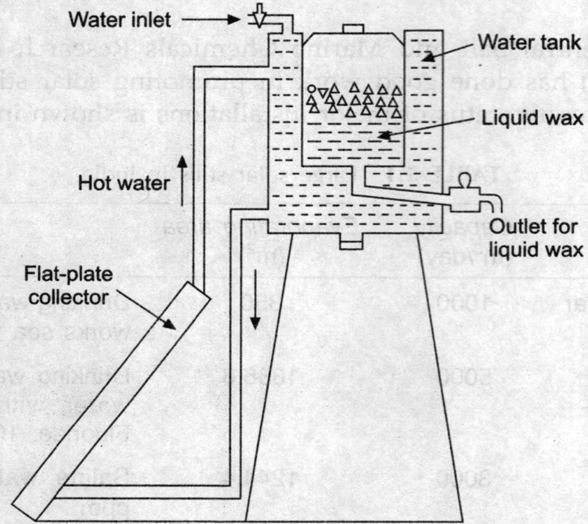

FIGURE 5.3 Liquid bath solar wax melter.

5.5 HEATING OF SWIMMING POOL BY SOLAR ENERGY

Solar energy can be used for maintaining swimming pool temperature during winter. It is necessary in extreme cold climate conditions to maintain water temperature between 20°C and 25°C. A swimming pool loses heat by conduction, convection, radiation and evaporation. An indoor swimming pool has the advantage of being maintained at the required temperature and being protected from dust, climate and birds. A schematic diagram of an indoor solar-heated swimming pool is shown in Figure 5.4.

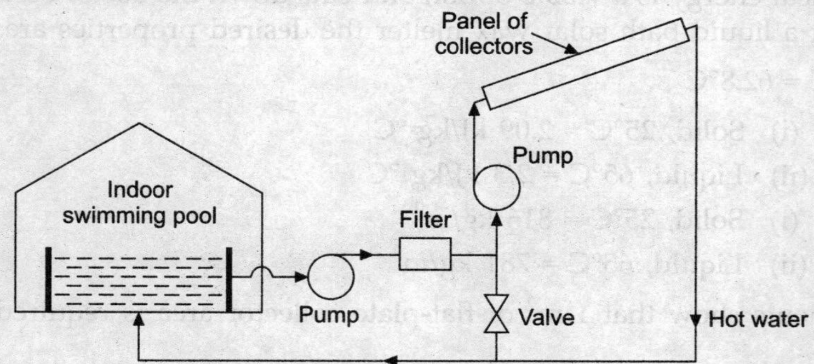

FIGURE 5.4 Schematic diagram of an indoor solar-heated swimming pool.

The area of the solar collectors required depends upon the capacity of the pool and the climatic conditions. Basically, the system has three sections: (i) solar collectors, (ii) water circulation pipes and pumps, and (iii) control system. The cost of the system depends on the availability of solar radiation, the type of collectors and atmospheric conditions. There are two types of collectors, namely unglazed and glazed. Unglazed collectors are economical and can be used where the requirement of pool temperature is small. For a larger requirement, glazed collectors are required.

5.6 THERMODYNAMIC CYCLES AND SOLAR PLANTS

Solar thermal energy can be converted into mechanical power by using any of the thermodynamic conversion cycles: the Rankine cycle, the Brayton cycle and the Stirling cycle. All steam power stations operate on Rankine cycle, where the working fluid, i.e., water is heated in the boiler and vapours thus produced are expanded in the turbine to perform mechanical work. The Brayton cycle is used to operate gas turbine plants working with a gaseous medium. The Stirling power cycle operates on a gaseous working substance and can use an external heat sources like solar, biogas or biomass.

Before discussing thermodynamic cycles, it will be relevant to mention the ideal Carnot cycle and the first and the second laws of thermodynamics. The first law can be expressed as: "In a system undergoing change, the energy can either be exchanged with the surroundings or be converted from one form to another with total energy remaining constant". The second law puts constraints on conversion of energy from one form to another with the fact that heat cannot be completely converted to work. This law recognizes different qualities of energy. "Total conversion of low quality energy to high quality energy in a cycle is impossible". Heat, work, and electrical energy are in ascending order of quality. Work can be fully converted to heat, while electrical energy being superior in quality can be converted fully into work or heat.

5.6.1 The Carnot Cycle

An ideal cycle was proposed by Sadi Carnot which operates at maximum efficiency within the purview of the second law of thermodynamics. The cycle constitutes four reversible processes as detailed in Figure 5.5. Water at point 1 is evaporated in the boiler at constant pressure to form steam at point 2 due to heat input. Steam is then expanded adiabatically doing work in turbine to be at point 3. After performing work, steam is partly condensed as heat is rejected to reach point 4. The cycle is completed as steam is compressed adiabatically to reach point 1. An adiabatic process does not exchange heat with its surroundings.

An isothermal process takes place at constant temperature. Entropy is the heat in the substance at the absolute temperature. Its unit is joule per kelvin. The enthalpy of a substance is $(U + PV)$ where U is the internal energy, PV is the product of pressure and volume of the system. If Q_H heat is supplied at temperature T_H; and after doing work, heat Q_L is rejected at low temperature T_L; then the efficiency of this cycle can be expressed as:

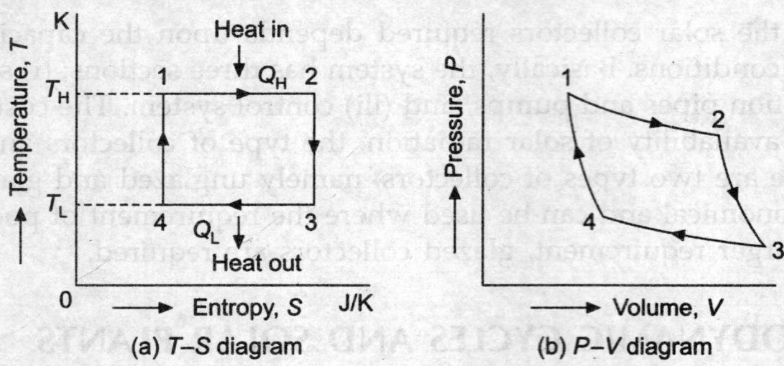

FIGURE 5.5   Ideal Carnot cycle on T–S and P–V diagrams.

$$\eta_{Carnot} = \frac{Q_H - Q_L}{Q_H} = \frac{T_H - T_L}{T_H}$$

$$= 1 - \frac{T_L}{T_H} \tag{5.1}$$

Equation (5.1) depicts that: (i) the machine efficiency improves as T_H of steam becomes higher, and (ii) the efficiency of the machine also increases if the sink temperature reduces. In an ideal condition if the sink temperature reduces to 0 K, then the theoretical Carnot efficiency becomes 100%, which is practically not possible to realize. The Carnot cycle cannot be realized in practice, hence it is useful for comparing with other thermal cycles only. The efficiency of a thermodynamic cycle is represented by the area of its entropy (T–S) diagram.

5.6.2   The Rankine Cycle

This cycle is used to operate steam power plants, where the water is heated in the boiler to produce steam (540°C). Steam so produced is allowed to expand in the turbine to perform work. Exhaust steam from the turbine is a mixture of steam and water droplets which is finally converted into water. Boiler feed pump (BFP) supplies this water at high pressure into the boiler drum to complete the cycle, as shown in Figure 5.6.

In the T–S diagram, 4–1a represents the compression of condensed water by BFP to boiler drum pressure, while 1a–1b is sensible heat, 1b–1c is latent heat, and 1c–2 is superheat addition into the cycle. Superheated steam is expanded performing work in turbine, represented by 2–3. Finally, 3–4 is the condensation of steam at constant pressure and temperature.

The efficiency of Rankine cycle is lower than that of Carnot cycle for the same temperature range. However, the work output is increased by superheating the steam and by partially expanding it and then reheating in several steps. Rankine efficiency is expressed by

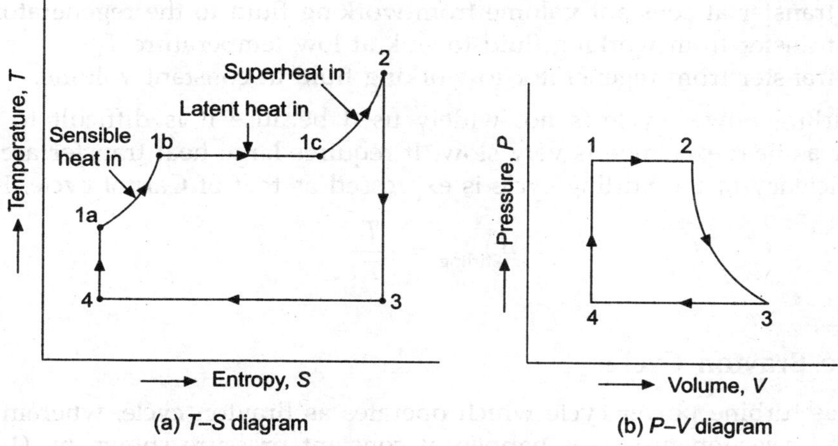

FIGURE 5.6 Ideal Rankine cycle with T–S and P–V diagrams.

$$\eta_{\text{Rankine}} = \frac{\text{net work done}}{\text{total heat addition}}$$

$$= \frac{(h_2 - h_3) - (h_{1a} - h_4)}{h_2 - h_{1a}}$$

where h_{1a}, h_2, h_3 and h_4 represent enthalpies at the respective points.

5.6.3 The Stirling Cycle

The Stirling cycle operates with hot air, where heat addition and rejection take place at constant temperature. It uses a definite mass of gas in a fully-sealed system, putting heat in and out from the working fluid (air or gas). The cycle carries four heat transfer processes, shown in Figure 5.7.

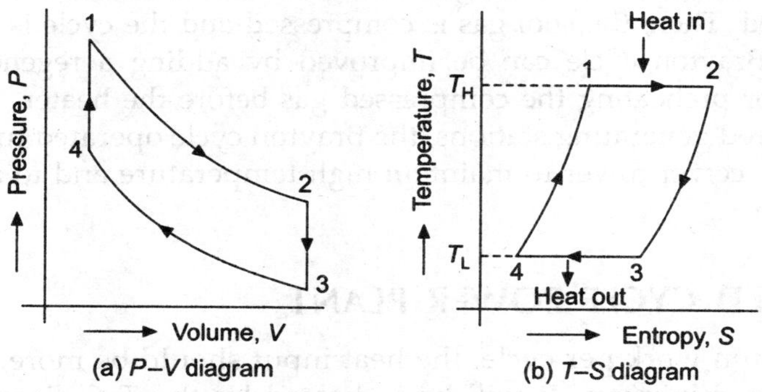

FIGURE 5.7 Ideal Stirling cycle on P–V and T–S diagrams.

1–2 Heat supplied to the working fluid at high temperature T_H with isothermal expansion.

2–3 Heat transfer at constant volume from working fluid to the regenerator.

3–4 Heat transfer from working fluid to sink at low temperature T_L.

4–1 Heat transfer from regenerator to working fluid at constant volume.

The Stirling power cycle is not widely used because it is difficult to design a regenerator as heat exchange is very slow. It requires large heat transfer areas.

The efficiency of the Stirling cycle is expressed as that of Carnot cycle, i.e.,

$$\eta_{\text{Stirling}} = -\frac{T_L}{T_H}$$

5.6.4 The Brayton Cycle

This is a gas turbine power cycle which operates as Brayton cycle, wherein the heat addition and rejection processes happen at constant pressure shown by (1–2), (3–4) (Figure 5.8).

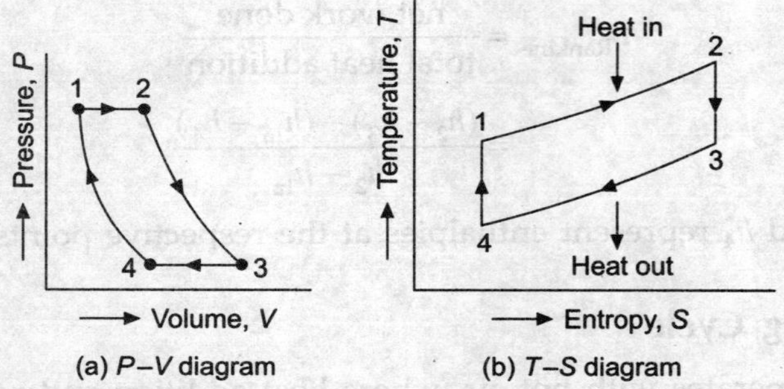

FIGURE 5.8 Ideal Brayton cycle on P–V and T–S diagrams.

Hot compressed gas at point 2 is allowed to expand through a turbine to perform work, represented by (2–3). Exhaust gas from the turbine enters the heat exchanger and heat is rejected. Then, the cool gas is compressed and the cycle is completed. The efficiency of the Brayton cycle can be improved by adding a regenerator after the turbine exhaust for preheating the compressed gas before the heater.

In solar-powered generating stations, the Brayton cycle operated turbine is located at the top of the receiver tower to maintain high temperature and to avoid excessive gas pressure drop.

5.7 COMBINED CYCLE POWER PLANT

To extract maximum work per cycle, the heat input should be more, while the heat rejected should be minimum. It will be indicated by the T–S diagram closed-loop area of the cycle. A steam turbine operating on Rankine cycle has the disadvantage of low-inlet temperature (540°C) compared with that of the gas turbine. On the other hand, the gas turbine has the disadvantage of high outlet temperature. In the 'combined cycle power plant', a gas turbine generator operating on Brayton cycle is used as a topping unit. Its high temperature exhaust is utilised in a steam turbine generator as a bottoming cycle unit. The resultant effect of combined cycle becomes an improved characteristic of high inlet temperature (T_H) and low exhaust temperature (T_L), resulting in high efficiency of 45%. The individual efficiency of a Brayton cycle gas turbine generator is about 22% and that operating on Rankine cycle is 35%.

5.8 SOLAR THERMAL POWER PLANT

Solar thermal power generation involves the collection of solar heat which is utilised to increase the temperature of a fluid in a turbine operating on a cycle such as Rankine or Brayton. In other methods, hot fluid is allowed to pass through a heat exchanger to evaporate a working fluid that operates a turbine coupled with a generator.

Solar thermal power plants can be classified as low, medium and high temperature cycles. Low temperature cycles operate at about 100°C, medium temperature cycles up to 400°C, while high temperature cycles work above 500°C. These cycles are discussed separately.

5.8.1 Low Temperature Solar Power Plant

A low temperature solar power plant uses flat-plate collector arrays shown in Figure 5.9. Hot water (above 90°C) is collected in an air insulated tank. It flows through a heat exchanger, through which the working fluid of the energy conversion cycle is also circulated. The working fluid is either methyle chloride or butane having a low boiling temperature up to 90°C. Vapours so formed operate a regular Rankine cycle by flowing through a turbine, a condenser and a liquid pump. As the temperature difference between the turbine outlet and the condensed liquid flowing out is small, i.e., about 50°C, the overall efficiency of the generating system is about 2% (8% Rankine cycle efficiency × 25% collector system efficiency). Finally, the organic fluid is pumped back to the evaporator for repeating the whole cycle.

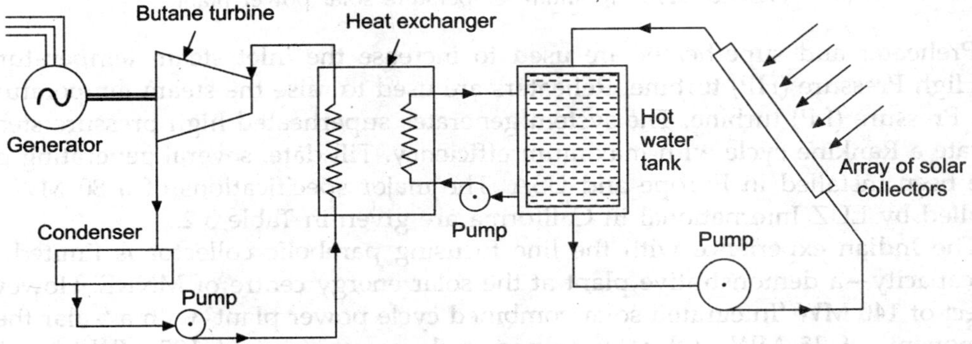

FIGURE 5.9 Low temperature solar power plant.

To reduce the capital cost, solar ponds are used instead of flat-plate collectors. Such plants up to 150 kW capacity are operative in Israel for the last 25 years.

5.8.2 Medium Temperature Solar Power Plant

Solar thermal power plants operating on medium temperatures up to 400°C, use the line focusing parabolic collector for heating a synthetic oil flowing in the absorber tube. A schematic diagram of a typical plant is shown in Figure 5.10. A suitable sun-tracking arrangement is made to ensure that maximum quantity of solar radiation is focused on the absorber pipeline.

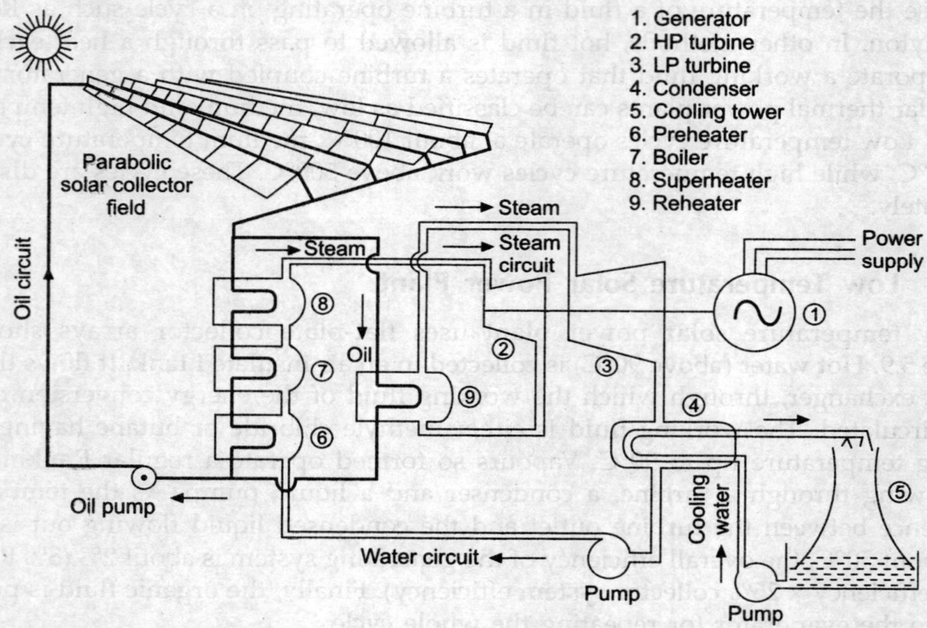

FIGURE 5.10 Medium temperature solar power plant.

Preheater and superheater are used to increase the inlet steam temperature for the High Pressure (HP) turbine. Reheaters are used to raise the steam temperature for Low Pressure (LP) turbine. The system generates superheated high pressure steam to operate a Rankine cycle with maximum efficiency. Till date, several generating plants have been installed in Europe and USA. The major specifications of a 80 MW plant installed by LUZ International in California are given in Table 5.2.

The Indian experience with the line focusing parabolic collector is limited to 50 kW capacity—a demonstrative plant at the solar energy centre of MNRE. However, a project of 140 MW 'Integrated solar combined cycle power plant' with a 'solar thermal component' of 35 MW and a 'combined cycle component' of 105 MW capacity at Mathania in Jodhpur district of Rajasthan is under active consideration. The 'solar thermal part' envisages parabolic collectors and the 'combined cycle power plant' is envisaged to run on 'regasified liquefied natural gas'. The total cost of the project is about ₹ 871 crore.

TABLE 5.2 Technical parameters of a 80 MW plant

Parameter	Value
Collector array area	464340 m^2
Type of collectors	Cylindrical parabolic
Axis orientation	North–South
Absorber tube	Steel body with selective surface enclosed in a glass cover
Collector's efficiency	70% for beam radiation
Fluid used	Synthetic oil up to 400°C
Overall efficiency	38%
Period of operation	8 hours a day
Generating cost	₹ 3.00/kWh

5.8.3 High Temperature Solar Thermal Power Generator

For efficient conversion of solar heat into electrical energy, the working fluid needs to be delivered into turbine at a high temperature. There are two possible systems—the 'paraboloidal dish' and the 'central receiver' to achieve high temperature.

With the paraboloidal dish, the concentrator tracks the sun by rotating about two axes and the solar beam radiations are brought to a common focus. A working fluid flowing through the focus is heated and the hot fluid is used to rotate a prime mover. In general, Sterling engines are installed for such systems to generate power having capacity of 10 to 100 kW with efficiency of about 30%. It is suitable as a standalone system to meet the local power needs of communities, away from the grid supply.

5.9 CENTRAL RECEIVER POWER PLANTS

In these power plants, solar radiations are reflected from arrays of mirrors (called heliostats) installed in circular arcs around the central tower. Reflected radiations concentrate on to the receiver. The array is provided with a tracking control system that focuses beam radiation towards the receiver as shown in Figure 5.11. Water is converted into steam in the receiver itself that operates a turbine coupled with a generator. Alternatively, the receiver may be utilised to heat a molten salt and this fluid is allowed to flow through a heat exchanger where steam is generated to operate the power cycle.

The 'central receiver' is an important part of the collection equipment. Typically, two receiver designs are in use—external type and cavity type. The external receiver is cylindrical in shape; the solar flux reaches the outer surface and heat is absorbed by the receiver fluid flowing through the tubes on the inner surface. In a 'cavity receiver', the solar flux enters through several apertures, where the radiant energy is transferred to the receiver fluid.

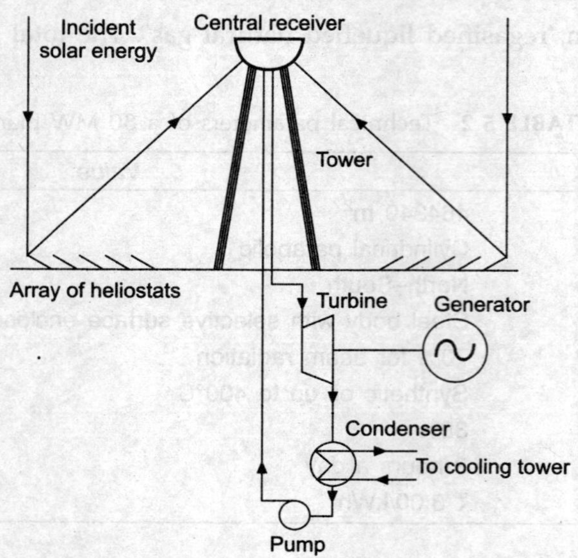

FIGURE 5.11 Central receiver power plant.

One of the biggest power plants installed during 1982 known as 'Solar one' at Barastow, US is a success story of this technology. Its technical parameters are given in Table 5.3.

TABLE 5.3 Technical parameters of a 10 MW plant at Barastow

Parameter	Value
Output	10 MW
Mirror field	1818 heliostats each having 39.3 m² area with total reflective area of 71447 m²
Receiver	Cylindrical 7 m in diameter and 13.5 m in height
Central tower	80 m high
Receiver fluid	Superheated steam at 510°C and 102 bar
Heat flux on absorber	Average 140 kW/m²; peak 350 kW/m²

5.10 SOLAR PONDS

The concept of solar pond was derived from the natural lakes where the temperature rises (of the order of 45°C) towards the bottom. It happens due to natural salt gradient in these lakes where water at the bottom is denser. In salt concentration lakes, convection does not occur and heat loss from hot water takes place only by conduction.

This technique is utilised for collecting and storing solar energy. An artificially designed pond filled with salty water maintaining a definite concentration gradient is called a 'Solar Pond'. A schematic diagram of a solar pond is shown in Figure 5.12. The top layers remain at ambient temperature while the bottom layer attains a maximum steady-state temperature of about 60°C–85°C.

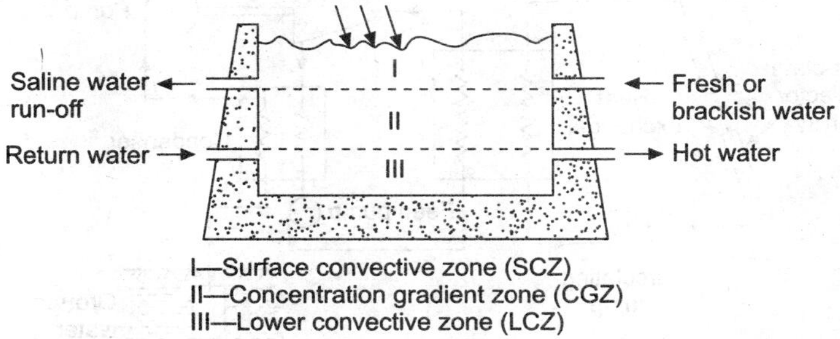

FIGURE 5.12 Schematic diagram of a solar pond.

For extracting heat energy from the pond, hot water is taken out continuously from the bottom and returned after passing through a heat exchanger. Alternatively, heat is extracted by water flowing through a submerged heat exchanger coil. As a result of continuous movement and mixing of salty water at the top and bottom, the solar pond can have three zones.

   (i) Surface Convective Zone (SCZ) having a thickness of about 10 cm–20 cm with a low uniform concentration at nearly the ambient air temperature.
  (ii) Non-Convective Zone (NCZ) occupying more than half the depth of the pond. It serves as an insulting layer from heat losses in the upward direction.
 (iii) Lower Convective Zone (LCZ) having thickness nearly equal to NCZ. This zone is characterized by constant temperature and concentration. It operates as the major heat-collector and also as the thermal storage medium.

The largest solar pond so far built is the 250,000 m^2 pond at Bet Ha Arava in Israel. Based on the Rankine cycle principle, this pond is used to generate 5 MWe of electrical power with an organic fluid.

In India, the first solar pond with an area of 1200 m^2 was built at the Central Salt Research Institute, Bhavnagar in 1973. Since then several solar ponds have been built and are in operation. The latest pond with an area of 6000 m^2, built at Bhuj (Gujarat) is the second largest in the world. It provides daily 90,000 litres of hot water at 80°C as process heat for can-sterilization. This pond maintains a stable salinity gradient with a maximum temperature of 99°C due to high radiation intensity and low thermal losses. The pond stores sufficient heat capable of generating 150 kW of power.

5.11 SOLAR PUMPING SYSTEMS

Water pumps can be driven directly by solar heated water or fluid which operates either a heat engine or a turbine. For low heads, the pump driven by vapour of a low-boiling point liquid heated by a flat-plate collector is used as shown in Figure 5.13. For larger heads, a parabolic trough concentrator or a parabolic bowl concentrator is installed to drive a steam turbine.

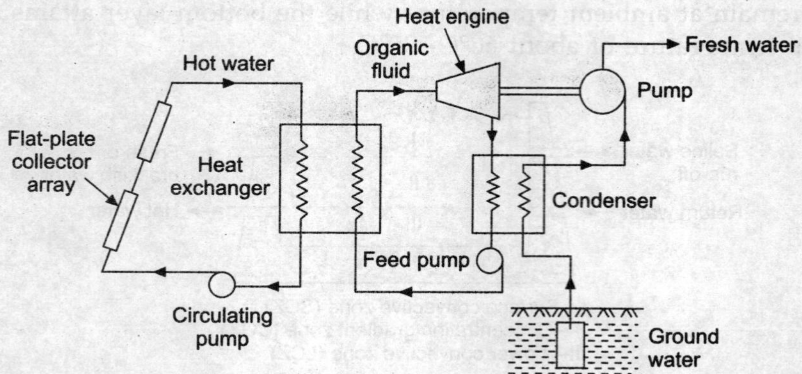

FIGURE 5.13 Schematic diagram of a solar pump.

Solar flat-plate collector arrays are installed to heat water or an organic fluid. Hot fluid then flows to a mixing tank/storage tank and then to a heat exchanger to convert the working fluid of the heat engine from liquid to vapour. It may be noted that R-115 is an acceptable working fluid as it gives high cycle efficiency besides its low cost.

Hot transport fluid or water is fed again into the collector circuit by a circulating pump. With heat engine cycle, discharged vapour from the turbine flows into the condenser where the vapour gets condensed. Working liquid is fed into the heat exchanger by a feed pump to complete the cycle. Pumped water is used as a coolant in the turbine condenser.

A higher temperature in heat exchanger or boiler, provides a high engine efficiency. An optimum range of operating temperature is used for a solar pumping system to attain maximum efficiency. Practically, energy efficiency, i.e., the percentage of solar energy collected with the quantity converted into useful work, is about 14%.

5.12 SOLAR AIR HEATERS

A solar air heater constitutes a flat-plate collector with an absorber plate, transparent cover at the top, a passage through which the air flows and insulation at the bottom and sides as shown in Figure 5.14. Air passage is only a parallel plate duct.

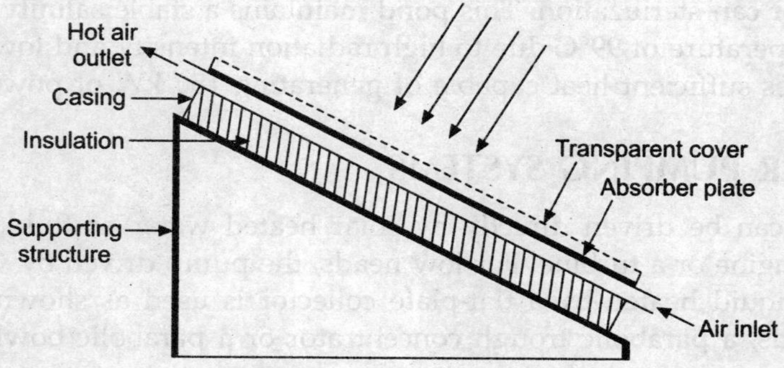

FIGURE 5.14 Solar air heater.

Air to be heated flows between the cover and the absorber plate which is fabricated from a metal sheet of 1 mm thickness. Cover is either made of glass or plastic of 4 mm to 5 mm thickness, glass wool of thickness 5 cm to 8 cm is used for bottom and side insulation. Full assembly is encased in a sheet metal box and kept inclined at a suitable angle. The face area of a solar heater is about 2 m^2, matching the heat requirement.

The value of heat transfer coefficient between the absorber plate and air is low and the operating efficiency of a simple air heater is also low. To boost heat transfer, the contact area of air with the absorber plate is increased either by adopting a V-shaped absorber plate or by designing two-pass air heaters as shown in Figure 5.15(a) and (b) respectively.

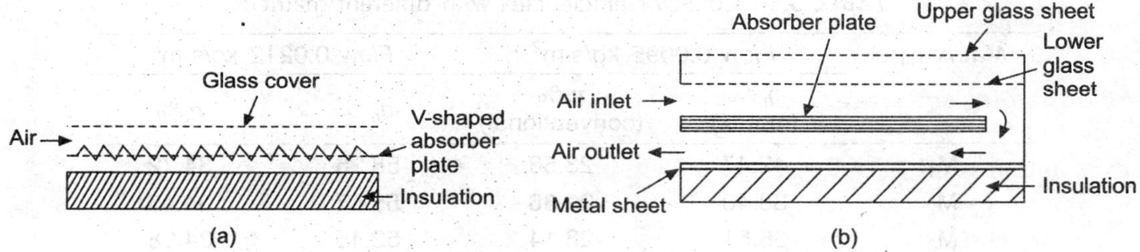

FIGURE 5.15   (a) V-shaped absorber plate, and (b) two-pass solar air heater.

The two-pass solar air heater carries two glass cover sheets, separated by an air gap which reduces heat losses.

In the matrix air heater, air flows through a porous metallic matrix which receives and absorbs solar radiation directly as detailed in Figure 5.16.

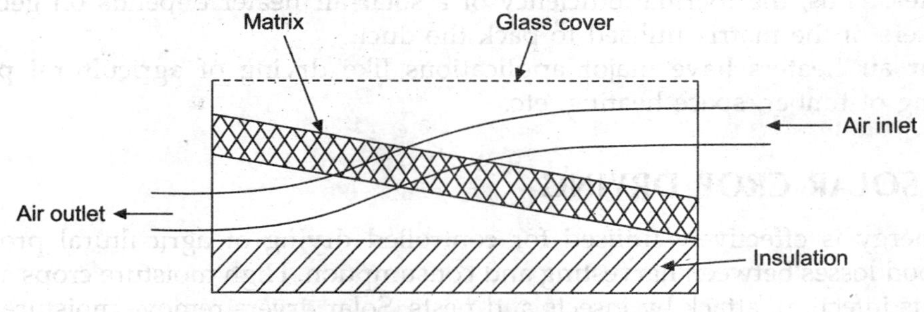

FIGURE 5.16   Matrix air heater.

Experiments were conducted at IIT Roorkee by L. Varshney and J.S. Saini during 1998 to determine the thermal performance of packed bed air heaters with wire-mesh screen matrix. Geometrical parameters of six matrices, i.e., wire diameter, pitch, number of layers, bed depth and pitch to diameter ratio, are given in Table 5.4.

Actual outdoor readings for flow rates of 0.0095 and 0.0212 kg/sm^2 were taken for thermal performance of these air heaters for comparison with conventional ones. The value of their efficiencies (η_m and η_c) at solar noon are detailed in Table 5.5.

TABLE 5.4 Geometrical parameters of wire-mesh screen matrices

Matrix type	Wire dia (mm)	Pitch (mm)	No. of layers	Bed depth (cm)	Pitch to diameter ratio
M_1	0.36	2.72	14	2.5	7.55
M_2	0.45	2.08	10	2.5	4.62
M_3	0.59	2.23	10	2.5	3.77
M_4	0.79	3.19	9	2.5	4.03
M_{4a}	0.79	3.19	7	2.5	4.03
M_{4b}	0.79	3.19	5	2.5	4.03

TABLE 5.5 Collector efficiencies with different matrices

Matrix type	Flow 0.0095 kg/s·m²		Flow 0.0212 kg/s·m²	
	$\eta_m\%$ (matrix)	$\eta_c\%$ (conventional)	$\eta_m\%$	$\eta_c\%$
M_1	42.47	23.58	58.26	34.72
M_2	38.40	22.36	55.47	34.45
M_3	36.51	23.14	52.45	34.28
M_4	40.83	22.97	56.08	34.46
M_{4a}	44.62	24.68	58.69	34.76
M_{4b}	47.07	24.30	59.60	34.74

It can be seen that solar air heaters with matrix M_{4b} give the best performance which is much higher compared with that of conventional air heaters at the given mass flow rates. Thus, the thermal efficiency of a solar air heater depends on geometrical parameters of the matrix utilised to pack the duct.

Solar air heaters have major applications like drying of agricultural products, seasoning of timber, space heating, etc.

5.13 SOLAR CROP DRYING

Solar energy is effectively utilised for controlled drying of agricultural products to avoid food losses between harvesting and consumption. High moisture crops are prone to fungus infection, attack by insects and pests. Solar dryers remove moisture with no ingress of dust, and the product can be preserved for a longer period of time.

A cabinet type solar dryer consists of an enclosure with a transparent cover as shown in Figure 5.17.

Openings are provided at the bottom and top of the enclosure for natural circulation. The material to be dried, is spread on perforated trays. Solar radiation enters the enclosure, is absorbed by the material and internal surfaces of the enclosure. Consequently, moisture from the product evaporates, the air inside is heated and natural air circulation starts. The temperature inside the cabinet ranges from 50°C to

75°C and the drying time for products like dates, grapes, apricots, cashew nuts and chillies varies from 2 to 4 days.

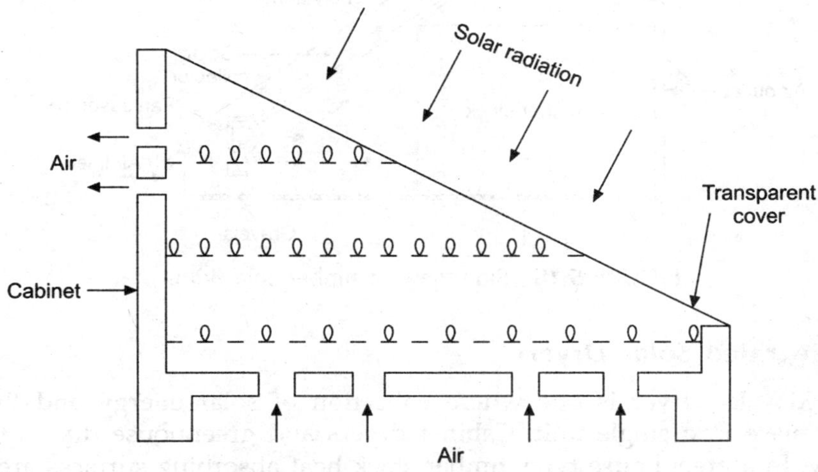

FIGURE 5.17 Cross section of a cabinet dryer.

For large-scale drying, i.e., seasoning of timber, corn drying, tea processing, tobacco curing, fish and fruit drying, solar kilns are in use.

5.14 SOLAR KILNS

In a solar kiln, heating and drying of products on a large scale, like tea, corn, fruits, timber, etc. is done by using solar energy. It operates on the principle that a transparent glass sheet or polythene sheathing allows solar radiation to pass through into the kiln and blocks long wavelength radiation emitted by products like timber back into the atmosphere. The important factors affecting the drying process are:

- Relative humidity and temperature of air
- Air flow rate
- Initial moisture content of the product
- Final desired moisture content of the product.

A solar kiln used for seasoning of timber consists of three major parts: (i) a wood seasoning chamber, (ii) a flat-plate collector, and (iii) a chimney seasoning chamber which is placed on a raised masonary platform. The chimney creates a natural draught in the seasoning chamber, causing hot air circulation around stacked wood as detailed in Figure 5.18.

Circulating air carries heat from the solar absorbing plate to timber logs and evaporates moisture. Drying is basically a heat and mass transfer process, i.e., the moisture from surface and inside the product is vaporized and removed by circulating hot air. Different types of solar dryers are discussed below.

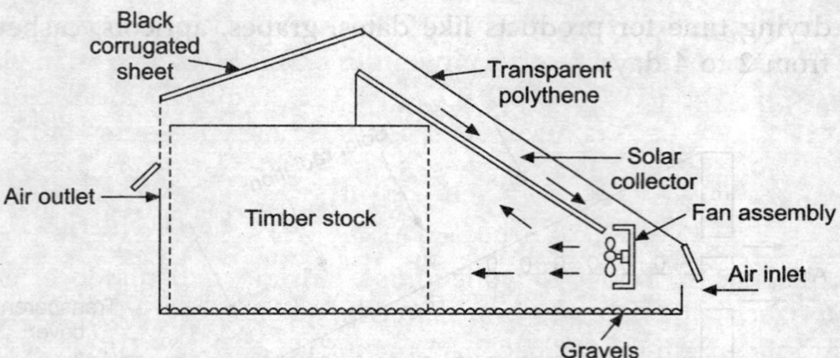

FIGURE 5.18 Side view of timber solar kiln.

5.14.1 Integrated Solar Dryers

An integrated solar dryer is one where collection of solar energy and the related drying take place in a single unit. Cabinet dryers and greenhouse dryers fall under this category. In a greenhouse type, timber stack heat absorbing surfaces are installed under one structure. The rate of drying is important and depends on the ratio of absorber surface to timber volume. The value of this parameter varies from 2 m^2 to 4 m^2 of absorber per cubic metre of timber volume. A low value of this parameter would indicate slow drying while a high value may cause twisting and splitting of the timber.

The calculation of kiln's performance compared to open drying is based on the efficiency of solar collector and kiln's capacity to season wood up to a given moisture level.

The efficiency η of the collector is expressed by an equation

$$\eta = \frac{T_K - T_A}{A_c I} \tag{5.2}$$

where T_K and T_A represent the kiln and ambient temperature respectively, A_c is the surface area of collector (m^2) and I is the radiation intensity (W/m^2) received on the collector. Pyronometer is used to monitor solar insolation and bulb thermometers are utilised to calculate relative humidity. Based on these readings, it has been found that the timber drying rate in a solar kiln is higher than that of open drying. The output is better as a kiln attains temperature up to 24°C above the ambient with operating efficiency of heat collector around 38%. Processing of timber in a solar kiln produces a quality product and ensures faster drying by 33%–57% compared to air drying.

Green timber contains a high proportion of moisture around 50% which has to be reduced to a value about 14% for long satisfactory use. Saw-woods (25 mm × 300 mm × 360 mm) of Mansonia Altissima were tested and dried from 46.16% to 15.02% moisture content in 12 days. Conventional drying causes defects like shrinking, warping, bending and infestation by insects. These defects have been eliminated by using solar kilns.

5.14.2 Distribution Solar Dryers

A distribution solar dryer has two parts: (i) a flat-plate air heater and (ii) a drying chamber. Air is heated in the flat-plate heater placed on the roof of a building. Hot air from the air heater is circulated in the drying chamber with a blower. These dryers may be designed in different sizes with various configurations depending upon the hot air temperature, the air flow rate and the type of products to be processed.

Solar drying systems are also economical for drying different industrial products such as chemical, leather, salt, plywood, and textiles.

5.15 SOLAR COOKERS

Cooking is a common application of solar energy in India. Several varieties of solar cookers are available to suit different requirements.

5.15.1 Box Solar Cooker

It consists of an outer box made of either fibre glass or aluminium sheet, a blackened aluminium tray, a double glass lid, a reflector, insulation and cooking pots as detailed in Figure 5.19. A blackened aluminium tray is fixed inside the box, and sides are covered with an insulating material to prevent heat loss. A reflecting mirror provided on the box cover increases the solar energy input.

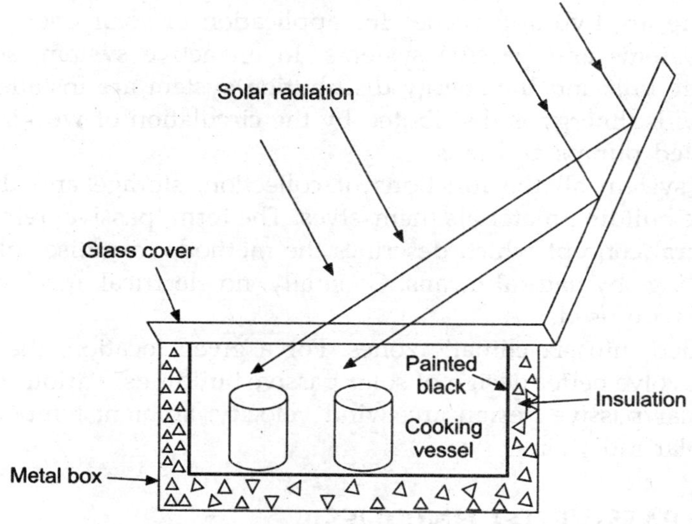

FIGURE 5.19 Box solar cooker.

Metallic cooking pots are painted black on the outer side. Food to be cooked is placed in cooking pots and the cooker is kept facing the sun to cook the food. An electric heater may also be installed to serve as a back-up during non-sunshine hours.

5.15.2 Dish Solar Cooker

A dish solar cooker uses a parabolic dish to concentrate the incident solar radiation. A typical dish solar cooker has an aperture of diameter 1.4 m with focal length of 0.8 m. The reflecting material is an anodized aluminium sheet having reflectivity of over 80%. The cooker needs to track the sun to deliver power of about 0.6 kW. The temperature at the bottom of the vessel may reach up to 400°C which is sufficient for roasting, frying and boiling. It can meet the requirement of cooking for 15 people.

5.15.3 Community Solar Cooker for Indoor Cooking

Like the dish solar cooker, the community solar cooker is a parabolic reflector cooker. It has a large reflector ranging from 7 m^2 to 12 m^2 of aperture area. The reflector is placed outside the kitchen so as to reflect solar rays into the kitchen. A secondary reflector further concentrates the rays on to the bottom of the cooking pot painted black. Temperature can reach up to 400°C and food can be cooked quickly for 50 persons.

5.16 ENERGY EFFICIENT BUILDINGS

Heavy energy demand for heating, cooling, ventilation and lighting is leading to depletion of precious environmental resources. Adopting an integrated approach, buildings can be constructed to meet the occupant's need for thermal and visual comfort with solar energy systems.

Basically, there are two approaches for application of solar energy to buildings, namely active systems and passive systems. In an active system, solar collecting panels, the storage unit and the energy distribution system are installed with one or more working fluids. Energy is distributed by the circulation of working fluids using electrically-operated pumps and fans.

In a passive system all the functions of collection, storage and distribution are carried out by the building materials themselves. The term 'passive' refers to the solar-related architectural concept which describes the methods to utilise solar heat that is available to buildings by natural means. Generally, no electrical, mechanical or power electronic controls are used.

India is divided into six climatic zones. For a given location, the knowledge of climate can help evolve better design of solar passive buildings. Various climatic factors that affect the solar passive design are: wind velocity, ambient temperature, relative humidity, and solar radiation.

5.17 SOLAR PASSIVE TECHNIQUES

There are two schemes for passive solar heating of energy efficient buildings.

5.17.1 Direct System Gain

Direct heat gain technique is generally used in cold climates. A direct gain passive solar heating system is shown in Figure 5.20 where the following techniques are used.

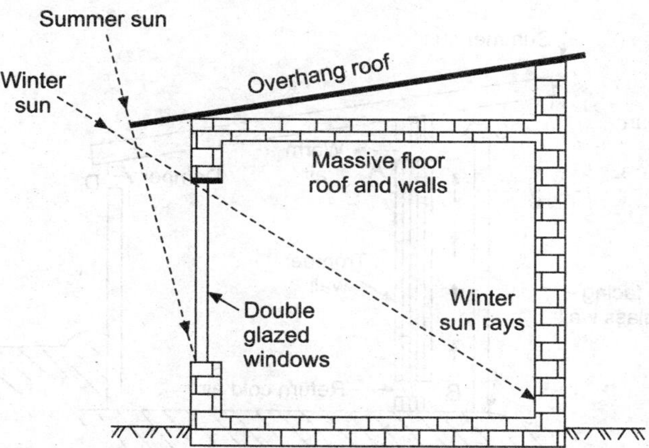

FIGURE 5.20   Direct heat gain solar passive system.

- Double glazed windows are located facing South to receive maximum sunlight during winter.
- An overhang above the windows or at the roof level is provided to give shade, during summer when the elevation of the sun is high.
- Insulating curtains are provided to cover the windows to reduces heat loss during night.
- Massive flooring and walls are used to increase the thermal mass to store heat during daytime; heat is released during the night to warm the interior.

5.17.2   Indirect Gain Systems (Thermal Storage Wall)

In direct heat gain passive-heated rooms, there are large variations in the room air temperature. To reduce variations in the room air temperature, a thermal storage wall is provided between the living space and the glazing. A diagram of such a system, designed by Professor Trombe, is shown in Figure 5.21.

In the Trombe wall passive system:

(i) The entire south-facing wall is double glazed by two sheets of glass or plastic with air-gap between the wall and the inner glazing. Hot air flows from bottom to top through this air-gap owing to natural convection.

(ii) A large blackened concrete thermal storage wall of 40 cm or more in thickness is constructed with the outer side facing the sun. The sunlight after penetration through the glazing is absorbed by the wall and thus the wall is heated.

Accordingly, the air between the glazing and the wall gets heated and flows into the room through the top vent. This circulation process continues and the cool air from the room enters into this gap through the bottom vent. In addition, the room is also heated by radiation and convection from the inner surface of the wall facing the room. During night, both vents are closed and heat transfer takes place only by radiation.

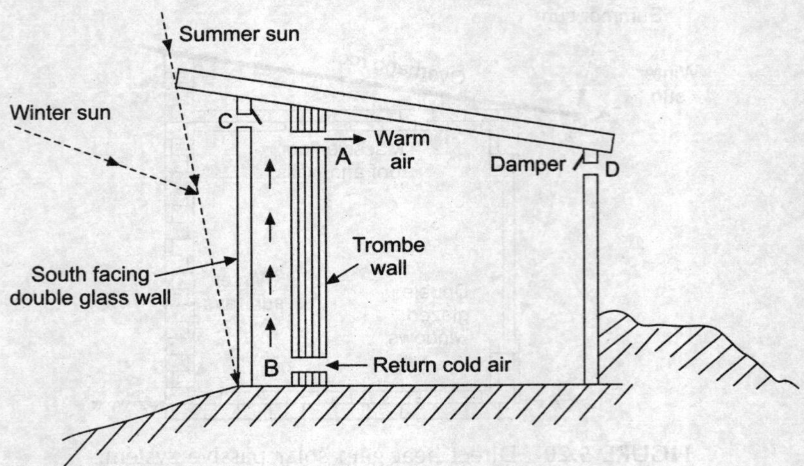

FIGURE 5.21   Trombe wall passive solar heating system.

During summer the vent A at the top of the south-facing wall is kept closed while the vents B, C and D are opened. The hot air between the glazing and the wall then flows out through the vent C and the air from the room flows in to fill this space. Simultaneously, the air is pulled into the room through the vent D which is located in a shaded cool area. The construction of the building is done in such a way that the overhanging roof prevents direct sun rays to heat glazing during summer.

5.18   SOLAR AIR-CONDITIONING AND REFRIGERATION

One of the thermal applications of solar energy is for cooling buildings (known as air-conditioning) or for refrigeration needed for preserving food. Solar cooling is advantageous in tropical countries where the cooling demand is the highest when the sunshine is the strongest. There are three modes of solar energy cooling: (i) evaporative cooling, (ii) absorption cooling, and (iii) passive desiccant cooling.

5.18.1   Evaporative Cooling

Evaporative cooling is a passive cooling technique, generally used in hot and dry climate. It works on the principle that when warm air is used to evaporate water, the air itself becomes cool and then it cools the living space of a building. Common techniques used for cooling are vapour absorption and vapour compression. Between these two, the absorption cooling system is considered to be more practical, since there is a seasonal matching between the energy needs of refrigeration system and the availability of solar radiation. The vapour absorption cooling system is discussed here.

5.18.2 Absorption Cooling System

A simple solar-operated absorption cooling system is shown in Figure 5.22. Water is heated in a flat-plate collector array and is passed through a heat exchanger called the generator. Suitable chemical solutions for absorption cooling are: (i) NH_3–H_2O where NH_3 is used as the working fluid, and (ii) LiBr–H_2O solution, where H_2O operates as the working fluid.

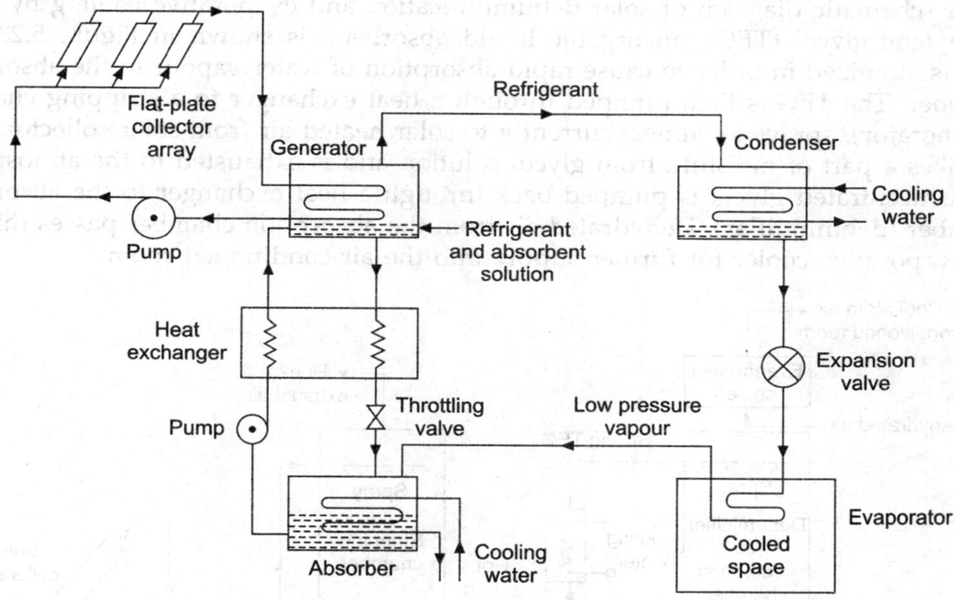

FIGURE 5.22 Solar absorption cooling system.

The whole system consists of four units: generator, condenser, evaporator, and absorber. The generator contains a solution mixture of absorbent and refrigerant, and this mixture gets heated with solar energy. Refrigerant vapour is boiled off at a high pressure and flows into condenser, where it gets condensed rejecting heat and becomes liquid at high pressure. Refrigerant then passes through the expansion valve and evaporates in the evaporator. The refrigerant vapour is then absorbed into a solution mixture taken from the generator in which the refrigerant concentration is quite low. The rich solution thus prepared is pumped back to the generator at a high pressure to complete the cycle. A heat exchanger is provided to transfer heat between solutions flowing between the absorber and the generator.

5.18.3 Passive Desiccant Cooling

The passive desiccant cooling method is effective in a warm and humid climate. Natural cooling of human body through sweating does not occur in highly humid conditions. The removal of moisture (dehumidification) from the room air using either the absorbent or the adsorbent, followed by evaporative cooling of air, is a workable air-conditioning method for use in a hot and humid climate.

Desiccant materials have a high affinity for water vapour which are used to dehumidify moisture. In solar air-conditioning, silica gel, molecular sieve and triethylene glycol are used as desiccant materials. In desiccant cooling, the hot and humid air from rooms is first dehumidified with a solid or liquid desiccant, then cooled by exchange of sensible heat and finally it is evaporatively cooled.

A schematic diagram of solar dehumidification and evaporative cooling by using triethylene glycol (TEG), an organic liquid absorbent, is shown in Figure 5.23. The TEG is atomized in order to cause rapid absorption of water vapour in the absorption chamber. The TEG is then pumped through a heat exchanger to a stripping chamber (regenerator), sprayed counter- currently to solar heated air from solar collectors. Hot air takes a part of moisture from glycol solution and is exhausted to the atmosphere. Hot concentrated glycol is pumped back through a heat exchanger to the absorption chamber (dehumidifier). Dehydrated air from the absorption chamber passes through the evaporative cooler for further supply into the air-conditioned room.

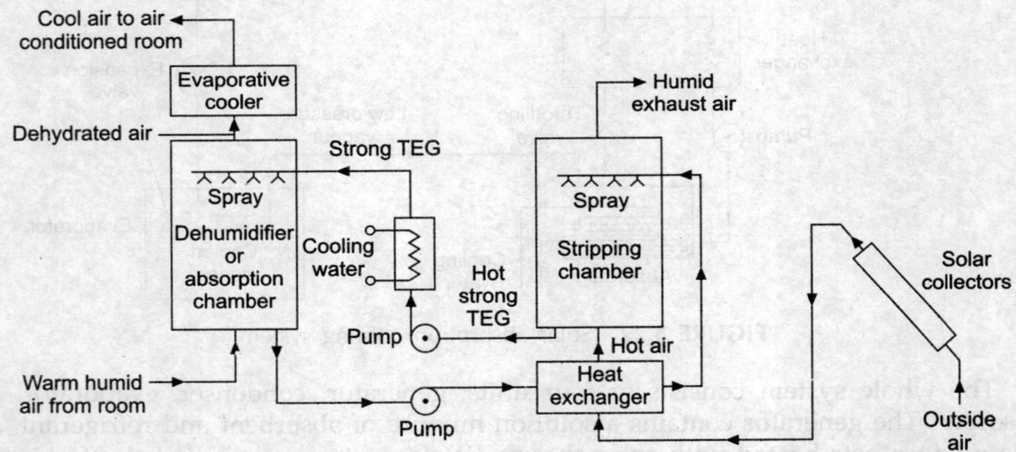

FIGURE 5.23 Schematic diagram of a solar dehumidification and evaporative cooling.

5.18.4 Energy-efficient Buildings in India

A number of buildings incorporating the solar passive architecture have been constructed in the country as offices, hostels and commercial centres and few of them are listed below.

1. Solar Energy Centre, Gwal Pahari, Gurgaon.
2. Himachal Pradesh Energy Development Agency building in Shimla.
3. A hostel for trainees in Leh.
4. West Bengal Renewable Energy Development Agency—office building in Kolkata.

5. Punjab Energy Development Agency, Chandigarh.

6. Centre for Wind Energy Technology, Chennai.

Such buildings help in reducing the requirement of conventional energy, besides providing comfortable conditions to the inhabitants in a eco-friendly manner.

5.19 SOLAR GREENHOUSES

Solar greenhouses are structures covered with glass or plastic sheets, suitable to grow vegetable and flowers under adverse climatic conditions (Figure 5.24). The basic requirements for a plant growth are: (i) light intensity, (ii) temperature, (iii) humidity, and (iv) amount of CO_2 in plant environment.

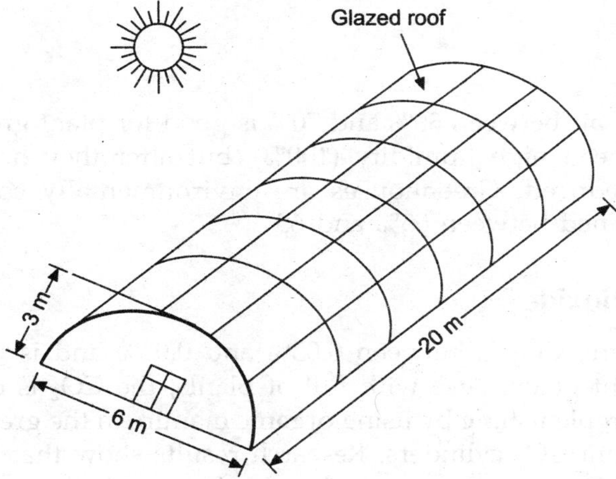

FIGURE 5.24 Schematic diagram of a pipe-framed greenhouse.

Plants manufacture their food by a process called photosynthesis which maintains a balance with respiration. In the respiration process, which is the reverse of photosynthesis, energy is liberated and used by the plant for nutrient uptake, division of cells and protein formation. Plants grow if photosynthesis is more than respiration and stop growing if both activities are equal. Plants will slowly perish if photosynthesis is less than respiration. The effects of various factors on plant growth in greenhouses are discussed in subsequent paragraphs.

5.19.1 Light Intensity

Visible radiations, CO_2 and water are used by plants to react and form carbohydrate and oxygen.

$$\text{Light Energy} + CO_2 + \text{Water} \xrightarrow{\text{Photosynthesis}} \text{Carbohydrate} + \text{Oxygen}$$

In the respiration process, carbohydrate reacts with oxygen to release energy which is utilised for the growth of plants.

$$\text{Carbohydrate + Oxygen} \xrightarrow{\text{Respiration}} CO_2 + \text{Energy + Water}$$

Minimum light intensity of 25,000 lux is sufficient for plant growth. A greenhouse structure, with two glazings, can have maximum light intensity up to 50,000 lux on a clear day.

5.19.2 Temperature

For plant survival, temperature is an important environmental factor. Temperature affects the movement of water, minerals and food in roots, stems and leaves. Ideal temperature range for winter crops is from 5°C to 15°C with a variation up to 3°C. For summer crops, the required temperature range is from 20°C to 30°C with a variation of 5°C.

5.19.3 Humidity

Relative humidity of air between 30% and 70% is good for plant growth. Saplings and germinating seeds need high humidity (100%), but after they have grown, relative humidity (RD) is reduced. Greenhouses are environmentally controlled chambers where RD is maintained between 55% and 65%.

5.19.4 Carbon Dioxide

In normal atmosphere, CO_2 is between 0.03% and 0.04% and is necessary for plant growth. In an airtight greenhouse with full of plants, the CO_2 is depleted in a short period and requires replenishing by using organic manure in the greenhouse or directly obtaining the gas from CO_2 cylinders. Research results show that by enriching air in a greenhouse with CO_2, crop matures early with higher yield.

Greenhouses are useful for growing vegetables and flowers during winter at high altitude places where the ambient temperature is below –5°C. In Ladakh, locations in higher reaches of Himachal Pradesh and Jammu and Kashmir (J & K), botanical gardens have been established in greenhouses where the crop yield is quite encouraging.

5.20 SOLAR FURNACE AND APPLICATIONS

A solar furnace is an optical equipment which concentrates solar radiation on a small area for creating a high temperature. To make a concentrated radiation in a small area from a large area receiving solar radiation, there are two ways: (i) refraction from a big single lens or multiple lenses, and (ii) paraboloidal reflector either single or heliostat type as shown in Figure 5.25.

With direct type solar furnaces, the highest heat flux density is obtained, but it is inconvenient as all the three components, i.e., both the lenses, the reflector and the target, have to be moved. Heliostat type solar furnaces are designed with optical axis either horizontal or vertical. From the operational point of view the horizontal axis furnace is most suitable.

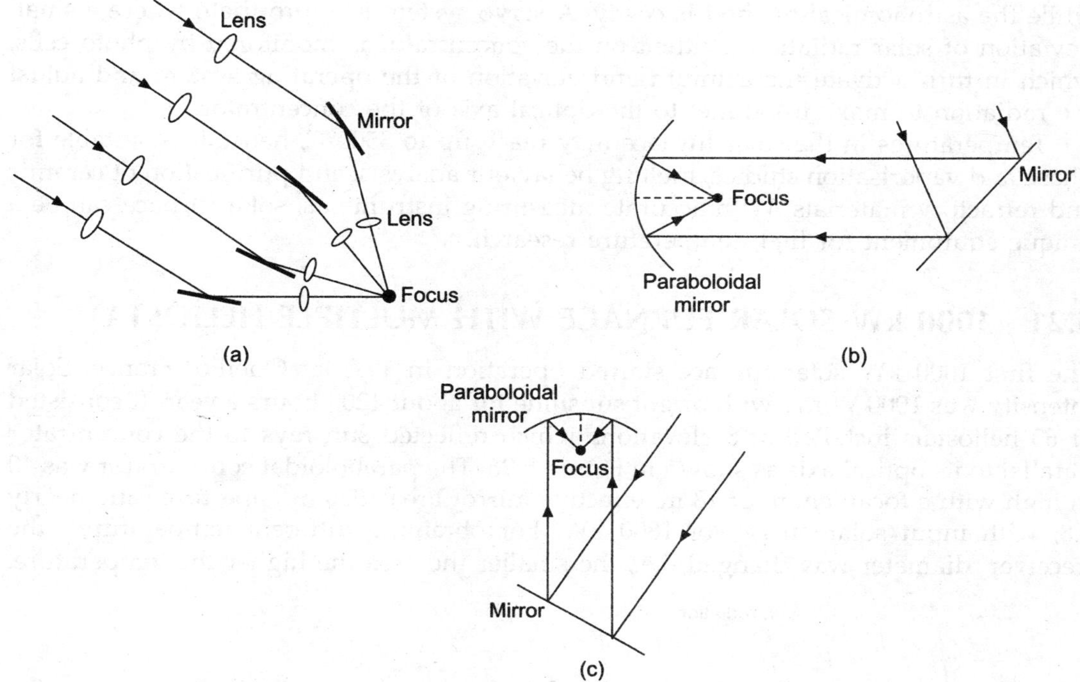

FIGURE 5.25 (a) Multiple lens type, (b) Heliostat type (optical axis horizontal), and (c) Heliostat type (optical axis vertical) solar furnaces.

Major components of a solar furnace

Concentrator: Solar furnaces use either a paraboloidal reflector concentrator or a spherical reflector concentrator. The paraboloidal reflector is considered superior due to unacceptable spherical aberration in a spherical reflector. However, a paraboloidal reflector needs to have optically smooth surface with correct geometry, high reflectivity of 0.93 to 0.94 and an aperture ratio less than 4. An electronically-polished aluminium sheet, finished with anodization, provides a better reflecting surface.

Heliostat: Heliostats in a solar furnace serve to orient solar radiation parallel to the optical axis of the concentrator. The shape and size of a heliostat is guided by the aperture of the concentrator as the heliostat has to reflect solar radiation over the full aperture of the concentrator, considering the latitude of location, the solar declination and the angular width of diverted rays from the heliostat. As a guide, the size of heliostat should be $1.4D \times 1.4D$ where D represents the size of aperture of the concentrator.

Sun tracking: For optimum functioning of a solar furnace, heliostats need to follow the sun from morning till evening. Such tracking can be performed manually, by an astronomical system or by a servomechanism system. Manual tracking is not accurate while the astronomical method is costly. A servo system is appropriate where a small deviation of solar radiation incident on the concentrator is monitored by photo cells, which in turn activate the azimuth and elevation of the operating system and adjust the radiation to make it parallel to the optical axis of the concentrator.

Temperatures in the solar furnace may reach up to 3500°C, hence it is suitable for phase and vaporisation studies, melting behaviour analysis, and purification of ceramic and refractory materials. With accurate measuring instruments, solar furnace can be a unique equipment for high temperature research.

5.21   1000 kW SOLAR FURNACE WITH MULTIPLE HELIOSTAT

The first 1000 kW solar furnace started operation in 1973 at Odeillo, France. Solar intensity was 1000 W/m^2, with bright sunshine for about 1200 hours a year. It consisted of 63 heliostats installed at 8 elevations which reflected sun rays to the concentrator parallel to its optical axis as shown in Figure 5.26. The paraboloidal concentrator was 40 m high with a focal length of 18 m, effective mirror area 1920 m^2, aperture ratio nearly 2.8, with input solar energy of 1800 kW. For obtaining different temperatures, the 'receiver' diameter was changed, i.e., the smaller the area the higher the temperature.

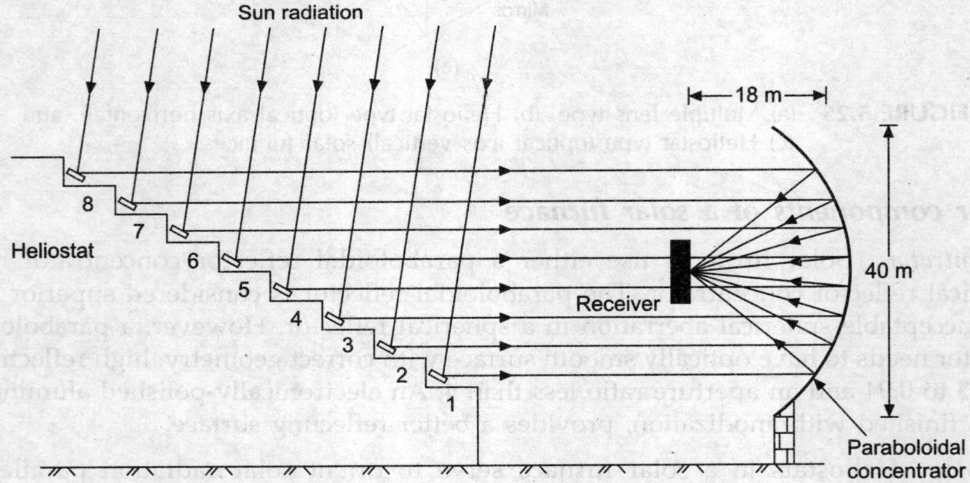

FIGURE 5.26   1000 kW solar furnace with heliostats.

REVIEW QUESTIONS

1. Discuss the uses of solar energy for heating purposes. How can solar thermal energy be used to supply potable drinking water in environmentally difficult places?

2. What are the different thermodynamic cycles? How are thermodynamic cycles useful for solar plants?

3. With the help of a schematic diagram, explain the
   (a) medium temperature solar plants, and
   (b) central receiver power plants.
4. What is solar pond? Discuss the principle of operation on which the solar pond works.
5. Discuss the methods by which agriculture crops and timber logs can be dried using solar energy.
6. Explain the difference between active and passive solar heating systems.
7. With the help of a schematic diagram, explain the "Trombe wall passive solar heating system".
8. Discuss and differentiate among evaporative cooling, absorption cooling and passive desiccant cooling methods, using solar energy.
9. What is greenhouse and how is CO_2 level maintained in it?
10. What are the important components of a solar furnace and what is the maximum temperature that can be obtained in such a furnace?

SOLAR PHOTOVOLTAIC SYSTEM

6.1 INTRODUCTION

Photovoltaic power generation is a method of producing electricity using solar cells. A solar cell converts solar optical energy directly into electrical energy. A solar cell is essentially a semiconductor device fabricated in a manner which generates a voltage when solar radiation falls on it.

In semiconductors, atoms carry four electrons in the outer valence shell, some of which can be dislodged to move freely in the materials if extra energy is supplied. Then, a semiconductor attains the property to conduct the current. This is the basic principle on which the solar cell works and generates power.

6.2 SEMICONDUCTOR MATERIALS AND DOPING

A few semiconductor materials such as silicon (Si), cadmium sulphide (CdS) and gallium arsenide (GaAs) can be used to fabricate solar cells. Semiconductors are divided into two categories—intrinsic (pure) and extrinsic. An intrinsic semiconductor has negligible conductivity, which is of little use. To increase the conductivity of an intrinsic semiconductor, a controlled quantity of selected impurity atoms is added to it to obtain an extrinsic semiconductor. The process of adding the impurity atoms is called *doping*.

In a pure semiconductor, electrons can stay in one of the two energy bands—the conduction band and the valence band. The conduction band has electrons at a higher energy level and is not fully occupied, while the valence band possesses electrons at a lower energy level but is fully occupied (Figure 6.1).

The energy level of the electrons differs between the two bands and this difference is called the band gap energy, E_g. Photons of solar radiation possessing energy E higher than the band gap energy E_g, when absorbed by a semiconductor material, dislodge some of the electrons. These electrons possess enough energy to jump over the band gap from the valence band into the conduction band. In this process, vacant electron positions or *holes* are left behind in valence band. These holes act as positive charges and can *move* if a neighbouring electron leaves its position to fill the *hole* site.

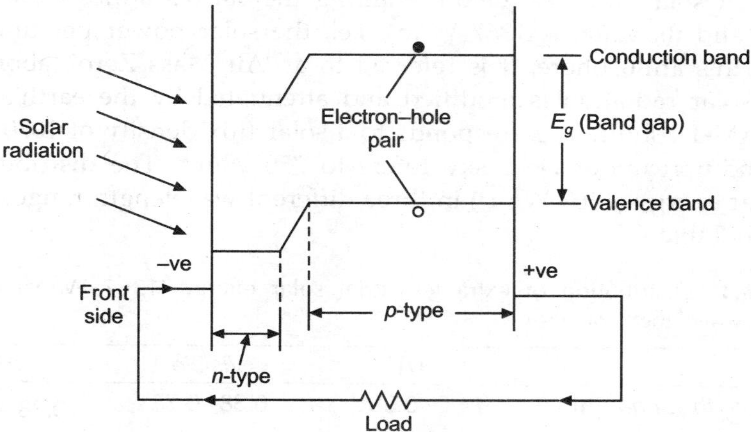

FIGURE 6.1 Semiconductor diode band structure.

Mobile electrons and holes can thus enable a current flow through an external circuit if a potential gradient exists in the cell material.

6.3 *n*-TYPE AND *p*-TYPE SEMICONDUCTORS

When a crystal of pure silicon with four valence electrons is doped with atoms having five valence electrons, for example, phosphorus, arsenic, antimony, the doped crystal carries excess electrons which can move freely, and silicon so treated is termed *n*-type semiconductor. If a pure silicon crystal is doped with atoms having three valence electrons, for example, boron, gallium, indium, a vacancy of one electron is created in the lattice, producing a *hole* with positive charge, which can freely move in the crystal. Silicon so treated makes a *p*-type semiconductor. Both *n*- and *p*-type doped semiconductors (called extrinsic semiconductors) have higher electrical conductivity than the pure (intrinsic) material.

6.4 PHOTON ENERGY

Sunlight is composed of tiny energy capsules called *photons*. The number of photons present in solar radiation depend upon the intensity of solar radiation and their energy content on the wavelength band. The solar spectrum constitutes three main regions.

1. Ultraviolet region ($\lambda < 0.4$ μm); 9% irradiance
2. Visible region (0.4 μm $< \lambda < 0.7$ μm); 45% irradiance
3. Infrared region ($\lambda > 0.7$ μm); 46% irradiance

Solar energy up to 98% is confined within the spectral region of wavelength from 0.25 μm to 2.5 μm. The far end of the infrared region, i.e., greater than 1.15 μm, has a big part of solar irradiance and this energy is not utilised by solar cells, which, in turn, constitutes the major cause of their low efficiency.

The level of solar intensity before entering the earth's atmosphere is called the *solar constant* and its value is 1367 W/m^2, i.e., the solar power per unit area at the top of the earth's atmosphere. It is referred to as 'Air Mass Zero', abbreviated AM-0. However, solar radiation is modified and attenuated by the earth's atmosphere. Further, the AM-1 condition corresponds to a solar flux density of 1070 W/m^2, while the AM-2 condition under clear sky refers to 750 W/m^2. The distribution of extra terrestrial solar energy (1367 W/m^2) in three different wavelength ranges (UV, Visible, IR) is given in Table 6.1.

TABLE 6.1 Distribution of extra terrestrial solar energy (1367 W/m^2) in three different wavelength ranges

	UV	Visible	IR
Wavelength range (μm)	0–0.38	0.38–0.78	0.78 → ∞
Energy (W/m^2)	88	656	623
Percentage in range	6.4%	48%	45.6%

When photons impinge on an atom of a semiconductor, they interact with electrons and are absorbed. This enhanced energy drives off electrons from the outer orbit. The major part of solar energy that reaches the earth's surface is in the visible region of the spectrum where photon energies vary from 1.8 eV deep red to 3.0 eV violet. In silicon, the band gap is about 1.1 eV; it therefore infers that photons with high energy are not effective in producing photovoltaic current. Because of limitations on collecting light and on absorbing photons, the silicon cells attain a theoretical maximum efficiency of 22%, cadmium telluride up to 25%, and gallium arsenide cells can go up to 25%.

6.5 FERMI LEVEL

Energy bands in a semiconductor are of two types—one which is filled with electrons known as *valence band* while the other which is empty is termed *conduction band*. The gap between the two bands is called the *band gap* as shown in Figure 6.2.

The Fermi energy level, E_f, is the energy position within the band gap from where a greater number of carriers, i.e., holes in p-type and electrons in n-type, get excited to become charge carriers. For an intrinsic semiconductor [Figure 6.2(a)], the Fermi level exists at the mid-point of the energy gap, whereas it moves closer to E_c (i.e., increases) in n-type semiconductors [Figure 6.2(b)]; similarly in a p-type semiconductor the Fermi level will lie close to E_v. In Figure 6.2(b), E_d represents the level of electrons from donor impurities, while in Figure 6.2(c) E_a represents the level of excess holes provided by acceptor impurities. Thermal energy kT (k is Boltzmann's constant = 1.38×10^{-23} J/K

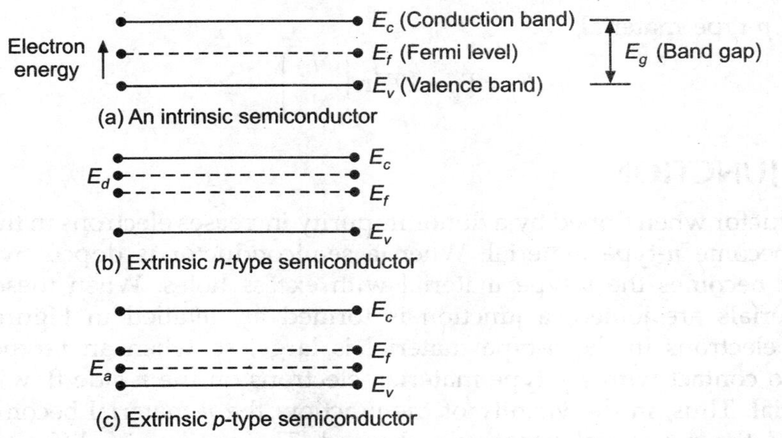

FIGURE 6.2   Diagram of energy levels in semiconductors.

and T is the absolute temperature) provides the energy differences $(E_c - E_d)$ and $(E_a - E_v)$ to excite the electrons. When thermal equilibrium is established, the number of electrons, n, per unit volume of crystal in conduction band is given by

$$n = N_c \exp\left(\frac{E_f - E_c}{kT}\right) \tag{6.1}$$

where N_c is the effective density of states in conduction band. Similarly, the density of holes p is expressed as

$$p = N_v \exp\left(\frac{E_v - E_f}{kT}\right) \tag{6.2}$$

where N_v is the effective density of states in valence band.

We have seen that the position of Fermi level is determined whether the semiconductor is doped with donors or acceptor atoms. If N_a is the concentration of acceptor atoms and N_d is the concentration of donor atoms, then

$$n = N_d = N_c \exp\left(\frac{E_f - E_c}{kT}\right) \tag{6.3a}$$

and

$$p = N_a = N_v \exp\left(\frac{E_v - E_f}{kT}\right) \tag{6.3b}$$

By solving Eqs. (6.3a) and (6.3b), for the n-type material the Fermi energy level is given by

$$E_f = E_c - kT \ln\left(\frac{N_c}{N_d}\right) \tag{6.4a}$$

and for the p-type material,

$$E_f = E_v + kT \ln\left(\frac{N_v}{N_a}\right) \tag{6.4b}$$

6.6 p-n JUNCTION

A semiconductor when doped by a donor impurity increases electrons in the conduction band and become n-type material. When a semiconductor is doped by an acceptor impurity, it becomes the p-type material with excess holes. When these n-type and p-type materials are joined, a junction is formed as detailed in Figure 6.3(a). The number of electrons in the n-type material is large; so when an n-type material is brought into contact with a p-type material, electrons on the n-side flow into holes of the p-material. Thus, in the vicinity of the junction, the n-material becomes positively charged and the p-material negatively charged. The process of diffusion of carriers continues till the junction potential reaches an equilibrium value at the time of equal flow of electrons and holes from both directions as shown in Figure 6.3(a). This is known as the unbiased condition of the p-n junction. In this condition, V is the contact potential (i.e., not an externally imposed potential) developed between the p-n junction. The contact potential so developed is a property of the junction itself.

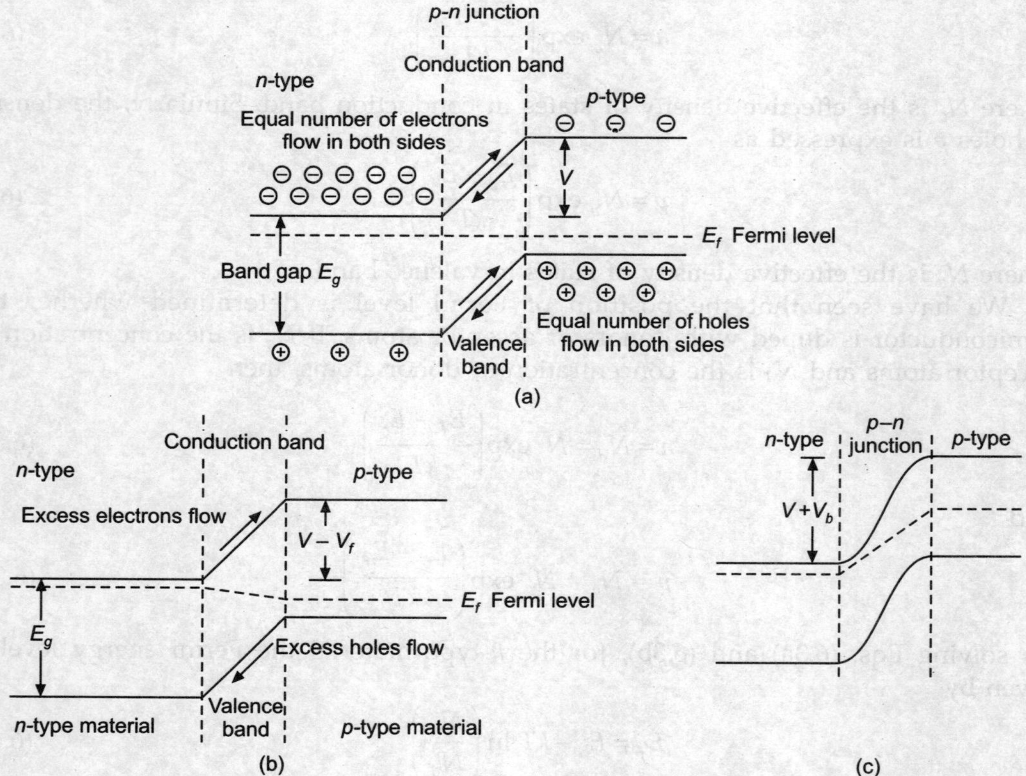

FIGURE 6.3 (a) A p-n junction, (b) a p-n junction with applied voltage V_f in forward bias, and (c) a p-n junction with reverse bias.

Now if an external voltage V_f is applied across the *p-n* junction in such a way that the magnitude of the potential difference across the *p-n* junction is *reduced* from V to $(V - V_f)$, the junction is said to be in the forward bias mode [Figure 6.3(b)]. Forward bias increases the flow of electrons in the *p*-material and the flow of holes in the *n*-material across the junction; thus the current flow across the *p-n* junction increases sharply.

Another condition arises, when a large reverse bias voltage V_b is applied across the junction as shown in Figure 6.3(c). In this case the potential difference across the *p-n* junction is *increased* from V to $(V + V_b)$. Now the current flow is only due to minority carriers, i.e., electrons from *p*-material to *n*-material and holes from *n*-material to *p*-material. The reverse bias make the in-built electric field stronger, resulting in negligible flow of current across the *p-n* junction.

When there is no illumination (dark) the flow of junction current I_j with imposed voltage V in a *p-n* junction is expressed by

$$I_j = I_0 \left[\exp\left(\frac{eV}{kT}\right) - 1 \right] \tag{6.5}$$

where I_0 is the saturation current (also called the dark current) under reverse bias and e is the electronic charge, and the other variables carry usual meanings.

6.7 PHOTOVOLTAIC EFFECT

When a solar cell (*p-n* junction) is illuminated, electron–hole pairs are generated and the electric current obtained I is the difference between the solar light generated current I_L and the diode dark current I_j, i.e.,

$$I = I_L - I_j \tag{6.6}$$

$$I = I_L - I_0 \left[\exp\left(\frac{eV}{kT}\right) - 1 \right] \tag{6.7}$$

This phenomenon is known as the photovoltaic effect.

6.8 EFFICIENCY OF SOLAR CELLS

Electrical characteristics of a solar cell are expressed by the current–voltage curves plotted under a given illumination and temperature conditions as shown in Figure 6.4.

The significant points of the curve are short-circuit current I_{sc} and open circuit voltage V_{oc}. Maximum useful power of the cell is represented by the rectangle with the largest area. When the cell yields maximum power, the current and voltage are represented by the symbols I_m and V_m respectively. Leakage across the cell increases with temperature which reduces voltage and maximum power. Cell quality is maximum when the value of 'fill factor' approaches unity where the Fill Factor (FF) is expressed as

$$FF = \frac{I_m V_m}{I_{sc} V_{oc}}$$

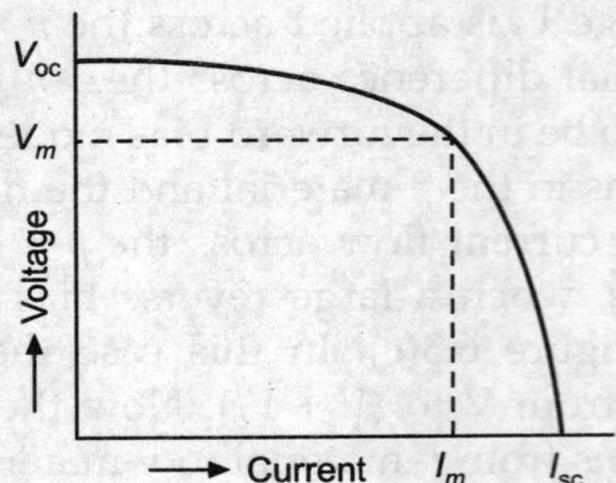

FIGURE 6.4 Current (I)–voltage (V) characteristic of a solar cell.

Maximum efficiency of a solar cell is defined as the ratio of maximum electric power output to the incident solar radiation. So,

$$\eta_{max} = \frac{I_m V_m}{I_s A_c} \tag{6.8}$$

where

I_s = incident solar flux

A_c = cell's area.

6.9 FACTORS LIMITING THE EFFICIENCY OF SOLAR CELLS

In a solar cell lot of energy gets lost, hence efficiency drops to about 20%. Various factors that causes energy losses and limit the conversion efficiency of the cell are:

1. Reflection losses: Nearly 30% of the incident radiation is lost through reflection from the surface of the cell. Reflection loss can be reduced to 3% by providing anti-reflection coating over the cell.

2. Incomplete absorption: Cell should be manufactured from material that is suitable for absorbing the energy of photons from solar radiation. Photons with less energy than the energy gap will generate heat in the cell, which is about 33%. The difference between conduction and valence band is called the band gap energy. The higher the energy gap the greater the wastage.

The energy of a photon in sun's light depends on its wavelength (λ) and represented by the equation

$$E = \frac{h \times v}{\lambda};$$

Substituting values of h and v

$$E = \frac{1.24}{\lambda}$$

where,

h = Planck's constant = 3×10^{-27} ergs

v = Velocity of light = 3×10^8 metre/sec.

Band gap of a semiconductor should match the solar spectrum and optimum band gap should be in the range of 1.1 to 2.3 eV for better efficiency. Material suitable for absorbing the energy of photons in sunlight are Silicon (1.12 eV), Gallium Arsenide GaAs (1.35 eV), Aluminium antimonide AlSb (1.49eV), Cadmium Telluride C_dT_e (1.5 eV), Gallium Phosphide GaP (2.24 eV).

Cadmium Sulphide-Cadmium Telluride cell has been commercialized in Japan. During last few years thin film of copper Indium Diselenide ($CuInSe_2$) has been made, which is widely used in space applications.

3. Partial utilization of Photon energy: Many photons in solar radiation generate electron hole pair having more energy than binding energy, which is likely to ejected as dissipated heat. therefore, combining two or more solar cells into a multifunction device is considered an effective method of making thin film PV device to reach 15% to 20% efficiencies.

4. Collection Losses: Only those electron-hole pair carriers that reach the junction before recombining are absorbed and contribute to the output current while others only generate heat. The collection efficiency depends upon

  (i) the absorption characteristic of semiconductors that determine the generation of electrons and hole pair in the crystal,
  (ii) the junction depth
  (iii) the width of the depletion layer
  (iv) the recombining rate of electrons and holes
  (v) the distance an electron will travel in the *p*-region and the hole in the *n*-region before recombining.

5. High Temperature Loss: Photovoltaic cell are exposed directly to the sun, as the temperature rises, leakage across the cell increases. This causes reduction in power output, for silicon the output decreases 0.5% per °C.

6. Series resistance losses: Electric current generated flows out of the top surface by a mesh of metal contacts provided to reduce series resistance losses. These contacts cover sufficient area which reduces the active surface and prove an obstacle to incident solar radiation.

7. Thickness of the cell: The cell must be thin in the 60 to 100 μm range.

8. Location of *p*–njunction: For higher efficiency, the *p*–*n* junction should be located near to the top surface (within 0.15 μm).

EXAMPLE 6.1

A solar cell (0.9 cm^2) receives solar radiation with photons of 1.8 eV energy having an intensity of 0.9 mW/cm^2. Measurements show open-circuit voltage of 0.6 V/cm^2, short-circuit current of 10 mA/cm^2, and the maximum current is 50% of the short-circuit current. The efficiency of cell is 25%. Calculate the maximum voltage that the cell can give and find the 'fill factor'.

Solution

$$\eta = \frac{V_{max} I_{max}}{P_{in}}$$

$$V_{max} = \frac{P_{in} \times \eta}{I_{max}} = \frac{0.9 \times 10^{-3} \times 0.25}{5 \times 10^{-3}}$$

$$= 0.045 \ V/cm^2$$

$$P_{max} = V_{oc} \times I_{sc} \times FF$$

$$FF = \frac{I_{max} \times V_{max}}{V_{oc} \times I_{sc}} = \frac{5 \times 10^{-3} \times 0.045}{0.6 \times 10 \times 10^{-3}}$$

$$= 0.0375$$

6.10 SEMICONDUCTOR MATERIALS FOR SOLAR CELLS

Solar cells are fabricated from semiconductor materials prepared in three physical states–single multicrystal, many small crystals (polycrystalline) and amorphous (noncrystalline).

6.10.1 Single Crystal Silicon

Silicon solar cells are commonly used for both terrestrial and space applications. The basic raw material is sand (SiO_2) from which silica (Si) is extracted and purified repeatedly to obtain the metallurgical grade silicon. It contains about 1% impurities and further processed to convert it to a purer semiconductor grade silicon. It is then finally converted into a single crystal ingot.

A single crystal ingot is a long cylindrical block of about 6 cm to 15 cm in diameter. Crystalline cells basically require 300 μm to 400 μm of absorber material; the ingot is sliced in wafers of 300 μm thickness as shown in Figure 6.5. These wafers are the starting material for a series of process steps such as surface preparation, dopants diffusion, anti-reflection coating, contact grid on the surface and base contact on the upper surface and on the lower one.

Solar cells are fixed on a board and connected in series and parallel combinations to provide the required voltage and power to form a PV module (Figure 6.6).

To protect the cells from damage a module is hermetically sealed between a plate of toughened glass and layers of Ethyl Vinyl Acetate (EVA). A terminal box is attached to the back of a module where the two ends of the solar string are soldered to the terminals. When the PV module is in use, terminals are connected directly to the load. Single PV modules of capacities ranging from 10 Wp (peak watt) to 120 Wp can provide power for different loads. Several panels of modules constitute an array, which is rated according to peak wattage it delivers at noon on a clear day. For higher outputs an 'array field' is created.

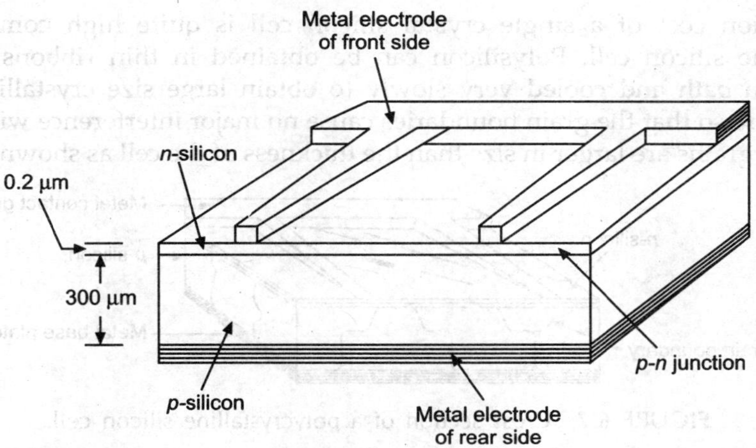

FIGURE 6.5 Cross section of a silicon cell.

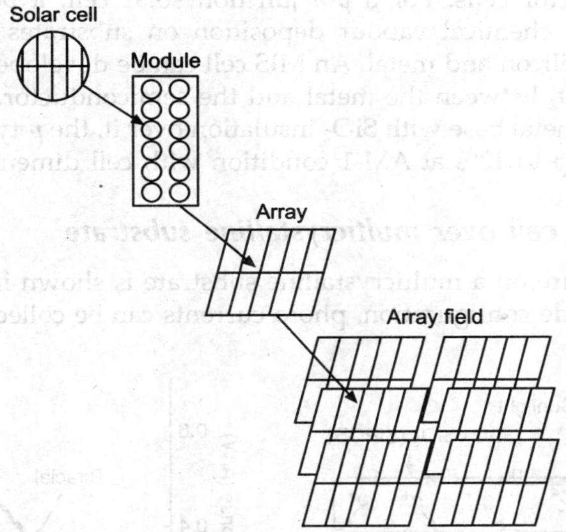

FIGURE 6.6 Solar cell, module, array and array field.

The size of an individual cell varies from 10 cm^2 to 100 cm^2 and a module contains about 20 cells to 40 cells. A standard module constituting 30 cells, each of 7.5 cm diameter, can provide electrical parameters of 12 volts, 1.2 ampere, and 18 watt peak power.

To reduce cost, methods have been developed to produce a ribbon of single crystal silicon from the molten pure silicon. The ribbon can be cut with minimum wastage into required sizes and processed directly to make solar cells.

6.10.2 Polycrystalline Silicon Cells

The production cost of a single crystal silicon cell is quite high compared to the polycrystalline silicon cell. Polysilicon can be obtained in thin ribbons drawn from molten silicon bath and cooled very slowly to obtain large size crystallites. Cells are made with care so that the grain boundaries cause no major interference with the flow of electrons and grains are larger in size than the thickness of the cell as shown in Figure 6.7.

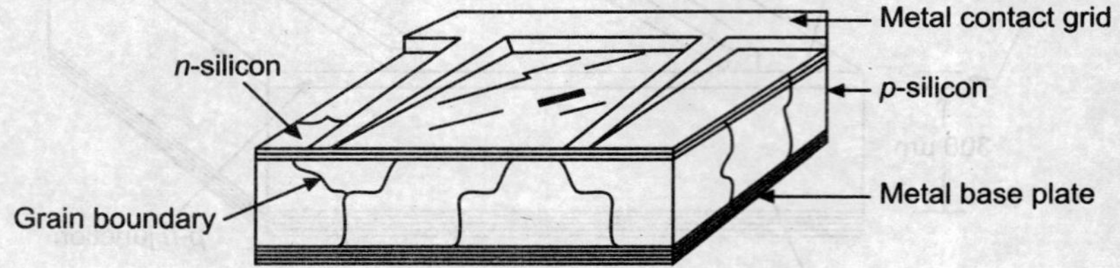

FIGURE 6.7 Cross section of a polycrystalline silicon cell.

The polycrystalline silicon solar cell can be fabricated in three designs, namely *p-n* junction cells, Metal Insulator Semiconductor (MIS) cells, and conducting oxide-insulator semiconductor cells. For a *p-n* junction solar cell, a polycrystalline silicon film is deposited by chemical vapour deposition on substrates like glass, graphite, metallurgical grade silicon and metal. An MIS cell can be developed by inserting a thin insulting layer of SiO_2 between the metal and the semiconductor. A nicely developed cell with chromium metal base with SiO_2 insulation over it, the *p*-type crystalline silicon can give efficiency up to 12% at AM-1 condition with cell dimension of 0.2 cm^2.

Bifacial crystalline cell over multicrystalline substrate

A bifacial cell structure on a multicrystalline substrate is shown in Figure 6.8(a). With a double-sided cathode configuration, photo currents can be collected from the nearest side of the cell.

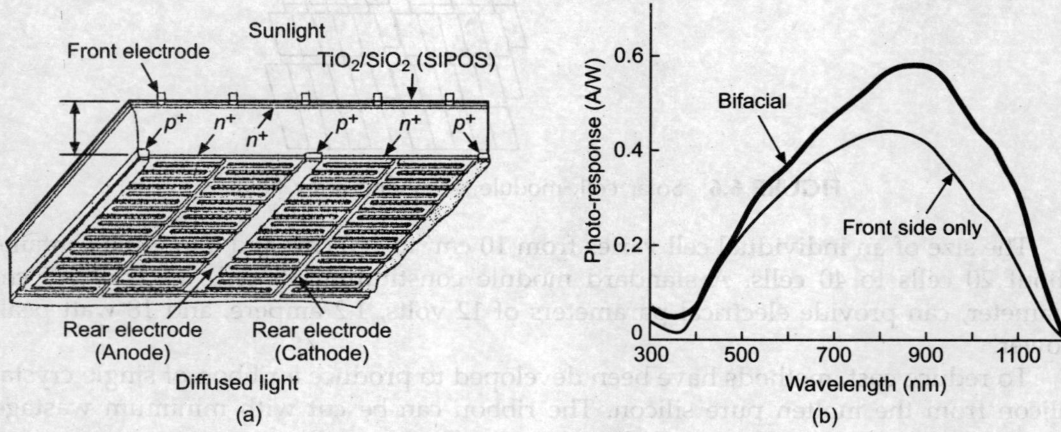

FIGURE 6.8 (a) Schematic of a bifacial cell and (b) spectral response curves of the bifacial cell.

The *p-n* junctions and electrodes are formed on both sides of a cell to collect the generated currents from both sides. The rear cathode acts as a current booster for the front cathode due to front sunlight and vice-versa. The spectral response [Figure 6.8(b)] of the bifacial cell is the summation of the independent front and rear cells. The spectral response improves in the long wavelength region due to effective collection of photo-currents by the rear cathode. The conversion efficiency of a bifacial cell developed by Hitachi Japan is reported to be up to 19%.

6.10.3 Amorphous Silicon Cells (Thin Film Cells)

Amorphous silicon cells are recently developed using thin film cells. Amorphous silicon is pure silicon with no crystal properties. It is highly light absorbent and requires only 1 µm to 2 µm of material to absorb photons of the incident light. Thin amorphous layers can be deposited on cheap substrates like steel, glass and plastic. Hydrogenated amorphous silicon (a-Si : H) is a suitable material for thin film solar cells, mainly due to its high photo-conductivity, high optical absorption of visible light with optical band gap of 1.55 eV. Thin films of nearly 0.7 µm can produce solar cells comparatively at low cost. Amorphous silicon cells can be fabricated in four structures: (i) metal, insulator–semiconductor (MIS), (ii) *p-i-n* devices, (iii) hetrojunction, and (iv) Schottky barriers.

The *p-i-n* junction, a-Si solar cells are beneficial for commercial production due to their good performance. A common type of *p-i-n* junction, a-Si solar cell, consists of a deposited layer of boron doped a-Si : H(200 Å) and above it, is a deposited layer of *n*-doped a-Si : H (80 Å). Then, a 70 Å thick layer of Indium Tin Oxide (ITO) is deposited over the *n*-type layer which serves in two ways, i.e., conducting electrode and anti reflective coating.

In a single junction (a-Si : H) solar cell, a part of solar radiation with less energy than band gap remains unutilized and wasted as heat, causing low cell efficiency. This drawback is solved by adopting a 'tandem structure' that involves stacked junctions where semiconductors having different energy gaps are erected on top of each other with decreasing band gap in the direction of light path.

Hitachi of Japan has developed tandem thin film solar cells consisting of three amorphous layers having different band gaps as shown in Figure 6.9(a). The top layer is of transparent conducting oxide and the first two cells are the standard a-Si:H cells serving as the intrinsic layer, and the third (last) layer is an alloy of silicon, germanium and hydrogen (a-Si Ge:H).

In this structure, the a-Si:H cells utilise the blue–green end of the spectrum, while a-SiGe:H cell utilises the red part of the spectrum. The spectral response of a tandem cell is shown in Figure 6.9(b), which shows the solar spectrum performance of each cell and the summation of tandem cells.

The spectral response is improved in long wavelength zones by the material provided with narrow band gap characteristic controlled by Ge contents. This three-layered tandem cell with band gaps of 2.0, 1.7 and 1.45 eV respectively can attain theoretical efficiency up to 24%.

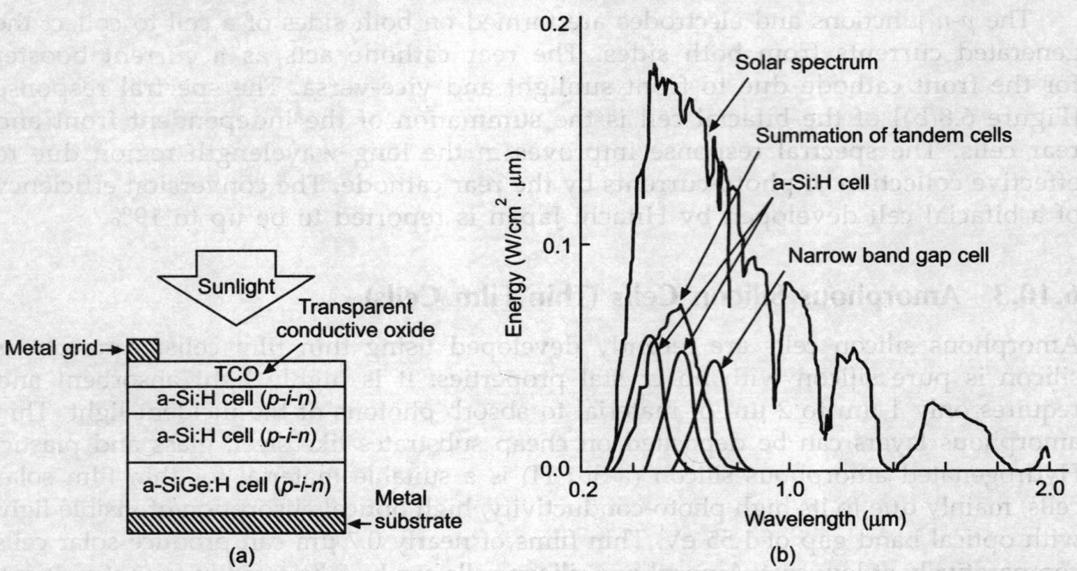

FIGURE 6.9 (a) Schematic of a three-layer tandem cell, and (b) spectral response of a tandem cell.

SPV technology has grown phenomenally, with cumulative installation of over 100 GW as of December 2012. Germany, Japan, US, and China leading the race, while India is poised to join with a target of 100 GW by 2022, a quantum jump from the present level of 48.556 GW (November 2021).

Type of SPV Solar Technology

There are two types of SPV technology:

(i) Crystalline silicon cells and
(ii) Thin film cells

They are based on the type of semiconductor and the process followed to manufacture them. Solar cells are made of various materials with different structure in order to reduce cost and achieve maximum efficiency. Thin film cells are less expensive than crystalline cells.

Principle of Working of a Solar Cell

Working of a solar cell depend on two steps:

1. Creation of pairs of negative and positive charges (known as electron -hole pairs) in the solar cell by absorbing solar radiation.
2. Separation of the negative and positive charges by creating potential gradient in the cell.

To achieve first step, the cell be manufactured of a material which can absorb the photons energy of sunlight. The photon energy E has relation with Wavelength (λ) represented by the equation.

$$E = HV/\lambda \text{ Joules}$$

where

H = Planck's constant = 6.62×10^{-34} Joules-second

V = Velocity of light = 3×10^8 m/s

λ = wavelength of photons in meters

Expressing λ in μm and energy in eV (electron-volt) (1eV = 1.6×10^{-19} Joules)

Substituting we get, $E = \dfrac{1.24}{\lambda}(eV)$

The energy in a photon must be higher than the semiconductor band-gap energy E_g in order to get absorbed in the cell material and create an electron hole pair to generate potential gradient in the cell.

A variety of compound semiconductors can be used to manufacture thin film solar cells.These compounds are Cadmium Sulphide(CdS),Cadmium Telluride (CdTe), Gallium Arsenide (GaAs).The combination of different band gap material in the tandem configuration lead to develop photovoltaic generator of higher efficiencies.

(a) **Gallium Arsenide cell:** Ordinary GaAs cells have thin films of n type and p type grown on a designed substrate. Other parts have cascading of many layers of p and n material formed using impurities of suitable elements. A GaAs has a direct band gap of 1.43 V which makes it suitable for PV thin film material. Single cell open circuit voltage is 0.8 V to 0.9 V. The efficiency is more than 20% which can be increased using more expensive GaAs subtracts. Due to high cost of production this cell is used for space crafts. Cost was $100 per peak watt.

(b) **Cadmium Sulphide:** Cadmium Telluride thin film cell: This cell is commercially used in Japan. Thin film heterogeneous junction with n-type CdS and p-type CdTe is fabricated as shown in Figure 6.10.

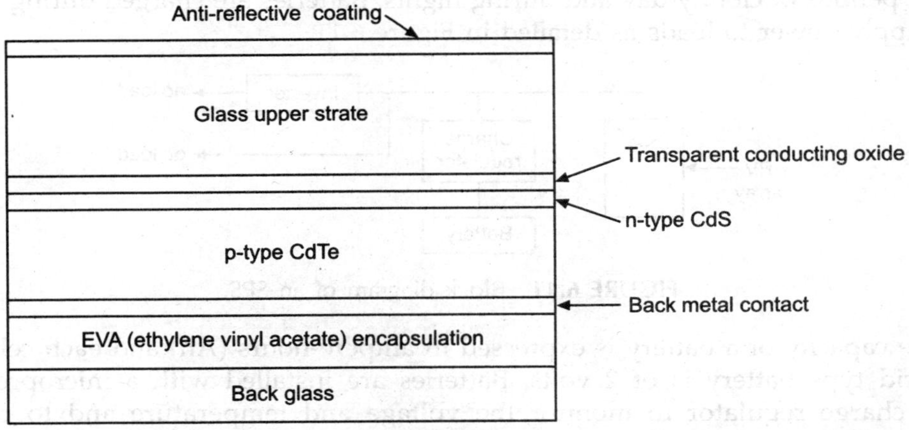

FIGURE 6.10 Magnified structure of CdTe cell.

CdTe has a direct band gap of 1.44 eV which is favorable to make a thin film cell. A 10 μm layer of CdTe is laid on a 12 μm layer of CdS. With this cell, a transparent conducting oxide layer is used instead of a metallic contact at the top of the n side. Encapsulation of ethylene vinyl acetate (a transparent material) is done to protect from environment.

Data for a module of 30 × 40 cm. Characteristic for incident flux of 1000W/m^2 having an efficiency of 8.7% is as:

Maximum output	10.5W
Open circuit voltage	52V (nearly)
Short circuit current	0.31 Amp

Its laboratory efficiency is 15% but commercial efficiency is lower than 10%. Increase for commercial efficiency and reduction in cost need R&D effort.

Cell Temperature

The temperature of a PV cell is obtained by an energy balance. The solar energy absorbed by a cell/module is converted partly into electrical energy and remaining into thermal energy. The electrical energy is used in the external circuit while thermal energy is dissipated. Heat transfer from the cell/module should be maximized, so as to make the cell to operate at the lowest temperature for higher efficiency.

6.11 SOLAR PHOTOVOLTAIC SYSTEM (SPS)

A PV module produces dc power. To operate electrical appliances used in households, inverters are used to convert dc power into 220 V, 50 Hz, ac power. Components other than PV module are collectively known as Balance of System (BOS) which includes storage batteries, an electronic charge controller, and an inverter.

Storage batteries with charge regulators are provided for back-up power supply during periods of cloudy day and during nights. Batteries are charged during the day and supply power to loads as detailed in Figure 6.11.

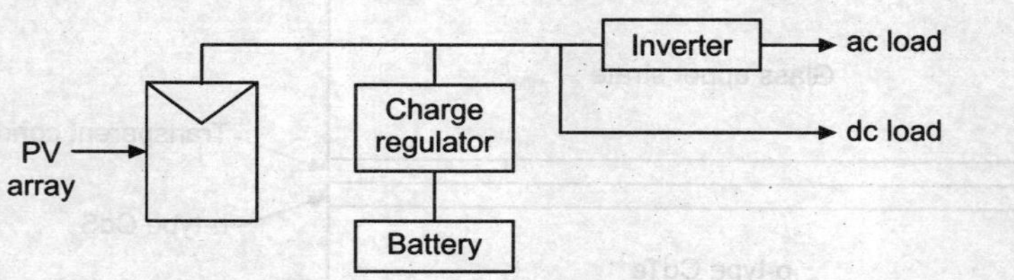

FIGURE 6.11 Block diagram of an SPS.

The capacity of a battery is expressed in ampere-hours (Ah) and each cell of the lead-acid type battery is of 2 volts. Batteries are installed with a microprocessor-based charge regulator to monitor the voltage and temperature and to regulate the input and the output currents to obviate overcharging and excessive discharge, respectively.

An inverter is provided for converting dc power from battery or PV array to ac power. It needs to have an automatic switch-off in case the output voltage from the array is too low or too high. The inverter is also protected against overloading and short circuit.

6.12 STANDARDS FOR SPV

Photovoltaic standards have been established in India by the Bureau of Indian Standard (BIS). For electrical safety and system reliability, PV devices need to conform to IS-12839 (1989) regulation regarding photovoltaic parts. Measurement of current and voltage is covered by IS-12762 (1989) and IS-12763 (1989) which deal with electrical characteristics of crystalline silicon cells.

6.13 APPLICATION OF PV SYSTEMS

Solar PV power systems may be categorized into four classes—standalone, PV hybrid, grid connected and solar power satellite. The standalone systems are self-sufficient, unreachable by state grid but have a battery system for continuous supply. A PV hybrid system is installed with a back-up system of diesel generator. Such system are used in remote military installations, BSF border outposts, health centres, and tourist bungalows. In grid-connected systems, a major part of the load during the day is supplied from the PV array, and then from the grid when the sunlight is not sufficient. These three PV systems and the solar power satellite are discussed in the following subsections.

6.13.1 Standalone PV Systems

Solar Lantern

Photovoltaic system provides an alternative to grid-based power is remote and inaccessible areas such as Kaikhali Island of Sunderbans, Gangetic Delta, West Bengal and in villages not yet electrified due to inaccessible terrains. Solar PV lantern replaces the need of a Kerosene-based lamp.

A solar PV lantern comprises a small solar PV module, chargeable battery, a charge controller and a CFL (Compact Fluorescent Lamp) or a LED (Light Emitting Diode) as shown in Figure 6.12.

Charge controller is provided to protect the solar chargeable battery from overcharging. Normal life of battery is 4 to 6 years. A solar module lasts about 25 years.

A solar lantern with a 7 watt CFL light to operate for 6 hours per night need energy per day = 7 × 6 = 42 watt-hours/day.

Batteries that are used on solar PV lantern are of 6 volt rating.

$$\text{Battery rating should be} = \frac{42\ \text{volt} \times \text{Amp} \times \text{Hours per day}}{6\ \text{volt}} = 7\ \text{Ah/day}$$

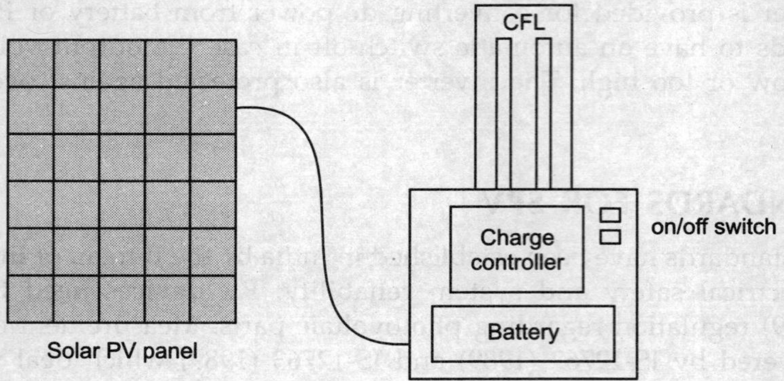

FIGURE 6.12   Solar PV lantern.

Thus, the size of the battery that is required must be 10 Wp solar PV module to cover up losses in the battery and electronics. Cost of this solar PV lantern is about ₹ 7000.

Solar street light

Solar street light as shown in Figure 6.13 describes a standalone PV power generating device. It comprises a compact fluorescent lamp, two 35 watt solar modules, and an 80 Ah tubular cell battery.

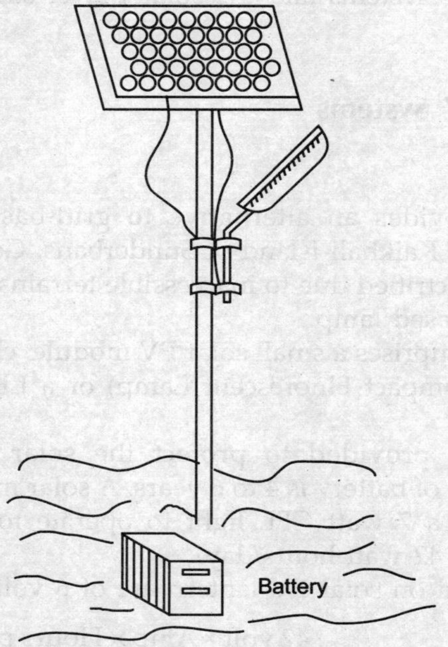

FIGURE 6.13   An SPV street light installation.

Home lighting system

Home lighting systems are the most popular solar PV units, typically designed to work with two light points and one TV point. When necessary, a small dc fan can also be run from this system.

6.13.2 SPV Water Pumping System

Individual farmers typically use an 1800 watt PV array to operate a 2 hp dc motor pumpset as shown in Figure 6.14. It can give water discharge of 140,000 litres per day from a depth up to 7 metres, sufficient to irrigate 5–8 acres of land holding several crops.

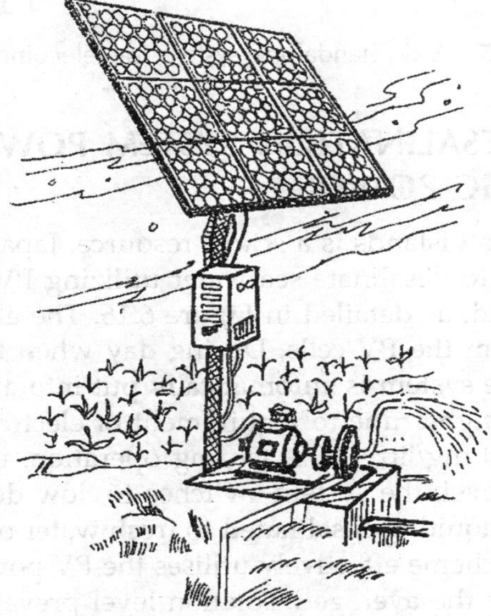

FIGURE 6.14 An SPV water pumpset.

6.13.3 SPV Cell for Communication Equipment in Snow-bound Areas

India's international boundary in Himalayas touches two neighbouring countries and hence reliable communication is a defence necessity. Grid power is not available; diesel generators are difficult to operate at sub-zero temperatures. Wind power generation is not feasible due to unsteady wind conditions with blizzards and snow storms occurring frequently. The only option is to use solar cells to produce electricity and charge the battery bank.

For a telecommunication network, PV modules of 4.5 kWp are sufficient to feed a battery bank of 48 V × 1200 A with a charge controller as detailed in Figure 6.15.

The SPV systems are virtually maintenance free with a reliable life of 25 years. They are noiseless and pollution free, suitable for remote locations with even fragile ecosystems.

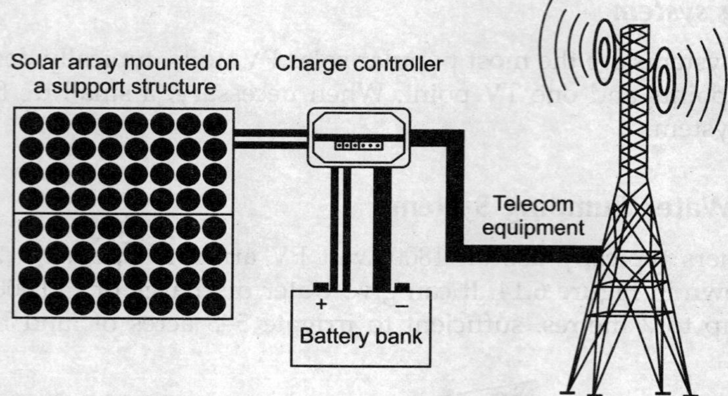

FIGURE 6.15 A dc standalone system for telecommunications.

6.14 SEA WATER DESALINATION SYSTEM POWERED BY A 25 kW PHOTOVOLTAIC POWER

Fresh water supply in small islands is a scarce resource. Japan installed a pilot project in an island during 1986, to desalinate sea water utilizing PV power 25 kWp with the support of M/s Hitachi Ltd. as detailed in Figure 6.16. The electrodialyzer is operated by dc current directly from the PV cells. During day when the solar energy is above a predetermined level, the system is automatically put into a 'high density operation', where the dc demand is higher due to requirement of electrodialization ranging from 3500 mg/litre TDS to 5000 mg/litre TDS. During operation, if the solar energy comes to a predetermined low level, the system switches to 'low density operation' and the stored 5000 mg/litre TDS liquid is desalinated to fresh water of 400 mg/litre TDS. Thus, the two-mode operation scheme effectively utilises the PV power to supply fresh water up to 5 m^3 daily, utilising the average insolation level prevalent in that area.

6.15 PV HYBRID SYSTEM

A dedicated PV power supply system is insufficient to maintain continuity of supply even with storage batteries. Standalone PV systems have a seasonal dependence and are not reliable during periods of low solar irradiance, cloudy days and nights. Thus, a hybrid energy system has been evolved to meet the load requirements without constraint. The most effective and economic solution is to install a PV system with a diesel generator along with storage batteries as shown in Figure 6.17. This system was installed in 1987 in an island isolated from main power grid. M/s Hitachi supplied 100 kW solar cell modules and the associated devices including the power conditioner. This design has two features: (i) standalone operation mode and (ii) a parallel operation mode with a diesel electric unit. Provisions have been made for installing an additional diesel generator in the system.

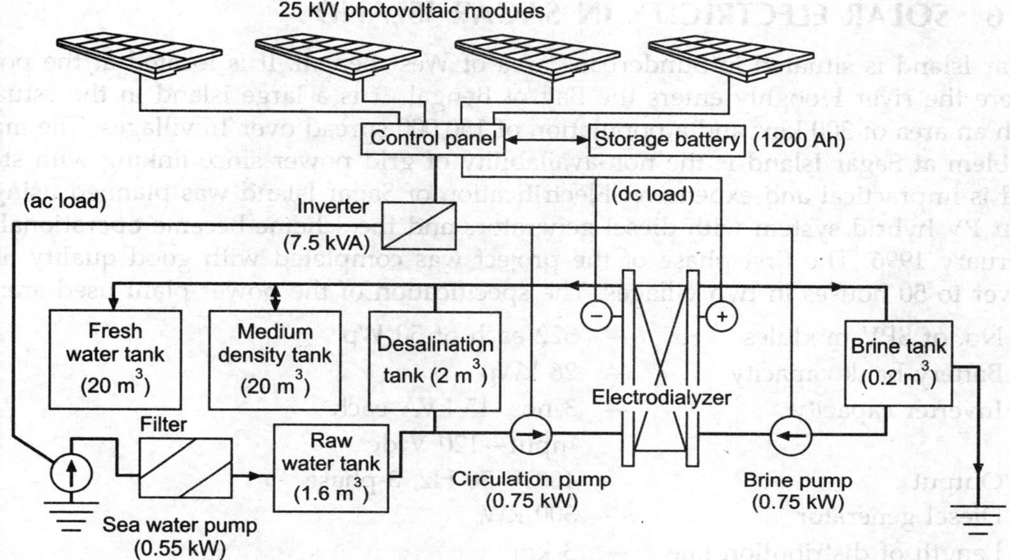

FIGURE 6.16 System flow diagram of a PV desalination system.

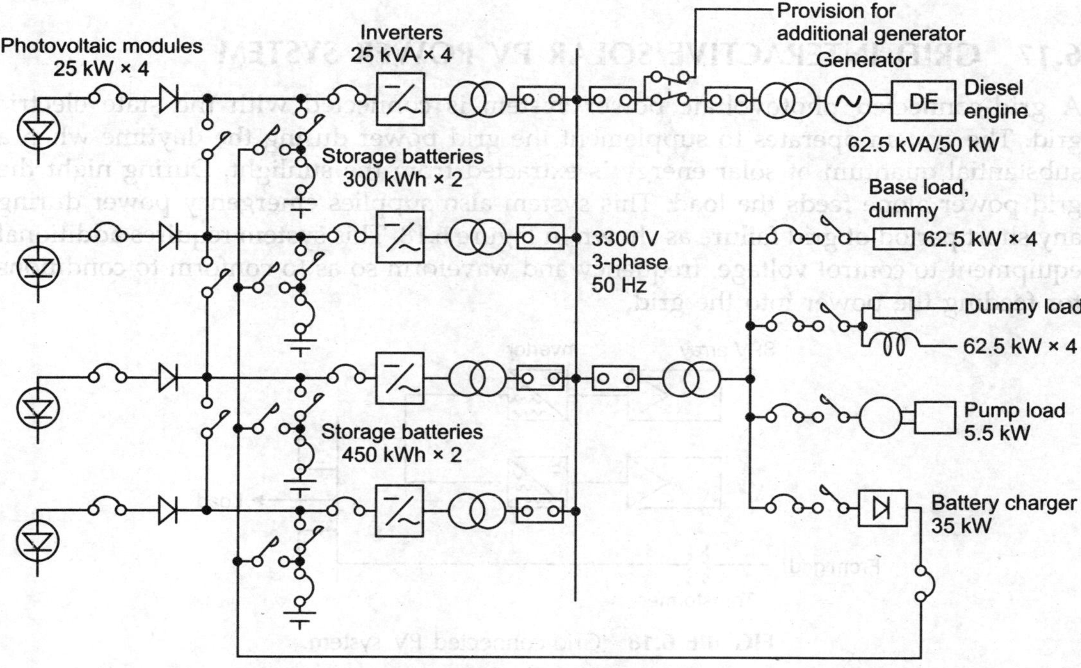

FIGURE 6.17 Single-line diagram of a 100 kW PV hybrid system with diesel generator.

6.16 SOLAR ELECTRICITY IN SAGAR ISLAND

Sagar Island is situated in Sunderbans area of West Bengal. It is located at the point where the river Hooghly enters the Bay of Bengal. It is a large island in the estuary with an area of 300 km^2 and a population of 150,000 spread over 16 villages. The main problem at Sagar Island is the non-availability of grid power since linking with state grid is impractical and expensive. Electrification of Sagar Island was planned using a solar PV hybrid system with diesel generators and the scheme became operational in February 1996. The first phase of the project was completed with good quality SPV power to 50 houses in two villages. The specification of the power plant used are:

No. of SPV modules	—	522 each of 50 Wp
Battery bank capacity	—	26 kWp
Inverter capacity	—	3 nos. 15 kVA each
		Input—120 V dc
Output	—	400 V 50 Hz, 3-phase
Diesel generator	—	300 kW
Length of distribution line	—	3 km

Sagar Rural Energy Development Cooperative is successfully managing the system and brings hope to thousands of villages in remote parts of the country.

6.17 GRID INTERACTIVE SOLAR PV POWER SYSTEM

A grid-connected photovoltaic power system is connected with the state electric grid. The system operates to supplement the grid power during the daytime when a substantial quantum of solar energy is extracted from the sunlight. During night the grid power alone feeds the load. This system also supplies emergency power during any short period of grid failure as shown in Figure 6.18. This system requires additional equipment to control voltage, frequency and waveform so as to conform to conditions for feeding the power into the grid.

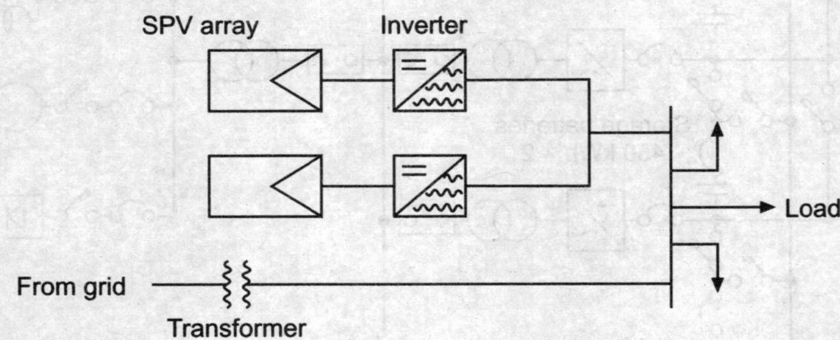

FIGURE 6.18 Grid-connected PV system.

A 25 kWp grid connected SPV power plant is functioning in Vidhyut Bhavan, the headquarters of West Bengal State Electricity Board at Salt Lake City, Kolkata. The salient features of the plant are:

Array capacity	—	25 kWp
No. of inverters	—	Two
Type of inverter	—	Pulse width modulation type, maximum point tracking
Rating of inverter	—	Grid interactive 15 kVA
Output of inverter	—	3-phase, 4-wire, 50 Hz, 400 V
Input to inverter	—	120 nominal dc voltage

6.18 SOLAR POWER PLANT USING A SATELLITE

Solar energy is a huge energy resource but difficult to utilise due to low density of the energy flux which is further decreased by atmospheric absorption and rotation of the earth. This constraint created an idea of a solar-powered generating satellite in space. It was proposed that solar-powered PV devices be arrayed in space as a circling satellite in a geo-synchronous orbit (36000 km away from the earth). Solar energy will then be received 24 hours a day and the efficiency of the system will not be hampered on account of the cloud cover over the earth. The schematic diagram is shown in Figure 6.19.

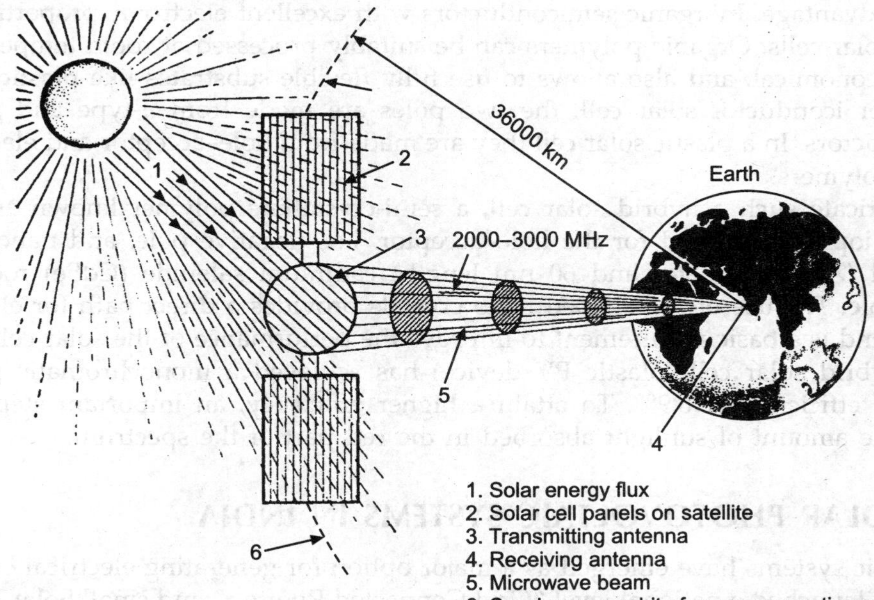

FIGURE 6.19 Schematic diagram of a solar power plant carried by a manmade satellite.

The power output from the solar array is converted to a narrow microwave beam (about 10 cm wavelength) by a magnetron, and transmitted to the earth to be received by an antenna and is then reconverted into commercial frequency electric power. The microwave beam sent from the satellite plant to the earth does not constitute any threat either to space, aircraft, or birds. At present such schemes are in the planning stage and yet to be implemented as a long-term solution to the energy shortage problem.

6.19 · PLASTIC SOLAR CELLS WITH NANOTECHNOLOGY

Photovoltaic devices will be used more and more in the near future as the production cost goes down. The fabrication of a simple semiconductor cell is a complex process and requires controlled conditions of high vacuum with temperature between 400°C and 1400°C.

Ever since the discovery of conducting plastic in 1977, there has been a constant quest to use these materials for the fabrication of solar cells. Plastic solar cells can be made in bulk quantities with lower cost, though their efficiency to convert solar radiation into electricity is low compared to semiconductor cells. A new generation solar cell that combines nanotechnology with plastic electronics has been launched with the development of a semiconductor polymer photovoltaic device. Such hybrid solar cells will be cheaper and easier to make in a variety of shapes.

Semiconductor nano-rods are used to fabricate energy efficient hybrid solar cells together with polymers. Hybrid materials, i.e., semiconductors and polymers provide a double advantage. Inorganic semiconductors with excellent electronic properties are good for solar cells. Organic polymers can be suitably processed at room temperature which is economical, and also allows to use fully flexible substrates like plastics.

In a semiconductor solar cell, the two poles are made from n-type and p-type semicon-ductors. In a plastic solar cell they are made from hole–acceptor and electron–acceptor polymers.

To fabricate such a hybrid solar cell, a semi-crystalline polymer known as poly (3-hexylth iophene) is used for the hole–acceptor, i.e., negative pole, and nanometre (nm) sized (7 nm diameter and 60 nm length) cadmium salenide (CdSe) rods for positive pole. The use of rod-shaped nano crystals provides a direct path for electron transport and is a basic requirement to improve the performance of the solar cell. This type of hybrid solar cell (plastic PV device) has achieved a monochromatic power conversion efficiency of 6.9%. To attain a higher efficiency, an important step is to increase the amount of sunlight absorbed in the red part of the spectrum.

6.20 SOLAR PHOTOVOLTAIC SYSTEMS IN INDIA

Photovoltaic systems have emerged as a major option for generating electrical energy. MNRE has launched a national level "Grid Connected Rooftop" and small Solar Power Plants Programme on 26th June 2014 which provides Central Financial Assistance (CFA) upto 30% of the capital cost for plants ranging from 1 kWp to 500 kWp in residential, commercial and institutional sectors.

This programme was envisaged to give boost to Rooftop PV and Small Solar Power Plants connected to distribution network either at 11kV or 440V three phase lines or of 220V single phase line. Huge potential is available for generating solar power using unutilized space on rooftops. Required quantities of power generated by each building can be used by the building occupants and surplus power can be fed into the grid.

Government of India also launched program with a target of 100 MW Grid Connected Solar PV Power Plants by generating 50MW power on Canal Tops and 50 MW on Canal Banks. Solar energy is land intensive, but it does not require extra land if it is installed on buildings and existing structures like canals. Some of the highlights of the completed projects during the year 2014–15 are given below:

- 24552 Solar street lights have been installed at various villages of Himachal Pradesh.
- Solar power plants having capacity of 3500 kWp installed for industries in Chhattisgarh.
- 18844 solar street lights installed at Lohiya Villages of Uttar Pradesh.
- 10,000 solar lanterns distributed in cyclone affected areas of Andhra Pradesh.
- A SPV power plant of 100 kW capacity installed at Legislative Assembly of Meghalaya.
- By 2022, 20 million solar lamps are expected to be installed.

Other Achievements are:

- 1.00 MW grid connected Rooftop 5 SPV Power Plant at Punjab Engineering College at Sector 12 Chandigarh.
- World's largest solar park of capacity 600 MW operate over 72,600 hectares in Charanka village Rann of Kutch Gujarat. It generated 343.33 million units in the year 2012–13 at a plant load factor of 18.3% and expected to produce 30 GW of power and 15 GW by 2024.
- Indian Renewable Energy Development Agency Ltd. (IREDA) as a part of Corporate Social Responsibility (CSR) supported Meghalaya Non-Conventional and Rural Energy Development Agency (MNREDA) to electrify 18 villages by providing 362 LED based solar PV lighting system. It was a joyous and proud moment for the communities of these villages when their houses glowed with clean/green energy during November 2015.
- India's total installed solar power generation capacity rose to 40,085 MW as of 2021.

Punjab to harness canal network for generation of solar power

The state have plans to harness solar PV energy over canal network of 5000 km, which has a potential for generation of 1000 MW. In first phase 5 MW canal top projects are under process and bids have been invited for another 15 MW canal based solar PV power projects.

Punjab has offered incentives to attract investors for solar energy projects. It includes exemption from pollution clearance, CLU charges, entry tax on equipment, registry and stamp duty charges on purchase and lease of land.

Punjab state has identified canal top areas that will be leased out at a very reasonable rate of ₹ 1.5 lakh/MW/annum.

Solar Photovoltaic Installation in India

Solar power has grown exponentially in India in the recent past. The country's cumulative capacity grew from mere 2.12 MW in 2007–08 to 40,085 in MW in January 2021. Such growth can be attributed due to two major mission policy namely Gujarat Solar Power Policy (GSPP) and Jawaharlal Nehru National Solar Mission (JNNSM) introduced in 2009 and 2010 respectively.

(i) Grid connected Solar PV Power projects taken up by NTPC and other PSUs got a good results and commissioning status as on February 2021 was 39,104 MW.

(ii) Solar Thermal Systems under special package for Leh and Kargil Region. MNRE initiated a project entitled "Ladakh Renewable Energy Initiative" to minimize dependence on conventional fuel in cold Leh and Kargil region in June 2010. The aim of the project was to meet energy/power requirement through local renewable sources. Achievement as on 31st December 2014 for various activities is in Table 6.2.

TABLE 6.2

S.No.	Device	Target achieved in Leh Region (in Nos.)	Target achieved in Kargil Region (in Nos.)
1.	Dish Solar Cooker (DSC)	4500	600
2.	Steam Cooking System (SCS)	15	—
3.	Domestic Green House (DGH)	2500	3000
4.	Commercial Green House (CGH)	750	150
5.	Solar Water Heating System (SWHS) sq. in area	20,384	1098

All the devices used are simple because required material except the polythene sheet and angle iron were locally available. Green houses are designed to maximise the capture solar energy during the day while minimizing heat loss at night.

6.21 JAWAHARLAL NEHRU NATIONAL SOLAR MISSION (JNNSM)

India receives nearly 3000 hours of sunshine every year, i.e. about 300 sunny day for 10 hours a day. Average solar radiation incident over the land is in the range of 4–7 kWh per day. Solar energy can be utilized in two ways

(i) solar photovoltaic technology which directly converts sunlight into electric energy and

(ii) solar thermal technology to use heat content into useful application.

The three phase JNNSM has setup an ambitious road map to achieve 1,00,000 MW of solar power by 2022, divided in 3 phases as detailed:

Phase 1 of JNNSM (2010–13): It will span remaining period of the 11th plan and first year of 12th plan upto 2012–2013. Summary of targets and achievements during Phase I of JNNSM (2010–13) are given in Table 6.3.

TABLE 6.3

Application segment	Target of Phase 1 (2010–13)	Achievement till March 2013
Grid Solar Power (large plants, rooftop and distribution grid plants)	1100 MW	1686.44 MW
Off-grid solar application	200 MW	252.5 MW
Solar Thermal Collectors (SWHS, Solar cooking, Solar cooling, Industrial process heat)	7 million sq. metres	7.01 million sq. metres

Achievements of Phase 1 have exceeded the targets set for the period.

Phase 2: It will cover remaining four years of the 12th plan from 2013–2017.

In 2021, installed capacity was 46,265 MW and rooftop solar installed capacity 6.1 GW.

Phase 3: 13th plan from 2017–2022 will cover Phase 3.
  Targets of Phase 2 and Phase 3 are given in Table 6.4.

TABLE 6.4

Application segment	Target for Phase 2 (2013–17)	Target for Phase 3 (2017–22)
Grid Solar Power (large plants, rooftop and distribution grid plants)	9000 MW	Decision is yet to taken
Off-grid solar application	800 MW	To be taken
Solar Thermal Collectors (SWHS, Solar cooking, Solar cooling, industrial process heat)	8 million sq. meters	-Do-

6.21.1 Solar-PV Cost is More and Reasons Thereof

Land is the most contentious of all projected cost. Requirement of laud is roughly 5 acre (2.2 ha) per MW.

Solar installations in India use crystalline cells made of silicon wafers having low efficiency of 12–18 per cent. The technology developed by the goverment-owned Central Electrical Laboratory in Ghaziabad is not commercially available. International companies are commercially producing solar cells with over 19 per cent efficiency.

However, future may not lie in silicon. Latest technology is multi-junction cells comprising several layers of semiconductors. Unlike single-layer crystalline silicon cells, which catches a fraction of the spectrum of sunlight, where several layers trap the entire spectrum. The mass over is using multi-junction cells. If the sunlight is concentrated on the multi-junction cells, their efficiencies reach up to 50 per cent. In India research on multi-junction cells needs to be taken up vigorously.

PV modules, consisting of solar cells, constitute 60 per cent of the total cost. At present cost of a PV module is US $ 2.2 per watt-peak. Considering above constraints, plus cost of mounting structures and power conditioning unit, the capital cost of a solar-PV project is ₹ 16.90 crore per MW. Accordingly solar-PV tariff is ₹ 17.91 per unit. To make solar mission a success, the challenge is to cut cost and improve efficiency.

6.21.2 Solar City Programme of UT Chandigarh

In Chandigarh where sun shines throughout the year, solar revolution is practiced.

With the interventions of UT Administration, the Honorable Joint Electricity Regulatory Commission (JERC) for Goa and Union Territories has notified Solar Tariff and Net Metering Guidelines 2015 which has provided a huge incentive to any electric consumers in Chandigarh through installation of Solar Power Plant on its roof. Under the guidelines any electric consumer has the option to either opt for "Net Metering" or can opt for 'Gross Metering' for consuming/selling solar power generated in its premises/rooftops.

Net Metering: The generated solar energy is fed to DISCOM energy meter on L.T. bus bar so that first it is used for self consumption and excess if any is exported to grid through Bi-directional energy meter (capable of measuring import and export of energy) which needs to be installed in place of existing Uni-directional electric meter installed by electricity department. The billing shall be done by Electricity Dept. of UT Chandigarh on the basis of import minus export basis. If export is more than import that can be carried forward to next billing cycle. This can continue till end of settlement period ending on 30th September and 31st March. In totality if export exceeds the import from the grid of the utility Punjab State Power Corporation Limited (PSPCL) the consumer can use it when s/he needs it without any charge. The Chandigarh rule is applicable throughout Punjab state. A capital subsidy of 30% is also offered to the eligible individual from the government for solar PV installation in the range of 1 kWp to 1000 kWp covering up to 80% of the sanctioned load.

Rooftop PV and Small Solar Power Generation Programme (RPSSGP)

Currently 71 projects have been commissioned under RPSSGP scheme with installed capacity 90.8 MW. The projects have been commissioned in 12 states as per detail given in Table 6.5.

TABLE 6.5

S. No.	State	No. of Projects	Capacity (MW)
1.	Andhra Pradesh	10	9.75
2.	Chhattisgarh	2	4
3.	Haryana	8	7.8
4.	Jharkhand	8	16
5.	Madhya Pradesh	3	5.25

S. No.	State	No. of Projects	Capacity (MW)
6.	Maharashtra	3	5
7.	Odiash	7	7
8.	Punjab	5	6
9.	Rajasthan	12	12
10.	Tamil Nadu	6	6
11.	Uttarakhand	3	5
12	Uttar Pradesh	4	7
	Total	71	90.8

All types of Module Technology used in the scheme are:

Monocrystaline = 4 No   Polycrystalline = 33 No
Thin film = 30 No   Mixed = 4 No

6.21.3 International Solar Alliance to Forge a Common Platform amongst 122 Solar-rich Countries Lying Fully or Partially between Tropic of Cancer and Capricorn

On 25th January 2016 French President and Indian Prime Minister laid the foundation stone for the International Solar Alliance (ISA) on the premises of National Institute of Solar Energy (NISE) Gurgaon. This was the culmination of India's announcement of the 21st meeting of Conference of Parties (COP21) of the UN Framework Convention on Climate Change (UNFCCC) in Paris during Dec. 2015. India's pledge to the Paris COP21 aims to draw 40% of electricity from renewable sources by 2030.

The ISA will supplement efforts of other institutions such as International Renewable Energy Agency (IRENA), Renewable Energy and Energy Efficiency Partnership (REEEP) International Energy Agency (IEA), Renewable Energy Policy Network for the 21st Century (REN21). India is set to play a pivotal role to massive solar energy generation. This solar initiative will be with zero emissions, most suitable when the world is battling with climate change. To achieve the target ISA has projected the mobilization of more than $1000 billion of investment by 2030.

In January 2016, Cabinet Committee on Economic Affairs has approved setting over 5000 MW grid connected solar PV projects on built, own and operate basis. This capacity would be developed in four trenches of each 1250 MW during four financial years 2015–16, 2016–17, 2017–18 and 2018–19. This would help in employment generation of about 30,000 people in rural and urban areas with reduction of about 8.525 million Tonnes of CO_2 emissions into environment every year.

Cumulative progress on PV solar system installed upto 31–12–2016.

Solar Lanterns Nos — 99,6841   (20 million expected by 2022)
Solar home light Nos — 13,96,036
Solar street light Nos — 4,42,936
Solar pumps Nos — 1,81,000 (2019)
Power plants MWp — 172

REVIEW QUESTIONS

1. Explain the current–voltage characteristic of solar cell. Also define the fill factor.
2. Discuss the reasons for low efficiency of solar cells.
3. Explain the different types of solar cells on the basis of material thickness and the type of junction structure.
4. Explain the different types of cells based on the material used for their fabrication.
5. Discuss the standalone type of PV systems.
6. With a neat sketch describe how photovoltaic power can be used to desalinate sea water for drinking purposes?
7. Write short notes on (i) PV hybrid system and (ii) grid-interactive solar PV system.
8. Discuss the concepts and feasibility of the futuristic solar power plant using a satellite.
9. Briefly explain how plastic solar cells with the help of nanotechnology can popularize the use of solar cells in the near future.
10. Survey and report the uses of SPV systems in India.

WIND ENERGY

7.1 INTRODUCTION

Wind is air in motion and it derives energy from solar radiation. About 2% of the total solar flux that reaches the earth's surface is transformed into wind energy due to uneven heating of the atmosphere. During daytime, the air over the land mass heats up faster than the air over the oceans. Hot air expands and rises while cool air from oceans rushes to fill the space, creating local winds. At night the process is reversed as the air cools more rapidly over land than water over off-shore land, causing breeze, as shown in Figure 7.1. On a global scale low pressure exists near the Equator due to greater heating, causing winds to blow from subtropical belts towards the Equator. Also, the axial rotation of the earth induces a centrifugal force which throws equatorial air masses to the upper atmosphere, causing deflection of winds.

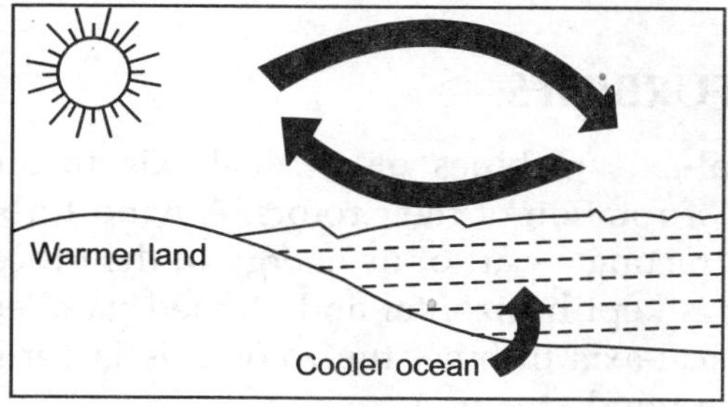

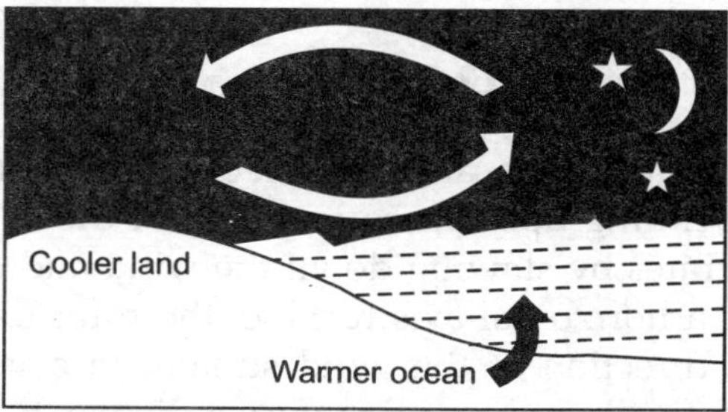

FIGURE 7.1 (a) Wind from ocean to land during daytime, and (b) wind from land to ocean during night.

7.2 HISTORICAL DEVELOPMENT

The concept of harnessing wind energy dates back to 4000 BC, when Egyptians used wind power to sail their boats in the Nile river. By the tenth century the wind mills were being used to grind grains in Iran and Afghanistan. Skilful technicians of Iran introduced the windmills to China where they gained popularity and were used to raise water for irrigation and sea water for production of salt.

The technology to harness wind energy reached western Europe via the Arabs. Wind machines became popular because the energy can be used in a number of ways. In 1854, Daniel Halladay in US introduced a wind pump. Windmills were in use for draining lakes, raising water for irrigation, industrial uses like sawing timber, extracting oil from oil seeds, and polishing stones. In West Indies, windmills were used for crushing sugarcane.

It was P. La'cour (Denmark), who in 1880 for the first time used the windmill as a source of electricity. A new era began after the First World War when experiments were carried out with windmills having sails of aerofoil section. A French engineer Darreius built an aerogenerator at Bourget in 1929 that had a tower 20 metres high with blades of the same diameter. In the late 1950s, Danish electrical companies successfully tested a 200 kW wind turbine with an asynchronous generator.

After the sudden price rise of fossil fuel in 1973, a number of countries were stimulated towards the development and use of renewable energy sources. In 1974, NASA constructed and operated a wind generator of 100 kW capacity with 38-m diameter rotor installed over a 30 m high tower. Success encouraged the US firms to manufacture a 2.5 MW generator in 1987. After 1990, the European and the Asian countries like Denmark, Germany, China and India encouraged private and cooperative sectors to install wind generators in capacities of 200 kW, and 500 kW to 1.5 MW.

The wind power programme in India is working quite satisfactorily. Provision of incentives instituted by the Ministry of New and Renewable Energy (MNRE), has made wind electricity competitive. As a result, wind electricity has emerged as an option for quality power. As on May 2004, India is ranked 4th in the world after Germany, USA, China in terms of wind power generation. Most of the capacity addition has been achieved through commercial projects by private investors.

7.3 CLASSIFICATION OF WIND TURBINES

Wind turbines are classified as horizontal-axis turbines or vertical-axis turbines depending upon the orientation of the axis of rotation of their rotors. A wind turbine operates by slowing down the wind and extracting a part of its energy in the process. For a horizontal-axis turbine, the rotor axis is kept horizontal and aligned parallel in the direction of the wind stream. In a vertical-axis turbine, the rotor axis is vertical and fixed, and remains perpendicular to the wind stream.

In general, wind turbines have blades, sails or buckets fixed to a central shaft. The extracted energy causes the shaft to rotate. This rotating shaft is used to drive a pump, to grind seeds or to generate electric power. Wind turbines are further classified into 'lift' and 'drag' type.

7.3.1 Lift Type and Drag Type Wind Turbines

Two important aerodynamic principles are used in wind turbine operations, i.e., lift and drag. Wind can rotate the rotor of a wind turbine either by lifting (lift) the blades or by simply passing against the blades (drag). Wind turbines can be identified based on their geometry and the manner in which the wind passes over the blades.

Slow-speed turbines are mainly driven by the drag forces acting on the rotor. The torque at the rotor shaft is comparatively high which is of prime importance for mechanical applications such as water pumps. For slower turbines, a greater blade area is required, so the fabrication of blades is undertaken using curved plates.

High-speed turbines utilise lift forces to move the blades, which phenomenon is similar to what acts on the wings of an aeroplane. Faster turbines require aerofoil-type blades to minimize the adverse effect of the drag forces. The blades are fabricated from aerofoil sections with a high thickness-to-chord ratio in order to produce a high lift relative to drag.

For electric power generation, the shaft of the generator requires to be driven at a high speed. For the same swept area, the energy extracted by a wind turbine operating on lift forces is several times greater than the energy from the drag-type turbine. Thus, the lift-type turbines are more suitable compared to drag-type turbines for electric power generation.

7.4 TYPES OF ROTORS

Different types of rotors used in wind turbines are: (i) multiblade type, (ii) propeller type, (iii) Savonious type, and (iv) Darrieus type. The first two are installed in horizontal-axis turbines, while the last two in vertical-axis turbines.

7.4.1 Multiblade Rotor

The multiblade rotor is fabricated from curved sheet metal blades. The width of blades increases outwards from the centre. Blades are fixed at their inner ends on a circular rim. They are also welded near their outer edge to another rim to provide a stable support. The number of blades used ranges from 12 to 18, as shown in Figure 7.2.

7.4.2 Propeller Rotor

The propeller rotor comprises two or three aerodynamic blades made from strong but lightweight material such as fibre glass reinforced plastic. The diameter of the rotor ranges from 2 m to 25 m as detailed in Figure 7.3. The blade slope is designed by using the same aerodynamic theory as for aircraft.

7.4.3 Savonious Rotor

The Savonious rotor comprises two identical hollow semi-cylinders fixed to a vertical axis. The inner side of two half-cylinders face each other to have an S shaped cross section as detailed in Figure 7.4.

FIGURE 7.2   Multiblade rotor installed on a tower.

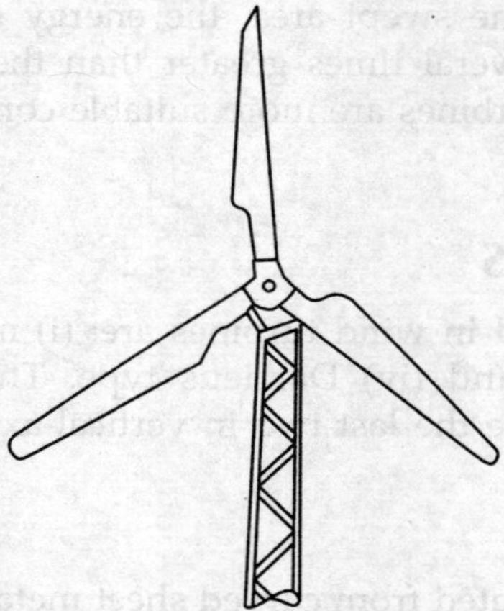

FIGURE 7.3   Propeller rotor installed on a tower.

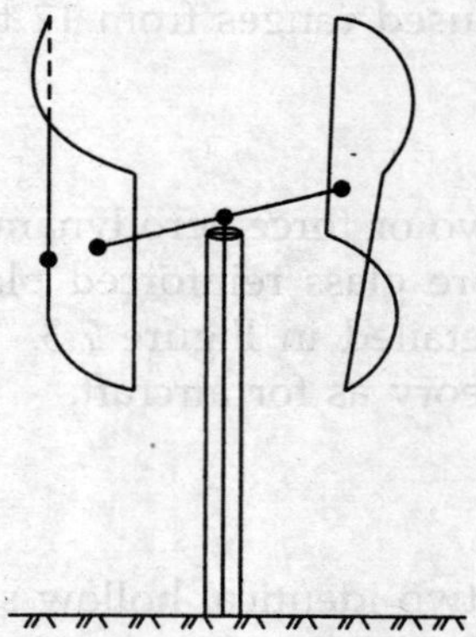

FIGURE 7.4   Savonious vertical-axis rotor.

Irrespective of wind direction, the rotor rotates due to pressure difference between the two sides. This vertical axis rotor was developed by an engineer Savonious of Finland in the year 1920. It is self starting and the driving force is mainly of drag type. The rotor possesses high solidity so as to produce a high starting torque and hence this rotor is suitable for water pumping.

7.4.4 Darrieus Rotor

This rotor has two or three thin curved blades of flexible metal strips. It looks like an egg beater and operates with the wind coming from any direction. Both the ends of the blades are attached to a vertical shaft as shown in Figure 7.5. It has an advantage that it can be installed close to the ground eliminating the cost of the tower structure.

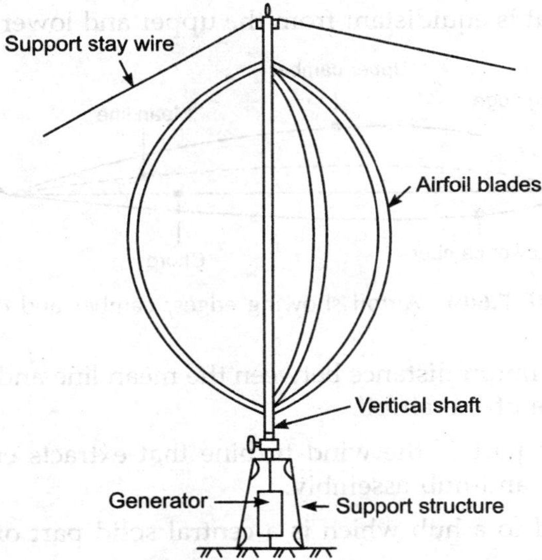

FIGURE 7.5 Darrieus rotor.

Lift is the driving force, creating maximum torque when the blade moves across the wind. This rotor was designed by a French engineer G.M. Darrieus in 1925. It is used for decentralized electricity generation.

7.5 TERMS USED IN WIND ENERGY

Airfoil (Aerofoil): A streamlined curved surface designed for air to flow around it in order to produce low drag and high lift forces.

Angle of attack: It is the angle between the relative air flow and the chord of the airfoil [Figure 7.6(a)].

Blade: An important part of a wind turbine that extracts wind energy.

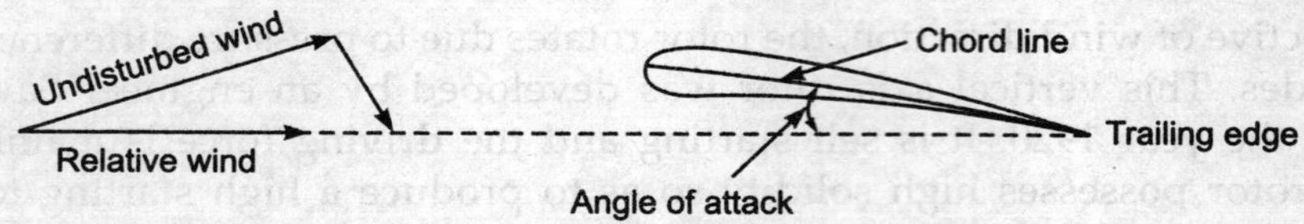

FIGURE 7.6(a)   Angle of attack of a wind turbine airfoil.

Leading edge: It is the front edge of the blade that faces towards the direction of wind flow [Figure 7.6(b)].

Trailing edge: It is the rear edge of the blade that faces away from the direction of wind flow [Figure 7.6(b)].

Chord line: It is the line joining the leading edge and the trailing edge [Figure 7.6(b)].

Mean line: A line that is equidistant from the upper and lower surfaces of the airfoil.

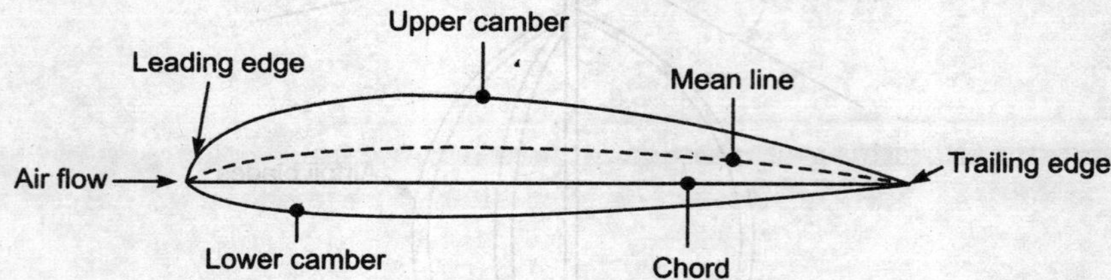

FIGURE 7.6(b)   Airfoil showing edges, camber and chord.

Camber: It is the maximum distance between the mean line and the chord line, which measures the curvature of the airfoil.

Rotor: It is the prime part of the wind turbine that extracts energy from the wind. It constitutes the blade-and-hub assembly.

Hub: Blades are fixed to a hub which is a central solid part of the turbine.

Propeller: It is the turbine shaft that rotates with the hub and blades and is called the propeller. Blades are twisted as per design. The outer profile of the blades conforms to aerodynamic performance while the inner profile meets the structural requirements.

Tip speed ratio: It is the ratio of the speed of the outer blade tip to the undisturbed natural wind speed.

Pitch angle: It is the angle made between the blade chord and the plane of the blade rotation.

Pitch control of blades: A system where the pitch angle of the blades changes according to the wind speed for efficient operation [Figure 7.6(c)].

Stall-regulated system: When the turbine blades are fixed at an optimum angle and the machine is stalled during high winds either by mechanical or hydraulic systems.

Swept area: This is the area covered by the rotating rotor.

Solidity: It is the ratio of the blade area to the swept area.

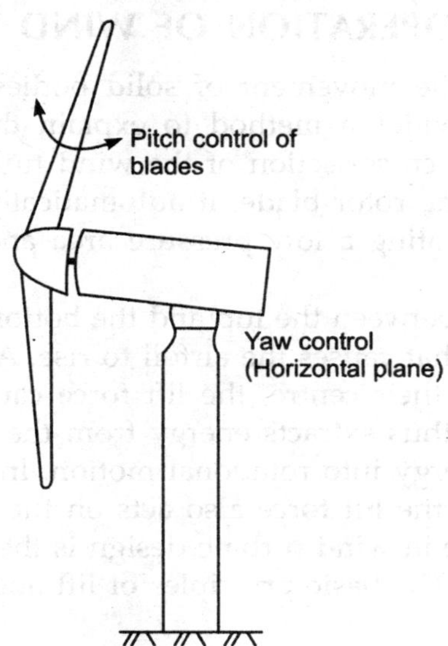

FIGURE 7.6(c) Pitch and yaw control of wind turbine.

Drag force: It is the force component which is in line with the velocity of wind.

Lift force: It is the force component perpendicular to drag force.

Nacelle: The nacelle houses the generator, the gear box, the hydraulic system and the yawing mechanism.

Yaw control: As the direction of the wind changes frequently, the yaw control is provided to steer the axis of the turbine in the direction of the wind. It keeps the turbine blades in the plane perpendicular to the wind, either in the upward wind direction or in the downward wind direction.

Cut-in speed: It is the wind speed at which a wind turbine starts to operate.

Rated wind speed: It is the wind speed at which the turbine attains its maximum output.

Cut-out speed: It is the wind speed at which a wind turbine is designed to be shut down to prevent damage from high winds. It is also called the *furling wind speed*.

Down wind: It is the opposite side of the direction from which the wind is blowing.

Up wind: It is the side of the direction from which the wind is blowing (in the path of the oncoming wind).

Wind rose: It is the pattern formed in a diagram illustrating vectors that represent wind velocities occurring from different directions.

Wind vane: A wind vane monitors the wind direction. It sends a signal to the controlling computer which activates the yaw mechanism to make the rotor face the wind direction.

7.6 AERODYNAMIC OPERATION OF WIND TURBINES

Aerodynamics deals with the movement of solid bodies through the air. In wind turbines, aerodynamics provides a method to explain the relative motion between airfoil and air. Airfoil is the cross-section of the wind turbine blade. When the wind passes over the surface of the rotor blade, it automatically passes over the longer or upper side of the blade, creating a low pressure area above the airfoil as shown in Figure 7.7(a).

The pressure difference between the top and the bottom surfaces results in a force called the aerodynamic lift that causes the airfoil to rise. As the blades can only move in a plane with the hub as their centre, the lift force causes rotation about the hub [Figure 7.7(b)]. The turbine thus extracts energy from the wind stream by converting the wind's linear kinetic energy into rotational motion. In addition to the lift force, a drag force perpendicular to the lift force also acts on the blade which impedes rotor rotation. The prime objective in wind turbine design is the desired lift-to-drag ratio of the blade (airfoil structure). The basic principles of lift and drag forces are dealt with in the next section.

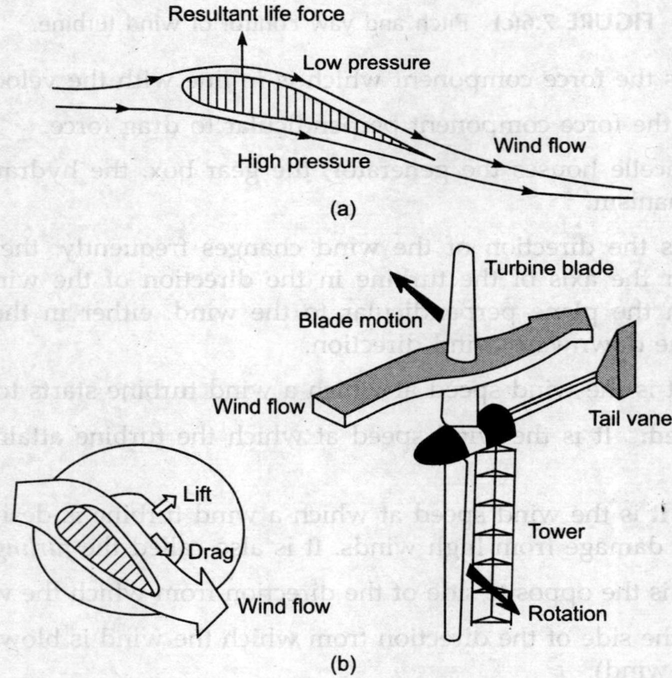

FIGURE 7.7 (a) Aerodynamic lift force on blade cross-section of wind turbine, and (b) the basic operating principle of wind turbine aerodynamic lift.

When air flows over solid bodies, several physical phenomena are noticed such as drag force acting on objects like trees and electric towers, the lift force developed by airplane wings, the lift force experienced by dust particles in a wind storm and the blade motion developed by a turbine. Either the fluid moves over a stationary body or a body moves through a standstill fluid; aerodynamically both activities are the same. The approach is to study the relative motion between the fluid and the body.

7.6.1 Drag

It is the resistance which a body experiences when a fluid moves over it. Flood water washes away animals, vehicles and buildings. Wind storm and hurricane knocks down transmission towers, trees, sweeps away catamaran and ships. These are a few undesirable examples of drag forces. The force that a flowing fluid exerts on a body in the direction of flow is called 'drag force'. Drag may bring an undesirable effect of friction, such as burning of space vehicles on entering into the earth's atmosphere. Reduction of drag is the basic engineering approach, associated with the reduction in fuel consumption in automobiles, aircraft and submarines. However, in certain engineering activities the drag produces a useful effect. A meteor from outer space burns due to friction with the earth's atmosphere, saving the inhabitants on earth from catastrophic impact.

Friction acts to help us as a 'life saver' in brakes of automobiles. Similarly, the drag force is useful in safe landing with a parachute.

7.6.2 Lift

When a body is immersed in a standstill fluid, only the normal pressure force is exerted on it. A flowing fluid in addition exerts tangential shear forces on the surface. Both these forces have two components, one is drag in the flow direction, the other is perpendicular to the fluid flow called 'lift'. It causes the body to move in the upward direction. The relative magnitudes of drag and lift forces depend completely on the shape of the object. Streamlined objects experience a smaller drag force than that experienced by blunt objects. Generation of lift always creates a certain amount of drag force.

Airfoils of a wind turbine are especially shaped to produce lift force on coming in contact with the moving air. It is achieved by fabricating the top surface of the airfoil as curved and the bottom surface nearly flat. Air flowing over the airfoil travels a longer distance to reach the tip-end of airfoil, in contrast to air flowing under the foil (Figure 7.8). It creates a pressure difference that generates an upward force which tends to lift the airfoil causing rotation of the wind turbine rotor. Good airfoils can have lift 30 times greater than drag.

7.7 WIND ENERGY EXTRACTION

Wind turbines extract energy from wind stream by converting the kinetic energy of the wind to rotational motion required to operate an electric generator. By virtue of the kinetic energy, the velocity of the flowing wind decreases. It is assumed that the

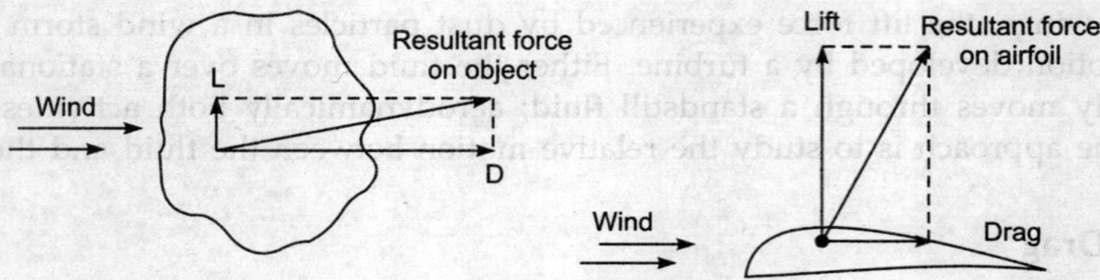

FIGURE 7.8 Relative magnitudes of lift and drag forces on a blunt object and a streamlined airfoil.

mass of air which passes through rotor is only affected and remains separate from the air which does not pass through the rotor. Accordingly, a circular boundary surface is drawn showing the affected air mass and this boundary is extended upstream as well as downstream as detailed in Figure 7.9.

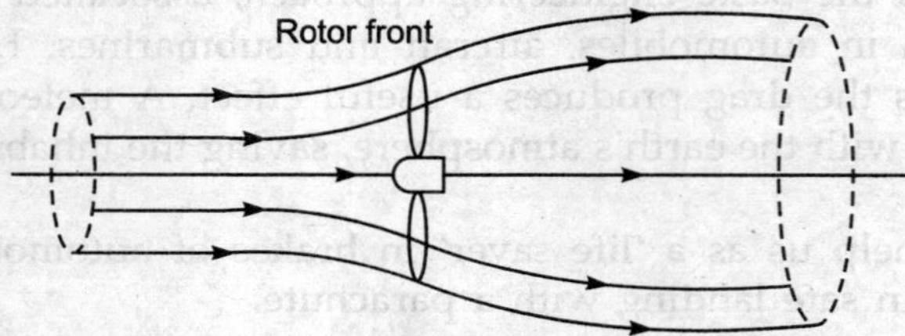

FIGURE 7.9 Representation of wind flow through turbine.

As the free wind (stream) interacts with the turbine rotor, the wind transfers part of its energy into the rotor and the speed of the wind decreases to a minimum leaving a trail of disturbed wind called **wake** [(Figure 7.10(a)]. The variation in velocity is considered to be smooth from far upstream to far downstream. However, the fall in static wind pressure is sharp as depicted in Figure 7.10(b). The wind leaving the rotor is below the atmospheric pressure (in wake region) but at far downstream it regains its value to reach the atmospheric level. The rise in static pressure is at the cost of kinetic energy, consequently further decreasing the wind speed.

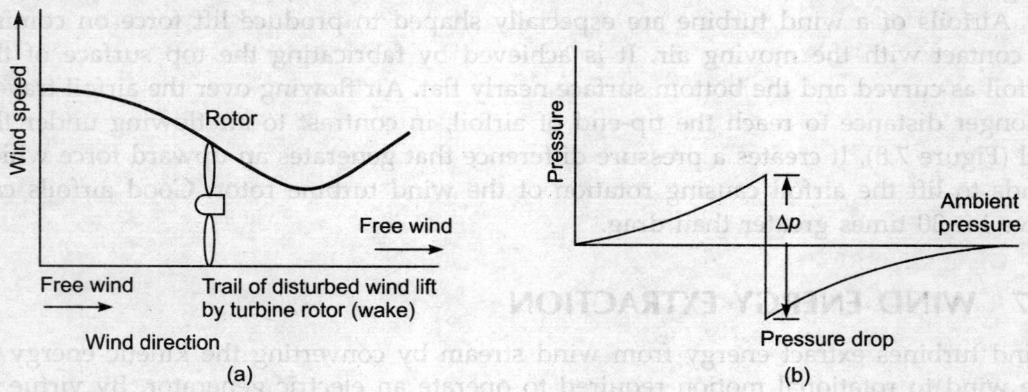

FIGURE 7.10 Change in wind speed and pressure in traversing the turbine rotor.

Wind flow is considered incompressible and hence the air stream flow diverges as it passes through the turbine. Also the mass flow rate of wind is assumed constant at far upstream, at the rotor and at far downstream. To compute the mathematical relationships, suppose:

P = atmospheric wind pressure
P_u = pressure on upstream of wind turbine
P_d = pressure on downstream of wind turbine
V = atmospheric wind velocity
V_u = velocity of wind upstream of wind turbine
V_b = velocity of wind at blades
V_d = velocity of wind downstream of wind turbine before the wind front reforms and regains the atmospheric level
A = area of blades
$\bar{M}$ = mass flow rate of wind
ρ = air density.

The kinetic energy of wind stream passing through the turbine rotor is

$$KE = \frac{1}{2}\bar{M}V_b^2$$

and

$$\bar{M} = \rho A V_b$$

Hence,

$$KE = \frac{1}{2}\rho A V_b^3 \qquad (7.1)$$

The force on the disc of the rotor can be expressed as

$$F = (P_u - P_d)A \qquad (7.2)$$

Force on the rotor can be expressed as change of momentum per unit time from upstream to downstream winds, i.e.,

$$F = \bar{M}(V_u - V_d) \qquad (7.3)$$

Applying the Bernaulli's equation to upstream and downstream sides,

$$P + \frac{1}{2}\rho V_u^2 = P_u + \frac{1}{2}\rho V_b^2 \qquad (7.4)$$

$$P_d + \frac{1}{2}\rho V_b^2 = P + \frac{1}{2}\rho V_d^2 \qquad (7.5)$$

Solving Eqs. (7.4) and (7.5), we get

$$P_u - P_d = \frac{1}{2}\rho(V_u^2 - V_d^2) \qquad (7.6)$$

Equating Eqs. (7.2) and (7.3), we get

$$(P_u - P_d)A = \bar{M}(V_u - V_d) = \rho A V_b(V_u - V_d) \qquad (7.7)$$

Solving Eqs. (7.6) and (7.7), we get

$$\frac{1}{2}\rho(V_u^2 - V_d^2) = \rho V_b(V_u - V_d)$$

or

$$V_b = \frac{V_u + V_d}{2} \tag{7.8}$$

In a wind turbine system "Steady Flow Work", W, is equal to the difference in kinetic energy between upstream and downstream of the turbine for unit massflow, $\bar{M} = 1$. Therefore,

$$W = (KE)_u - (KE)_d$$

$$= \frac{1}{2}(V_u^2 - V_d^2) \tag{7.9}$$

The power output P of wind turbine is the rate of work done, using the mass flow rate equation.

$$P = \bar{M}\left(\frac{V_u^2 - V_d^2}{2}\right)$$

$$= \rho A\left(\frac{V_u + V_d}{2}\right)\left(\frac{V_u^2 - V_d^2}{2}\right)$$

$$= \frac{1}{4}\rho A(V_u + V_d)(V_u^2 - V_d^2) \tag{7.10}$$

For maximum turbine output P, differentiate Eq. (7.10) with respect to V_d and equate to zero to obtain

$$\frac{dP}{dV_d} = 3V_d^2 + 2V_u V_d - V_u^2 = 0$$

The above quadratic equation has two solutions, i.e., $V_d = \frac{1}{3}V_u$ and $V_d = V_u$

For power generation $V_d < V_w$, so we can have only $V_d = \frac{1}{3}V_u$ $\tag{7.10a}$

Therefore,

$$P_{\max} = \frac{8}{27}\rho A V_u^3 \tag{7.11}$$

$$= \frac{16}{27}\left(\frac{1}{2}\rho A V_u^3\right)$$

$$= 0.593\left(\frac{1}{2}\rho A V_u^3\right)$$

Total power in wind stream is

$$P_{\text{total}} = \frac{1}{2}\rho A V_u^3 \qquad\qquad (7.11a)$$

Therefore, $\qquad\qquad\qquad\qquad P_{\text{max}} = 0.593 P_{\text{total}}$

Maximum theoretical efficiency η_{max} (also called the power coefficient C_p) is the ratio of maximum output power to total power available in the wind, i.e.,

$$\text{Power coefficient, } C_p = \frac{P_{\text{max}}}{P_{\text{total}}} = 0.593 \qquad\qquad (7.12)$$

The factor 0.593 is known as the **Bitz limit** (After the name of the engineer who first derived this relationship).

Available efficiency

Theoretically, the maximum power extracted by a turbine rotor is 59.3% of the total wind energy in the area swept by the rotor. Considering the rotor efficiency to be 70%, bearing, vibrations, friction losses and generator efficiency 90%, the available efficiency η is 60% of C_p, i.e.,

$$\eta_a = 0.6 \times 0.593$$
$$= 35.5\%$$

7.8 EXTRACTION OF WIND TURBINE POWER

Equation [7.11(a)] can be expressed as $P_{\text{total}} = \dfrac{\rho}{2} \cdot \dfrac{\pi}{4} D^2 V_u^3$. Accordingly, for a given wind speed at a site, P_{total} would increase four times if the rotor diameter is doubled. The designer of a wind turbine always tries to increase the rotor diameter to optimize the extraction of the wind energy. The cumulative effect of wind speed and rotor diameter on the availability of wind power can be observed in Figure 7.11.

While selecting the wind turbine it is necessary to know the energy needs and the availability of wind speeds at the given site. Economically, it is known that the wind system cost varies according to the rotor size. Referring to Figure 7.11, if the rotor diameter of 40 m is selected in lieu of 20 m at the proposed site having wind speed of 10 m/s, the available power rises up to 1 MW from a low value of 0.25 MW, i.e., becomes four times more.

7.9 WIND CHARACTERISTICS

Power in the wind is proportional to the cube of the wind speed [Eq. (7.11)] and is highly site specific. It is necessary to carry out wind measurements if the performance of wind turbines is to be estimated accurately. The highest wind speed sites are on exposed hill tops, offshore or on coastal sites. For developing wind energy at any site the different parameters required are:

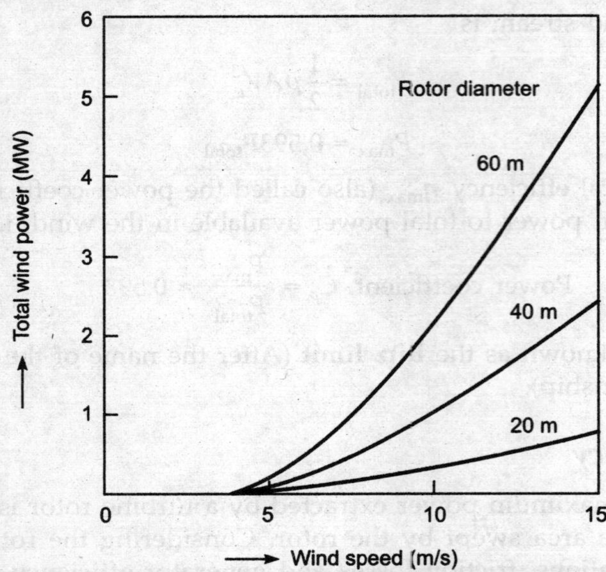

FIGURE 7.11 Variation of wind turbine power with rotor diameter and wind speed.

(i) Mean wind speed

(ii) Daily, seasonal and annual variations in wind speed. Wind speed frequency distribution, generally described by **Weibull** probability distribution.

(iii) Direction of wind by studying the 'Wind Rose' data for micro-siting of WEGs (Wind Energy Generator Systems). The map of wind roses gives the frequency distribution of main wind flow.

(iv) Wind speed variation with height above ground (power in the wind increases with height).

Definitive wind resource data can be obtained from the Indian Meteorological Department (IMD) where microprocessor-based anemometers are used at wind mapping stations. These instruments can make highly accurate wind measurements for the estimation of power production. These readings are analysed to assess the performance and economic viability of a Wind Energy Conversion System (WECS).

7.10 MEAN WIND SPEED AND ENERGY ESTIMATION

Wind is air-in-motion, a vector quantity represented by its magnitude and direction. For energy extraction from the wind, only the horizontal component of its magnitude is of use. Hence, the direction of the wind is not considered and its data is referred to as 'wind speed', a scalar quantity. Variation in wind speed is a perpetual occurrence and to access energy content for a given period 'mean wind speed' is calculated from a large number of readings observed over one year, with hourly, daily, weekly and monthly intervals.

Energy estimation

Hourly mean wind speed data are recast into a number of hours in the year for which the speed equals or exceeds the specified value. A graph of wind speed, as shown in Figure 7.12, against the duration (in hours over a year for which that speed is experienced) is plotted.

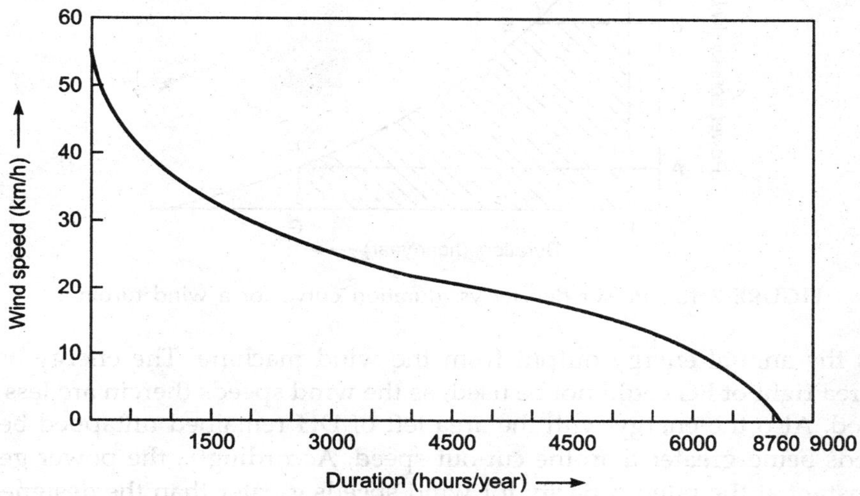

FIGURE 7.12 Wind speed vs. duration curve of a location.

The curve begins from x-axis which corresponds to the calm period of zero wind speed. The curve then rises with decreasing slope, reaches a point of inflection and then joins with the y-axis. It directly provides the total number of hours per year during which hourly wind speeds remain in the range of 18 to 20 kmph, this range being suitable for power generation.

7.11 POWER DENSITY DURATION CURVE

The energy content in the wind can be estimated by plotting a curve of power density against duration. As power $(P) \propto (\text{speed})^3$, this curve can be drawn by plotting $(\text{speed})^3 \times$ constant, on the y-axis and the duration on the x-axis, as shown in Figure 7.13. Three speeds associated are: cut-in-speed (V_c) below which the machine stops to rotate and no power is generated. The 'design speed' (V_d) is the wind speed at which the wind machine generates power. A governing mechanism is provided to maintain output power constant at rated value even for wind speeds higher than the designed speed. The cut-out speed (V_f) is the high wind speed, experienced during storms at which the wind machine is shut down to save it from mechanical damage. It is also called the 'furling speed'.

Referring to Figure 7.13, the points A, B and C are power densities (kW/m^2) pertaining to cut-in, design and cut-out speeds, respectively. The hatched area EFGHJ

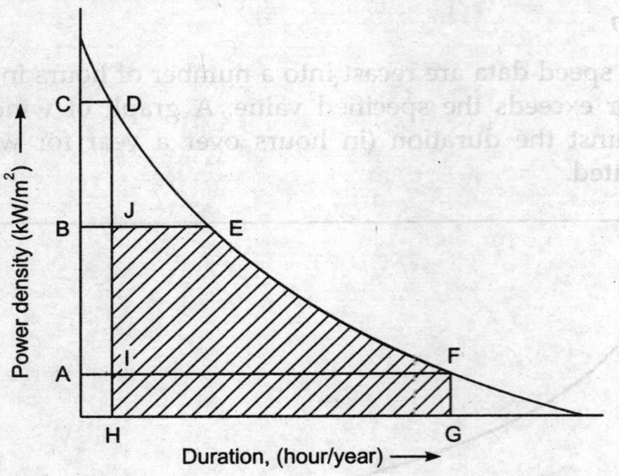

FIGURE 7.13 Power density vs. duration curve for a wind turbine.

represents the annual energy output from the wind machine. The energy associated with the area right of FG could not be used, as the wind speeds therein are less than the cut-in speed. Also the energy with the area left of DH remained untapped because of wind speeds being greater than the cut-out speed. Accordingly, the power generation is kept constant at the rated capacity for wind speeds greater than the designed speed; the energy represented by the area above the line JE is also not harnessed. Thus, the actual energy output of the wind machine is shown by the shaded area EFGHJ.

7.12 WEIBULL PROBABILITY DENSITY FUNCTION

Wind industry designers require variation of wind speed to be visualized accurately in order to optimize the design of their turbines. Evaluation of field data on wind speeds gives the probable wind energy availability at the site. Wind energy extracted at a location cannot be expressed accurately as wind speeds are highly variable. For correct estimation, the wind energy at a site is also decided by the wind profile at the site. The wind regime at a site can be mathematically described by 'Weibull probability density function' which is given as

$$f(V) = \frac{K}{C}\left(\frac{V}{C}\right)^{K-1} \exp\left\{-\left(\frac{V}{C}\right)^{K}\right\} \qquad (7.13)$$

It has two parameters: C, the scale parameter (has a value of about 1.1 times the average wind speed) and K, the shape parameter which determines the shape of function $f(V)$. The expression depicts that the wind speed V is available at the site for duration of $f(V) \times 8760$ hours per annum. Thus, the energy contribution by wind speed V will be $P \times f(V) \times 8760$ at the respective site.

The parameters C and K of a wind regime at a site are determined from the annual mean wind speed averaged over few years, based on wind measurements at the site at an interval of 10 minutes. A value of K greater than 3 indicates more regular, steadier winds. However, for $K = 2$, the distribution is called a cumulative 'Rayleigh distribution'; it is used in evaluating a wind resource. If P_{rt} is the rated output, then P_{out}, i.e., the power output of a wind turbine is a function of wind speed and expressed as

$$P_{out} = P_{rt} g(V)$$

The average power output P_{avg} of a wind turbine is given by

$$P_{avg} = P_{rt} \int f(V)\, g(V)\, dV \tag{7.14}$$

Equation (7.14) depicts the power output of a wind turbine at a given speed and frequency at which that speed occurs, summed over all possible speeds. Thus,

$$\frac{P_{avg}}{P_{rt}} = \int f(V)\, g(V)\, dV \tag{7.15}$$

$$\text{Capacity Factor (CF)} = \frac{\text{average power output during a period}}{\text{rated power output}}$$

Thus, the integral part of Eq. (7.15) is also called the Wind Turbine Capacity Factor (WTCF).

7.13 FIELD DATA ANALYSIS

Capacity factor projects economics of a wind station

Considering that a 250 kW wind turbine with a hub height of 30 m is to be installed, the annual mean capacity factor needs to be calculated for a few wind monitoring stations. The cut-in, the rated and the furling wind speeds for a 250 kW generator taken are 14.4 kmph, 56 kmph and 100 kmph, respectively. Wind speeds are normally measured at 10 m height, extrapolated to 30 m height and the appropriate Weibull parameters are used in computations. The results based on 2 years to 5 years data over 1986–1991 for six stations are given in Table 7.1.

TABLE 7.1 Capacity factors of WTGS (Wind Turbine Generation System)

Station	January	May	July	August	Annual
Kakulakonda (Andhra Pradesh)	0.042	0.243	0.466	0.500	0.229
Surajbari (Gujarat)	0.012	0.186	0.257	0.150	0.130
Chalkewadi (Maharashtra)	0.058	0.054	0.354	0.114	0.157
Jaisalmer (Rajasthan)	0.025	0.151	0.203	0.146	0.113
Talayathu (Tamil Nadu)	0.017	0.194	0.365	0.486	0.190
Muppandal (Tamil Nadu)	0.080	0.290	0.298	0.268	0.218

Source: MNRE (Govt. of India, Urja Bharti, 1993)

The mean annual CF is found to be 23% at Kakulakonda (AP) and 22% at Muppandal (TN). Stations where CF is less than 12% are not considered suitable for power generation. One important feature of wind climatology of India is that 70%–80% of wind energy available in a year is confined to just five months of the year, i.e., May–September. It so happens that favourable winds occur in association (i) with South-West monsoon current which blows during June–September and (ii) with strong circulation of dry wind in May, associated with a thermally-induced low pressure area over the country. During the other seven months, the amount of wind energy is hardly sufficient to operate large capacity WEGs.

7.14 ANNUAL PERCENTAGE FREQUENCY DISTRIBUTION OF WIND SPEED

The annual percentage frequency distribution of hourly wind speeds at Harshad in Gujarat, Chalkawadi in Maharashtra and Kakulakonda in Andhra Pradesh is shown in Figure 7.14. At Harshad the curve is sharply peaked at about 20 kmph speed.

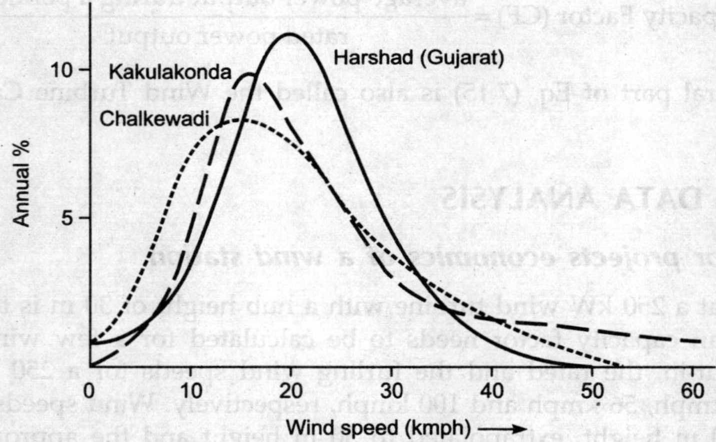

FIGURE 7.14 Annual percentage frequency distribution of wind speed.

At Chalkewadi there is a broad maximum at 14–16 kmph, while at Kakulakonda even though the frequency is maximum at 14–16 kmph, there are several occasions when a higher percentage is observed at wind speeds ranging higher than 30–50 kmph. Consequently the wind power density at Kakulakonda is quite high, i.e., 388 W/m^2 for a year.

For stations having the same monthly average wind speed, the wind power density could be different, because of the differences in the shape of frequency distribution curves. The sharper the curve, the lower will be the average power. The broader the peak, the greater will be the mean power. As a quantitative measure the Weibull shape parameter K is higher in value, for the sharper peak in frequency distribution curve.

7.15 DIRECTION OF WIND AND WINDROSE DATA

Meteorological parameters such a wind direction, wind speed, temperature and rainfall are recorded hourly for micro-siting of WEGs at the site. From the data obtained, 'Windrose' is plotted which is the pattern formed showing vectors representing wind velocities from different directions. The windrose diagrams of two stations for winter and summer are shown in Figure 7.15 as an illustration.

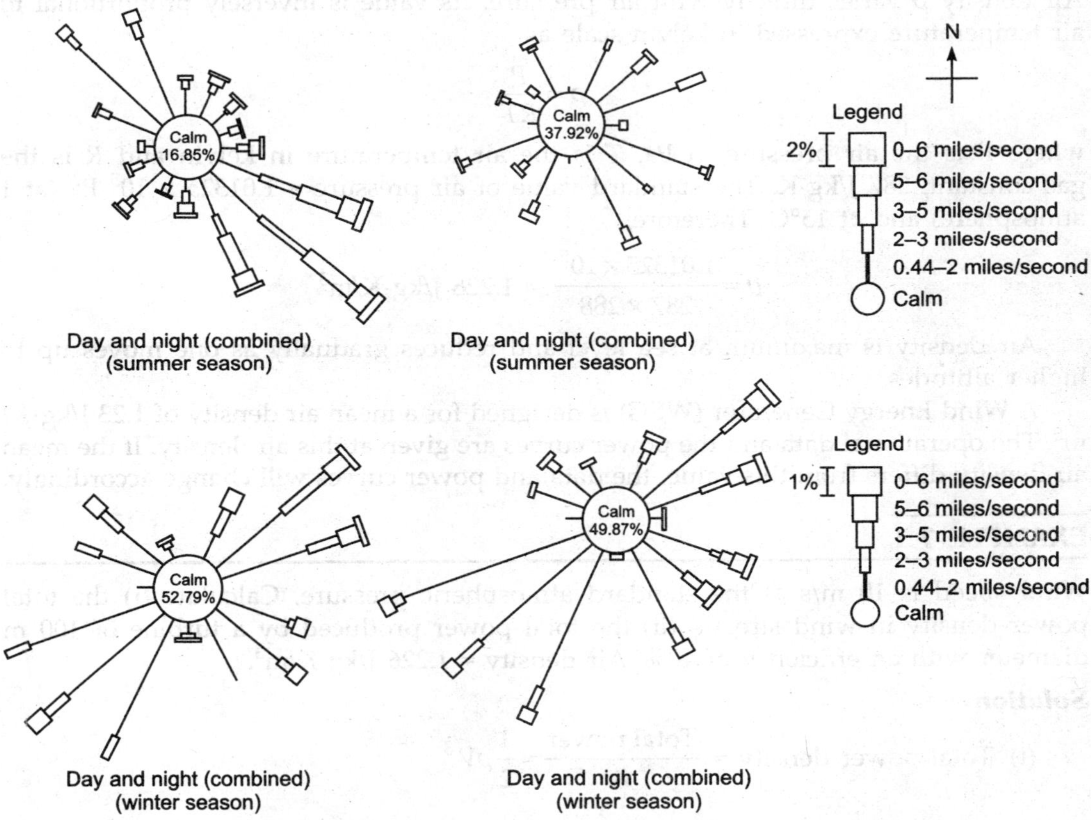

FIGURE 7.15 Windrose diagram.

The main direction of the wind flow at a given location is determined by magnetic compass and wind vane. The map of windroses gives the frequency distribution of main wind flow.

The wind vane sends signals about the changes in wind direction to the controlling computer. This activates the yaw mechanism located in the nacelle to turn the rotor face in the wind direction.

The wind direction plays an important role in deciding the siting of individual WEGs. Variation in 'wind speed' and 'wind direction' on an annual basis is one of the main criteria on which the "wind power plant" layout is finalized.

A straight coastline site where the 'windrose' is uniform or a land surface with a few wind breaks (roughness class I) are the favourable sites.

While analyzing the wind resource potential, it should be ensured that the prevalent wind is not turbulent. It should also be confirmed that the wind is not gusty.

7.16 CALCULATION FOR AIR DENSITY

Air density ρ varies directly with air pressure. Its value is inversely proportional to air temperature expressed in kelvin scale as

$$\rho = \frac{P}{RT}$$

where P is the air pressure in Pa, T is the air temperature in kelvin and R is the gas constant, 287 J/kg·K. The standard value of air pressure = 1.01325×10^5 Pa (at 1 atmosphere) and at 15°C. Therefore,

$$\rho = \frac{1.01325 \times 10^5}{287 \times 288} = 1.226 \text{ J/kg·K/m}^3$$

Air density is maximum at sea level and reduces gradually as one moves up to higher altitudes.

A Wind Energy Generator (WEG) is designed for a mean air density of 1.23 J/kg·K/m^3. The operational data and the power curves are given at this air density. If the mean air density differs from this value, the data and power curves will change accordingly.

EXAMPLE 7.1

Wind speed is 10 m/s at the standard atmospheric pressure. Calculate (i) the total power density in wind stream, (ii) the total power produced by a turbine of 100 m diameter with an efficiency of 40%. Air density = 1.226 J/kg·K/m^3.

Solution

(i) Total power density = $\dfrac{\text{Total power}}{A} = \dfrac{1}{2}\rho V^3$

$$= \frac{1}{2} \times 1.226 \times 10^3 = 613 \text{ W/m}^2$$

(ii) Total power produced = Efficiency × Power density × Area

$$= \frac{40}{100} \times 613 \times \frac{\pi}{4}(100)^2 \times \frac{1}{1000}$$

$$= 1924.8 \text{ kW}$$

7.17 VARIATION OF WIND SPEED WITH ELEVATION

The wind speed increases with height above the ground. Increase in wind speed with elevation h (above ground level) is called *wind shear*. The wind speed at the ground is zero due to the friction between the ground surface and air. Increase in wind speed with height is due to temperature gradient and it depends on the type of terrain (ground roughness) over which the wind has blown and the atmospheric stability. Sites can be divided into four types with yearly output from 225 kW WEG.

1. More than 10 km offshore (roughness class zero) with yearly output of 779,000 kWh.
2. Open landscape with few wind breaks, also sea-shore sites (roughness class 1) with yearly output of 577,000 kWh.
3. Suburban areas with farms, gardens and with medium wind breaks (roughness class 2) with yearly output of 454,000 kWh.
4. Urban districts, high trees, buildings and structures with many wind breaks (roughness class 3) with yearly output of 301,000 kWh.

Based on the data from several locations, for sites of low ground roughness, the change in wind speed with height can be expressed by an equation

$$\frac{V_2}{V_1} = \left(\frac{H_2}{H_1}\right)^{\alpha} \tag{7.16}$$

where V_1 and V_2 are wind speeds at levels H_1 and H_2, respectively. This is known as power law index α which depends on the roughness of terrain. Its value taken as is 1/7 for open land and 0.10 for calm sea area. For a particular site, the value of power law index is obtained from the measured wind speed at two heights, i.e.,

$$\alpha = \frac{\log V_2 - \log V_1}{\log H_2 - \log H_1} \tag{7.17}$$

The ideal wind energy sites have a low value of α. Generally, wind measurements are carried out at an elevation of 10 m. However, modern wind turbines are installed at a hub height of 25 m to 50 m. Wind speed at the required height is extrapolated from Eq. (7.16) with $\alpha = 1/7$.

Extrapolation of power density

If power density is to be extrapolated to a higher height, say 40 m above ground from the value at 20 m, it is found to vary logarithmically and expressed by an equation

$$\frac{P_{40}}{P_{20}} = \left(\frac{40}{20}\right)^{3\alpha} \tag{7.18}$$

7.18 ENERGY PATTERN FACTOR IN WIND POWER STUDIES

To design a wind energy system for a given site, it is necessary to know the characteristic of natural wind and estimate the wind resource. Wind power is expressed by the equation

$$P = K_E \cdot \frac{1}{2}\rho\bar{V}^3 \ \text{W/m}^2$$

where ρ is the air density = 1.2 kg/m^3 at sea level, $\bar{V}$ is average wind speed in m/s and K_E is the 'Energy Pattern Factor'. K_E is a ratio defined as

$$K_E = \frac{\text{actual wind energy available for the month}}{\text{calculated energy from the mean monthly wind speed}}$$

$$= \frac{\text{mean power density for the month}}{\text{mean power density for the monthly mean speed}}$$

$$= \frac{(1/2)\rho \sum V_h^3 / N_m}{(1/2)\rho(\bar{V}_m)^3}$$

where V_h is the hourly wind speed during the month, N_m is the number of hourly wind speed values during the month, and

$$\bar{V}_m = \sum \frac{V_h}{N_m}$$

Thus,

$$K_E = \frac{\text{mean of cubes of hourly wind speeds for the month}}{\text{cube of mean hourly wind speed for the month}}$$

As the numerator is always greater than the denominator, so K_E is always greater than unity.

For a natural wind region, the values of K_E have been determined as given in Table 7.2.

TABLE 7.2 Values of K_E

K_E	Type of wind regime
6	Polar
2.70	Continental, irregular
1.57–1.92	Coastal, maritime
1.22–1.36	Trade winds

For a given region, the concept of 'Energy Pattern Factor' (K_E) is useful to calculate the available wind energy from annual speed. It helps to choose a location with limited wind data.

7.19 BEAUFORT WIND SCALE

A numerical scale, graded from 0 to 12, devised by Admiral Beaufort in England, which matches natural indicators of wind speeds is given in Table 7.3.

TABLE 7.3 Beaufort wind scale

Code number	Wind speed (mph)	Visible wind effects
0	0–1	Calm; smoke rises vertically
1	1–3	Light air; wind direction by smoke draught
2	4–7	Light breeze; leaves rustle
3	8–12	Gentle breeze; leaves in motion
4	13–18	Moderate breeze; raises dust, tree branches move
5	19–24	Fresh breeze; small trees sway
6	25–31	Strong breeze; large branches in motion
7	32–38	Moderate gale; whistling in telegraph wires
8	39–46	Gale; trees in motion
9	47–54	Strong gale
10	55–63	Storm
11	64–74	Cyclone (violent storm)
12	Over 74	Hurrican

It is useful to codify the wind pattern about an area for the safety of WEG from the extreme environmental conditions.

7.20 LAND FOR WIND ENERGY

For a broad overview of the wind energy resource around the world, let us divide the globe into a number of regions according to their wind climate. The characteristics of these regions are described to predict the wind resources.

7.20.1 Regions

The Tropic: Tropical regions are at 30° North and South of the Equator, dominated by seasonal wind systems, like the monsoon and the trade winds. These regions are high pressure belts. These regions are characterized by high growth rate of population, resulting in an increasing interest in all kinds of energy, including wind energy. India is dominated by monsoon type flow and has a comprehensive database of meteorological measurements.

The Equator: This is the high temperature and humidity region due to 'low pressure' belt around the equator. Winds blow from areas of high atmospheric pressure, i.e., sub-tropical belts, towards the equator, and are known as 'trade winds'.

7.20.2 Areas

There are well-defined areas in above regions with rich wind resource, such as open sea, coastal areas, hills, valleys, terrace, saddle and khals (low depression).

Open seas: Open sea is in general characterized by a very high wind potential. An overview of the offshore resources given in the map helps to decide the wind energy potential of islands (like Andaman and Nicobar in Bay of Bengal). Ocean–atmospheric data set, wind speed and direction, reported from ships crossing the oceans is compared with coastal measurements to decide the application of wind energy.

Coastal areas: Land sites close to the coastline experience stronger winds compared to (flat) inland sites in the same wind regime. Sea breeze usually has diurnal pattern due to temperature difference between the sea and the land. During daytime, land is hotter than sea, while at night the situation reverses. Heat lowers the pressure, winds thus blow from sea to land during day and from land to sea during night. Thus, coastal offshore sites, become attractive wind resource zones with about 10 km width.

Hills: In hilly areas the topography enhances the wind potential. It is due to basic laws of continuity of fluid flow with conservation of momentum and energy of the flowing air mass in certain geomorphological features. Rounded hills and ridges experience higher wind speeds due to acceleration over the hill as shown in Figure 7.16. Acceleration of wind over a ridge depends on the height and its slope profile.

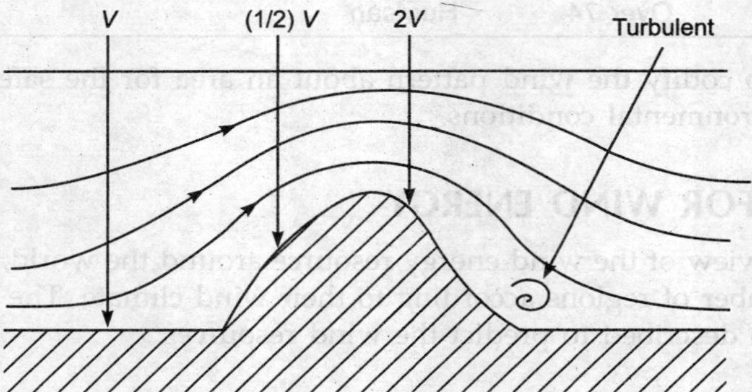

FIGURE 7.16 Schematic diagram showing the wind acceleration factors over a hill.

Monsoon winds from the Arabian Sea and the Bay of Bengal strike the Himalayas and dissipate their energy in mountain ranges. Fairly strong winds exist even during winter on well-exposed peaks. There are many potential windy hill ridges in the Himalayas.

Valley: When two steep slopes meet each other, a 'valley' is formed. An ideal location for an aero generator in a medium-depth valley is at the bottom of the valley along a river bed. For deep valleys, the location should be on 'ridge', as the wind profile totally shatters due to wall effects at the bottom where eddy currents also persist.

Terrace: When one or both escarpments instead of rising smoothly, try to level out half way up and then again start climbing, a terrace is formed. Aerodynamically, a terrace is an ideal location for wind turbine generators.

Saddle: The saddle actually resembles a horse saddle. When the mountain range dips shallow between two escarpments, a saddle is formed. It is a commonly occurring wind potent land formation. High wind power potential is at the location on the saddle where the escarpment just starts to climb upwards at either ends or at the 'seat' of the saddle.

7.20.3 Khals (Low Depressions)

Low depression saddles and water divides, having suitable aerodynamic conditions, are common sites in rural Garhwal Himalayas. These are known as *Khals* in local dialect. It is observed that river valleys, separated by a smaller water divide, have a change in their altitude and a change in air pressure. During the day, the air in these valleys is heated, after sunset the temperature decreases, consequently the pressure of the cool air on mountain summits becomes more and more. As a result, the cool air drains from summits to valleys during night. After sunrise the surface air of the valley again warms up. Typically, a temperature difference of 0.5°C produces winds of 5 m/s at 10 m above ground which is a good wind energy potential. Survey records indicate that about 135 Khals have been formed in between the Alaknanda catchment in the North and the Ramganga catchment in the South.

7.21 DESIGN OF WIND TURBINE ROTOR

There are two forces that operate on the blades of a propeller type wind turbine. One is the axial thrust which acts in the same direction as that of the flowing wind stream. The other is the circumferential force acting in the direction of wheel rotation that provides the torque.

7.21.1 Thrust on Turbine Rotor

A turbine extracts wind energy, causing the difference in momentum of air streams between the upstream and downstream sides as shown in Figure 7.17.

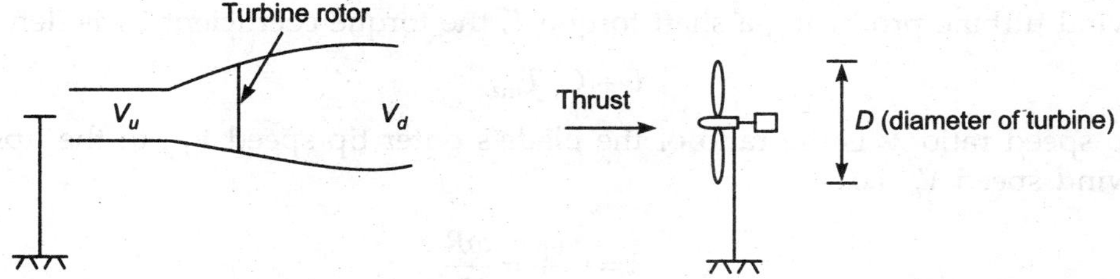

FIGURE 7.17 (a) Wind flow across turbine rotor, and (b) axial thrust on rotor.

$$F_x = \frac{1}{2}\rho A (V_u^2 - V_d^2)$$

$$= \frac{1}{2}\rho \frac{\pi}{4} D^2 (V_u^2 - V_d^2)$$

$$= \frac{\pi}{8}\rho D^2 (V_u^2 - V_d^2) \tag{7.19}$$

where D is the rotor diameter.

For maximum output, $V_d = \dfrac{1}{3} V_u$ [Eq. 7.10(a)]

Therefore, $F_{x(\max)} = \dfrac{\pi}{8} \rho D^2 \left(V_u^2 - \dfrac{1}{9} V_u^2 \right)$

$$= \dfrac{\pi}{9} \rho D^2 V_u^2 \qquad (7.20)$$

For designing a WEG, a large axial force can be obtained using large diameter turbines. The upper limit of the diameter needs to be optimized by matching structural design with economy.

7.21.2   Torque on Turbine Rotor

Maximum torque T on a turbine rotor would occur when maximum thrust can be applied at the blade tip farthest from the axis. A propeller turbine of radius R experiences

$$T_{\max} = F_{\max} \cdot R$$

From Eq. (7.19), F_x becomes maximum when $V_d = 0$. That is,

$$F_{\max} = \dfrac{1}{2} \rho A V_u^2$$

Hence, $T_{\max} = \dfrac{1}{2} \rho A V_u^2 R \qquad (7.21)$

For a wind turbine producing a shaft torque T, the torque coefficient C_T is defined by

$$T = C_T T_{\max} \qquad (7.22)$$

The 'tip speed ratio λ' is the ratio of the blade's outer tip speed V_{tip} to the upstream (free) wind speed V_u, i.e.,

$$\lambda = \dfrac{V_{\text{tip}}}{V_u} = \dfrac{\omega R}{V_u} \qquad (7.23)$$

where   is the angular velocity of the rotor and R is the blade radius. Substituting the value of R in Eq. (7.21),

$$T_{\max} = \dfrac{1}{2} \rho A V_u^2 \left(\dfrac{V_u \lambda}{\omega} \right) = P_{\text{total}} \dfrac{\lambda}{\omega} \qquad (7.24)$$

because, $\dfrac{1}{2} \rho A V_u^3 = P_{\text{total}}$ (wind power in upstream side).

Maximum shaft power, $P_{\max}$, is the power obtained from the turbine and is given as

$$P_{\max} = T \cdot \omega = C_T T_{\max} \omega \qquad (7.25)$$

Equating Eqs. (7.12) and (7.25),

$$C_p\, P_{\text{total}} = C_T\, T_{\text{max}}\, \omega = C_T\, P_{\text{total}}\, \frac{\lambda}{\omega} \cdot \omega$$

Therefore,
$$C_p = \lambda C_T \qquad (7.26)$$

$C_p = 0.593$ as per Eq. (7.12), the Bitz limit.

So, ideally
$$(C_T)_{\text{max}} = \frac{0.593}{\lambda} \qquad (7.27)$$

7.21.3 Solidity

Solidity σ is defined as the ratio of the blade area to the circumference of the rotor. Solidity determines the quantity of blade material required to intercept a certain wind area. Hence,

$$\sigma = \frac{Nb}{2\pi R} \qquad (7.28)$$

where N is the number of blades, b is the blade width and R is the blade radius.

For example, if a 3-metre radius rotor has 24 blades, each 0.35 m wide, the solidity is

$$\sigma = \frac{24 \times 0.35}{2\pi \times 3} \times 100 = 44.6\%$$

Solidity represents the fraction of the swept area of the rotor which is covered with metal.

Variation of solidity σ with tip speed ratio λ is shown in Figure 7.18.

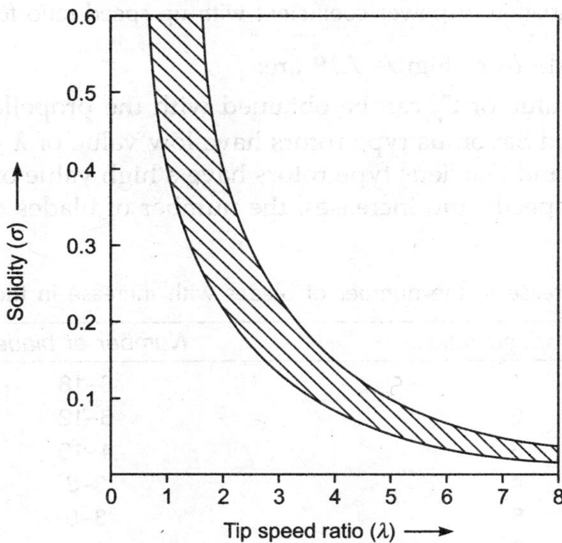

FIGURE 7.18 Variation of solidity σ with tip speed ratio λ.

The following observations can be made from Figure 7.18.

(i) A two or three-bladed wind turbine has a low solidity and so needs to rotate faster to intercept and capture wind energy with aerofoil blades like aircraft. Otherwise the major part of wind energy would be lost through the large gaps between the blades. High speed wind turbines have a low starting torque.

(ii) Rotors having a high value of solidity, like the multibladed wind water pump turbine, operate at low tip speed ratio. Such rotors need a high starting torque.

A high solidity rotor rotates slowly and uses the drag force while a low solidity rotor uses lift forces. The rotor will foil to rotate if the solidity is less than 0.1.

Also, the variation of power coefficient C_p with tip speed ratio λ as obtained from field experiments, has been plotted for several horizontal and vertical axis rotors and is shown in Figure 7.19.

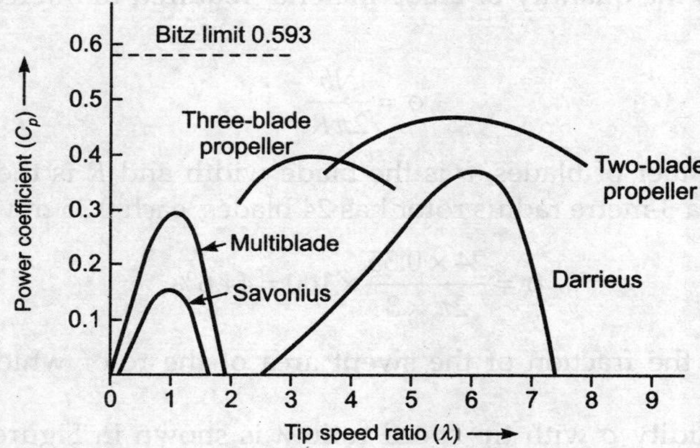

FIGURE 7.19 Variation of power coefficient with tip speed ratio for different rotors.

The observations made from Figure 7.19 are:

(i) The highest value of C_p can be obtained with the propeller type rotors.

(ii) Multiblade and Savonius type rotors have low value of λ (around unity) while the propeller and Darrieus type rotors have a high value of λ varying from 3 to 7. As the tip speed ratio increases, the number of blades decreases as detailed in Table 7.4.

TABLE 7.4 Decrease in the number of blades with increase in the tip speed ratio

Tip speed ratio	Number of blades
1	8–18
2	6–12
3	4–10
4	3–8
5	3–5
8	3–4
10 above	1–2

(iii) Propeller rotor curves are not sharp while for other rotors the rise and fall of power coefficient is quite fast around the maximum value.

(iv) The maximum value of power coefficient under ideal conditions is 0.593 (Bitz limit)

Thus, the rotor of a Wind Energy Conversion (WEC) system can be designed for a specific application by studying the characteristics from Figures 7.18 and 7.19. It is obvious that Savonius and multiblade rotors are suitable for low-speed operation such as water pumps to irrigate fields and to meet the drinking water requirements in rural areas. Modern three-blade or two-blade propeller turbines and Darrieus vertical turbines are suitable for high-speed operation, more appropriate to generate electrical power. It can be seen that a two-blade turbine attains peak power coefficient close to the theoretical maximum value of 59.3%.

The rotor of a wind turbine is a vital part and its blade (say a 500 kW wind turbine) is designed as detailed in Figure 7.20.

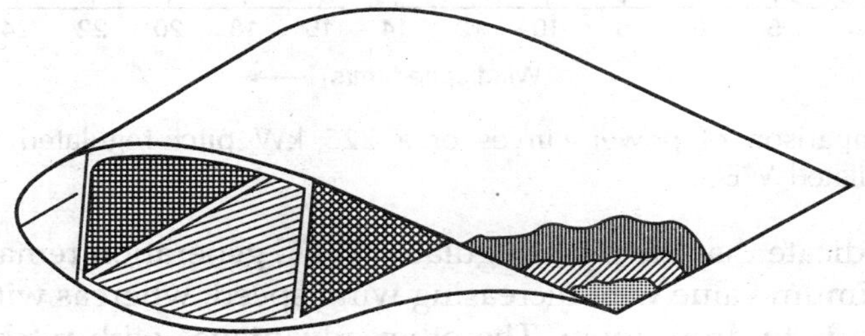

FIGURE 7.20 Section of a 19-metre blade with 10,000 mm radius.

The layout of the rotor is based on a design philosophy that ensures high annual production with minimum structural loads, characterized by a low solidity with aggressive profiles. The outer profiles of the blade are selected for high aerodynamic performance, whereas the inner profile is designed to meet structural requirements. The blade skin carries high-profile stiffness to prevent buckling.

To assess the fatigue life of the blades, full-scale tests are performed both flapwise and cordwise, to 10 million cycles in each direction at a load equivalent to 20 years of lifetime.

7.22 DESIGN OF REGULATING SYSTEM FOR ROTOR

The rotor blades are designed to withstand centrifugal force for every wind load. Centrifugal forces tend to exert pull on blades whereas wind loads cause bending stress on blades. To prevent overstressing the turbine during high wind and to maximize power generation, a regulating system is required. Two basic types of WEGs have been designed namely: (i) Stall Regulated and (ii) Pitch Regulated.

The blades of stall-regulated wind turbines are designed with a suitable blade profile, and the thickness and chord distribution with a calculated blade twist. The blades are rigidly fixed to the rotor. At high wind speeds, less torque is produced over the rotor shaft limiting the power output. Actually for 'stall-regulated' turbines, the pitch angle distribution along the blades is constant for all wind speeds. Hence, these turbines are unable to cope variations in wind speed, and changes in wind direction. So 'stall regulation' is used for small capacity turbines only.

The above shortcomings are solved in pitch-regulated wind turbines where the blades can rotate about the length of the axis to regulate the power. A comparison of power curves of a 225 kW pitch-regulated WEG vis-a-vis stall-regulated WEG is shown in Figure 7.21.

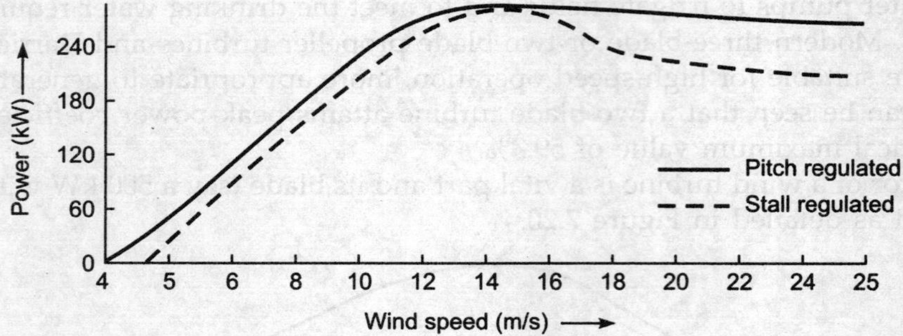

FIGURE 7.21 Comparison of power curves of a 225 kW pitch-regulated WEG vis-a-vis stall regulated WEG.

The curves indicate that the pitch-regulated WEG generation remains constant after reaching the maximum value with increasing wind speed, whereas with stall-regulated, the generation starts to drop down. The other advantages with pitch regulation are:

(i) Pitch regulation makes it possible to rotate the blades to a position which stops and starts the rotor at any wind speed.

(ii) With low wind speed, a pitch-regulated WEG can generate maximum torque to start the rotor.

(iii) Pitch control system is not affected by change in air density and change in temperature due to placement of WEG at certain heights above sea level.

(iv) It is easy to get a greater capacity factor for a pitch-regulated WEG.

(v) With pitch-regulated WEGs, the blades always experience laminar flow across the profile and turbulent, chaotic flow has no influence.

7.23 WIND POWER GENERATION CURVE

The power curve of a wind turbine indicates power output as a function of wind velocity at hub height as shown in Figure 7.22. The curve shows the steady idealized characteristic, but in practice the wind speed constantly varies.

A wind turbine develops less power than the wind's stream power, due to friction and spillage and the curve in Figure 7.22 shows the following limiting speeds:

(i) *Cut-in speed* (V_{in}):   It is the wind speed (14 km/h or 4 m/s) at which the turbine output begins. It is higher than the speed at which the turbine starts rotating. Before starting to rotate, the turbine remains in the braked position.

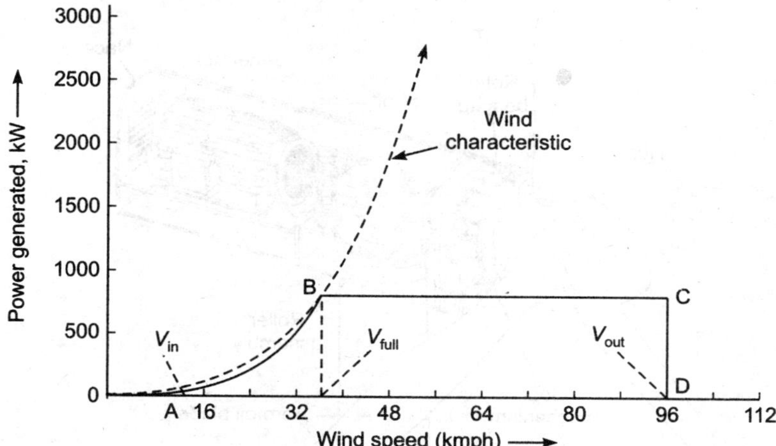

FIGURE 7.22 Wind turbine power curve.

(ii) *Rated speed* (V_{full}): It is the wind speed at which the turbine is designed to generate the rated power. When the wind speed is more than the 'cut-in speed' but less than the rated speed, the pitch angle of blades is selected to deliver maximum power. Pitch angle is controlled to maintain constant rated power above the rated wind speed.

(iii) *Cut-out speed* (V_{out}): When the speed reaches the upper limit (90 kmph or 25 m/s) the turbine stops to generate power as a safety measure in order to protect the turbine and the generator.

As the wind reaches the cut-in speed V_{in} the WTG starts generating power; it then moves up to the point B to deliver the rated power. The blade pitch control operates at B to maintain a constant power output BC. At C, the cut-out wind speed is reached and the turbine is stopped to avoid structural damage.

7.24 SUB-SYSTEMS OF A HORIZONTAL AXIS WIND TURBINE GENERATOR

Wind energy, extracted by blades, rotates the shaft which, by using the gear and coupling mechanism, operates the generator housed inside a nacelle. A roller assembly links the tower with the nacelle to permit its rotation about a vertical axis to keep the rotor in wind direction. Large wind turbine generators use pitch regulation and run at a fixed speed (50 cycles/second) to facilitate synchronization with the grid supply. A 225 kW WEG having a rotor diameter of 27 metres with swept area of 573 sq. m, installed on a tubular tower, is shown in Figure 7.23 with its various subsystems as follows:

Blades: Wind turbine blades need to be lightweight and possess adequate strength and hence require to be fabricated with aircraft industry techniques. The blades are made of glass fibre reinforced polyster with a suitable structural geometrical shape to create *lift* as the air flows over them.

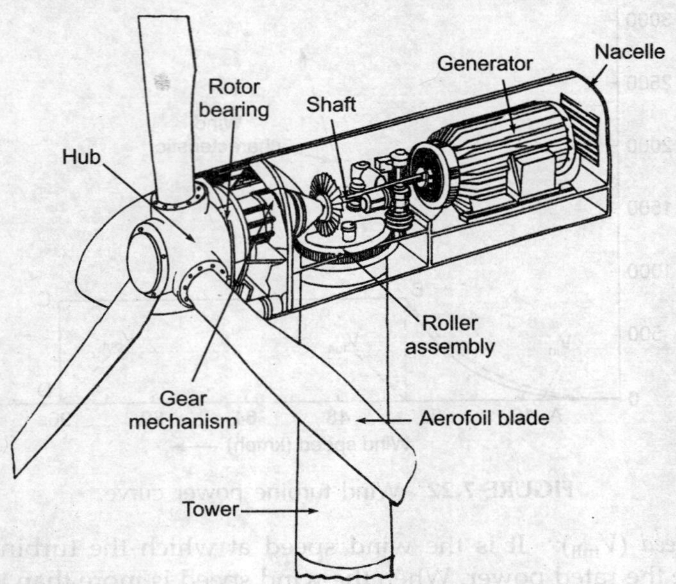

FIGURE 7.23 Subsystems of a horizontal axis wind turbine generator.

Nacelle: It houses the generator, the gear box hydraulic system and the yawing mechanism. Nacelle is placed at the top of the tower and is linked with the rotor.

Power transmission system: Mechanical power generated by rotor blades is transmitted to the generator through a two-stage gear box. From the gear box, the transmission shaft rotates the generator with a built-in friction clutch. The gear box is provided to increase the generator speed to 1500 rpm.

Generator: Generally the large WTGs, used with grid-connected systems, have induction generators. They use reactive power from grids and feed the generated power to boost the grid supply. Medium capacity WTGs use synchronous generators installed to electrify villages, and provide industrial power supply to remote places. Small capacity WTGs use permanent magnet dc generators which supply power to microwave stations and illuminating lighthouses.

Yaw control: Yawing is done by two yawing motors, which mesh with a big-toothed wheel mounted on top of the tower. Yaw control continuously tracks and keeps the rotor axis in the wind direction. During high speed wind, i.e., more than the cut-out speed, the machine is stopped by turning the rotor axis at right angles to the wind direction.

Brakes: Braking of WEGs is done by full feathering. An emergency STOP activates the hydraulic disc brakes fitted to the high-speed shaft of the gear box.

Controllers: WEGs are monitored and controlled by a microprocessor-based control unit. A controller monitors the parameters in the nacelle besides controlling the operation of the pitch system. Variations in the blade position are performed by a hydraulic system, which also delivers pressure to the brake system.

Tower: Modern wind turbine generators are installed on tubular towers. Large turbines use lattice towers designed to withstand gravity loads and wind loads. The height of the tower is decided for obtaining the designed value of wind speed and dimensions of the rotor (the higher the turbine capacity the larger the rotor).

7.25 MODES OF WIND POWER GENERATION

By nature, wind is not a steady source of energy, therefore, it cannot on its own meet the needs of consumers at all times. Necessarily, it has to be integrated with some other sources to provide a constant backup. Wind Electric Generators (WEGs) operate in one of the following three modes.

  (i) Standalone mode
 (ii) Backup mode like wind–diesel
(iii) Grid-connected mode.

7.25.1 Standalone Mode

This type of aero-generator represents decentralized application of wind energy and is characterized by the situation where an individual energy consumer or a group of consumers install their own wind turbine. The generating capacity of the WTG is matched with the energy requirement. The two most promising applications of the wind energy conversion system are:

  (a) Power supply for domestic use and battery charging.
  (b) Windmill water pump for irrigation and drinking purposes.

A WEG with a capacity of 2.5 kW to 5 kW is useful for domestic power supply. It operates independently with a battery and its charging equipment is as detailed in Figure 7.24. Such installations are useful for remote mountainous regions where the extension of grid or supply of oil is a remote possibility. Special benefit accrues where the wind speeds are suitable for power generation. It is preferred to have electric power at controlled frequency. As the wind changes speed, the pitch of the blades is adjusted to control the frequency of turbine rotation.

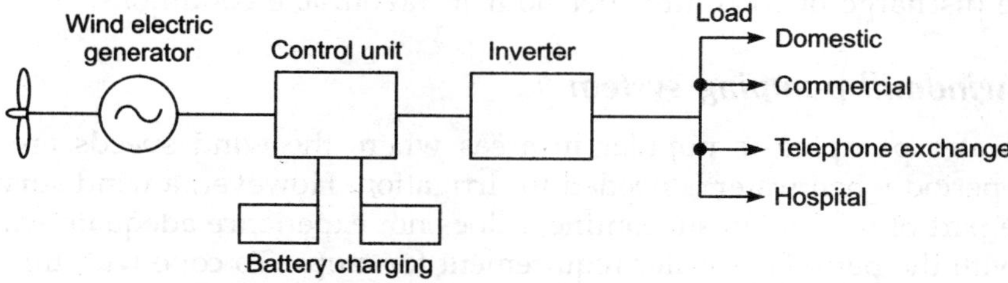

FIGURE 7.24 A standalone 5 kW wind electric generator.

A windmill water pump comprises a wheel with pressed steel blades secured to a shaft. The windmill is mounted on the top of a lattice steel tower. The rotary motion of the windmill is converted into reciprocating motion to pump water as illustrated in Figure 7.25.

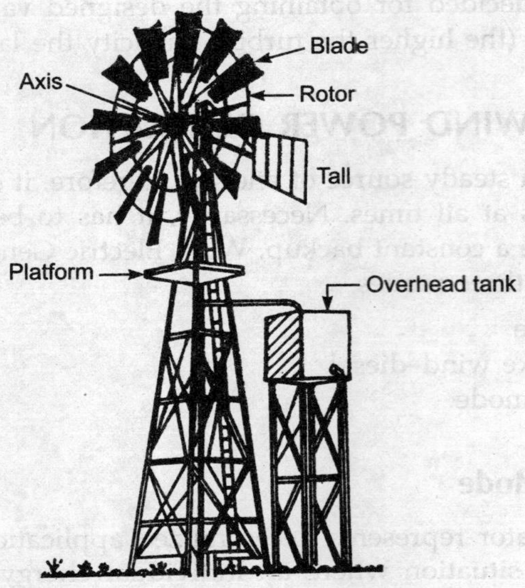

FIGURE 7.25 Windmill water pump.

Sample technical specifications of water pumping windmill are:

Cut-in wind speed	—	6 km/h (1.7 m/s)
Rated wind speed	—	13 km/h (3.6 m/s)
Cut-out wind speed	—	35 km/h (9.7 m/s)
Rotor rpm	—	85 at the rated wind speed
No. of blades	—	18
Rotor diameter	—	3.25 m
Tower height	—	10 m

A system like the above is capable of lifting water from a depth of 20 m to 50 m, with an effective discharge of 2300 litres per hour in favourable conditions.

Versatile windmill pumping system

Windmill water pumping is popular in areas where the wind speeds are adequate during the period when water is needed for irrigation. However, a wind survey shows that a large part of the Indian subcontinent does not experience adequate wind speeds matching with the periods of water requirement for crops. To cope with this situation, the Water Technology Centre of the Indian Agriculture Research Institute, New Delhi, has designed a versatile windmill pumping system as shown in Figure 7.26. The wind turbine has a Savonious rotor, which can also be operated employing a pair of bullocks when the wind speed is inadequate during periods of urgent water requirement.

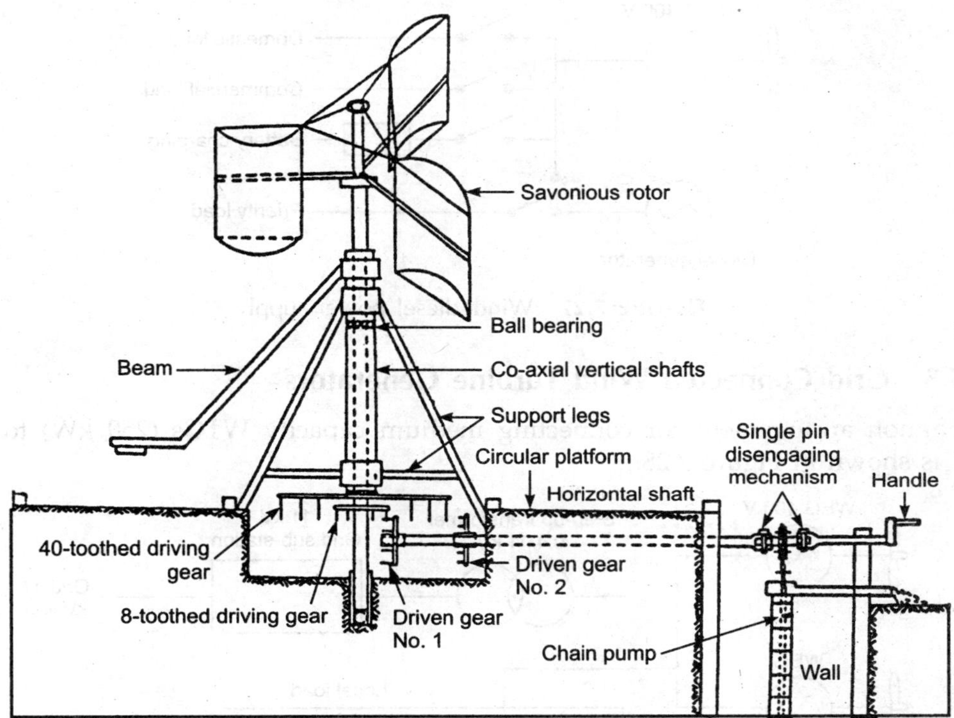

FIGURE 7.26 Constructional details of a versatile windmill pumping system.

Coupling wind pump with drip irrigation system

A superior approach to use water, pumped by wind energy, is to couple the drip irrigation system with the windmill pumping system. It is an innovative approach for optimal agriculture production with renewable energy, in semi-arid areas not covered by grid power.

7.25.2 Backup Mode Like Wind–Diesel

Wind energy, being intermittent, requires a backup of diesel generator to maintain a 24-hour power supply. In areas inaccessible to grid power, the emergency loads of hospitals, defence installations and communication services are met with a wind–diesel hybrid system, while the general loads of domestic and commercial establishments are fed by WTG, as detailed in Figure 7.27.

As the wind speed drops, low tariff loads are automatically switched off to reduce the demand. During the period of no wind, priority loads are fed by the diesel generator. Load management allows the full capital value of WTG to be used at all times, besides maximum utilisation of free wind energy.

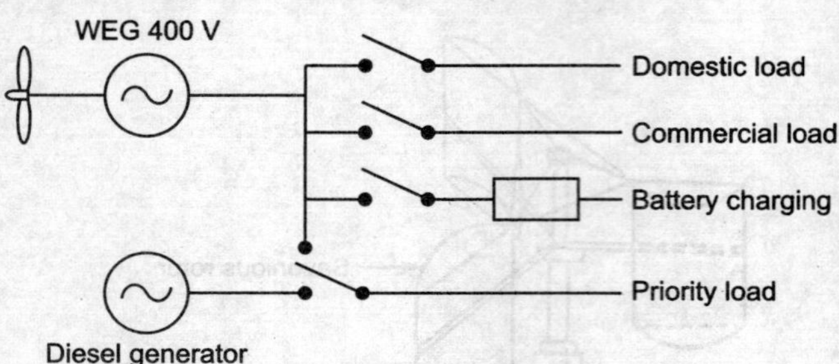

FIGURE 7.27 Wind–diesel power supply.

7.25.3 Grid Connected Wind Turbine Generators

A common arrangement for connecting medium capacity WTGs (250 kW) to 'state grid' is shown in Figure 7.28.

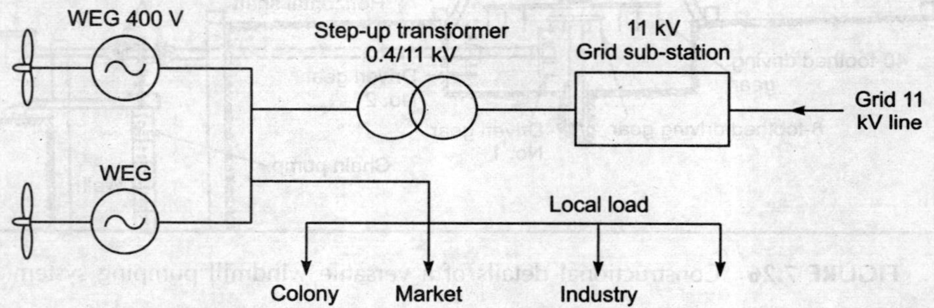

FIGURE 7.28 Grid-connected wind turbine generators.

WTGs generate electric power at 400 V; it is then stepped up to make this voltage compatible to the grid (11 kV). In India, grid-connected WEGs constitute wind farms where the generated power is distributed among the nearby consumers and the excess power is exported to the grid. Electrical energy is purchased (imported) from the grid during periods of no wind.

7.26 ADVANTAGES AND DISADVANTAGES OF WIND ENERGY SYSTEM

The advantages of WEC systems are:

1. Wind energy is a renewable source of energy and can be tapped, free of fuel cost.
2. The WTG produces electricity in an environmentally friendly way.
3. It can supply electric power to remote inaccessible areas like the Upper Himalayan range (Ladakh to Sikkim), Andaman and Nicobar islands, remote desert locations in Rajasthan, coastal areas of Kutch (Gujarat) and deep forest tribal settlements in Madhya Pradesh.

4. Public opinion is in favour of WTGs rather than fossil fuel and nuclear power generation. People do accept a wind turbine closer to their homes (2 km to 5 km). In contrast, the minimum acceptable distance to a nuclear power plant is 60 km.
5. Wind power generation is cost effective.
6. It is economically competitive with other modes of power generation.
7. Wind energy development is dynamic and an exciting addition to the landscape which increases public awareness of energy generation.
8. It is reliable and has been used for ages.

The disadvantages of WEC systems are:

1. Wind energy has low energy density and normally available at only selected geographical locations away from cities and load centres.
2. Wind speed being variable, wind energy is irregular, unsteady and erratic.
3. Wind turbine design is complex and needs more research and development work due to widely varying atmospheric conditions where these turbines are made to operate.
4. Large units have less capital cost per kWh, but require capital intensive technology. In contrast, small units are more reliable but have higher capital cost per kWh.
5. Wind energy systems require storage batteries which contribute to environmental pollution.
6. Wind farms are established in locations of favourable wind. These locations are in open areas away from load centres. Consequently, the connection to state grid is necessary.
7. Wind energy systems are capital intensive and need government support.

EXAMPLE 7.2

Wind at one standard atmospheric pressure and 15°C has a speed of 10 m/s. A 10-m diameter wind turbine is operating at 5 rpm with maximum efficiency of 40%. Calculate (i) the total power density in wind stream, (ii) the maximum power density, (iii) the actual power density, (iv) the power output of the turbine, and (v) the axial thrust on the turbine structure.

Solution

Total power density $= \dfrac{1}{2}\rho V_u^3 = \dfrac{1}{2} \times 1.226 \times 10^3$

$$= 613 \text{ W/m}^2$$

Maximum power density [Eq. (7.11)] $= \dfrac{P_{max}}{A} = \dfrac{8}{27}\rho V_u^3$

$$= \dfrac{8}{27} \times 1.226 \times 10^3 = 363 \text{ W/m}^3$$

Actual power density = Efficiency × Total power density

$$= 0.40 \times 613 = 245.2 \text{ W/m}^2$$

Power output, $P = 0.245 \times \dfrac{\pi D^2}{4} = \dfrac{0.245 \times \pi \times 10^2}{4} = 19.33 \text{ kW}$

Axial thrust [Eq. (7.20)], $F_{x(max)} = \dfrac{\pi}{9} \rho D^2 V_u^2 = \dfrac{\pi}{9} \times 1.226 \times 10^2 \times 10^2$

$$= 4277.40 \text{ N}$$

EXAMPLE 7.3

Design the rotor radius for a multiblade wind turbine that operates in a wind speed of 3 kmph to pump water at a rate of 6 m³/h with a lift of 6 m. Also, calculate the angular velocity of the rotor. *Given:* water density = 1000 kg/m³, g = 9.8 m/s, water pump efficiency = 50%, efficiency of rotor to pump = 80%, C_p = 0.3; λ = 1.0, and air density = 1.2 kg/m³.

Solution

Power required to pump water $= \dfrac{6 \times 1000}{3600} \times 9.8 \times 6 = 98$ W

Power required at rotor $= \dfrac{98}{0.5 \times 0.8} = 245$ W

From Eq. (7.12), $C_p \times P_{\text{total}} = P_{\text{max}}$

Hence, $\qquad 0.3 \left[\dfrac{1}{2} \times 1.2 \times \pi R^2 \left(\dfrac{36 \times 1000}{3600} \right)^3 \right] = 245$

where R is radius of the turbine rotor.

Thus, $\qquad\qquad\qquad\qquad\qquad R = 0.66$ m

As λ = 1 the number of blades in a multiblade turbine varies from 8 to 18.

From Eq. (7.23) the angular velocity of rotor, $\omega = \dfrac{\lambda V_u}{R}$.

Therefore, $\qquad\qquad \omega = 1.0 \times 36 \times \dfrac{1000}{3600} \times \dfrac{1}{0.66}$

$$= 15.1 \text{ rad/s}$$

$$= 144 \text{ rpm}$$

EXAMPLE 7.4

A WEG generates 1500 watts at rated speed of 24 kmph at the atmospheric pressure and temperature of 20°C. Calculate the change in output if the wind generator is operated at an altitude of 1800 m, temperature 10°C, wind speed 30 kmph, and air pressure 0.88 atmosphere.

Solution

Air density,
$$\rho = \frac{P}{RT}$$

where P is the air pressure in Pa, T is temperature in kelvin and R is gas constant (= 287 J/kg·K).

Air pressure at 1 atmosphere = 1.01325×10^5 Pa

Air density at 1800 m = $\dfrac{0.88 \times 1.01325 \times 10^5}{287 \times 283}$ = 1.10 kg/m^3

Now,
$$P = \frac{1}{2}\rho A V^3$$

or
$$1500 = \frac{1}{2} \times 1.2 \times A \times \left(\frac{24 \times 1000}{3600}\right)^3$$

or
$$A = 8.44 \text{ m}^2$$

Power generated at 1800 m, $\quad P = \dfrac{1}{2} \times 1.10 \times 8.44 \times \left(\dfrac{30 \times 1000}{3600}\right) = 2686$ W

REVIEW QUESTIONS

1. Explain how local winds are created during daytime and night.
2. Discuss the different types of wind turbines used to extract wind energy.
3. With the help of a neat sketch, discuss the different types of rotors used in wind turbines.
4. Explain the terms: camber, nacelle, solidity, cut-in speed, cut-out speed, windrose, and wind vane.
5. Prove that the maximum turbine output can be achieved when $V_d = (1/3)V_u$.
6. Derive an expression for energy that can be extracted from wind.
7. Discuss the favourable sites for installing windmills.
8. Write short notes on advantages and disadvantages of WEC systems.
9. Calculate the maximum power output of a 15-m diameter wind turbine at one atmospheric pressure and wind speed of 12 m/s.
10. For an 8-blade wind turbine, calculate the angular speed of the rotor to lift water from 6-m depth if the radius of the turbine rotor is 1 m and the wind speed is 10 m/s (given $\lambda = 1$).

WIND ENERGY FARMS

8.1 INTRODUCTION

Wind energy is the first among renewable energy resources to become an economically viable source of power generation. Technological improvements have brought down the cost of wind power equal to that of coal-fired thermal power plant.

India is blessed with many natural meteorological and topographical settings that are conducive to high speed winds suitable for power generation. Energy content in wind in different regions varies with latitudes, land sea dispositions, altitudes and seasons. In India, the prime factor governing the availability of wind energy at a particular site, is its geographical locations with reference to monsoon winds. A site is considered suitable where the wind speed is 18 km/h (5 m/s). Maximum wind energy can be tapped from a windy site by installing several wind turbines. The generated power is fed into a network. The whole system is called a 'Wind Farm'.

8.2 WIND RESOURCE SURVEYS

A fundamental prerequisite for determining the feasibility of wind power generation is the availability of wind data at a given location. Three types of wind survey projects were undertaken during 1985 by MNRE with the Indian Institute of Tropical Meteorology, Bangalore.

- The first category was of a wind monitoring project to determine windy locations, using a 20-m mast and a microprocessor-based measuring instrument, to generate data for wind power development.

- The second category constituted wind mapping projects, based on 5-m mast, to establish wind regime in a given area of a state on an extensive basis.

- The third category projects covered complex terrain studies in hilly and mountainous regions to determine the wind flow in mountain passes and over undulated terrains.

Data collected from survey projects identified major windy sites in the coastal areas and also confirmed that several interior locations in Kerala, Tamil Nadu, Andhra Pradesh, Madhya Pradesh, Karnataka, Maharashtra and Rajasthan have wind energy potential. The characteristic feature of large scale wind flow over India is the monsoon circulation. Wind flows are generally high from April to September and low during the rest of the period.

8.3 ASSESSMENT OF WIND AVAILABILITY FROM METEOROLOGICAL DATA

Meteorological data is used to evaluate the following:

1. To identify the areas where the highest wind speeds are available.
2. To measure Mean Annual Wind Speeds (MAWS) and their variability from year to year.
3. To record Monthly Mean Wind Speeds to indicate wind regimes for the area.
4. Measurement of Daily Mean Wind Speeds in order to understand their variation during different seasons of the year.

The wind climatology of India is determined by two extensive monsoon systems. Strong South-West monsoon winds during June to September over Western parts of Indian peninsula provide bulk wind energy potential. During November to February, North-West monsoon winds are relatively weaker and have lesser wind potential. For economic utilisation of wind power, a MAWS of at least 18 km/h is required.

The MAWS is an approximate index of the wind potential at a site. To evaluate the energy potential, another statistical parameter, i.e., mean monthly wind speed is required. It provides a comprehensive pattern about variability in wind energy during the course of the year. Windy areas of peninsular India experience annual maximum wind speeds (30–35 km/h) during May/June to September.

To gain further insight into the variability of winds, the daily mean wind speed is analysed which is the average of the winds during the 24 hours of the day. Higher winds suitable for power generation persist during June to September with some ups and downs which reveal the quasi-oscillatory character of monsoon current. A threshold speed of 15 km/h is the lowest speed needed to operate the wind electric generators.

Wind power classification

Wind power is classified depending on resource potential at 30 m height and is given in Table 8.1.

TABLE 8.1 Wind power classification

Resource potential	Wind power density (W/m²)	Wind speed (m/s)
Fair	100–150	4.3–5.0
Good	150–200	5.0–5.5
	200–300	5.5–6.3
Excellent	300–400	6.3–7.0
	400–600	7.0–8.2
	600–1000	8.2–10.1

8.4 ESTIMATION OF WIND ENERGY POTENTIAL

Wind speed extrapolation

Wind speed data are recorded by data loggers at a height of 10 m and 20 m. Wind speed increases with height as per the power law equation. Since WEGs are installed at a greater height, it is necessary to extrapolate the mean wind speed measured at one level to higher levels.

8.4.1 Methods of Calculations

The estimation of wind energy potential is based on the following methods:

  (i) Based on wind data of a specific site using frequency distribution.
 (ii) Based on the type of wind energy generator.
(iii) Based on Weibull factors of the wind data and WEG's characteristics.

8.4.2 Equations Used for Calculations

(a) Based on wind data

Annual energy generation and other factors are calculated at a specific site based on the following equations and characteristics of WEGs.

1. Power law index α as per Eq. (7.16)

$$\frac{V_2}{V_1} = \left(\frac{H_2}{H_1}\right)^{\alpha}$$

2. Wind power density (P_d) as per Eq. [7.11(a)]

$$P_d = \frac{1}{2}\rho V_u^3 \text{ W/m}^3$$

(b) Based on wind energy generator (WEG)

1. Capacity factor (CF)

CF is defined as the ratio of average power output of a turbine during a month or a year to the rated power output.

2. Capacity utilisation factor (CUF)

$$\text{CUF} = \frac{\text{actual energy generated}}{\text{theoretical energy generated}}$$

(c) Capacity factor on the basis of WEG characteristics and using Weibull factors as per Eq. (7.13) where C is the scale factor and K is the shape factor.

Energy likely to be generated is calculated using the power curve of the given WEGs and the above equations, based on frequency distribution and the results are then tabulated for analysis.

8.5 WIND RESOURCE ASSESSMENT IN INDIA

The Centre for Wind Energy Technology (C-WET), Chennai, recently conducted a wind resource assessment programme in co-ordination with the state nodal agencies. Accordingly, an annual mean wind power density greater than 250 W/m^2 at 50 m height, was recorded at 211 wind monitoring stations, covering Andaman and Nicobar Islands, Andhra Pradesh, Gujarat, Karnataka, Kerala, Lakshadweep, Madhya Pradesh, Maharashtra, Orissa, Rajasthan, Tamil Nadu, Uttarakhand and West Bengal.

8.5.1 Wind Power Potential and Achievements in India

India's wind power potential has been assessed at 302 GW, at a hub height of 100 m as detailed in Figure 8.1. However, the potential for grid-interactive wind power is less, i.e., around 15,000 MW (sites having wind power density in excess of 300 W/m^2 at 50 m hub height are considered).

As of March 31, 2021, India **Installed wind power capacity** was 39.25 GW, the fourth largest **wind power producer** in the world after China, USA and Germany. The global installed capacity of wind power had reached over 733 GW.

The installation of wind power has always exceeded government targets. In the 10th (2002–07) and 11th (2007–12) Five-Year Plans, against targets of 1500 MW and 9000 MW, 5427 MW and 10,260 MW of wind power was installed. In the 12th plan (2012–17) it was 15000 MW. Further, MNRE has made a mission to develop a long-term sustainable policy so as to attain a high target of 100,000 MW wind power total installation by 2022.

India's Wind Power Potential

Potential of wind power in the Indian sub-continent is estimated as:

- At a hub height of 50 metres = 43130 MW
- At a hub height of 80 metres = 102778 MW
  (assessed by centre for wind energy technology)
- At a hub height of 120 metres = 695.50 GW
  (estimated by Lawrence Berkeley National Laboratory in US)
- At a hut height of 100 m = 302 GW.

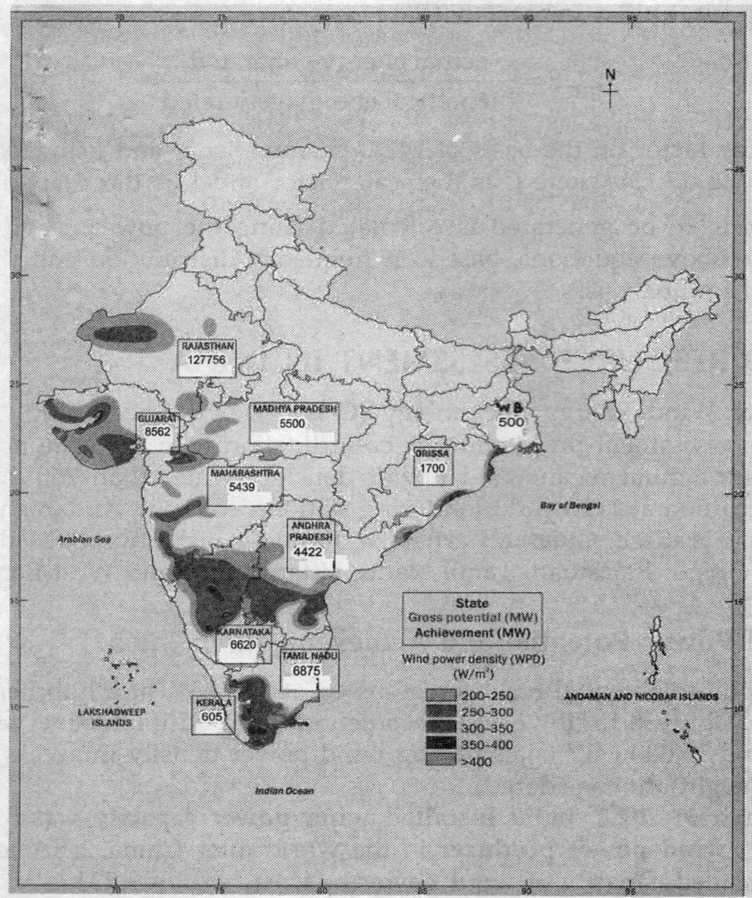

FIGURE 8.1 India's wind power potential and achievements as on 31st March 2021 in MW.

Phase I: land availability, characteristic location and landscape profile

The proposed capacity of a project determines the land requirement. Considering 250 kW wind turbines (rotor diameter D = 25 m) the distance between turbines in the same row is taken 5 times the rotor diameter and the distance between turbine rows is kept 10 times the rotor diameter. However, as per the norms of MNRE, a wind farm requires 12 hectares/MW in sites having wind power density in excess of 200 W/m² at 50 m hub-height.

Accessibility to wind project site

Approach roads are required from the main road to the Wind Power Project site for transportation of wind turbines with generators, electrical equipment, civil construction material and erection equipment.

Soil characteristic

Soil investigation of the proposed site, e.g. sand, loose/hard soil rock, etc. has to be carried out for foundation design. Earthing design has also to be worked out based on soil characteristics.

State grid

Grid proximity to the site has to be studied as a stable grid must be available to pump the generated electricity to the Electricity Board grid.

Ambient conditions at the proposed site

It is necessary to know the ambient conditions at the proposed site, as under special conditions the ambient conditions would affect the performance of WEGs. In areas where the conditions could be rough, the following data is required:

   (i) Temperature conditions (°C)
  (ii) If the turbine is to be installed near the sea then:
   • salt concentration in the air
   • sand concentration in the air
 (iii) Relative humidity of the air
 (iv) Corrosion factor.

Phase II: Micro-siting of Wind Electric Generators (WEGs)

Once the site selection is done, specific siting of WEGs is necessary to optimize the power output.

   (i) A visual inspection of the land helps in understanding the topography of the terrain. WEGs are located at the highest level of the land in the region of least turbulence.
  (ii) Array efficiency should normally be above 95%, which depends on specific configuration and orientation. Minimum loss due to 'shadow effect' should be ensured.

A schematic layout of a 10 MW 'Wind Power Plant', having 50 nos. of 225 kW WEGs, is shown in Figure 8.2.

Annual energy output

The power curve of a 225-kW WEG is shown in Figure 8.3 as a function of wind speed distribution pattern (assumed to follow the Weibull probability density function):

Calculations are based on Weibull probability distribution given in Eq. (7.13) where C and K are scale and shape factor respectively. The parameter C has dimensions of velocity and is given by $C = 1.1 \times$ average wind velocity. Parameter $K > 3$ depicts regular and steadier winds but $K = 2$ is used for evaluating a given wind resource.

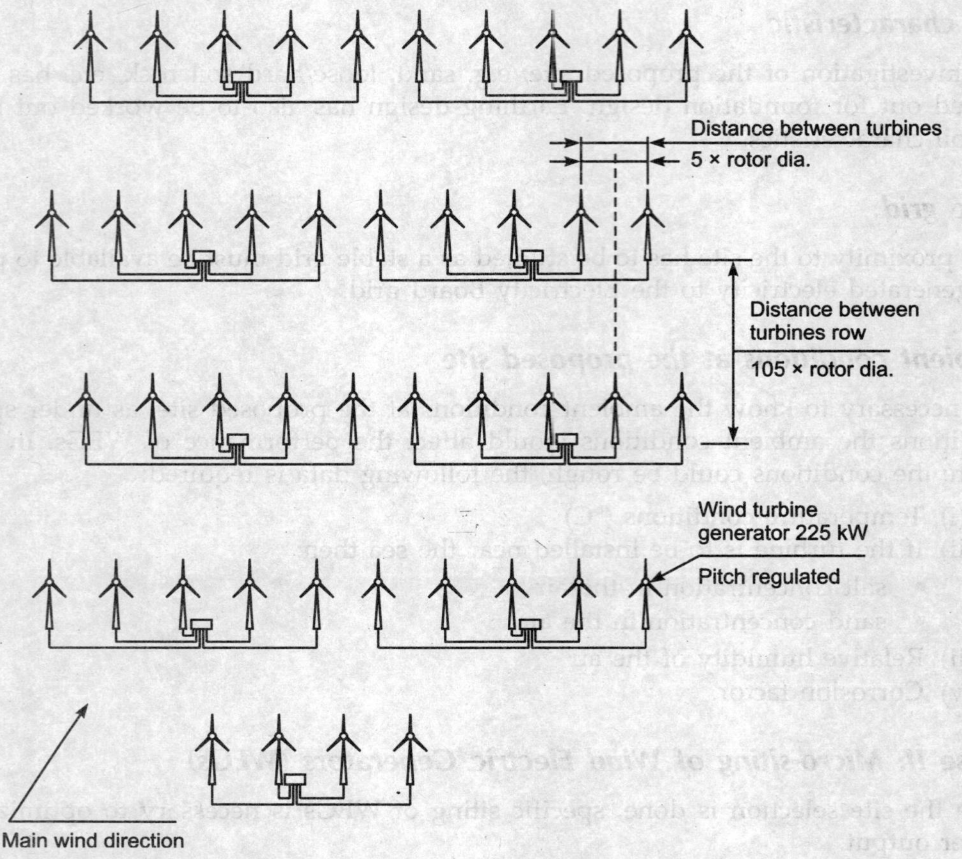

FIGURE 8.2 Schematic layout of a 10-MW wind power plant.

Weibull function is used to describe the wind regime at any site. Annual generation at wind farms with different wind speeds is given in Table 8.2.

TABLE 8.2 Annual generation at wind farms with different wind speeds

Average wind speed (m/s)	Units generated (kWh)
8	795,000
7	636,000
6	465,000
5	296,000

The capacity of a wind generator is optimized to suit the site by having theoretical energy projections. A right choice of a WEG reduces the generation cost so as to be able to compete economically with fossil fuel energy.

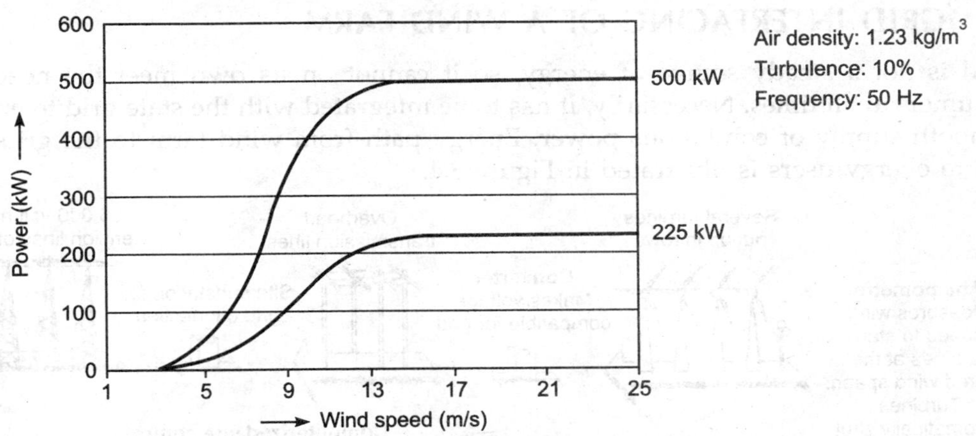

FIGURE 8.3 Power curve of a 225-kW wind energy generator.

8.6 SELECTION OF OPTIMUM WIND ENERGY GENERATOR (WEG)

Many types of WEGs are commercially available in the capacity range of 1 kW to 3 MW. It is necessary to select the best suited WEG for a particular site, for generating maximum energy. Production of electricity depends on several factors, i.e., mean wind speed of the site, characteristic of WEG, hub height, cut-in speed, rated and furling wind speed of the machine.

Methodology adopted is based on comparative statement of the following two modes of evaluation.

(i) Calculated values of the annual energy generation using the power curve of WEG with the annual capacity factor and the annual capacity utilisation factor.

(ii) Power density is calculated from the wind data over a period of few years to select an optimum WEG using Weibull parameters.

Technical data of 225 kW and 500 kW WEGs from an Indian manufacturer (Vestas) is shown in Table 8.3.

TABLE 8.3 Technical data of WEGs

Generator rating (kW)	Rotor diameter (m)	Hub height (m)	Voltage (V)	Cut-in speed (m/s)	Cut-out speed (m/s)	Brake aerodynamics	Controls
225	27	32	400	3.5	25	Full feathering of blades	Microprocessor based
500	47	50	690	4	25	Full feathering of blades	Microprocessor based

Turbine blades are flexible, generator is asynchronous, nacelle cover is of fibre glass reinforced polyester. Lightning protection for total installation is by a shielding system.

8.7 GRID INTERFACING OF A WIND FARM

Wind is not a steady source of energy, so it cannot on its own meet the needs of consumers at all times. Necessarily, it has to be integrated with the state grid to ensure a smooth supply of continuous power. Energy path from wind farm to the grid and then to energy users is illustrated in Figure 8.4.

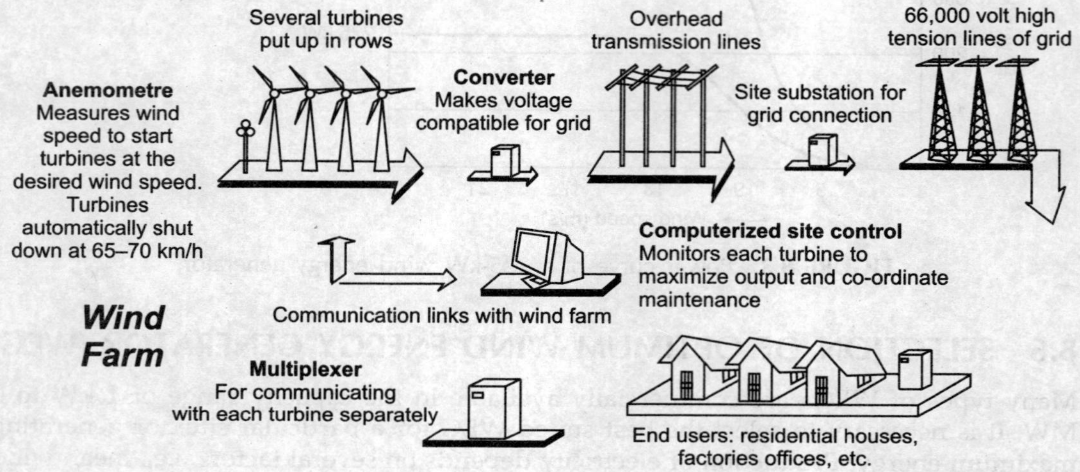

FIGURE 8.4   Energy path from wind farm to the user.

WEGs generate at 400 V which is stepped up to 11 kV, then the overhead transmission lines connect to the substation for grid connection. Power evacuation may be on 11 kV or 22 kV or 33 kV lines depending on (a) the availability of distribution network, (b) the number of WEGs installed in the wind farm. A rule of thumb as given in Table 8.4 is adopted.

TABLE 8.4   Type of interfacing for different capacities of WEGs

Capacity	Type of interfacing
Up to 2 MW	Existing 11 kV/22 kV line
2 to 6 MW	Separate 11 kV/22 kV line
6 to 10 MW	Separate 33 kV line
Above 10 MW	Locate 66 kV or 110 kV substation at the wind farm site

8.8 METHODS OF GRID CONNECTION

For a single-row layout of wind farm, one transformer connected to two WTGs is the most economical solution, whereas for a multi-array wind farm, one transformer is connected to four turbines, as shown in Figures 8.5(a) and (b).

Generated power of wind farm, collected at 11 kV or 33 kV, is then further stepped up to the appropriate class of voltage while integrating with the state power grid, as shown in Figure 8.6.

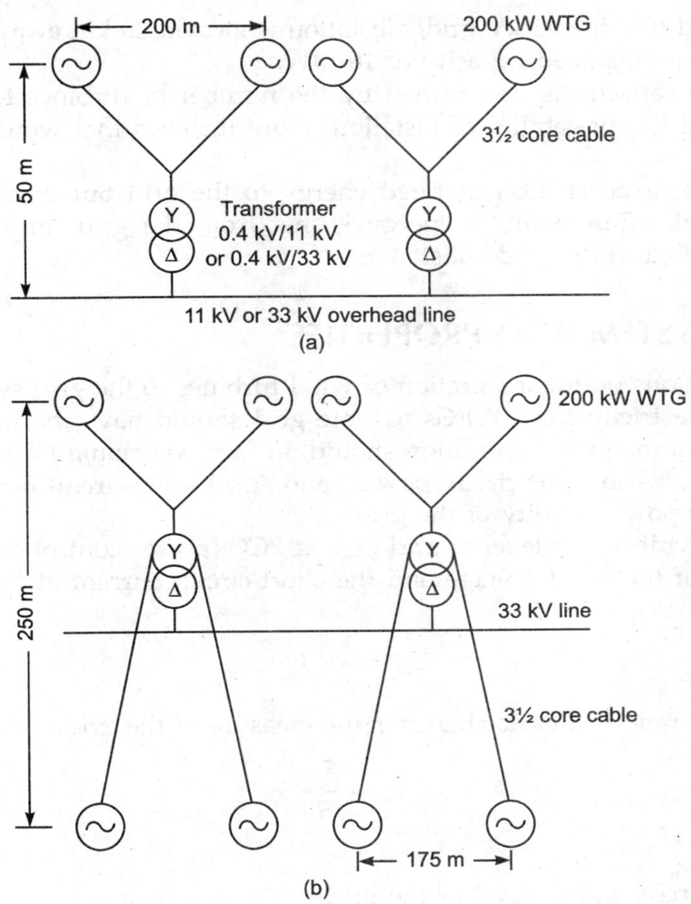

FIGURE 8.5 (a) Electrical system for a single-array wind farm, and (b) electrical system for a multi-array wind farm.

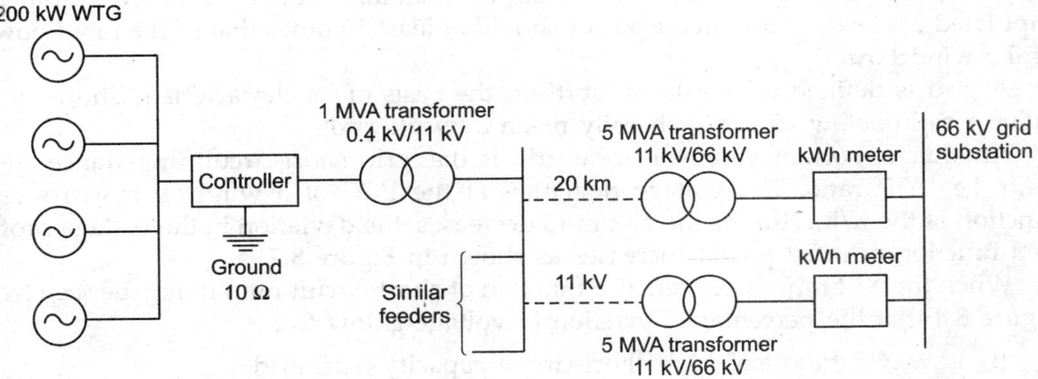

FIGURE 8.6 Electrical connection of a wind farm with the grid.

It is assumed that the 66 kV grid/substation is located 20 km away from the wind farm which has an installed capacity of 10 MW.

Transformer capacity is determined by the number of turbines to be connected, keeping in mind the possibility of installing more turbines that would be connected at a later date.

A wind farm exports the generated energy to the grid but during the no-wind periods the local requirement of energy is met from the grid. Import-export kWh meters are installed in the grid substation.

8.9 GRID SYSTEM AND PROPERTIES

There are limitations on the integration of wind turbines to the grid system. Pumping the generated electricity from WEGs to state grid should have the minimum power quality impact on the grid. Integration should, in fact, contribute to the improvement in power quality. "Grid short-circuit power" and "grid short-circuit ratio" have a great influence on the power quality of the grid.

The short-circuit power level of grid (S_{sc}) at PCC (power control centre) is defined as the product of the rated voltage and the short-circuit current at PCC of the wind farm, i.e.,

$$S_{sc} = \frac{3}{2} V_r I_{sc}$$

The short-circuit ratio is a basic characteristic measure of the grid which is defined as

$$R_{sc} = \frac{S_{sc}}{S_r}$$

where

S_{sc} = short-circuit power level of the grid
S_r = rated turbine power level.

Short-circuit power affects the voltage deviation at the terminals of the wind turbine generating the rated power. Thus, to maintain voltage deviation within the stipulated 2% level, short-circuit power should at least 50 times that of the rated power of the wind farm.

A grid is defined as 'weak' or 'stiff' on the basis of its characteristic short-circuit ratio. A ratio below 20 may generally mean a 'weak grid'.

Another important parameter of a grid is the 'grid short-circuit impedance angle ratio' i.e., X/R ratio. The voltage deviation at the PCC of a wind farm varies as a function of the X/R ratio. As the X/R ratio decreases, the deviation in the voltage profile as a function of rated power increases as shown in Figure 8.7.

When the X/R ratio is plotted as a function of short-circuit ratio it may be seen from Figure 8.8 that the percentage deviation in voltage is low for:

(i) Low X/R ratio and high short-circuit capacity (stiff grid)
(ii) High X/R ratio and low short-circuit capacity (weak grid).

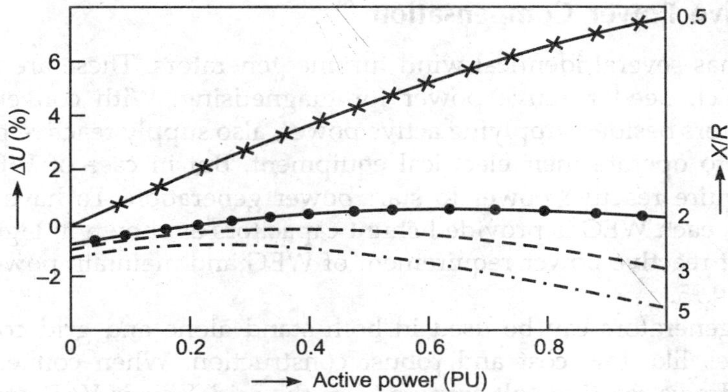

FIGURE 8.7 Graph of X/R between deviation in voltage (DU) and rated power.

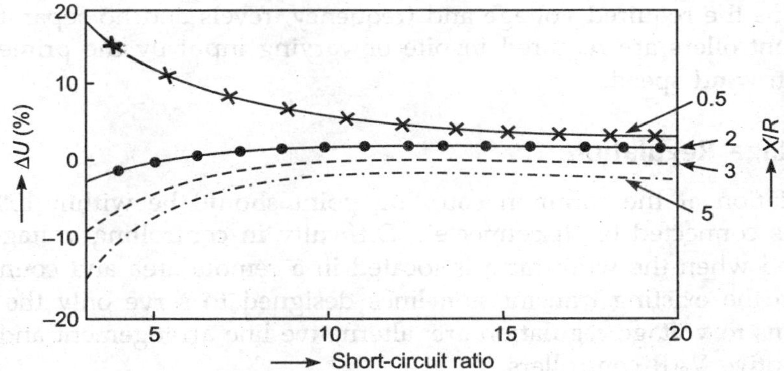

FIGURE 8.8 Graph of X/R between deviation in voltage (DU) and short-circuit ratio.

Thus, it can be concluded that a low X/R ratio calls for a high short-circuit capacity of the grid. Once the grid parameters are known, the generator which suits the grid based on its reactive power requirement, can be opted for.

Wind Electric Generators are designed to operate satisfactorily within the following grid parameters

Voltage   —   400/440 V ± 13%
Frequency   —   50 Hz, –3 Hz, +1 Hz
Asymmetry current ±12.5%

The interface issues with the grid include the following:

1. Reactive power supply
2. Voltage regulation
3. Frequency control.

The above issues are separately discussed below:

8.9.1 Reactive Power Compensation

A wind farm has several identical wind turbine generators. These are induction type generators which need reactive power for magnetising. With conventional energy system, generators besides supplying active power, also supply reactive power required by consumers to operate their electrical equipment. But in case of WEGs (induction type), they require reactive power to start power generation. To have availability of reactive power, each WEG is provided shunt capacitors as shown in Figure 8.10. These capacitors meet reactive power requirement of WEG and maintain power factor at the rated value of 0.95.

Induction generators can be used in both stand alone and grid connected mode with advantages like low cost and robust construction. When connected with grid system, WEG draws reactive volt ampere from the grid. This is VAR drain on the grid system which is compensated by the use of terminal capacitors as explained above. Generated power is fed to local 11 kV grid through 0.415/11 kV transformer. The grid maintains the required voltage and frequency, revels and no separate voltage or frequency controllers are required inspite of varying input by the prime mover due to changes in wind speed.

8.9.2 Voltage Regulation

Voltage variation at the common coupling point should be within 15% when the wind farm is connected or disconnected. Difficulty in controlling voltage regulation is accentuated when the wind farm is located in a remote area and connected to the grid through the existing transmission lines designed to serve only the load in the area. Solutions to voltage regulation are: alternative line arrangement and addition of static or adaptive VAR controllers.

8.9.3 Frequency Control

Utilities operating wind power plants connected to a weak, isolated grid, can have difficulty in maintaining the normal system frequency of 50 Hz. The system frequency shows fluctuations when gusting winds cause the power output of wind plants to change rapidly. Low frequency operation affects the output of WEGs in two ways:

- Several WEGs do not get cut in when the frequency is less than 48 Hz, thus resulting in loss of output.
- The output of WEGs at low frequency operation is reduced due to low speed of the rotor.

Power generation of a 250 kW WEG at various frequencies is shown in Figure 8.9.

8.10 CAPACITY OF WIND FARMS FOR PENETRATION INTO GRID

Power supply from the wind farm snaps with the drop in wind speed. Therefore, the grid must have the requisite spinning/reserve capacity to keep the continuity of power

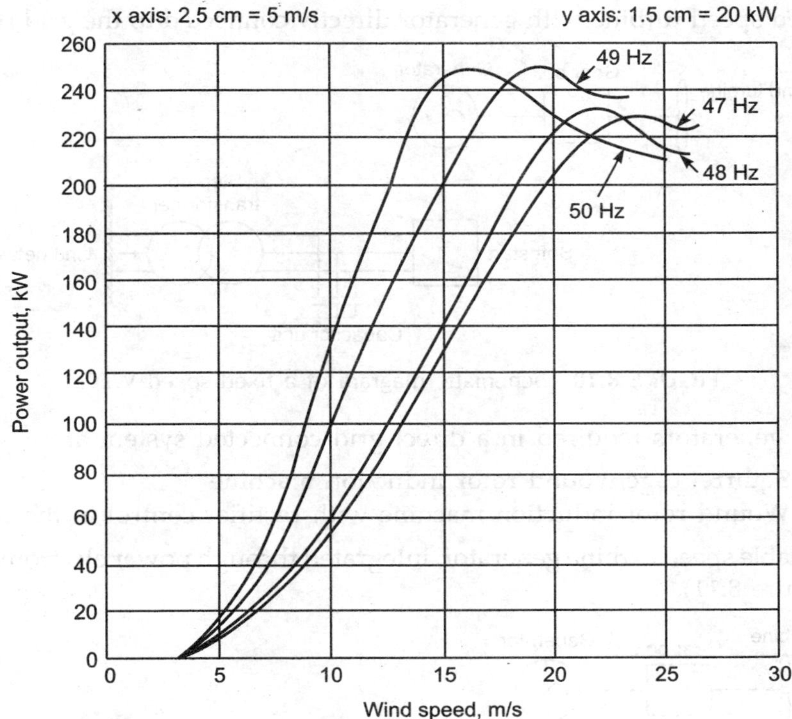

FIGURE 8.9 Power curve of a 250 kW WEG at different frequencies.

supply. Studies conducted in different countries found feasible wind penetration with minimum wind speed and spinning power to the utility as given below:

Holland	below 5%	Australia	20 to 30%
Denmark	over 19%	India	25%
Spain & Portugal	over 11%	Germany & Ireland	over 6%

Global scenario

2020 was the best year in history for the global wind industry with over 93 GW of new capacity (53% more than the preview year). Total wind power installed capacity of the world stood at 743 GW by 2020. 127 countries now have commercial wind power installations. In US and Europe it is planned to carry power from remote wind resources to load centre through EHV lines.

Electrical system of generators with wind turbines

For interaction of wind turbines with the grid, there are two broad classifications as follows:

(a) Fixed speed turbine with generator directly connected to the grid (Figure 8.10).

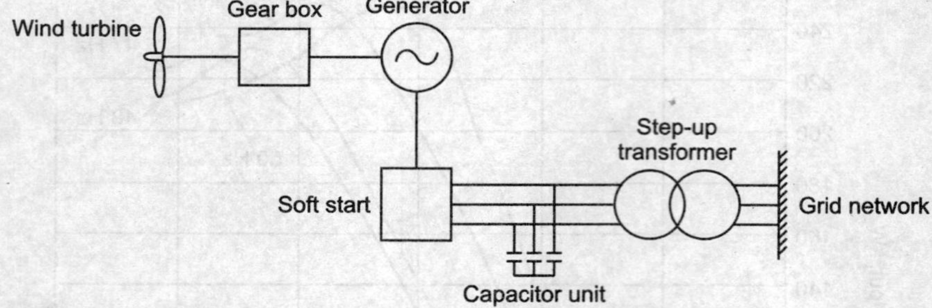

FIGURE 8.10   Schematic diagram of a fixed speed WEG.

The generators required in a direct grid connected system are:

(i) Squirrel cage/wound rotor induction machine

(ii) Wound rotor induction machine with rectifier control in the rotor

(b) Variable speed turbine generator, integrated through power electronic converters (Figure 8.11).

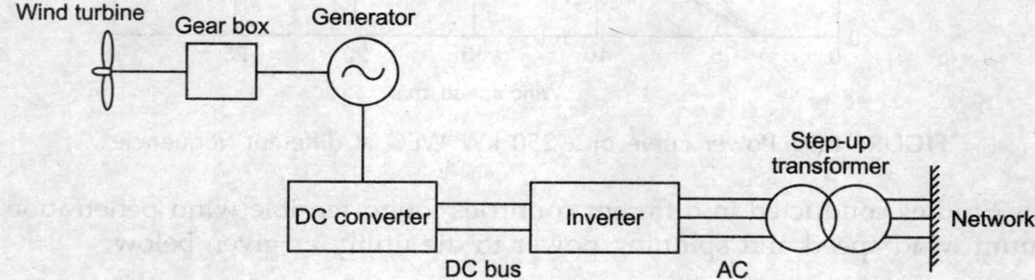

FIGURE 8.11   Schematic diagram of a variable speed WEG.

Under systems integrated through power electronic converters, the generators required are:

(i) Wound rotor induction machine with thyristor/IGBT bridge on the rotor for reduced converter size.

(ii) Permanent magnet synchronous machine with thyristor bridge and permanent magnet/wound rotor machine with IGBT bridge on the stator for full converter size.

8.11   MICROPROCESSOR-BASED CONTROL SYSTEM FOR WIND FARMS

Large wind farms need a fast and an accurate control system to obtain optimum output. A microprocessor-based control system, as shown in Figure 8.12, is used with the grid-connected wind farms. It is equipped with remote control and automatic call facility. The controller can communicate with the wind farm through a PC and a modem on a telephone line. The control system is equipped to change settings and adjust parameters for optimal output.

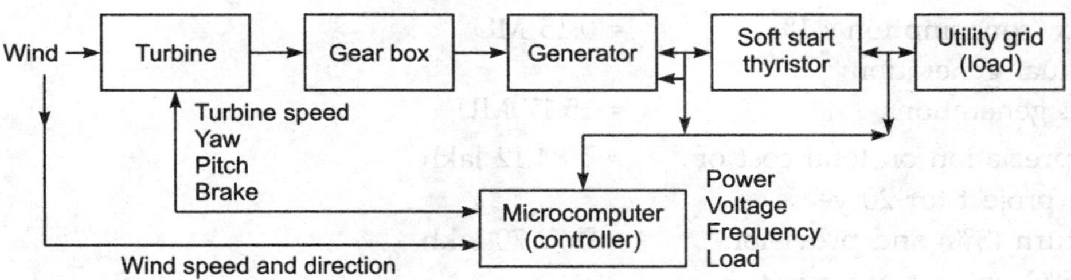

FIGURE 8.12 Block diagram of a wind farm with microprocessor-based control system.

The microcomputer receives the input of wind speed and direction along with load requirement of voltage and frequency. It sends signals to the turbine to establish proper yaw (direction control), blade pitch and to activate the brakes in high winds. The microcomputer may turn on optimal loads in strong winds and can also adjust the power conditioner to change the load voltage and frequency.

8.12 ECONOMICS OF WIND FARMS

Cost economics gives the total cost of installation along with the cost per unit of energy delivered to grid. Methodology to calculate the cost of energy, as adopted by Central Electricity Authority (CEA), which includes the cost of all capital equipment, electrical and civil works is given as:

Project Details

Installed capacity	= 5 MW
Project cost	= ₹ 2000 lakh
Interest rate	= 14%
Construction period	= 6 months
Depreciation amount	= 0.9 project cost × depreciation factor @ 4.5%
Return and provision	= 3.5%
Loan repayment	= 10 years
Moratorium period	= 1 year
Operation and maintenance	= 1% of project cost

Cost Analysis

Cost of the project	= ₹ 2000 lakh
Interest during construction on capital cost	= ₹ 77 lakh
Total cost of the project	= ₹ 2077 lakh

Cost per MW	= ₹ 415.4 lakh
Annual generation (million units)	= 15.32 MU
Aux. consumption @1% annual generation	= 0.15 MU
Net generation	= 15.17 MU
Depreciation on total cost of the project for 20 years	= ₹ 84.12 lakh
Return (3%) and provision (0.5%) on cost of project for 20 years	= ₹ 72.70 lakh
Benefit cost ratio	= 1.86
Cost of generation for 1 year	= ₹ 3.09/kWh
Levalised cost of generation	= ₹ 2.49/kWh
Levelised selling tariff (SEB)	= ₹ 4.17/kWh
Selling price for first year	= ₹ 3.02/kWh
Payback period	= 6 years
Twenty years average cost of generation	= ₹ 2.12/kWh
Twenty years average cost of selling price to SEB	= ₹ 4.99/kWh

Techno-economic analysis confirms wind energy potential, and suitability of the selected WEG for a particular site. This assessment indicates that the project is economically viable.

Concluding remarks

Cost-wise the WEG is nearly ₹ 4.0 crore/MW against ₹ 6.5 crore for a large hydro and ₹ 5.50 crore for thermal power. Annual maintenance of thermal plants is 15% of project cost while wind farms need only 1%. Thus, the economic and the environmental characteristics favour suitability of wind power.

REVIEW QUESTIONS

1. Compile the meteorological data of wind speeds in India and classify the sites into Fair, Good and Excellent wind energy potential sites.
2. Discuss the various methods of estimating the wind energy potential.
3. Explain the planning stages of installing wind power projects.
4. Evaluate the selection of optimum wind energy generation (WEG).
5. What is grid interfacing and how can the grid connection be made?

6. Discuss the issues related to grid interconnection of wind farms.
7. With neat sketches, discuss:

    (a) Fixed speed WEG
    (b) Variable speed WEG

3. Discuss the control system that can be used to monitor a large wind farm interconnected to the grid.
4. Write a review on the potential, the present installed capacity and the future of wind power generation in India.
5. Write a short note elaborating the environmental impact of wind energy.

9

SMALL HYDROPOWER

9.1 INTRODUCTION

Falling water as a source of energy is known from ancient times. It was used to turn water wheels for grinding corn. With industrial development during the 19th century, wooden water wheels were replaced by turbines. With the invention of electricity, water turbines were coupled with generators to produce electrical energy. In India, the first hydropower station of 130 kW was commissioned during 1897 in the hills of Darjeeling in West Bengal. Subsequently, many small hydropower stations were set up utilising canal falls. After independence in 1947, India is marching ahead to develop hydropower as part of multipurpose projects which also provide benefits of irrigation water, industrial and drinking water supply, flood control, and so on.

Hydropower projects essentially harness energy from flowing or falling water in rivers, rivulets, artificially created storage dams or canals. Potential energy in water is converted into shaft work utilising a hydraulic prime mover. Electrical energy is obtained from an electric generator coupled to the shaft of the prime mover.

Attention has also been focused on smaller size and dispersed sites oriented hydro-electric power plants. Small hydro is environmentally benign, operationally flexible, useful for stand-alone applications in isolated remote areas.

9.2 POWER EQUATION

There are two main parameters, i.e., the quantity of water flow per unit time and the vertical fall of water for the determination of the generating potential for a hydro-electric power station. Vertical fall (or head) of water may be available due to topography of the site or may be created by constructing a dam. Water flow is available in perennial rivers, canal or rain-fed systems.

The amount of electric power generated (measured in kilowatts) is proportional to the product of net head (metre) and flow in cubic metre per second. Power generated in kW is expressed by

$$P = 9.81QH\eta \tag{9.1}$$

where

Q = discharge through turbine, in m^3/s
H = net head, in m
η = system efficiency, in %.

Water head in hydraulic systems is defined in several ways as follows:

Gross head: It is the difference in level from the upper surface of water at the highest usable point to the lowest level at the discharge side of the turbine when no water is flowing.

Net head: It is head of water available for doing work on the turbine. It is the gross head less the hydraulic losses occurred in carrying water to the entrance of the turbine.

Rated head: It is the head at which the turbine produces the rated output at the rated speed.

9.3 CLASSIFICATION OF SMALL HYDROPOWER (SHP) STATIONS

The Central Electricity Authority (CEA) and the Ministry of New and Renewable Energy (MNRE) have classified SHPs depending on capacity range and available head. The classifications are as follows:

Based on capacity (MNRE Report 2005)

Category	Unit size
Micro	Up to 100 kW
Mini	101–1000 kW
Small	1–25 MW

Depending on head

Ultra low head	Below 3 metre
Low head	Above 3 metres and up to 40 metre
Medium/high head	Above 40 metre

Field analysis of several small hydro-electric projects revealed a range of suitable net head (m) with water discharge (m^3/s) to generate optimal power as shown in Figure 9.1.

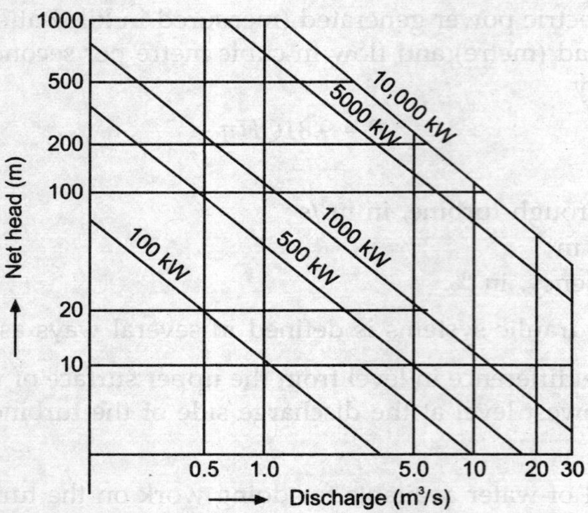

FIGURE 9.1 Graph for hydropower units—net head (m) vs discharge (m3/s)—to generate power.

9.4 CLASSIFICATION OF WATER TURBINES

Water turbines are classified based on the action of flowing water on turbine blades, the existing head and the quantity of water available, the direction of water flow on turbine blades, and the name of the inventor.

Broadly, water turbines are divided into two classes—reaction and impulse turbines with further sub-divisions as low, medium, and high head turbines (Figure 9.2).

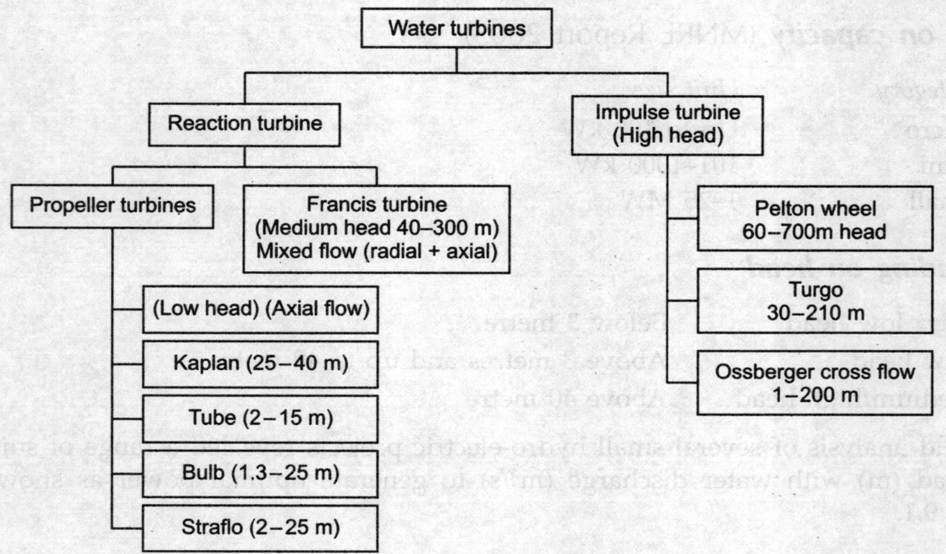

FIGURE 9.2 Classification of water turbines.

Small hydro is characterized with low head and nominal water flow. Net head available to the turbine leads to the selection of the type of turbine, and the rate of water flow determines the capacity of the turbine.

9.4.1 Reaction Turbines

The essential features of medium- and low-head turbines shall be covered by enumerating the details of Francis turbine. Francis turbine blades are joined to two rims 1 and 2 as shown in Figure 9.3(a) and are especially shaped [Figure 9.3(b)] to ensure maximum extraction of energy from water.

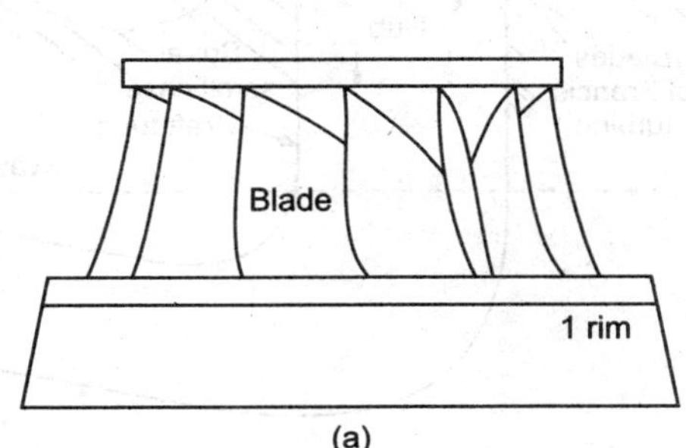

FIGURE 9.3 Francis turbine: (a) front view and (b) bottom view.

The major parts of a Francis turbine system are:

1. Penstock pipe from high water level to scroll casing.
2. Scroll casing provided around turbine welded with penstock on upper side and draft tube on lower side.
3. Guide vanes installed on pivots to control water entering the runner.
4. Turbine wheel with blades, i.e., a runner.
5. Draft tube.

Penstock pipe is provided as passage for water under pressure. It terminates as a spiral scroll case around the turbine, and is welded with draft tube on the discharge side. Guide vanes are arranged on pivots around the turbine, and their degree of opening controls the quantity of water entering the turbine and consequently the power output can be adjusted. The runner of a Francis turbine consists of a number of fixed curved blades, arranged evenly along the circumference of the runner.

Water under pressure enters the runner from the guide vanes towards the centre in radial direction and discharges out of the runner axially. Francis turbine is thus an inward mixed flow (radial + axial) type. Water completely fills the passages between the blades. Energy partly in the kinetic form and partly in the pressure form is imparted to the runner to rotate it as shown in Figure 9.4(a).

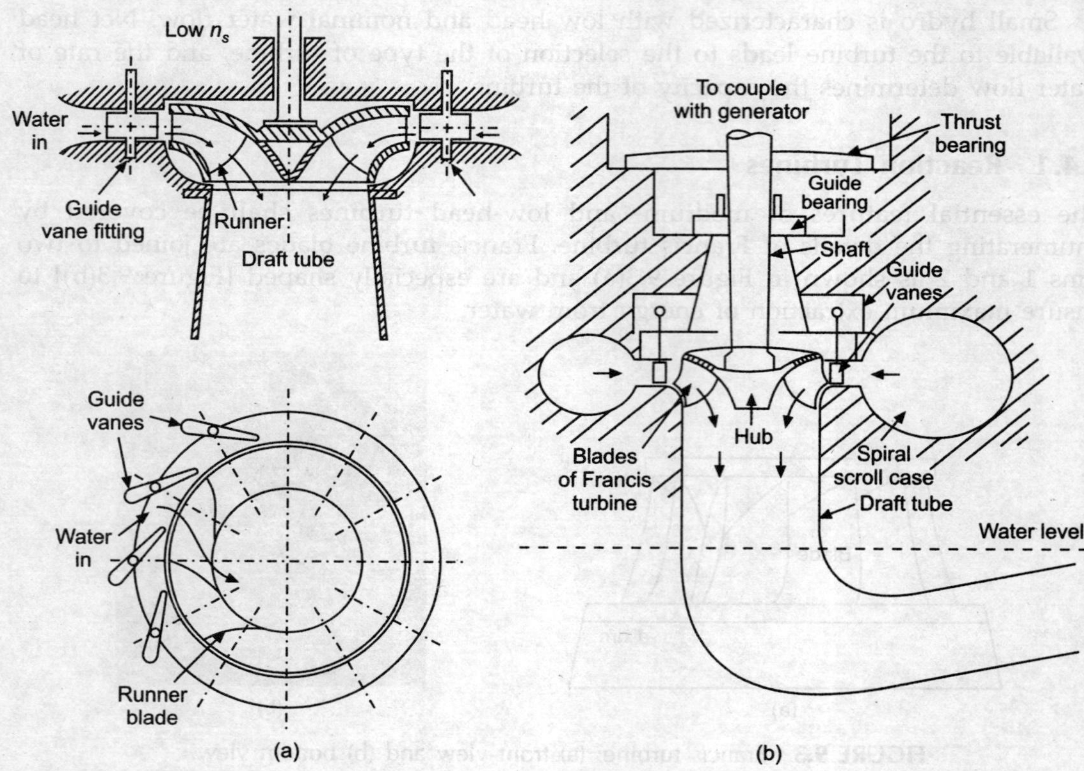

FIGURE 9.4 (a) Flow of water through the guide vanes and runner of Francis turbine, and (b) cross section of Francis turbine.

The draft tube is an outflow bend and an upper taper pipe fabricated of steel plates. It enables the turbine to be installed above the tail race level without losing the head below the runner. Water leaving the runner at certain velocity at low pressure possesses kinetic energy. A large proportion of this energy is recovered by giving the draft tube a suitable taper. The draft tube operates under condition of suction and submerging the lower end of the tube prevents air from entering and destroying the vacuum. The draft tube tapers towards the outlet and the velocity of water gradually reduces, facilitating its discharge smoothly into the tail race as shown in Figure 9.4(b).

9.4.2 Axial Flow Turbines

Axial flow reaction turbines are suitable for low heads and, therefore, need a large quantity of water. These are sub-divided into three types.

(i) *Propeller type:* Propeller turbines are with fixed blades and adjustable guide vanes. Turbine discharge and generator output can be only controlled over a limited range.

(ii) *Semi-Kaplan:* Turbines with adjustable runner blades and fixed guide vanes are called semi-Kaplan. This design offers high efficiencies at several operating points.

(iii) *Kaplan turbine:* Named after the Austrian engineer, V. Kaplan who designed it with adjustable runner blades and guide vanes as shown in Figures 9.5(a) and (b). Runner blades and guide vanes are regulated to variable flow rates. It offers good efficiency even at partial load.

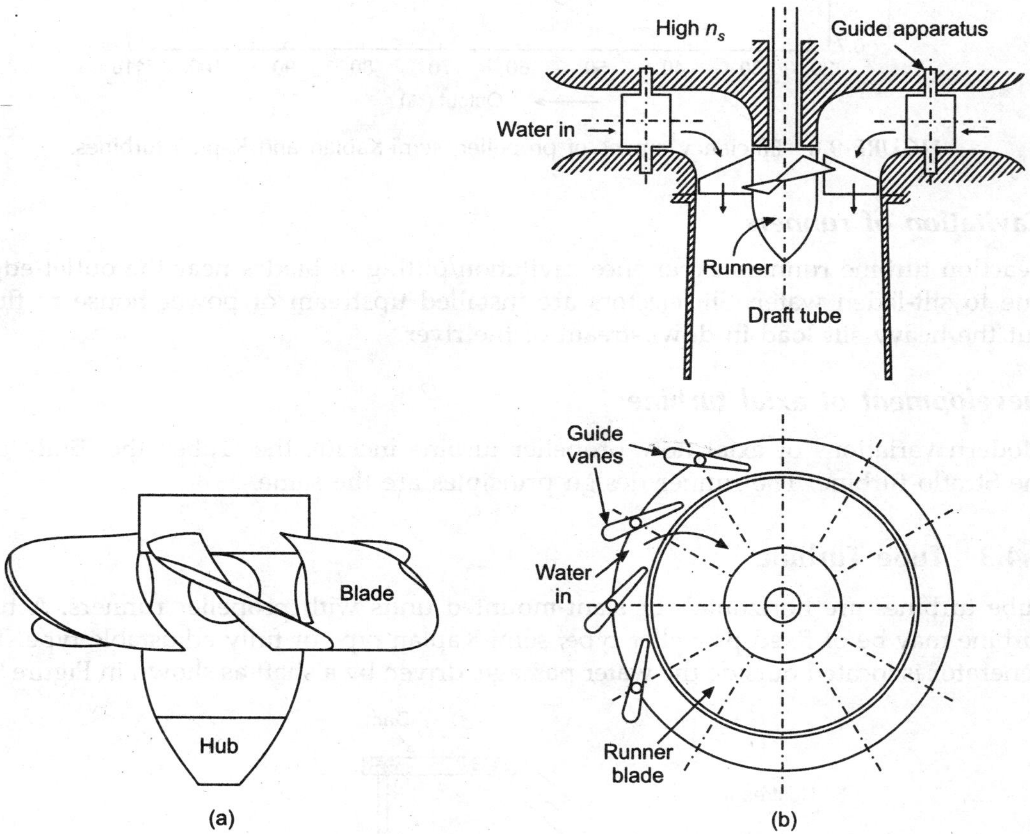

FIGURE 9.5 (a) Kaplan turbine—front view, and (b) flow of water through Kaplan turbine guide vanes and runner.

Figure 9.6 shows the efficiency curve of propeller, semi-Kaplan and Kaplan turbines.

Speed regulation

With bigger units the guide vanes are actuated by hydraulic servomotors while for smaller units the electric motor is used. The governor regulates the speed by control of the guide vane opening, changing the pitch angle of the turbine blade, and matching the load requirement.

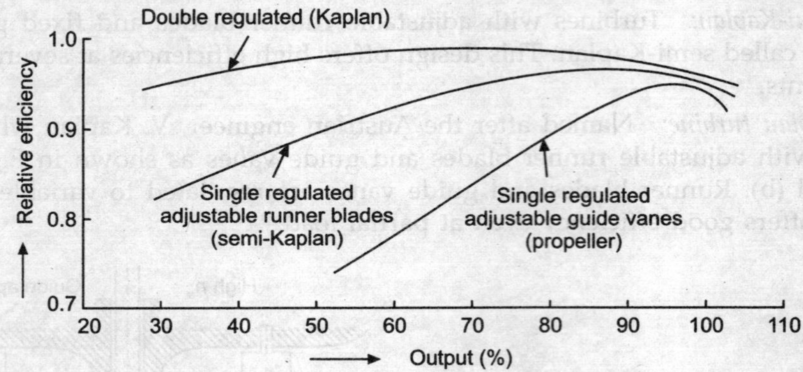

FIGURE 9.6 Efficiency curves of propeller, semi-Kaplan and Kaplan turbines.

Cavitation of runners

Reaction turbine runners experience cavitation/pitting of blades near the outlet edges due to silt-laden water. Silt ejectors are installed upstream of power house to flush out the heavy silt load in downstream of the river.

Development of axial turbine

Modern variations of axial flow propeller turbine include the 'Tube', the 'Bulb' and the Straflo turbine. The runner design principles are the same.

9.4.3   Tube Turbine

Tube turbines are horizontal- or slant-mounted units with propeller runners. A tube turbine may be of fixed propeller type, semi-Kaplan type or fully adjustable type. The generator is located outside the water passage, driven by a shaft as shown in Figure 9.7.

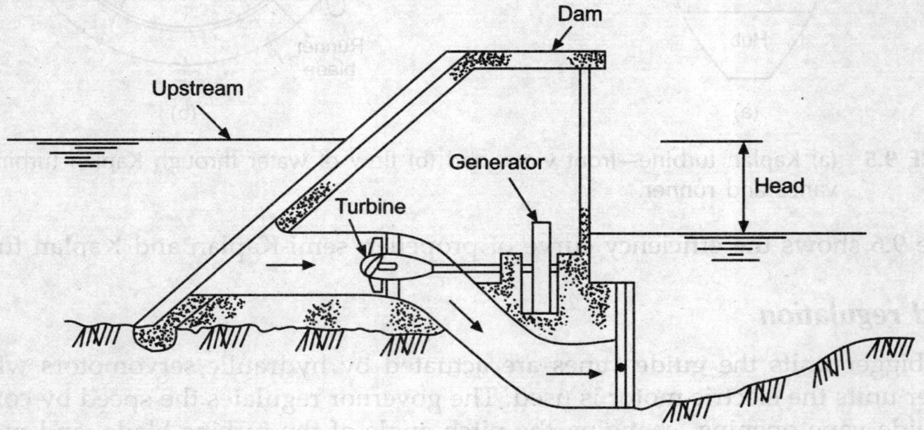

FIGURE 9.7 Tube-turbine installation at a low-head dam.

The performance range of a tube turbine with movable blade runner and fixed guide vanes is good. It operates efficiently between the head range of 2–15 metre especially where the discharge is heavy compared to the head. As the stream flow approaching the runner is axially symmetrical, a higher 'specific speed' can be used with reduction in turbine and generator size. Tube-turbines are available in the range from 5 kW to 700 kW for heads up to 20 metres. A tube-turbine can also be used as a pump.

The requirements of civil works in a powerhouse are reduced as the height and the width required are 60% of the dimensions needed for a conventional turbine and generator.

9.4.4 Bulb Turbine

Bulb turbines are horizontal units that have propeller runners directly connected to the generator. The generator is enclosed in a watertight bulb shaped enclosure. The bulb unit is placed horizontally, completely submerged in the water passage (Figure 9.8).

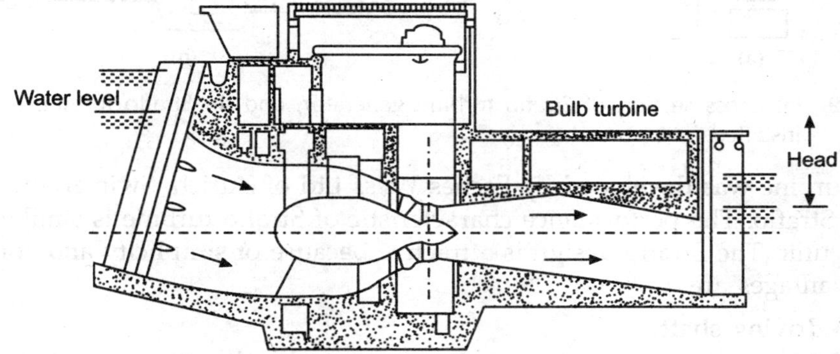

FIGURE 9.8 Bulb turbine installation.

Bulb turbines are available with fixed or adjustable runner blades. The performance characteristics are similar to the vertical adjustable propeller turbine. Bulb units operate efficiently between the head range of 1.25 m to 25 m with a discharge of 3 cumecs to 70 cumecs. Being compact in design, the powerhouse floor space and the height for the bulb turbine installations are minimized. Other advantages over a Kaplan unit are:

- No spiral case
- Friction loss is minimum due to straight draft tube
- Less civil works construction
- Less affected by cavitation
- Higher specific speed.

Bulb units can be used as reversible pump turbine units. This function cannot be performed by conventional units.

9.4.5 Straflo Turbine

A Straflo turbine is one where the generator rotor is mounted at the periphery of the turbine runner, thereby providing minimum obstruction to the flow as shown in Figures 9.9(a) and (b).

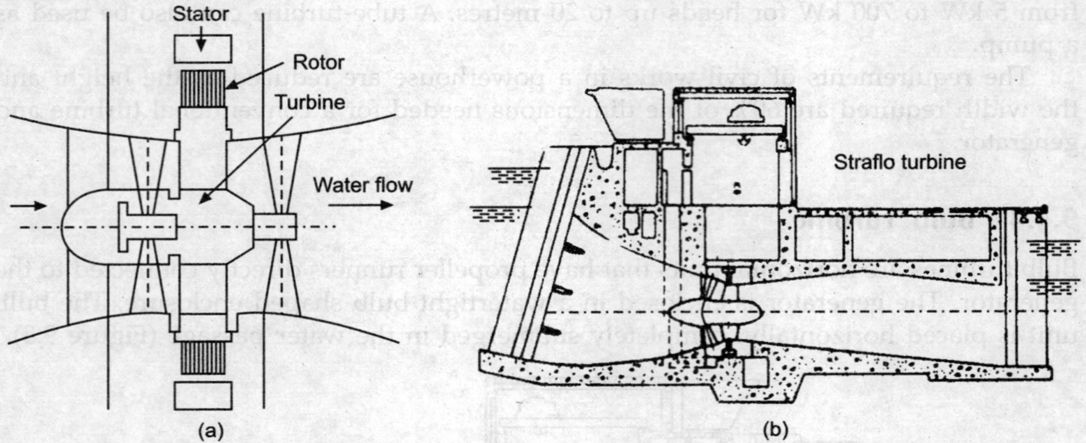

FIGURE 9.9 (a) Cross section of Straflo turbine generator, and (b) Straflo turbine and generator installed in a powerhouse.

This turbine was developed by Esches Wyss Ltd of Zurich, Switzerland and given the name Straflo. The performance characteristic of Straflo turbine is similar to that of the 'Bulb' unit. The Straflo design is attractive because of simplicity and compactness. Other advantages are:

- No driving shaft
- A higher output generator can be accommodated as the same is mounted on the outer periphery
- A larger inherent inertia ensures better stability compared to bulb turbines of the same capacity.

The Straflo unit is suitable for the head range of 2 m–50 m and water flow of 3–20 cumecs. Capacities ranges from 100 kW to 1900 kW.

9.5 IMPULSE TURBINES

An impulse turbine consists of a wheel or runner, with a number of buckets around its periphery. High velocity water, issuing from one or two nozzles, impinges on the buckets causing the wheel to rotate.

The pressure of water before the nozzle causes the energy to be converted into kinetic form that is imparted to the wheel. The turbine is set above the tail water level; water leaving the buckets falls into a pit below the runner and escapes by the tail race as shown in Figure 9.10. The head between the tail race and the nozzle is ineffective for producing power.

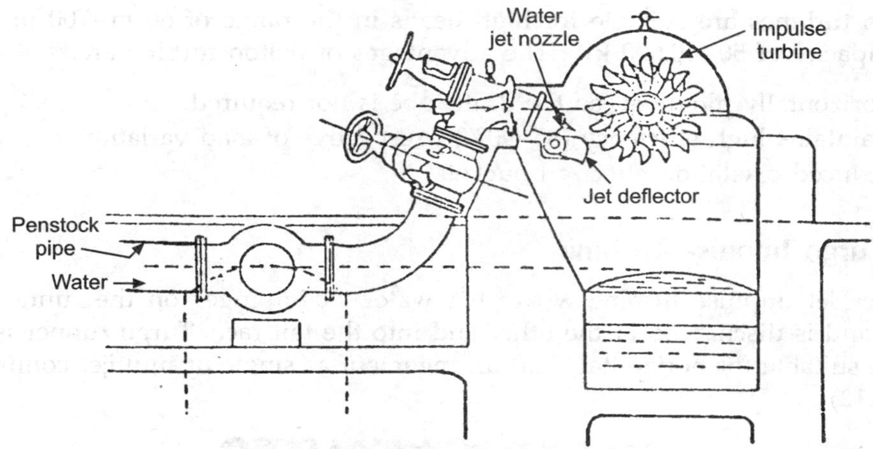

FIGURE 9.10 Impulse turbine mounted horizontally.

Various types of impulse turbines are:

1. Pelton turbine (for high head)
2. Turgo impulse turbine (for medium head)
3. Ossberger crossflow turbine (for low head).

9.5.1 Pelton Turbine

Pelton turbine is installed with a horizontal shaft as shown in Figure 9.10. Buckets are shaped like two spoons placed side by side with a knife edge between them. A jet striking the knife edge gets divided into two equal parts and water is diverted through 180° by the bucket thus transferring energy to the turbine wheel as shown in Figures 9.11(a) and (b).

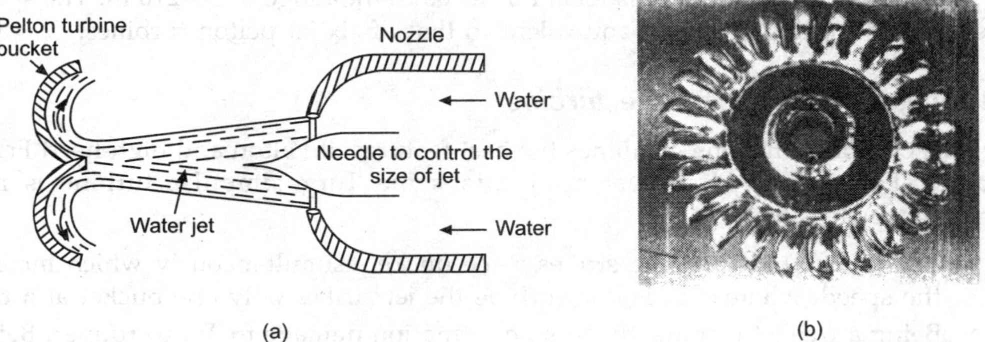

FIGURE 9.11 (a) Water jet and control needle on pelton turbine bucket, and (b) Pelton turbine.

Control of the turbine is maintained by hydraulically operated needle nozzles in each jet. In addition, a jet deflector is provided for emergency shutdown. The deflector diverts the water jet from the buckets to the wall of the pit-liner.

Pelton turbines are suitable for high heads in the range of 60 m–700 m with an output capacity of 50–10,000 kW. The advantages of pelton turbine are:

- Horizontally mounted, so the draft tube is not required.
- Maintains high efficiency (85–90%) irrespective of load variation.
- Reduced cavitation effect on buckets.

9.5.2 Turgo Impulse Turbine

It is a free jet impulse turbine where the water jet impinges on the runner cup at one side and is discharged at the other end into the tail race. Turgo runner is cast in one piece suitable for horizontal shaft arrangement as single or multijet configuration (Figure 9.12).

FIGURE 9.12 Turgo impulse turbine.

The Turgo impulse turbine is ideal for heads in the range of 30–210 m. The specific speed of this turbine is almost equivalent to that of six jet pelton turbines.

Advantages of turgo impulse turbine

The Turgo impulse turbine combines the best features of Pelton and high-head Francis turbines. For small hydropower applications the Turgo impulse turbine is more advantageous because:

- The jet in Turgo turbine strikes three buckets simultaneously which increases the speed, whereas in Pelton turbine the jet strikes only one bucket at a time.
- Being a free jet turbine, there is no cavitation damage to Turgo runner. Being a reaction turbine the cavitation damage in Francis turbine is high particularly for small hydropower applications, where part load operations are predominant.
- Governing the Turgo impulse turbine with a long penstock is possible without making a provision for surge tank/relief valve.

- The efficiency curve of the Turgo impulse turbine is almost flat for a wide range of loads, resulting in efficient part load operation. This does not happen with the Francis turbine.

- Horizontal split-casing of Turgo impulse facilitates easy inspection and repairs as only the top-half of the turbine needs to be removed.

9.5.3 Ossberger Crossflow Turbine

The crossflow turbine is another form of impulse wheel that can be used in low head applications. It was designed by Ossberger Falirik Co. of Germany.

The turbine carries the horizontal shaft, the runner in rotor form has a number of blades; and the length of blades can be changed matching with the output. Blades are curved only in the radial direction, hence no axial thrust is experienced which feature obviates the need of a thrust bearing.

Water enters through a rectangular jet into a cylindrical runner and passes from periphery towards the centre, then after crossing the open centre it moves outwards. As the water passage physically crosses the runner, hence the name given is crossflow.

An exploded view of the typical Ossberger turbine assembly is shown in Figure 9.13(a) and the flow pattern in Figure 9.13(b).

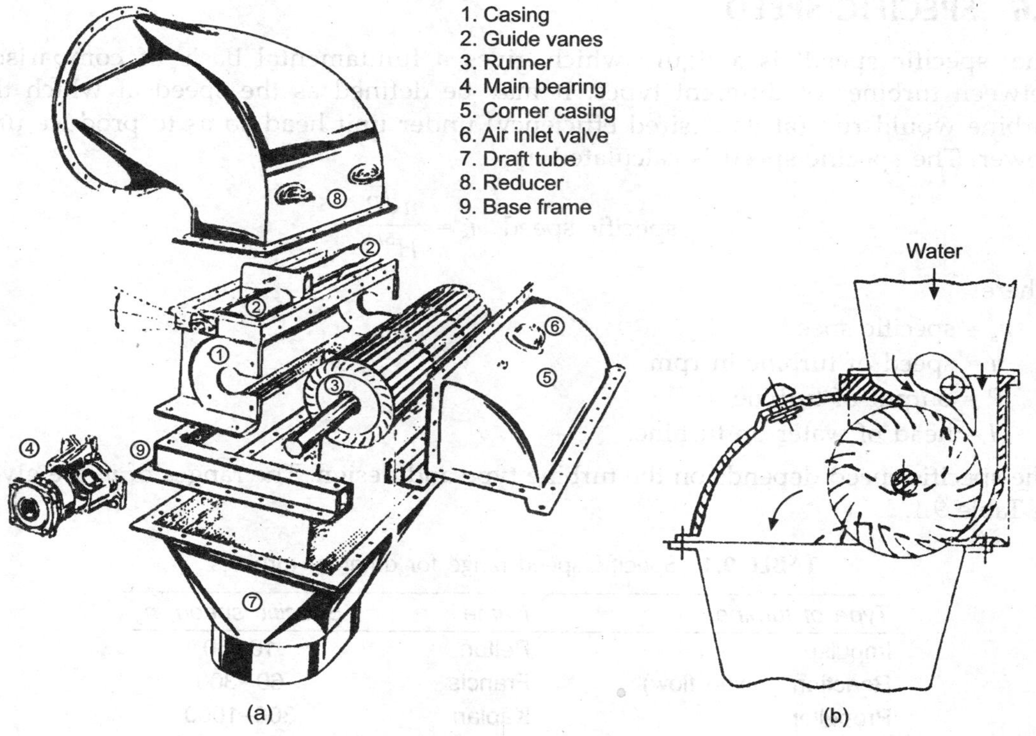

FIGURE 9.13 (a) Exploded view of typical Ossberger turbine assembly, and (b) vertical inlet flow pattern of Ossberger turbine.

For obtaining higher efficiency at part load, the turbine guide vane is split into two valve sections—one covering two-third and the other balance one-third of the runner. At maximum flow conditions, both sections are open. At moderate rates, the two-third section is open and at reduced flow rates only the one-third section of the guide vane is open. The expected peak efficiency of the crossflow turbine is 85%. The allowable head range is from 1–200 m for flow of 0.03–9 cumecs.

Crossflow turbines are equipped with a conical draft tube creating a pressure below atmosphere in the turbine chamber. Therefore, the difference between the turbine centre line elevation and the tail water is not lost to a crossflow turbine as in the case of an impulse turbine. Air is admitted into the chamber through an adjustable air inlet valve which is used to control the pressure. Other advantages are:

- Crossflow turbines are free from cavitation
- Crossflow turbines are suitable for low-head applications where flow is fluctuating
- The efficiency curve is flat over a wide range of flow and head conditions
- Runners are self cleaning
- Crossflow turbines have a less complex structure, hence there is savings in cost.

9.6 SPECIFIC SPEED

The 'specific speed' is a figure which gives a fundamental basis of comparison between turbines of different types. It may be defined as the speed at which the turbine would run (at its desired efficiency) under unit head so as to produce unit power. The specific speed is calculated as

$$\text{specific speed, } n_s = \frac{n\sqrt{P}}{H^{5/4}}$$

where

n_s = specific speed
n = speed of turbine in rpm
P = output of turbine
H = head of water on turbine.

The specific speed depends on the turbine type and design. The ranges of n_s are given in Table 9.1.

TABLE 9.1 Specific speed range for different turbines

Type of turbine	Name	Specific speed, n_s
Impulse	Pelton	10–50
Reaction (mixed flow)	Francis	60–300
Propeller	Kaplan	300–1000
	Bulb	>1000

It infers that high-head operational turbines have a low value of specific speed while low-head turbines have a high value of n_s.

EXAMPLE 9.1

It is required to develop 15,000 kW at 214 rpm under a head of 100 metre with a single runner. What type of turbine should be installed?

Solution

$$n_s = \frac{n\sqrt{P}}{H^{5/4}}$$

$$= \frac{214\sqrt{15000}}{100^{5/4}}$$

or
$$n_s = 83$$

Hence, a reaction turbine should be used.

9.7 RANGE OF APPLICATION OF VARIOUS TYPES OF TURBINES FOR A SMALL HYDRO PROJECT

Having explained the various turbines suitable for small hydropower, it is necessary to select a suitable turbine for a given project. The types of turbines that would be useful at various combinations of head and desired power output are plotted in Figure 9.14 over a range of heads and power from 3–300 m and 10–1000 kW.

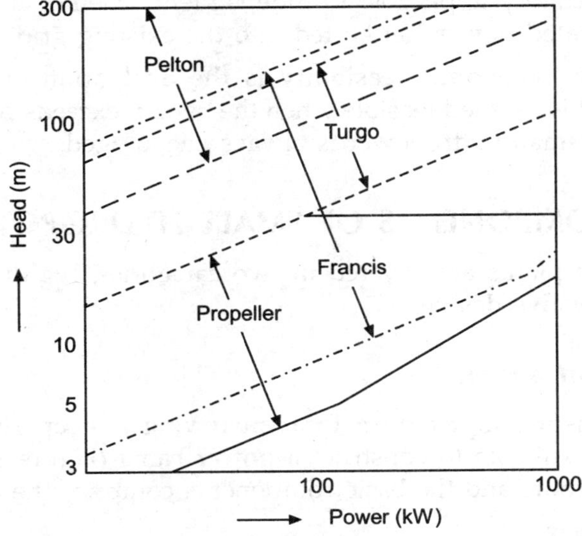

FIGURE 9.14 Range of applications of various turbines.

Figure 9.14 is constructed with the following assumptions:

- The rpm is in the range of 600–3600
- Direct drive
- Specific speed in the range of optimum efficiency for a given design
- At constant n and n_s the head is related to the power $H_V(n/n_s)^{4/5}\,(P)^{2/5}$.

Thus the upper limit represents the maximum rpm and minimum n_s and the lower boundary is determined from the lowest rpm and maximum n_s without cavitating. Cavitation limits are based on a net positive draft head of one atmosphere.

9.8 CIVIL WORKS FOR SMALL HYDROPOWER FACILITIES

Small hydropower projects have distinct attractive features, simplicity in design, short gestation period, environmental friendly with no submergence, resources available locally and suitable for decentralized application.

Having identified a potential site, a feasibility study is conducted to decide whether the project be designed and constructed after considering four major parameters.

Water flow record: Water flow in river or canal should be available round the year to make the project viable to the users.

Available head: High water head schemes (in hilly areas) need smaller quantity of water to produce the desired power. Low-head schemes (in plains) have to handle large quantities of water, consequently the civil structures and the generating plant tend to be comparatively costly.

Location: The project site should be located within a reasonable distance from the users. This is particularly applicable to independent schemes in isolated hilly areas. In plains, the generated power can be fed into the existing grid.

Economic analysis: Economic feasibility is the evaluation of project's costs and benefits. The project is deemed feasible when the benefit exceeds cost. In Indian context of power shortage, small hydropower is always encouraged.

9.9 MAJOR COMPONENTS OF SMALL HYDROPOWER PROJECTS

Small hydro-electric plants are covered in two categories, high/medium head design and low-head innovative design.

High/medium head design

A typical arrangement adopted to a location having a steep river and topography of available land is suitable to construct a power canal (Figure 9.15). A hilly stream traverses the canal route and the basic components comprise the following structures:

- (i) Diversion weir
- (ii) Desilting tank

(iii) Water conductor system
(iv) Forebay
 (v) Penstock, thrust block and surge tank
(vi) Spillway
(vii) Powerhouse
(viii) Tail race.

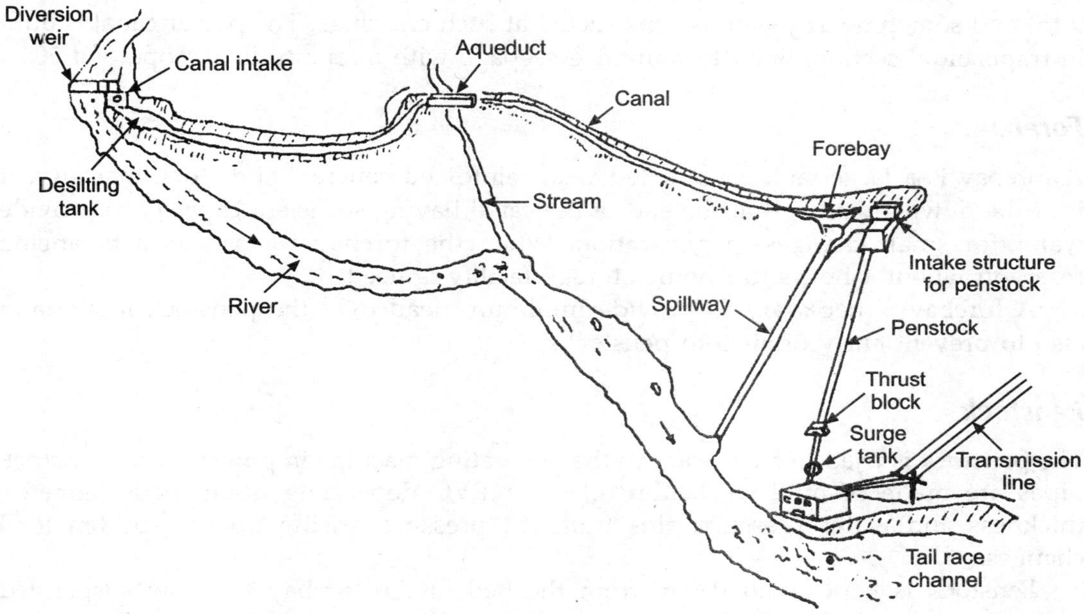

FIGURE 9.15 Typical arrangement of a high/medium head small hydro powerhouse.

Diversion weir and intake

A diversion weir is used to divert river water to intake channel. A trench type diversion weir is used where rock is not available in the river bed. A trench with a grating of iron bars is provided at the bed of stream/river. Water falls in the trench, and large-size sediments roll down the stream. The weir is constructed such that it diverts all the lean season flows and the structure stands safe during monsoon floods.

The intake structure is located at the end of the trench weir and is provided with a gate to control the release of water in power canal.

Desilting tank

A desilting tank is constructed to remove pebbles and coarse suspended material in order to prevent erosion of turbine blades. The abrasion effect increases as the head increases. For high-head turbines, silt size more than 0.2 mm is removed. For medium-head turbines, silt particles more than 0.5 mm size are trapped. The speed of water flow is maintained within 40–60 cm/s to control cavitation. The desilting tank is periodically flushed to remove sediments.

Water conductor system

A water conductor system from intake to forebay is designed with minimum head loss and little water seepage. An **aqueduct** is provided to cross a hilly stream (Figure 9.15) and sometime a syphon is constructed at such crossings. The power canal is built in trapezoidal section, lined to minimize seepage with a longitudinal slope of 1:500.

Forebay

A forebay is a large tank constructed with reinforced concrete and stone masonry. It is at the downstream and at the end of the canal having sufficient capacity to provide water for small changes in generation. When the forebay is used as a balancing reservoir, about 4 hours to 6 hours storage facility is needed.

A forebay is necessary to provide minimum head over the penstock intake and also to prevent entry of air into penstock.

Penstock

The penstock is used to feed water to the generating machine in powerhouse. Penstock pipes are made of mild steel, fibre glass or PVC depending upon their diameter, thickness and water pressure plus transient pressure arising due to sudden load changes.

Penstock is kept about 0.6 m from the bed of the forebay to allow suspended matter to settle which is flushed occasionally. A bell-mouth entry is adopted to reduce head loss and ensure smooth entry of water from the forebay tank into the penstock.

Thrust/anchor blocks are provided wherever penstock changes direction to counteract unbalanced pressure and forces of momentum change as shown in Figure 9.15.

In medium-head powerhouses, where the water conductor pipe length is more than five times the head of the machine, a 'surge tank' is located nearest to the turbine. It consists of a vertical tank reaching to level above that of high water in the reservoir. The lower end of the tank is connected to the penstock. When the load on the turbine is reduced, the governor closes the gates, water level in the surge tank rises and the excessive pressure is prevented, when the load increases the sudden demand of water is met from the surge tank.

Spillway

A spillway arrangement is provided at the penstock intake and it does not allow the water level to rise and flood the area during sudden load rejection. An opening is provided in the forebay at the maximum water level and the spilled water is discharged in the river (Figure 9.15).

Powerhouse

In the powerhouse, turbine generator control panels and auxiliary equipment are installed and operated. A firm foundation for the turbine and the generator is essential. Centre-to-centre distance between machines depends upon the runner diameter. The height of the powerhouse side walls from the floor may be 3 m to 5 m.

Tail race

A tail race is a water channel, used to drain down the water discharged from the draft tube to the river. The tail race must maintain a proper tailwater elevation so as to prevent cavitation and inefficient operation of propeller turbine. From the hydraulic point of view, the water level should be maintained to keep the turbine and the draft tube submerged, otherwise the draft tube vacuum may break and stop the turbine.

9.10 LOW-HEAD SMALL HYDRO PROJECTS

Low-head small hydro projects are situated on perennial run-of-rivers and canal drops. Different types of low-head small hydro powerhouses are discussed below:

9.10.1 Run-of-river Small Hydro Powerhouse

A run-of-river plant is one where a rock-filled dam is constructed across the river with an overflow spillway in one abutment and turbines with generators installed opposite to the abutment as shown in Figure 9.16.

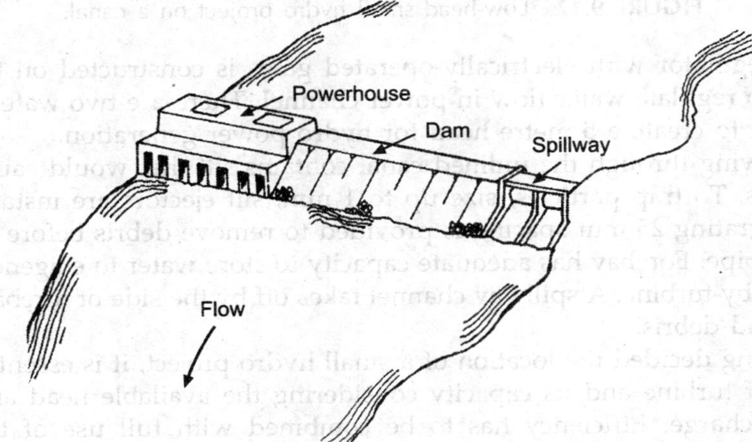

FIGURE 9.16 Typical layout of a dam, spillway and powerhouse for low-head small hydro projects.

The bulb, tube and straflo turbines may be adopted due to their straight-through flow characteristics. The flow of water in the river/stream varies, being minimum during winter and maximum during rainy season. The turbine parameters are selected to match with the minimum water flow conditions in order to operate the turbine uninterrupted round the year.

Flow of flood water tends to produce a negative pressure at the outlet of the turbine draft tube, thus, helping to counteract the tendency of reduction of generating head (due to rise in tail water elevation) during flood flows.

9.10.2 Low-head Small Hydro Project on a Canal

The existing canals constructed for irrigation usually have 2–3 m fall after every 7 km to 10 km as per the topography of the area. These are attractive sites for small hydro projects either with a single fall or adding two falls by constructing a power channel parallel to the main canal (Figure 9.17).

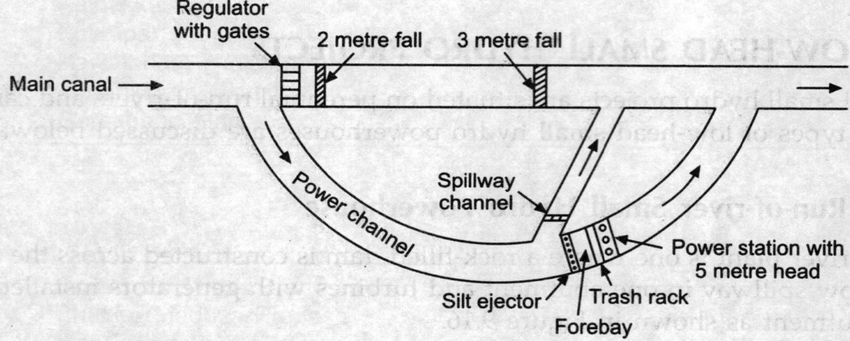

FIGURE 9.17   Low-head small hydro project on a canal.

A head regulator with electrically-operated gates is constructed on the left bank of the canal to regulate water flow in power channel. There are two water falls on the canal, utilized to create a 5 metre head for hydro power generation.

Water flowing through the unlined canal contains silt that would cause pitting of turbine blades. To trap particles, size up to 1 mm, silt ejectors are installed. A trash rack of steel grating 25 mm spacing is provided to remove debris before water enters the penstock pipe. Forebay has adequate capacity to store water in exigency of sudden load rejection by turbine. A spillway channel takes off by the side of forebay to remove flushed silt and debris.

After having decided the location of a small hydro project, it is essential to decide on the type of turbine and its capacity considering the available head and the rated minimum discharge. Efficiency has to be combined with full use of the available hydraulic energy throughout the operation period.

Equation (9.1) gives the approximate discharge requirement as

$$Q = K \frac{P}{H} \text{ cumec}$$

where $K = 0.12$–0.13.

Consider a canal discharge as 50 cumec; then H (head) vs. P (power) diagram shown in Figure 9.14 helps to select propeller turbines. Irrigation canals on annual basis carry variable water flows. A Kaplan turbine with controllable blades and guide vanes will be the correct choice with power output as

$$P = \frac{QH}{K} = \frac{50 \times 5}{0.12} \cong 2000 \text{ kW}$$

9.11 ELECTRIC GENERATORS

The choice of the generator, synchronous or induction, depends on the application. If the small hydro project is to operate in isolation (i.e., away from the state grid), synchronous generators are installed to meet the requirement of consumers, i.e., local community, defence department, telecommunication equipment, tea estates, cottage industries in hilly areas. Indicative parameters of a 1000 kW brushless, three-phase, 50 Hz self-exciting and self-regulating alternator are:

Voltage	:	415 V (380–450 V)
Voltage regulation	:	±1.5%
P.F.	:	0.8
Insulation class	:	F/H
Transient response	:	Less than one second
Excitation system	:	Electronic automatic voltage regulator

Synchronous generators can vary the power factor and contribute reactive power to the system. Two or more generators can be synchronized to feed the full load.

9.11.1 Induction Generator

When a small hydro project is in the vicinity of the state grid, both are interfaced and then less costly induction generators are installed, which obtain their reactive excitation from the grid. During the off-grid time the small hydro plant operates in isolation; a capacitor bank installed specially supplies the excitation power.

A general method of getting the plant on-line is to start the induction generator as a motor with the turbine runner spinning 'dry' and then open the guide vanes of the turbine to load the unit. The unit begins to operate as a generator.

9.12 EXAMPLES OF SMALL HYDRO-ELECTRIC PROJECT INSTALLATION WITH UNIQUE FEATURES

The following are the descriptions of the some small hydro-electric projects in India.

   (i) Micro hydro-electric project in Shansha (Keylong) (Lahaul and Spiti Valley) HP

   (ii) Micro hydel project in Kakroi (Sonepat) Haryana

(iii) Western Yamuna canal hydro-electric project (Powerhouses A, B, C and D) in Yamunanagar (Haryana)

9.12.1 Micro Hydro-electric Project Shansha (Keylong)

This project was envisaged (in 1960) at Shansha (12 km from Keylong), a border tribal area of Himachal Pradesh. The geographical features of the site were:

- It is a snow-bound area across Rohtang pass (4281 metre high). The working period is 4 months only (July to October).

- There was no regular road, so steel and cement were transported on mules. Turbines and generators were taken to site on power wagons.

- Skilled labour was arranged from Punjab area.

- Water supply to the project was obtained by diverting a perennial glacier fed tributary of river Chenab. Thus, there is no upstream reservoir, and the powerhouse operates as run-of-river plant.

The Shansha hydro-electric project was constructed and tested by the author, Er. K.C. Singal, from 1964 to 1966. It was commissioned and inaugurated by the then Governor of Punjab Shri Dharam Vir ICS (Retd) on 24th October, 1966. Two nearby villages were electrified, tribal people felt rejoiced to see electric lights in their houses.

An overview of the project, water channel along the mountain side, bye-pass channel, head tank, pen-stock pipe line, powerhouse building for two turbines and generators and an electric pole are shown in Figure 9.18(a).

(a)

(b)

FIGURE 9.18 (a) An overview of the micro hydro-electric project at Shansha, and (b) Shri Dharam Vir ICS (Retd) the then Governor of Punjab welcomed by Er. K.C. Singal before inauguration.

The major technical parameters are:

Water Conductor System

Flume length	= 198 metre
Penstock steel pipe 61 cm diameter from head tank to powerhouse	= 63.7 metre

Turbine

Francis vertical turbine capacity	= 75 hp
Head	= 23 metre
Water quantity	= 350 litre/s
Speed	= 1000 rpm

Generator

Capacity	= 50 kW
Voltage	= 415 V
Phase	= 3
Poles	= 6
Frequency	= 50 cycles

Governor

The governor consists of a servomotor, the pressure tank oil pump and the guide vane operating mechanism.

Control Panel

The control panel consists of an air circuit breaker, a frequency meter, energy meters and voltage regulating equipment.

9.12.2 Micro Hydel Project in Kakroi (Sonepat) Haryana

The other micro hydel project is located on the Western Yamuna Canal in Kakroi village near Sonepat. The available head is 1.6 metre, this being the lowest head project of Asia. The project was approved by MNRE. Design, procurement and erection was done by the Alternate Hydro Energy Centre, IIT Roorkee while the Micro Hydel Project Kakroi was tested and commissioned by the author, Er. K.C. Singal, the then Superintendent Engineer during October–November 1988. Dr. Maheshwar Dayal, the then Secretary, Govt. of India, MNRE, inspected the Kakroi project on 20th November, 1988 (Figure 9.19(b)). A power channel constructed to utilise the canal fall is shown in Figure 9.19(a).

The technical parameters of the project are:

1. Discharge of power channel is 31.6 cumec. Each turbine requires 10.2 cumec water. Generation is at 415 V and plant is synchronized with the HSEB grid.

2. **Units**

    First Unit: Split-type tubular turbine coupled with a synchronous generator of Voest Alpine (Austria)

Second Unit: Tubular turbine coupled with an induction generator of BHEL (India)

Third Unit: Bulb turbine coupled with a synchronous generator of ESSEX (USA)

The cost of the project was about ₹ 1.62 crore and its generation capacity is 1.5 lakh units per month.

(a)　　　　　　　　　　　　　　　　(b)

FIGURE 9.19　(a) Power channel and control gates of Kakroi hydel project, and (b) Dr. Maheshwar Dayal, and Er. K.C. Singal discussing the project.

9.12.3　Western Yamuna Canal Small Hydro-electric Project

The Western Yamuna Canal Hydro-electric Project is a low-head small hydro project constructed in Yamunanagar district of Haryana. The Western Yamuna Canal had a number of small falls between Hathnikund and Dadupur where it enters the plains. The terrain has a good natural slope affording over 52 m of difference in elevation. The project was planned to develop power by utilising this total fall with installation of the following three power stations in cascade.

Stage I

Powerhouse A (2 × 8 MW)　—　(RD-3000 m) Fall of 12.8 m
Powerhouse B (2 × 8 MW)　—　(RD-7600 m) Fall of 12.8 m
Powerhouse C (2 × 8 MW)　—　(RD-11600 m) Fall of 12.8 m

Stage II

Powerhouse D (2 × 7.2 MW)　—　(Upstream of Tajewala head work) Fall of 10 m

Hydel canal built from Hathnikund barrage up about 1 km from Tajewala Head Works, then runs parallel to Western Yamuna Canal (WYC) up to Dadupur where the two steams rejoin (Figure 9.20).

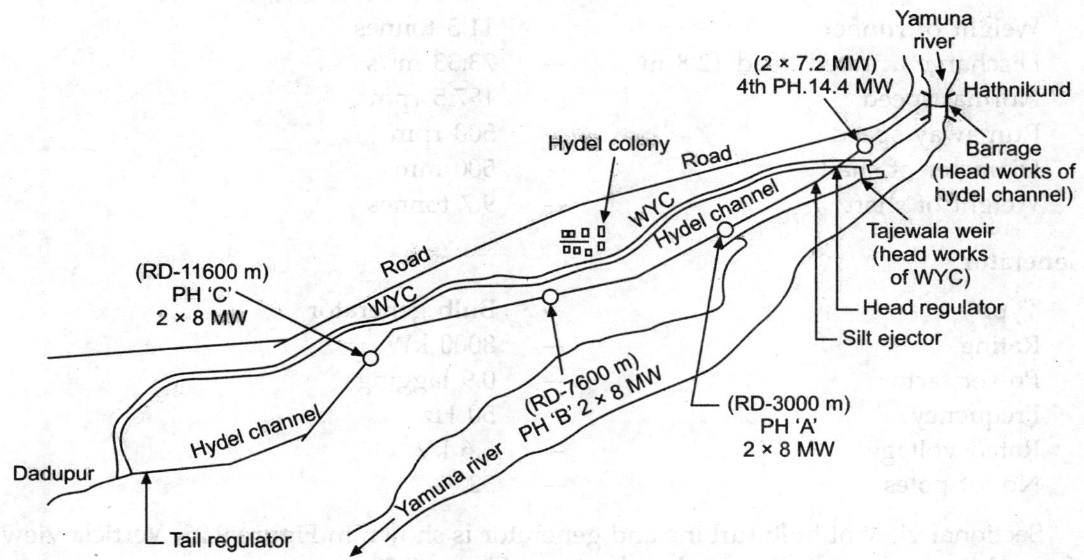

FIGURE 9.20 Layout plan of Western Yamuna Canal—small hydro-electric project.

Powerhouses A, B and C built on power channel, incorporating two identical bulb type generating units each of 8 MW rating, are similar in design. Each powerhouse complex comprises the following features:

- A forebay of the canal and trash racks
- Bypass channel take off from the side of forebay chambers
- Intake gate for each generating unit (6.45 m × 9.75 m)
- Draft tube gate provided for each unit. Intake gates and draft tube gates are operated by 60 T and 20 T gantry cranes respectively. Draft tube gate is first follow up inflamble gate is lighted to releases the water into the turbine.

Powerhouse Structure

Powerhouse is of standard indoor type designed for bulb units comprising an integral concrete structure from intake to draft tube output.

Generating Units

Generating units are of upstream bulb type with adjustable guide vanes and runner blades, manufactured by Fuji Electric Company of Japan. Ratings and other particulars of turbine and generator are:

Turbine

Type	—	Bulb turbine having capacity 9350 kW
No. of blades	—	4
Inlet diameter	—	3.15 metre

Weight of runner	—	11.3 tonnes
Discharge at rated head 12.8 m	—	73.33 m^3/s
Normal speed	—	187.5 rpm
Run-away speed	—	500 rpm
Diameter of shaft	—	500 mm
Weight of shaft	—	9.7 tonnes

Generator

Type	—	Bulb generator
Rating	—	8000 kW
Power factor	—	0.9 lagging
Frequency	—	50 Hz
Rated voltage	—	6.6 kV
No. of poles	—	32

Sectional view of bulb turbine and generator is shown in Figure 9.21. Verticla view of a mchine in power plant is also shown in Figure 9.22

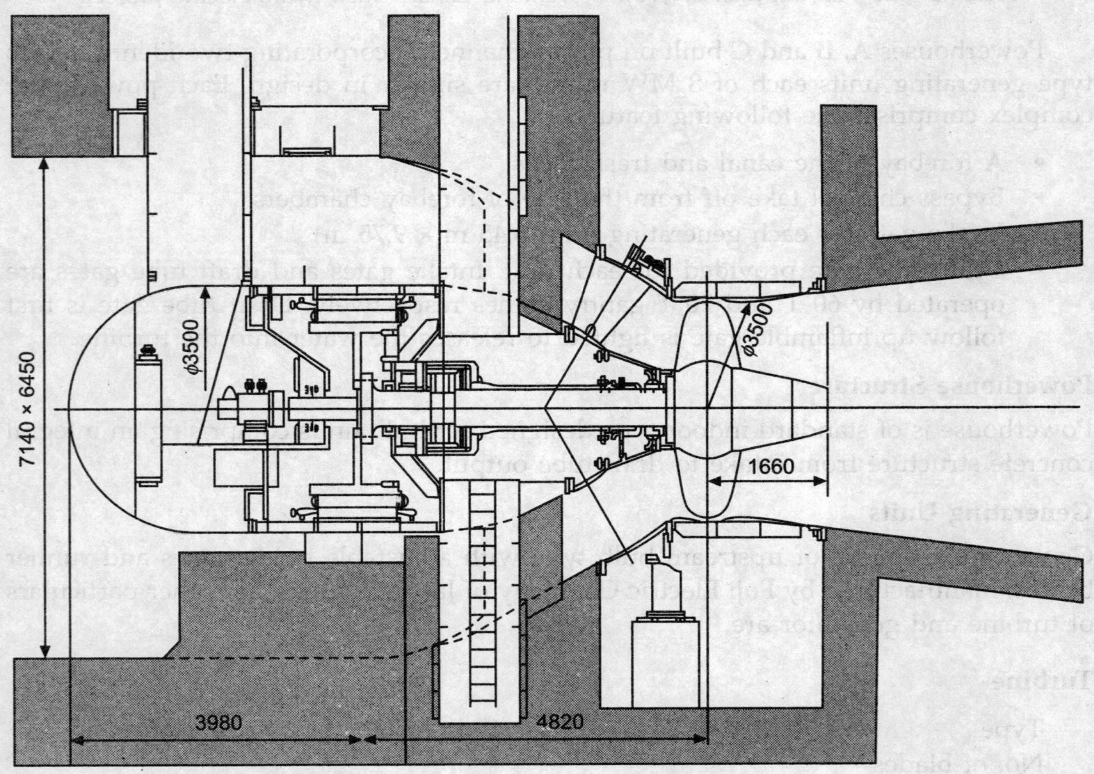

FIGURE 9.21 Sectional view of bulb turbine and generator.

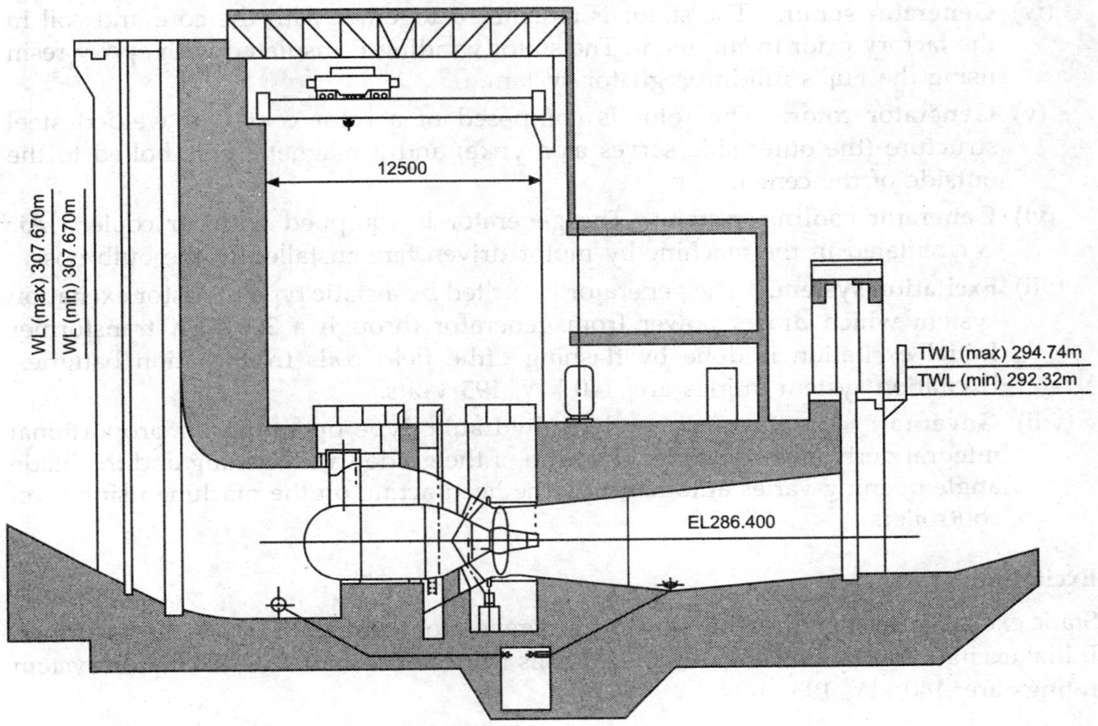

FIGURE 9.22 Vertical sectional view of a machine in a power plant.

9.13 CONSTRUCTION OF BULB TURBINE AND GENERATOR

(i) **Bulb Support System:** The bulb in which the generator is mounted is supported by two stay vanes. The stay vanes are designed to maintain the hydraulic thrust and the generator short-circuit torque applied to the bulb. The upper stay vane is used as a passage for inspection of turbine bearing, shaft, seal, etc. The lower stay vane is provided with oil and water drain pipes.

(ii) **Bearing:** Turbine guide bearing and the generator guide bearing of cylindrical type are mounted on the each side of the shaft. The generator rotor is mounted on the shaft in the manner of overhung. The thrust bearings, both in the normal and counter directions are of Mitchel segment type and are housed in an oil tank together with the generator guide bearing and supported by bearing bracket.

(iii) **Turbine runner:** The turbine runner has 4 blades made of 13% chrome stainless steel. The runner servomotor is housed in a runner boss. The hydraulic oil for the runner servomotor is fed from the oil supply device at the end of the generator auxiliaryshaft.

(iv) **Generator stator:** The stator is assembled together with the core and coil in the factory prior to shipment. The stator winding is insulated with epoxy resin using the Fuji's full-impregnator system.

(v) **Generator rotor:** The rotor is composed of a rotor center of welded steel structure (the outer side serves as a yoke) and a magnetic pole bolted to the outside of the center.

(vi) **Generator cooling system:** The generator is equipped with air coolers. Air is circulated in the machine by motor driven fans installed in the bulb nose.

(vii) **Excitation system:** The generator is excited by a static type thyristor excitation system which draws power from generator through a 300 kVA transformer Initial excitation is done by flashing ofthe field coils from station batteries. Excitation system ratings are: 140 kW, 195 Volts.

(viii) **Governor:** The governor is electrohydraulic type operating on proportional integral derivative principle. The ratio of the guide vane opening and the blade angle opening varies automatically the head acting on the machine using level controllers.

Excitation System

Static excitation system drawing power from generator through a 300 kVA transformer. Initial excitation is by flashing of the field coils from station batteries. Excitation system ratings are: 140 kW, 195 volts.

Governor

The governor is of electro-hydraulic type operating on proportional integral derivative principle. The ratio of the guide vane opening and the blade angle opening varies automatically the head acting on the machine using level controllers.

Power Evacuation

Each generating unit is provided with a 6.6 kV/66 kV, 10 MVA step-up transformer. Power from this project (5.5 lakh units per day from all the four powerhouses) is fed into the Haryana State Electricity Board grid which is connected to the Northern Regional grid.

	Stage I *Powerhouse A, B & C*		Stage II (4th unit)	
Cost of project	₹ 45.71	crore	₹ 12.47	crore
Annual Energy Generation	336	GWh	78	GWh
Cost of generation (1980 estimate)	19	paise/kWh	19.7	paise/kWh

9.14 GLOBAL SCENARIO OF SMALL HYDRO POWER (SHP)

There is no globally acceptable definition of SHP. In India SHP is considered up to 25MW, in Brazil it is 30MW while in China SHP is to 50 MW.

By the end of year 2019, total world SHP installed capacity rose 78 GW. 54% of this was in China (65 GW) followed by Japan (3.5GW), the United States (3 GW) and India (2.4GW).

Starting in 1958 China has continued a focused policy on small-scale hydropower development at the local level. At present, this is by far the largest capacity in any country. Over 300 million people in China enjoy the benefit of electrification through SHP.

Environmental Impact

Small hydro projects are environment friendly and non-polluting. Due to the environmentally benign nature of SHP, MOEF has exempted hydel projects of less than ₹ 500 million from environmental clearance.

However, ecological flow is to be maintained in run-off SHP, especially during ban months, i.e. when snow melting reduces and water flow in the river reduces. Ecological flow is the minimum water flow that should be released from the hydropower project in the river to maintain the ecosystems and sustain livelihoods dependent on the river. These include drinking water, irrigation need and fish harvesting by the people living in the area around.

9.14.1 Small Hydropower (SHP) in India

The Ministry of New and Renewable Energy (MNRE) is assigned with responsibility of development and promotion of SHPs. The estimated potential of the country for small hydropower covering up to 25MW capacity is 19749MW from 6474 identified sites. The aggregate installed capacity till June 2019, was nearly 4,604.81 MW. 116 SHP projects totaling 589.99 MW are under construction. The cost of SHP projects varies between ₹ 8-11 crore per MW depending upon the location and its site topography.

Case study 1: The Ministry at present is implementing a project titled 'Ladakh Renewable Energy' to minimize dependence on diesel in the Ladakh cold region and to meet power requirement through renewable energy sources locally available. The project envisages setting up of 22 small/mini hydel projects with an aggregate capacity of 23.8 MW at a total cost of ₹ 267 crore.

Case Study 2: Ministry approved a package of ₹ 550 crore to electrify 1483 border villages of Arunachal Pradesh. Work has been completed and all households have been provided with home lighting systems.

9.14.2 Economic Viability of SHPs

Small hydro power could be developed economically by simple design of turbines, generators and civil works. Its economic viability can be gauged from the following:

- Small hydro electric schemes have short gestation period.
- Minimum financial needs as compared to other sources of energy.

- SHP projects being low cost development schemes do not cover transmission system for interconnection to the grid. So, these projects are developed to supply a local load connected at distribution voltage.

- With Kyoto Protocol coming into operation, development of energy from SHP become a very important tool in reduction of CO_2 emission as it fetch additional sums of money through sale of carbon credits under Clean Development Mechanism (CDM).

- Minimum operation and maintenance expenses.

- There is a large scope to harness abundant potential of SHP in mountains for developing high and medium head run-of-river schemes, which are connected to the grid.

9.14.3 Advantages of Small Hydro Schemes

1. Small hydro is more consistent compared to other renewables like wind and solar with respect to its availability for power generation. With run-off river a small pond is used to meet the daily peak requirement of power.
2. Small hydro is useful for off-grid, rural in far flung isolated communities having no hope of grid extension for years to come.
3. The SHP programme in India now is largely private investment driven. Generally SHP projects are economically viable and private sector has been showing interest to set up SHP projects. Electrifying villages and providing job opportunities in remote area.
4. Cheap commercial energy in the hills from micro-hydel protect forests, otherwise villagers are largely meet their energy demand from forest wood.
5. Small hydel projects do not encounter the problems as associated with large hydel projects such as deforestation and resettlement of villagers due to submergence.

REVIEW QUESTIONS

1. Briefly discuss the different types of small hydro-power generating plants.
2. Discuss and differentiate between reaction and impulse turbines.
3. Explain the working principle of "Ossberger Cross Flow Turbine".
4. Define specific speed. Find the specific speed when 150 kW power is to be generated under a head of 100 m at 300 rpm. Also, suggest the type of turbine to be used based on specific speed.
5. Describe the major components required for the high/medium head hydro-power projects.
6. Explain the working of a low-head small hydro project on a canal.
7. Write a short note on micro hydro-electric power plants.

8. What types of electric generators are used in small hydro power projects? Discuss in brief.

9. Define the terms: gross head, net head, rated head, total head, cavitation of runners, penstock and spillway.

10. Write short notes on:
    (a) Francis turbine
    (b) Pelton turbine
    (c) Kaplan turbine
    (d) Straflo turbine

8. What types of electric generators are used in small hydro power projects? Discuss in brief.
9. Define the terms gross head, net head, rated head, turbine runners, penstock and spillway.
10. Write short notes on:
(a) Francis turbine
(b) Pelton turbine

10 GEOTHERMAL ENERGY

10.1 INTRODUCTION

The earth is a great reservoir of heat energy in the form of molten interior. Surface manifestation of this heat energy is indicated by hot water springs and geysers discovered at several places. Heat can be experienced from the temperature rise of the earth's crust with increasing depth below the surface. Radial temperature gradient increases proportionally to depth at a rate of about 30°C per km. At a depth of 3–4 km, water bubbles up; while at a depth of 10–15 km the earth's interior is as hot as 1000° to 1200°C. The core of the earth consists of a liquid rock known as **'Magma'** having a temperature of about 4000°C.

This geothermal heat is transferred to the underground reservoir of water which also circulates under the earth's crust. Its heat dissipates into the atmosphere as warm water and the steam vents up through the fissures in the ground as hot springs and geysers. Limitless heat content in magma plus the heat generated by radioactive decay of unstable elements such as K_{40}, Th_{232} and U_{235} which are abundant in the earth's crust are forms of geothermal energy and considered as a renewable energy resource.

10.2 STRUCTURE OF THE EARTH'S INTERIOR

The earth consists of a series of concentric shells. Its internal structure can be divided into three parts—Crust, Mantle and Core—as shown in Figure 10.1.

The crust

The solid crust of the earth is 70–100 km thick and can be divided into continental crust 20–65 km under the continents and oceans crust 7 km under the ocean basins. The study of seismic waves has indicated that the earth's crust underneath the continents

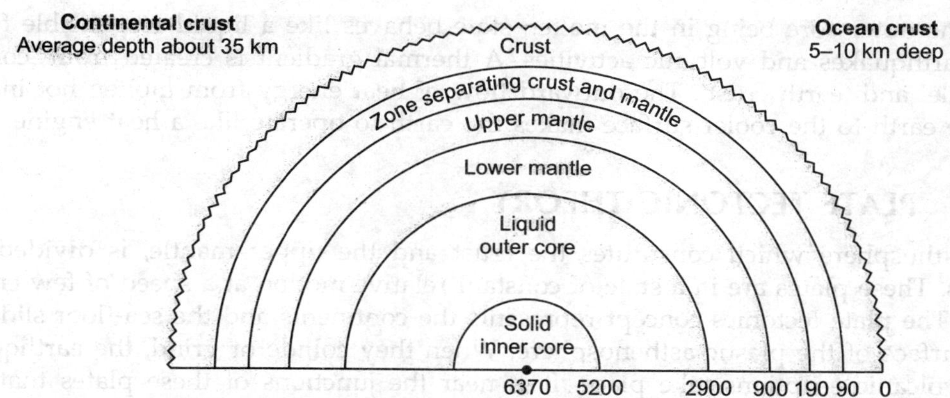

FIGURE 10.1 Half-cross section of the earth. Distances are from the surface in km.

is thicker than that underneath the oceans as seismic waves travel faster in oceanic crust than in continental crust. The oceanic crust consists of low-density rocks (basalt) whereas the continental crust largely contains the granite.

The mantle

The upper rigid part of the mantle extends up to 100 km below the separating crust and contains mainly iron and magnesium. The crust and upper mantle form the 'lithosphere'. The lower mantle extending up to 2900 km below the earth's surface is less rigid and is hotter. This is known as the 'asthenosphere' and is capable of being deformed. The phenomena of plate tectonics, i.e., the movement of the earth's crust is caused by the movement of the lithosphere over the asthenosphere as shown in Figure 10.2.

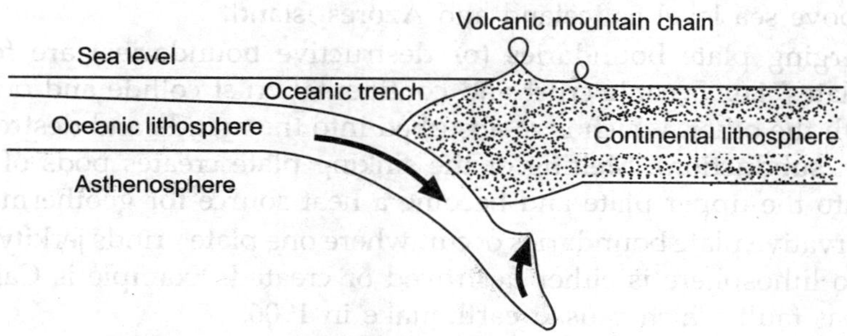

FIGURE 10.2 Movement of the lithosphere over the asthenosphere.

The core

It forms about 33% of the earth's mass and has a radius of 3500 km. The outer core is molten or liquid while the inner core (radius 1170 km) is believed to contain nickel-iron alloy. The hot molten rock of the mantle is called 'Magma'.

The outer core being in the molten state behaves like a liquid responsible for all the earthquakes and volcanic activities. A thermal gradient is created from 'core' to 'mantle' and 'earth crust'. The outward flow of heat energy from molten hot interior of the earth to the cooler surface makes the earth to operate like a heat engine.

10.3 PLATE TECTONIC THEORY

The lithosphere which constitutes the crust and the upper mantle, is divided into plates. These plates are in a state of constant relative motion at a speed of few cm per year. The plate tectonics concept represents the continents and the sea-floor slides on the surface of the plastic asthenosphere. When they collide or grind, the earthquakes and volcanic eruptions take place. It is near the junctions of these plates that heat travels rapidly from the interior magma to surface volcanoes. The active volcanoes are due to geothermal activity.

Most volcanoes and the earthquakes occur in narrow bands along the major dislocations in the earth's substructure that mark the edges of the crustal plates. The oceanic survey revealed the world-wide system of mid-oceans ridges along which new crust is continuously formed, and at the same time, seismographic observations showed that many earthquakes had their focal points beneath the oceans trenches. Movements that produce earthquakes were analysed and it was observed that the plate generated at a mid-oceans ridge was far away, plunging down an earthquake-ridden subduction zone. In this way the boundaries of the plates are identified.

The boundaries between the plates are of three types: (i) diverging plate boundaries, (ii) converging plate boundaries, and (iii) conservative plate boundaries.

(i) Diverging plate boundaries (or constructive boundaries) are formed when two plates move apart, allowing up-welling of molten magma from asthenosphere to create new lithosphere. Thus, mid-oceanic ridges are formed which sometime rise above sea level as Iceland and Azores island.

(ii) Converging plate boundaries (or destructive boundaries) are formed when two plates, i.e., oceanic crust and continental crust collide and one plate sinks beneath the other, which is re-absorbed into the mantle and destroyed (process known subduction). Melting of the sinking plate creates pods of magma that rise into the upper plate and become a heat source for geothermal reservoir.

(iii) Conservative plate boundaries occur, where one plate grinds jerkily past another and no lithosphere is either destroyed or created. Example is California's San Andreas fault which caused earthquake in 1906.

10.4 GEOTHERMAL SITES, EARTHQUAKES AND VOLCANOES

Geothermal resources are associated with tectonic activity, as it allows ground water to be heated with the subsurface heat source. Geothermal fields require a combination of three geological conditions—a natural underground source of water; an impermeable layer that traps water and allows formation of steam; and, a large mass of hot rock in the vicinity of water system. In the plate boundaries, earthquakes, volcanoes and regions of heat flow are largely located.

Most of the world's volcanic activities and geothermal sites are located in the circum-pacific belt known as 'rim of fire'. It starts from New Zealand, encompasses Philippines, Japan, West coasts of North America and Mexico. Another belt runs from Iceland touching the British Isles, through Azores across the Atlantic to the West Indies, with a branch running through the Mediterranean Sea (Figure 10.3).

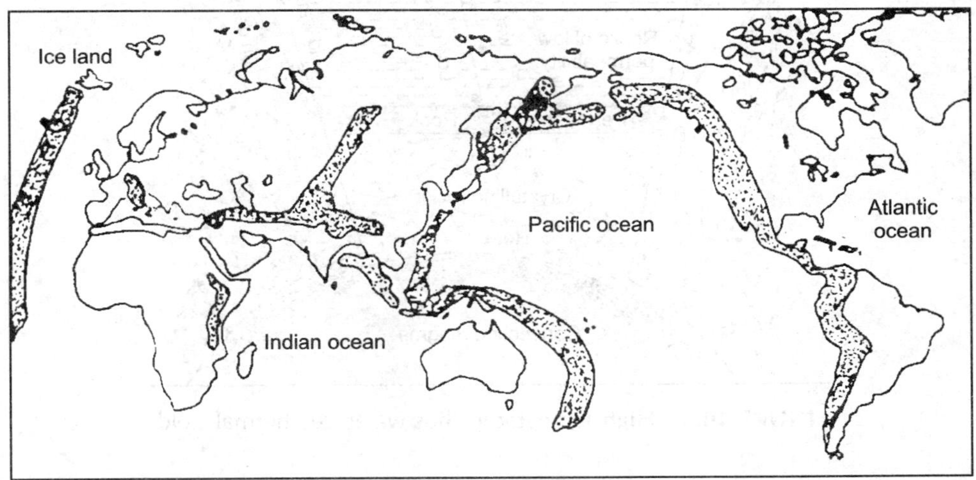

FIGURE 10.3 Regions of geothermal sites, earthquakes and volcanic activity.

Geothermal fields exist in subduction zones where oceanic plateau bend downwards under a continental plate near Japan, Indonesia, New Zealand and Central America. Geothermal sites are also found where collision of continental plates occurs as north west area of Indian–Australian and Eurasian plates. Himalayan geothermal fields on Indian and Chinese side are due to this same reason.

10.5 GEOTHERMAL FIELD

A typical geothermal field is shown in Figure 10.4. Cool rainwater percolates underground from a large surface area (1000 sq. km) and then circulates downwards. At depths of 2 km to 6 km, water is heated by conduction from hot rocks, which in turn are heated by molten rocks.

Water expands on heating and flows buoyantly upwards in a restricted cross-sectional area ($1–50$ km^2). If rocks have many inter-connected fractures or pores, heated water rises rapidly to the surface in the form of hot water springs or shows up as geysers. However, if the upward movement of heated water is impeded by rocks with few fractures and pores; geothermal energy is stored in the reservoir rock below the impeding layers.

Whenever such a site is drilled, steam and hot water gush out through the drilled hole and become a source of geothermal energy for use in a power plant.

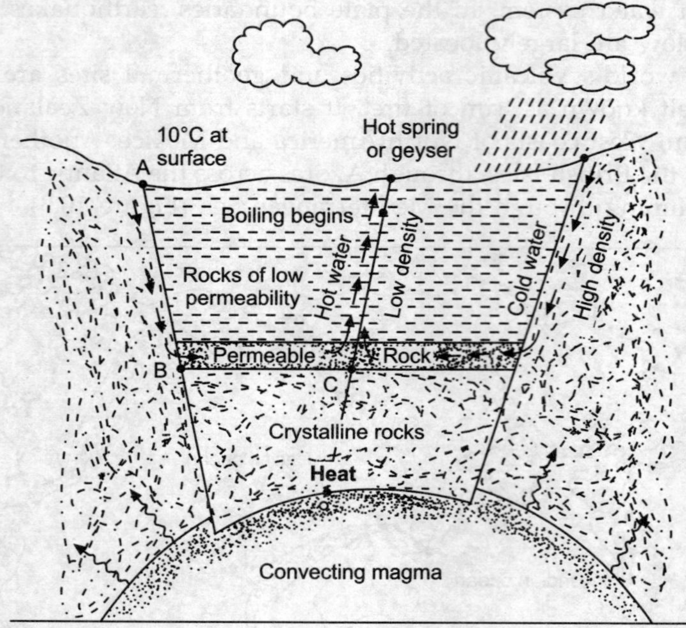

FIGURE 10.4 High temperature hot water geothermal field.

10.6 GEOTHERMAL GRADIENTS

To utilise geothermal energy, a steady rise of the earth's temperature with increasing depth is necessary. It is called geothermal gradient as represented in Figure 10.5.

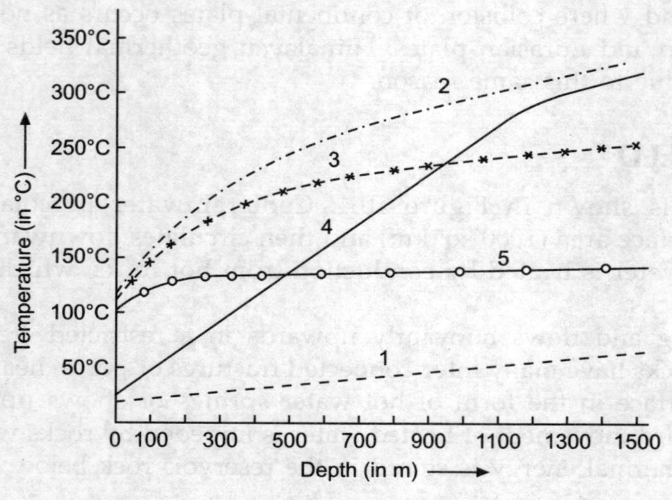

Detail of curves
1. Average gradient 30°C/km
2. Increase in boiling point of water due to rise in pressure
3. Temperature of water in upflowing spring
4. Impermeable rock insulates geothermal reserve
5. Leak in rocks to create springs of hot water or steam gysers

FIGURE 10.5 Geothermal gradients.

The figures are based on measurements within a few km of the earth's surface. The average gradient near the surface is about 30°C/km, as represented by curve 1. Boiling temperature of water is expressed by curve 2 which goes above 100°C with increase in depth, causing a rise in pressure. At locations where the crust is fractured, water percolates downwards, gets heated and gushes upwards in the form of hot springs. It is represented by curve 3, such manifestations of hot water springs exist in Iceland. Curve 4 depicts the effect of impermeable rock which locks up geothermal fluid and does not allow heat flow towards the earth's surface. There are locations with leaks in impermeable rocks, where water generates steam which is released to surface in the form of geysers represented by curve 5. Such phenomena are seen at Lardarello in Italy and geysers of California in the USA. The geothermal gradient is expressed in °C and heat flow in mW/m^2.

10.7 GEOTHERMAL RESOURCES

Geothermal resources are of five types:

1. Hydrothermal

   (a) Hot water
   (b) Wet steam (superheated water from highly pressurized underground reservoirs)

2. Vapour dominated resource
3. Hot dry rock resource
4. Geo-pressured resource
5. Magma resource.

10.7.1 Hydrothermal Resource

Hydrothermal resources (geothermal reservoirs) are hot water or steam reservoirs that can be tapped by drilling to deliver heat to the surface for thermal use or generation of electricity. Such fields exist in zones of structural weakness as given in Figure 10.6.

It may be seen that only a part of the rock is permeable constituting the geo-fluid reservoir, so the field is able to produce commercially a viable resource. Sites of these resources adopt the geographical name of their locality such as Larderallo field in Italy, Wairakei field in New Zealand and Geysers geothermal field in California.

Hot water fields

At these locations hot water below 100°C gushes out as hot spring. The geothermal aquifers being covered by confining layers keep the hot water under pressure. Generally the geothermal water contains sulphur in colloidal form widely used as medicated curative water for skin diseases. In northern India, such a spring exists at Tatapani on the right bank of river Sutlej 54 km from Shimla. Other locations are 'Sahestra Dhara' near Dehradun, sacred kund at Badrinath in Uttarakhand, Sohna sulphur water tank in

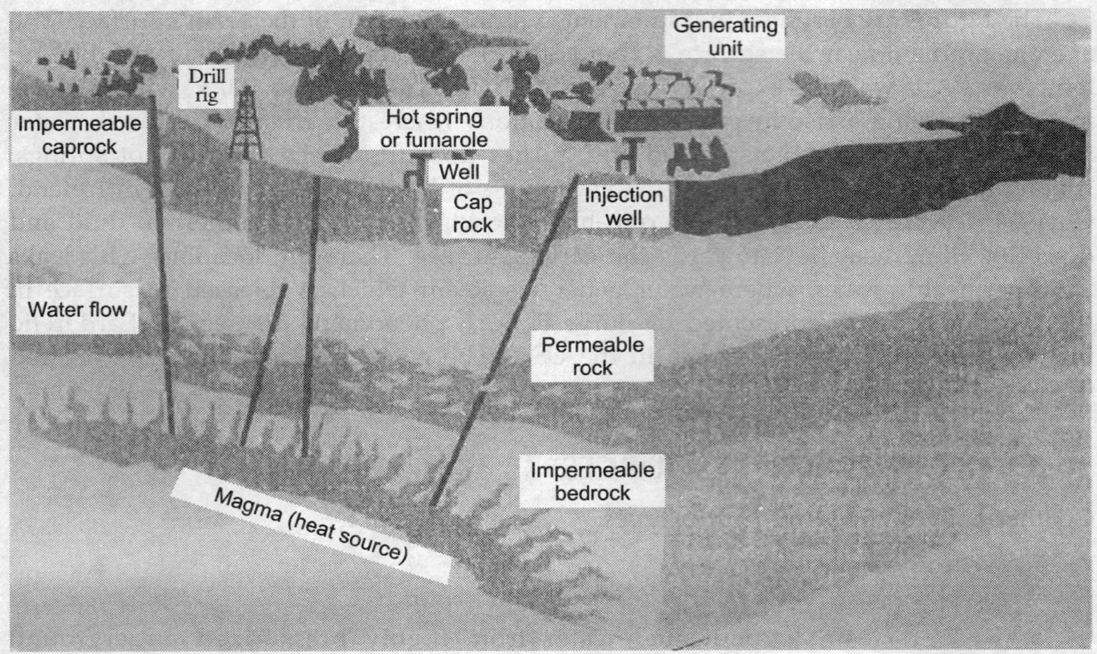

FIGURE 10.6   Cross-section showing the characteristics of a hydrothermal geothermal site.

Gurgaon (Haryana) and Manikaran in Kulu Valley (Himachal Pradesh). Internationally known fields are Pannonian basin (Hungary), Po river valley (Italy) and Klamath Falls Oregon (USA).

Wet steam fields

The pressurized water is at more than 100°C and contains small quantities of steam and vapour in the geothermal reservoir (370°C). With this formation, liquid is in dominant phase that controls pressure in the reservoir. Steam occurs in the form of bubbles surrounded by liquid water. Sites where the steam escapes through cracks in the surface are called 'fumaroles'.

An impermeable cap-rock prevents the fluid from escaping into the atmosphere. Drilling is carried out to bring the fluid to the surface. The fluid is used to produce steam and boiling water in predominant phase.

Examples of wet steam fields generating electrical energy are: Los Azufre (Mexico), Puna (Hawaii, USA), Dieng (Indonesia), Azores (Portugal), Latera (Italy) and Zunil (Guatemala).

10.7.2 Vapour-dominated Resource

Vapour dominated reservoirs produce dry saturated steam of pressure above the atmosphere and at high temperature about 350°C. Water and steam co-exist, but steam is in dominant phase and regulates pressure in the reservoir. Steam obtained from such a geothermal field directly drives a turbine. Major geothermal power plants in the world are: Malsukawa (Japan), The Geysers California (USA), Mt. Amiata (Italy) and Kamojang (Indonesia).

A hot dry rock field also comes under this category. This is the geological formation with high temperature rocks at 650°C, heated by conductive heat flow from magma but contains no water. To tap its energy the impermeable rock is fractured and water is injected to create an artificial reservoir. Water circulates and hot fluid returns to the surface through the other drilled well as steam and hot water which are used to generate electricity.

10.7.3 Geopressured Resource

Geopressured resources contain moderate temperature brines (160°C) containing dissolved methane. These are trapped under high pressure (nearly 1000 bar or 987 atmosphere) in a deep sedimentary formation sealed between impermeable layers of shale and clay at depths of 2000 m–10,000 m. When tapped by boring wells, three sources of energy are available—thermal, mechanical (pressure) and chemical (methane).

Technologies are available to tap geopressured brines as investigated in off-shore wells in Texas and Louisiana at the US Gulf Coast zone up to a depth of nearly 6570 m but have not proved economically competitive. Extensive research is yet to confirm the long-term use of this resource.

10.7.4 Magma

Magma is a molten rock at temperatures ranging from 700°C to 1600°C. This hot viscous liquid comes out at active volcanic vents and solidifies. It may form reservoirs at some depth from the earth's surface. Magma Chambers represent a huge energy source, but the existing technology does not allow recovery of heat from these resources.

10.8 GEOTHERMAL POWER GENERATION

Electric power from geothermal resources can be developed in the following manner.

1. Liquid-dominated resource

    (a) Flashed steam system
    (b) Binary cycle system

2. Vapour-dominated resource

10.8.1 Liquid-dominated Resource

Geothermal fluid is either available from natural outflow or from a bored well. The drilling cost increases greatly with depth and the technically viable depth is 10 km.

Thus, only the geothermal wells of maximum output at shallow depths offer the best prospects for power generation.

Flashed steam system

The choice of geothermal power plant is influenced by brine characteristics and its temperature. For brine temperatures more than 180°C, the geothermal fluid is used. This flashed steam system is suitable for power generation as detailed in Figure 10.7. Geothermal fluid is a mixture of steam and brine, it passes through a flash chamber where a large part of the fluid is converted to steam. Dry saturated steam passes through the turbine coupled with the generator to produce electric power. Hot brine from the flash chamber and the turbine discharge from the condenser are reinjected into the ground. Reinjuction of the spent brine ensures a continuous supply of geothermal fluid from the well.

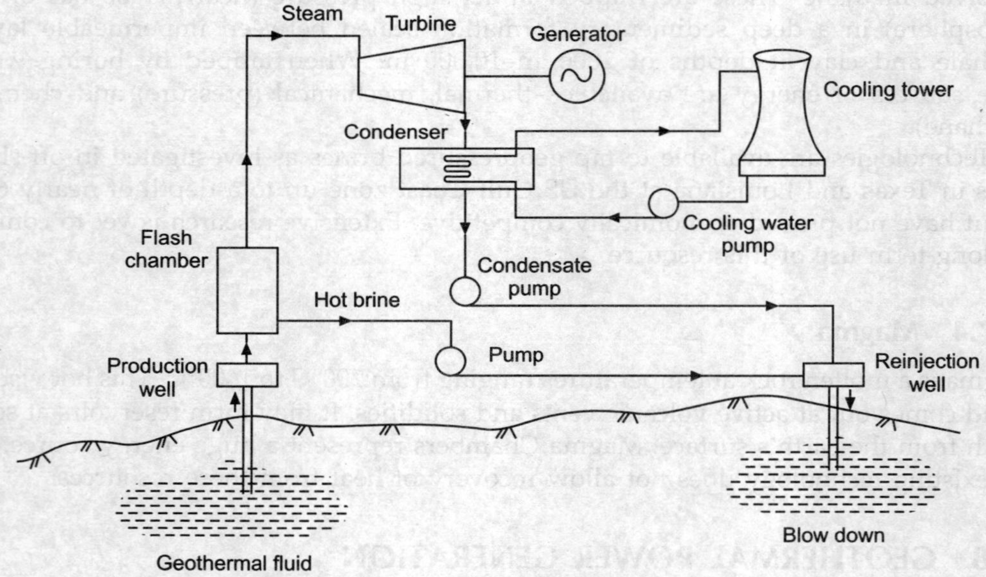

FIGURE 10.7 Flashed steam geothermal power plant.

Commercially available turbogenerator units in the range of 5–20 MWe are in use. To improve the total efficiency of the system, hot water is utilised for poultry farming in cold regions.

Binary cycle system

A binary cycle is used where geothermal fluid is hot water with temperature less than 100°C. This plant operates with a low boiling point working fluid (isobutane, freon) in a thermodynamic closed Rankine cycle. The working fluid is vaporized by geothermal heat in a heat exchanger as shown in Figure 10.8.

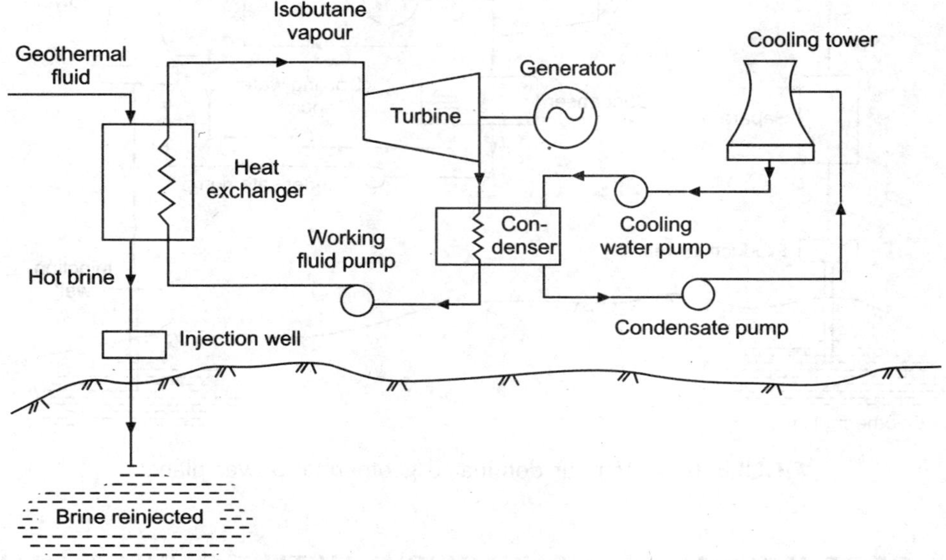

FIGURE 10.8 Binary cycle geothermal power plant.

Vapour expands as it passes through the turbine coupled with the generator. Exhaust vapour is condensed in a water-cooled condenser and recycled through a heat exchanger. Power plants of 11 MW in California and 10 MW at Raft River Idaho USA operate on binary cycle.

10.8.2 Vapour-dominated Geothermal Electric Power Plant

In a vapour-dominated plant, steam is extracted from geothermal wells, passed through a separator to remove particulate contents and flows directly to a steam turbine (Figure 10.9).

Steam that operates the turbine coupled with the generator is at a temperature of about 245°C and pressure 7 kg/cm^2 (7 bar) which are less than those in conventional steam cycle plants (540°C and 130 kg/cm^2). Thus, the efficiency of geothermal plants is low, i.e., about 20%.

Exhaust steam from the turbine passes through a condenser and the water so formed circulates through the cooling tower. It improves the efficiency of the turbine and controls environmental pollution associated with the direct release of steam into the atmosphere. Waste water from the cooling tower sump is reinjected into the geothermal well to ensure continuous supply.

At present such a system is being operated to generate power at Larderallo Italy, and at the Geysers in California.

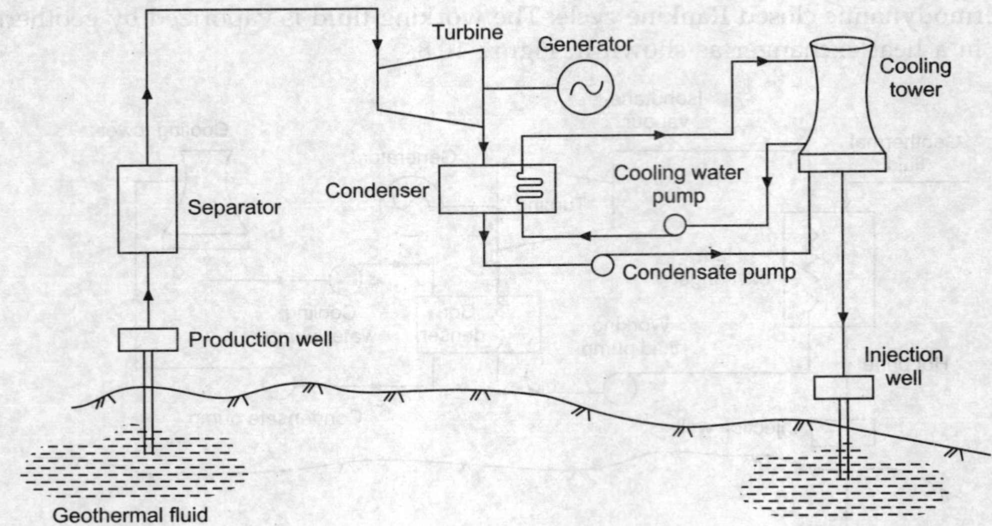

FIGURE 10.9 Vapour dominated geothermal power plant.

10.9 GEOTHERMAL—PREHEAT HYBRID WITH CONVENTIONAL PLANT

Geothermal brine at low temperature is usefully utilised to heat feed water in conventional fossil-fuelled power plants as shown in Figure 10.10. Geothermal heat replaces all low-temperature feed water heaters used ahead of the deaerating heater. Then, the boiler feed pump (BFP) takes over prior to high pressure (HP) feed water heaters which receive heat from the steam bled from the high pressure (HP) turbine. Feed water then flows into the economizer before entering the boiler drum.

10.10 IDENTIFICATION OF GEOTHERMAL RESOURCES IN INDIA

India has good potential for geothermal energy. Govt. of India Ministry of New and Renewable Energy (MNRE) have estimated the capacity to produce 10,600 MW of power. However, geothermal power has not yet been exploited and India does not appear on the geothermal power map of the world.

National Geophysical Research Institute (NGRI) conducted surveys and have identified several sites which are suitable for power generation as well as for direct use. The geothermal springs are clustered in seven provinces given below, and detailed in Figure 10.11.

1. The Himalaya
2. Cambay
3. West coast
4. Son-Narmada-Tapi (SONATA)

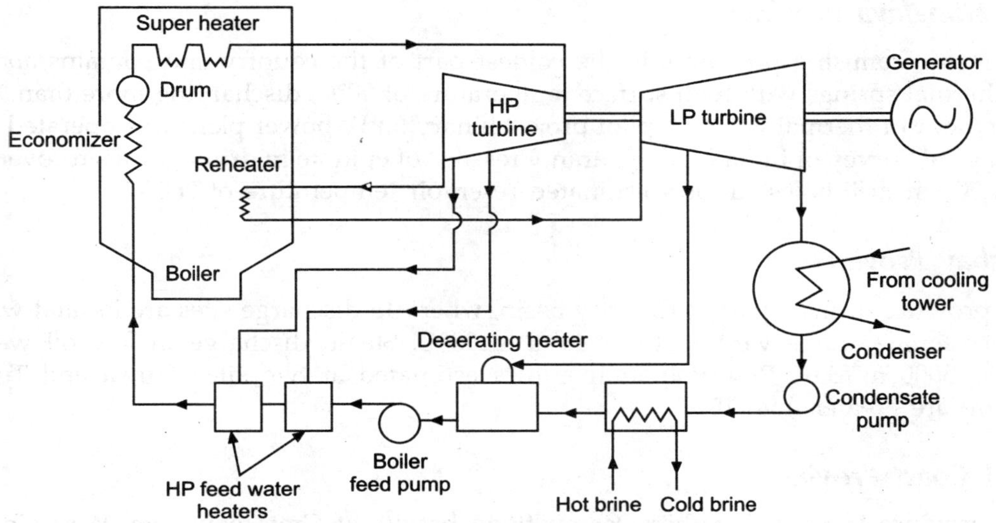

FIGURE 10.10 Schematic diagram of geothermal preheat hybrid with conventional plant.

5. Bakreswar
6. Godavari
7. The Barren Island

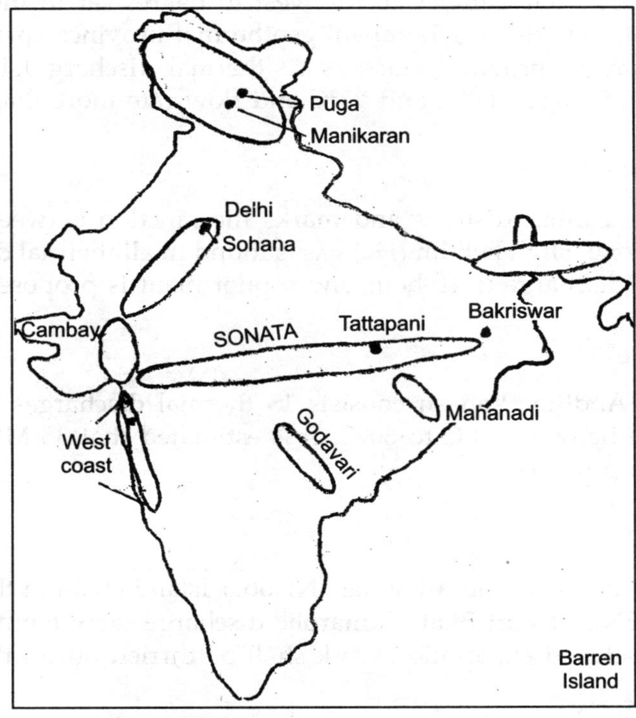

FIGURE 10.11 Geothermal provinces of India.

The Himalaya Province

It is most promising provinces in the coldest part of the country and contains about 100 thermal springs with high surface temperature of 90°C, discharging more than 190 tonne/hour of thermal water. A pilot project binary 5 kW power plant was operated by Geological Survey of India at Mainkaran. Presence of epidote in drill-cuttings recovered from 500 m drill-holes support estimated reservoir temperature of 260°C.

Cambay Province

This province forms a part of Cambay basin, where 15 discharge sites are located with surface temperatures varying from 40°C to 90°C. Steam discharge in few oil wells exceed 3000 m³/day. Reservoir temperature estimated at two sites (Tuwa and Tulsi Shyam) are greater than 15°C.

West Coast Province

This province is located within Deccan flood basalts of Cretaceous age. West Coast province enjoys a thin lithosphere of 18 km thickness, thereby making this province a most promising sites for exploitation.

SONATA Province

This province extends from Cambay in the west to Bakreswar in the east with high geothermal gradient and encloses Tattapani geothermal province spread over an area of 80,000 sqm. Tattapani province encloses 23 thermal discharge sites with surface temperature varying between 60°C and 95°C and flow rate more than 4000 L/min.

Bakreswar Province

It falls in Bengal and Bihar districts and marks the junction between SONATA and Singh bhum shear zone. High helium (He) gas is found in all thermal discharges (water and gases). The He discharge is 4L/hour and a pilot plant is proposed to recover it.

Godavari Province

Godavari valley in Andhra Pradesh consists 13 thermal discharges having range of surface temperature between 50°C to 60°C. It is estimated that 38 MW power can be generated from this province.

The Barren Island

This province forms a part of the Andaman–Nicobar Island chain in the Bay of Bengal and is located 116 ENE of Port Blair. Fumarolic discharge carry temperature between 100°C and 500°C. Detailed exploration work shall be carried out in this province.

10.11 UTILISATION OF GEOTHERMAL ENERGY IN INDIA

Geothermal energy available in India is at low temperature (150°C) and is used for different projects including pilot power plants.

Power generation

A 5 kW pilot geothermal power plant has been installed at Manikaran by the GSI and National Aeronautical Laboratory (NAL), Bangalore. This plant operates on a closed loop Rankine cycle utilising Freon–113 as the working fluid, have been designed and fabricated by NAL.

The National Geophysical Research Institute of Hyderabad conducted Magnets Telluric (MT) studies in Tattapani geothermal field in Chhattisgarh. Based on these findings the installation of a demonstration power plant of 300 kW capacity is under consideration. Similar studies for Puga geothermal fields in Jammu and Kashmir (J&K) by NGRI Hyderabad are in progress.

Indian government has set a target of generating 10 GW electric power through Geothermal by 2030. Intensive survey carried out to find suitable site where geothermal source is in abundance. As per reports of Ministry of New and Renewable Energy Govt. of India 340 sites have been identified suitable for Geothermal power generation. These sites exist in 11 states namely, J&K, Himachal Pradesh, Haryana, Bihar, Jharkhand, Madhya Pradesh, Chhattisgarh, Gujarat, Odisha, Andhra Pradesh and West Bengal. Financial involvement to set up geothermal power generation are:

1. Tentative cost to install 1 MW plant = ₹ 25–30 crore.
2. Expected cost of one unit Energy = ₹ 12.00 kWh (which is high)

Other use of Geothermal Energy

Space heating

Puga (J&K) being at high altitude, experiences low ambient temperatures up to −35°C during winter. Here, a 62.5 m^3 hut is heated with geothermal water, which helps to maintain the inside temperature at 20 ± 2°C.

Extraction and refining of borax and sulphur

Geothermal hot water in Puga valley is used for refining the locally occurring borax and sulphur and for processing of Tsokar lake salt. The extraction plant has the capacity to handle 2 tonne/day of borax ore, while the refining plant can process 500 kg/day of borax. The pilot plant for sulphur refining can process 100 kg sulphur per day.

Greenhouse heating

Geothermal water is used at Chumathang (J&K) for greenhouse cultivation. A suitable temperature (20–25°C) for agriculture production is maintained inside the greenhouse during winter where the outside temperature dips down to −25°C. Several varieties of vegetables and flowers are grown in the greenhouse—a boon to local population.

Refrigeration

A geothermal energy-based absorption refrigeration system operates a 7.5 tonne capacity cold storage plant at Manikaran (HP). The plant uses ammonia as the refrigerant and geothermal water at 90°C.

10.12 GLOBAL STATUS OF ELECTRICITY GENERATION FROM GEOTHERMAL RESOURCES

The geothermal-based electrical energy generation capacity in the world stands at approximately 10715 MWe. A global-level study of renewables in the year 2000 showed that geothermal energy ranked third after small hydro and biomass. There are several countries where geothermal energy is dominant.

Iceland began to use natural hot water in 1930 for greenhouses and domestic space heating. The island is situated on an exposed segment of mid-Altantic ridge, which is a boundary between the Eurasian and the American continental plate. It is rich in geothermal resource—an entire city building of Reykjavik and Hveragerdi town are heated by natural hot water with a distribution pipe line of 64 km. Steam from one of its large geothermal reservoirs was used during 1969 to feed a 17 MWe power generating plant. The total installed geothermal generating capacity stood at 155 MW during the year 2018.

New Zealand is another country that has exploited geothermal energy since early 1960s, where a 192 MWe plant was installed at Wairakei in North Island. In addition, a 110 MW (thermal) plant at Kawerau feeds natural steam to Tasman pulp and paper mill. The present geothermal generating capacity of the country is nearly 1027 MW.

Philippines uses geothermal energy for several low-grade heat (200–250°C) industrial processes. Its installed capacity of geothermal power stood at 1931 MW in the year 2020.

Italy is the first country where steam from Larderallo field was used to produce electricity in 1904. Its installed capacity rose to 127 MW in 1944. The total capacity of geothermal electric power reached 1086 MW in the year 2020.

In Japan, geothermal power production in mid-1960s was 13 MWe at Otak, and 20 MWe at Matsukawa. The installed capacity rose to 546.9 MW in the year 2000 using the Hot Dry Rock (HDR) technology. Currently at 535 MW.

The United States of America started late in geothermal energy extraction and installed 420 MWe near the Geysers field on the West coast. This site is in proximity to the tectonic plate boundaries that gives rise to high temperature gradients, permitting both power generation and direct applications. In the year 2005, the geothermal generating capacity in the USA rose to 2544 MWe—the highest in the world. Many towns in the USA, namely California, San Bernardino, Colorado and Oregon use geothermal energy. Currently at 3.673 GW (in 2020).

There are a few more countries who have done dominant work in installing geothermal generating units as detailed in Table 10.1.

At present, 70 countries of the world use 15.4 GW geothermal energy for space heating, industrial and agricultural applications whereas 26 countries utilise geothermal energy for electricity generation as shown in Table 10.1.

In 2020, United States led the world in geothermal production with 3763 MW of installed capacity from 77 plants. The largest group of geothermal power plants in the world is located at The Geysers, a geothermal field in California. 2nd largest is Indonesia with 2133 MW in 2019. The Philippines is the 3rd highest producer with 1931 MW of capacity.

TABLE 10.1 Installed geothermal electric capacity

Country	Capacity (MW) *in 2019*	Country	Capacity (MW) *in 2019*
United States	3676	El Salvador	204
Indonesia	2133	Costa Rica	166
Philippines	1918	Nicaragua	88
Turkey	1526	Russia	82
New Zealand	1005	Papua-New Guinea	56
Mexico	962.7	Gautemala	52
Italy	944	Portugal	29
Kenya	861	China	24
Iceland	755	France	16
Japan	601	Ethiopia	7.3
Iran	250	Germany	6.6
		Austria	1.4
		Australia	1.1
		Thailand	0.3
		Total	15,406 MW

Geothermal electric plants are built on the edges of tectonic plates where high temperature geothermal resources are available near the surface. However, development of binary cycle power plant and improvements in drilling and extraction technology enable enhanced geothermal system.

10.13 ADVANTAGES OF GEOTHERMAL ENERGY

Various advantages associated with electricity generation from geothermal energy are:

- Electricity generation from geothermal source is pollution free and does not contribute to green house effect.
- It is economical as power stations need small space.
- No fuel is needed, so recurring expenditure is small.
- Once geothermal power station is built, the energy is almost free.
- Geothermal energy is renewable. It is a constant energy source and also ubiquitous, its cost will not rise with time.

- Geothermal electric power plants are on line 97% of the time, whereas nuclear plants average only 65% and coal plants only 75% online time.
- Geothermal plants are modular, and can be installed in increments as required.
- Construction time is only 6 months for plants in the range 0.5 MW to 10 MW, and as little as 2 years for cluster of plants.
- Geothermal plants can be used both as base line and peaking power.
- No CO_2 is generated from geothermal power plant, boon to protect environment.
- Geothermal energy is unlimited and is available 3 to 4 km below the earth crust. Inside temperature varies from 84°C to 98°C (Chhattisgarh) which is sufficient to run a power plant.
- Geothermal plant requires only small land, i.e. 0.75–1.2 acre per MW. In contrary a solar plant needs 5–8 acre land per MW.

REVIEW QUESTIONS

1. What is geothermal energy? What is plate tectonic theory and how is it related to geothermal energy?
2. What do you understand by geothermal fields?
3. How are geothermal sites, earthquakes and volcanoes related?
4. How can geothermal energy be extracted for useful purposes?
5. What are the various types of geothermal resources available?
6. Define and discuss geothermal gradients.
7. Discuss the various ways of geothermal power generation.
8. Discuss the indirect utilisation of geothermal energy.
9. Write short notes on the environmental impacts of geothermal energy.
10. Discuss the global status vis-à-vis the current status of geothermal energy in India.
11. Enumerate advantages of geothermal power plants.

ELECTRIC POWER GENERATION BY OCEAN ENERGY

11.1 INTRODUCTION TO TIDAL ENERGY

All forms of energy available on the earth are, in the first instance, derived from solar energy, with the exception of nuclear, geothermal and tidal energy. Wind, ocean waves, and rivers are driven by the energy from the sun. Coal, oil, gas, wood and grasses are formed by solar energy, which splits carbon dioxide with water to produce cellulose which has either been fossilized (to form coal, oil and gas) or been turned to starch and sugar to produce biomass. In view of the rising prices of fossil and nuclear fuels, combined with adverse environmental impacts with their use in electric power generation, of late *there has been an increased interest in the exploitation of tidal energy.*

11.2 TIDAL CHARACTERISTICS

The tides are caused by the combined attraction of the sun and the moon on the waters of the revolving globe. The effect of the moon is about 2.6 times more than that of the sun, influencing the tides of the oceans. Thus, tide is a periodic rise and fall of the water level of the ocean. Twice during a lunar* day (i.e., within 24 hours 50 minutes) the water in oceans and seas rises and falls. The excess of 50 minutes over the solar day results in the maximum water level, occurring at different times on different days. The amplitude of water level variations at different points on the earth depends on the latitude and the nature of the shore. The rotation of the earth causes two high tides and two low tides to occur daily at any place.

The revolution of the moon around the earth increases the time interval between two successive high tides from 12 hours to about 12 hours and 25 minutes. As the moon revolution takes about 28 days, the three bodies, i.e., the sun, the moon and the earth are in alignment every two weeks at new and full moon. During these periods the sun and the moon act in combination to produce tides of maximum range as shown in Figure 11.1.

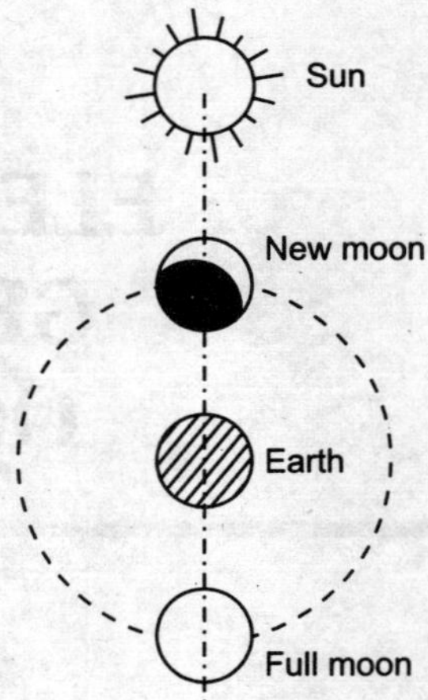

FIGURE 11.1 The sun and the moon acting in combination to creat spring tides.

The solar pull comes in line with the lunar pull at 'New Moon' and 'Full Moon', causing greater flow and ebb, known as spring tides. On the other hand, if the two pulls act at right angles to each other, as at waxing and waning 'Half Moons', i.e., in the first and the third quarters, we get low tides called 'Neap Tides' as shown in Figure 11.2.

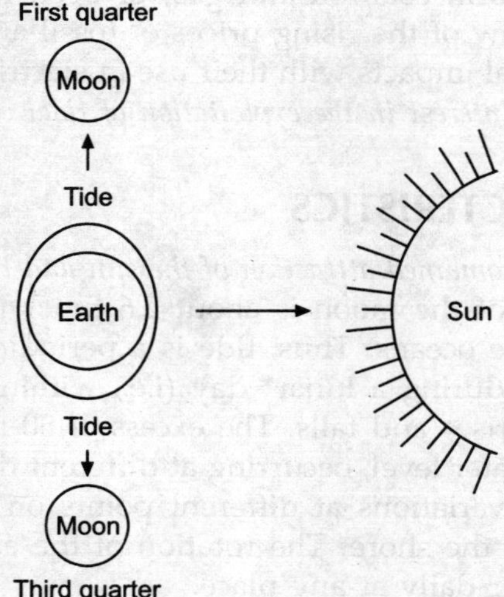

FIGURE 11.2 The sun and the moon are at right angles to each other, causing neap tides.

The spring tide is particularly great when the moon is 'New' and 'Full' at which time it is at the closest point of its orbit to the earth. The revolution of the earth and the moon together around the sun gives rise to further variation, and due to this effect the highest spring tide occurs at the equinoxes in March and September as shown in Figure 11.3. It has an important bearing on the design of a tidal power plant. A high tide is experienced at a point which is directly under the moon. At the same time, at a diameterically opposite point on the earth's surface, there also occurs a high tide due to dynamic balancing of the ocean water over the globe. In the course of the earth's rotation the water buldges out as shown in Figure 11.4.

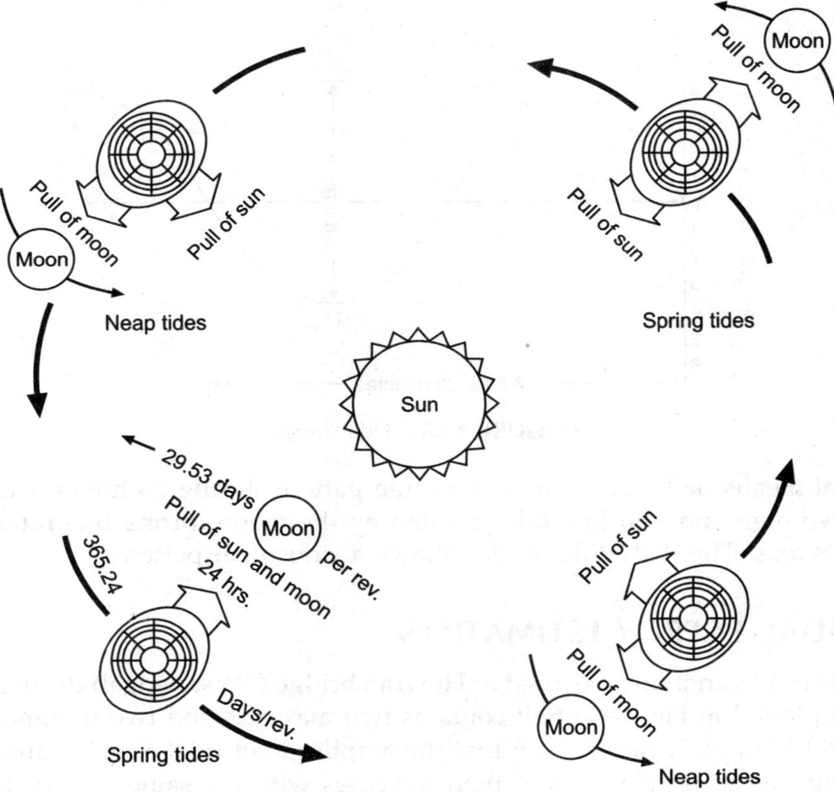

FIGURE 11.3   Origin of tides.

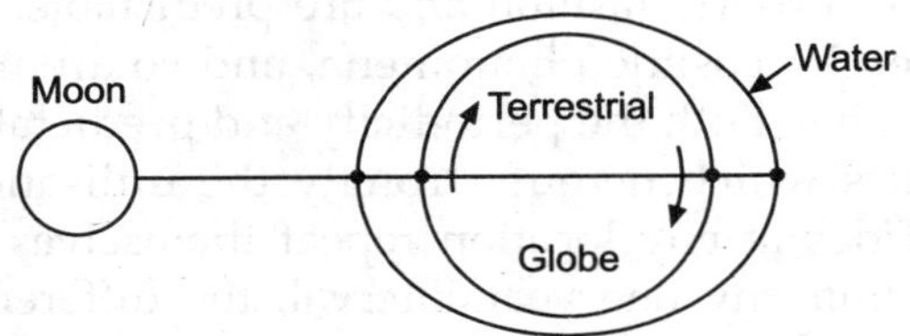

FIGURE 11.4   Distribution of water over the earth's surface under the effect of the moon.

11.3 TIDAL RANGE

The 'tidal range' is expressed as the difference in water levels between two consecutive high tides and low tides. The rise and fall of water level in the sea during tides can be represented by a sine curve shown in Figure 11.5. The figure shows the point B, a position of high tide, while the point D represents a position of low tide. One tidal day is of 24 hours and 50 minutes and there are two tidal cycles in one tidal day. The normal tide is a semi-diurnal tide with a period of 12 hours and 25 minutes.

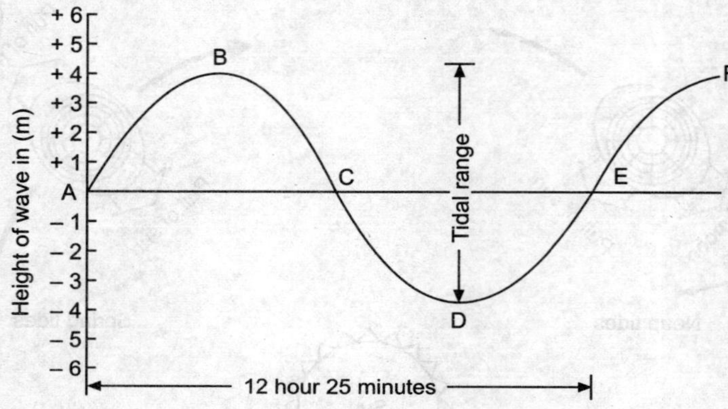

FIGURE 11.5 Tidal range.

Diurnal means daily, i.e., activities of tide pattern during 24 hours. Diurnal tides indicate two high and two low tides created by the moon during one rotation of the earth on its axis. The daily tidal cycle follows a sinusoidal pattern.

11.4 TIDAL ENERGY ESTIMATION

Tidal water level variations recorded at Howrah bridge (West Bengal) during one lunar month are plotted in Figure 11.6. It contains two maximas and two minimas during a period of 29.33 days. It can be seen that the amplitude of tidal range reduces steadily from spring tide to neap tide, and then increases with the same pattern to the next spring tides. This monthly cycle occurs due to one revolution of the moon around the earth. Both the tide cycles, namely the daily and the monthly cycles at a particular location repeat in a most orderly fashion and are predictable.

The tides are caused by cosmic phenomena, and so are not affected by weather conditions and yearly rains. Both the periodicity and predictability of tidal action are important characteristics which favour strongly the utilisation of this phenomenon as an energy source. Tides at any location repeat themselves almost identically in a cycle of 19.0 years. Within any one year interval, the differences are small, and the available energy is practically the same from year to year. Precisely for the semi-diurnal tides, there is a relation between the tidal range and the hours of the high and low tides. Thus, at a particular location, the tidal range at a given time during the day shall always be within limits of the known maximum value. This data proves useful in deciding the location of a tidal power plant.

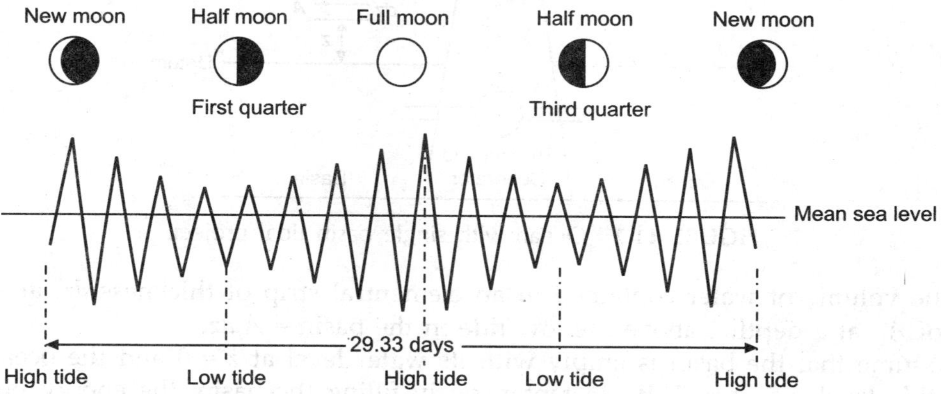

FIGURE 11.6 Water level variation caused by tides during one lunar month.

The tide range varies and depends upon the land situation against the sea. On open, exposed head lands, tides are moderate which may range from 1 m to 2 m. In a gulf, bay or creek, tides are greatly amplified, in certain cases several times than those occuring at a nearby open coast. The amplification is maximum if the bay is funnel shaped.

The tidal regime differs completely from the regime of river. Once the availablity of tidal energy is investigated, it will not be influenced by summer or monsoon and there are no dry or wet years. The variation of the tides and the time of occurrence of high and low tides can be predicted with great accuracy years ahead. This makes it possible to determine the energy and the dependable peak available from a tidal power plant.

11.5 ENERGY POTENTIAL ESTIMATION FOR A TIDAL POWER PROJECT

In the tidal power scheme, a barrage is used to create the water head considering the variation of tidal height in the basin. The barrage is used to impound water during rising tide in one or more basins, which is then released through hydraulic turbines installed in the barrage during the period the tide recedes. Thus, the potential energy of water is converted into electrical energy. For optimal output of tidal power plant, an estuary or creek is the best choice for constructing the barrage. It provides high tidal range besides large storage of water. By using a reversible turbine, electric power can be generated during the rising tide, when the basin is filled, and again during the falling tide when the basin is emptied.

Consider a basin of surface area A m^2 at the maximum basin level. Let R be the range of the tide and V the volume of water stored from the low level to high tide level as shown in Figure 11.7.

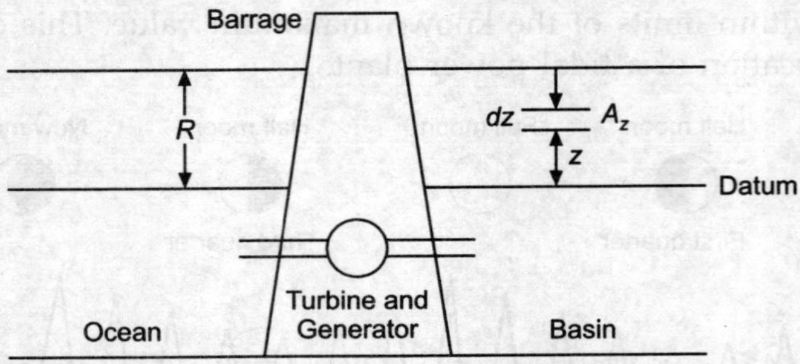

FIGURE 11.7 Ocean with single basin tidal project.

The volume of water contained in an elementral strip of thickness dz, at surface area of A_z, at a depth z above the low tide in the basin $= A_z dz$.

Assume that the basin is empty with its water level at $z = 0$ and the ocean is at high tide level, i.e., $z = R$. By instantaneously filling the basin, the energy potential available is E_f. Then,

$$E_f = \rho g \int_{z=0}^{z=R} z\, A_z\, dz$$

where

ρ = sea water density in kg/m^3

   = 1025 kg/m^3

g = gravitational constant

   = 9.81 m/s^2

For the particular case where A_z is constant and independent of z,

$$E_f = A_z\, \rho g \int_{z=0}^{z=R} z\, dz$$

$$= A\rho g\, \frac{R^2}{2} \tag{11.1}$$

The above equation provides energy conversion from a single basin type with single effect, i.e., either filling the basin or emptying the basin. The duration of time for single effect is 6 hours and 12.5 minutes which is equal to 22350 seconds.

The average theoritical power P generated by the water is W in watts during a semi-diurnal tide of 6 hours and 12.5 minutes (22350 seconds)

$$\text{Average power, } P = \frac{W}{\text{Time in seconds}} = \frac{A\rho g R^2}{2 \times 22350}$$

or

$$\frac{P}{A} = \frac{1}{44700} \times 1025 \times 9.81\, R^2 \text{ W/m}^2$$

Average power generated during one filling or emptying process

$$= 225 \, AR^2 \text{ kW} \tag{11.2}$$

where A is the area of the basin in m^2 and R is the range of the tide in m.

The average power generated is calculated based on average operating head of $R/2$ which is available only for a limited period under a single basin emptying operation. There are friction losses, conversion efficiencies of turbine and generator that reduce the power output. Tidal engineering practice based on studies at various tidal sites has revealed that the optimal annual energy production is only 30% of the average theoritical power calculated above.

EXAMPLE 11.1

A simple single-basin type tidal power plant has a basin area of 22 km^2. The tide has a range of 10 m. The turbine stops operation when the head on it falls below 3 m. Calculate the average power generated during one filling/emptying process in MW if the turbine-generator efficiency is 74%. Take specific gravity of sea water as 1.025.

Solution

$$\text{Energy potential, } E_f = A\rho g \int_{3}^{10} z \, dz$$

$$= \frac{1}{2} A\rho g \, (10^2 - 3^2)$$

where

$$A = 22 \times 10^6 \text{ m}^2$$
$$\rho = 1025 \text{ kg/m}^3$$
$$g = 9.81 \text{ m/s}^2$$

$$\text{Average power, } P_f = \frac{W}{\text{Time}} = \frac{1}{2 \times 22350} \times 22 \times 10^6 \times 1025 \times 9.81(10^2 - 3^2) \text{ W}$$

$$= \frac{1}{44700} \times 22 \times 1025 \times 9.81 \times 91 \text{ MW}$$

or

$$P = 450.3 \text{ MW}$$

Turbine-generator efficiency is 74%.

Thus, power output $P = 450.3 \times \dfrac{74}{100}$ MW

$$= 333.22 \text{ MW}$$

11.6 ENERGY AND POWER IN A DOUBLE CYCLE SYSTEM

In a double effect system as shown in Figure 11.8 the energy available in tide sea water is converted into electrical energy during flood tide (rising tide) when the basin is filled and also during the ebb tide (falling tide) when the basin is emptied.

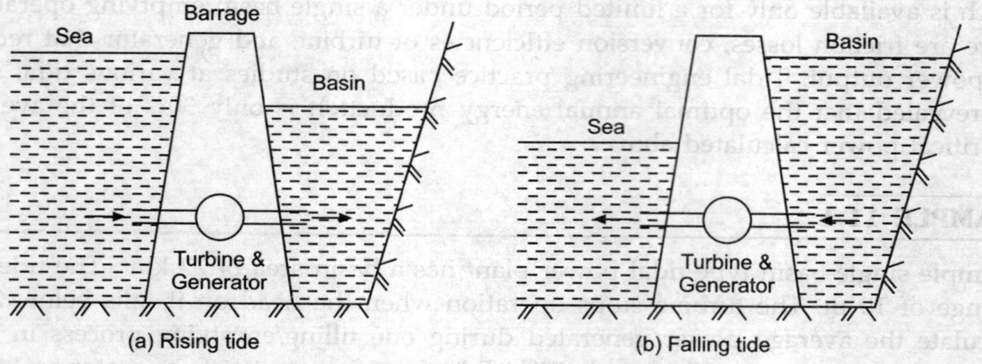

FIGURE 11.8 Double cycle system.

The flow of water through the turbine during rising and falling tides is in opposite directions. For this reason, a reversible water turbine is used, which acts as a turbine for either direction of flow. During rising tides, a large quantity of water flows into the basin through the turbine by opening the sluice gate. Filling of the basin continues along with the generation of electric power until the tide water levels of the sea and the basin become equal. At this position the sluice gate is closed. Subsequently, during falling the tide water from the basin flows into the sea through the turbine and electric power is generated. As the water level in the basin drops, a point is reached when the difference in water levels between the sea and the basin becomes too small to generate power. At this point of time the generating units are shut down. The basin is again filled during rising tide and the cycle repeats to convert tidal energy into electrical power.

It was seen in Eq. (11.2) that the average power generated during one filling of the basin is

$$P_f = 225AR^2 \text{ kW}$$

During the emptying process of the basin, the same amount of power is generated. Thus, the energy potential available during the filling and emptying operations is given by

$$E_f + E_e = A\rho g \frac{R^2}{2} + A\rho g \frac{R^2}{2}$$

$$= A\rho g R^2 \tag{11.3}$$

The theoritical average power generated in a double effect, single-basin system is given by

$$P_f + P_e = 2 \times 225AR^2 \text{ kW}$$

$$= 450AR^2 \text{ kW} \tag{11.4}$$

The double effect tidal plant generates double energy per tidal cycle, so it is 100% more efficient than the single effect plant.

11.7 YEARLY POWER GENERATION FROM TIDAL PLANTS

To harness tidal energy for maximum efficiency, a double cycle system is most suitable. In real sense, the tidal energy is the potential energy of sea water. For filling the basin, sea water gains potential energy due to lunar gravitational pull; while for the emptying process, the basin water flows out due to gravity action of the earth. The energy available from a tidal plant depends on two factors, namely the tidal range and the volume of water accumulated in the basin. Tidal energy is the slowly-increasing hydro energy during filling of the basin, and after a period of nearly three hours it attains its peak value. When the tide recedes, water is allowed to flow from basin to sea; it is then slowly-decreasing hydro energy and attains its lowest value when the turbine stops after a period of three hours. Thus, the energy available from a tidal plant can be calculated in a similar way as for as hydro-electric plant.

Assuming the following:

H = tidal range, i.e., the difference between the maximum and minimum water levels in the basin, expressed in m

V = volume of water that can be contained in the basin, expressed in $(\text{m})^3$

A = mean base area of the basin, expressed in $(\text{m})^2$

So, $V = AH$.

Let Q be the average quantity of water in cubic metre per second that flows in or flows out from the basin. Therefore,

$$Q = \frac{AH}{t}$$

where t is the total time in seconds required for filling or emptying the basin.

Theoritical work done by Q quantity of water falling through H' metres is given by

$$W = \rho QH' \text{ kg-m}$$

Power generated at any point of time

$$P = \frac{\rho QH'}{75}\eta \text{ hp}$$

where

$\rho = 1025 \text{ kg/m}^3$ for sea water

1 hp = 75 kg-m/s

η = efficiency of the system.

Hence,
$$P = \frac{\rho QH'}{75}\eta \times 0.736 \text{ kW} \qquad (\because \ 1 \text{ hp} = 736 \text{ W})$$

$$\text{Total energy per tidal cycle} = \int\limits_0^t P\,dt = \int\limits_0^t \frac{\rho QH'}{75}\eta \times 0.736\,dt$$

There are, on average, 705 tidal cycles in a year.

$$\text{Yearly power generation from a tidal project} = \int\limits_0^t \frac{\rho QH'}{75}\eta \times 0.736 \times 705\,dt$$

Therefore,

$$P_{\text{year}} = \frac{1025 \times 0.736 \times 705}{75}\int\limits_0^t QH'\eta\,dt \tag{11.5}$$

$$= 10.06 \times 705\eta \int\limits_0^t QH'\,dt$$

EXAMPLE 11.2

For Rann of Kutch the basin area of a tidal project is 0.72 sq. km, with a difference of 6 m between the high and low water levels. The average available head is 5 m and the system generates electric power for 4 hours in each cycle. Assuming the overall efficiency as 80%, calculate the power in kW at any point of time and the yearly power output. Density of sea water is 1025 kg/m^3.

Solution

$$\text{Volume of the basin, } AH = 0.72 \times 10^6 \times 6 \text{ m}^3$$

$$\text{Average discharge, } Q = \frac{AH}{t}$$

$$= \frac{0.72 \times 10^6 \times 6}{4 \times 3600}$$

$$= 300 \text{ m}^3/\text{s}$$

$$\text{Power at any point of time, } P = \frac{\rho QH'}{75} \times \eta \times 0.736 \text{ kW}$$

$$= \frac{300 \times 1025 \times 5}{75} \times 0.736 \times 0.8$$

$$= 120.95 \times 10^2 \text{ kW}$$

$$\text{Energy generated per tidal cycle} = 120.95 \times 10^2 \times 4 \text{ kWh}$$

$$= 483.80 \times 10^2 \text{ kWh}$$

$$\text{Total number of tidal cycles in a year} = 705$$

$$\therefore \quad \text{Yearly energy generation} = 483.80 \times 10^2 \times 705 \text{ kWh}$$

$$= 341.08 \times 10^5 \text{ kWh}$$

11.8 DEVELOPMENT OF A TIDAL POWER SCHEME

11.8.1 Site Selection

For a favourable tidal power development, a site must have a large tidal range and must be capable of storing a large quantity of water for energy production with minimum dam and dyke construction. For achieving a high storage capacity, the site should be located in an estuary or a creek. The site should be near to a load centre to minimize the transmission requirements.

The suggested approach to the development of a tidal power scheme leading to the construction of a tidal power plant is as follows:

1. *Pre-feasibility study:* Acquisition of data such as tides, local topography, infrastructure, etc. is the first requirement.
2. *Feasibility study:* It comprises mathematical modelling, preliminary energy computation, foundation investigations, hydraulic model studies, detailed analysis of various modes of operation.
3. Detailed design, preparation of specifications and tender documents.
4. Construction of the plant.

11.8.2 Pre-feasibility Study

The following maps, charts, data and information about the scheme site need to be collected during the pre-feasibility study.

   (i) Local land area map, survey of India map and hydrographic charts
  (ii) Historical data on tides and tidal currents
 (iii) Geotechnical properties of the sea bed and coastal region in the study area
  (iv) Typical weather conditions, rainfall wind and wave data
   (v) Nearest high voltage substation for connecting the generated electric power with the state grid

11.8.3 Types of Tidal Power Plants

Tidal power plants can be broadly classified into the following four categories:

   (i) Single-basin single-effect plant
  (ii) Single-basin double-effect plant
 (iii) Double-basin with linked-basin operation
  (iv) Double-basin with paired-basin operation

11.8.4 Single-basin Single-effect Plant

It is the oldest form of tidal power development and the basis of many tide mills. A tidal power plant is simply a barrage (dam or dyke) across an estuary or creek, whose principal elements are a powerhouse and a sluice as shown in Figure 11.9

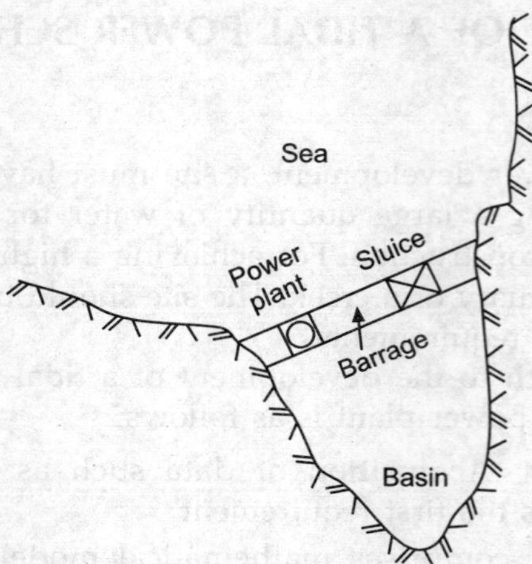

FIGURE 11.9   Single-basin single-effect tidal plant.

The basin is filled through the sluice by the rising tide. The water level in the basin reaches the highest level of the tide. It povides the water head of tidal range to the turbine. The sluice gate is closed. The turbine is started only when the water in the sea is at falling tide level. As the tide continues to fall a hydraulic head is formed at the barrage and at an appropriate time water is released from the basin through the generating unit installed in the powerhouse. Electric power generation continues until the head is reduced to the minimum turbine operating level. It normally occurs after the tide has reached its lowest point and has begun to rise again. At this stage the turbine water passage is closed and all discharge from the basin is stopped. When the rising tide reaches the basin level, the filling sluice is opened, refilling of the basin starts and the cycle is repeated. The cycle of operation showing the water level, the generating and the refilling periods, is depicted in Figure 11.10.

The above cycle of operation offers different output characteristics depending on the time with respect to the tidal cycle at which generation starts and stops and on the turbine capacity. There are three variables which affect the power characteristics of ebb tide at a particular tidal site, namely:

(i) The turbine capacity

(ii) Minimum head under which the turbine will operate efficiently

(iii) Time at which generation starts and stops.

These three variables need to be adjusted to produce the best possible results. In general the aim should be to get as long a period of operation as possible, and with this objective, the turbines would commence and stop operating at the minimum head consistent with high efficiency.

In a single-basin single-effect tidal plant with ebb tide operation, the generation period is only for 3.5 hours during every tide cycle. There are two tide cycles per day, so the energy available is intermittent and fluctuates from a maximum at spring tides to a minimum at neaps.

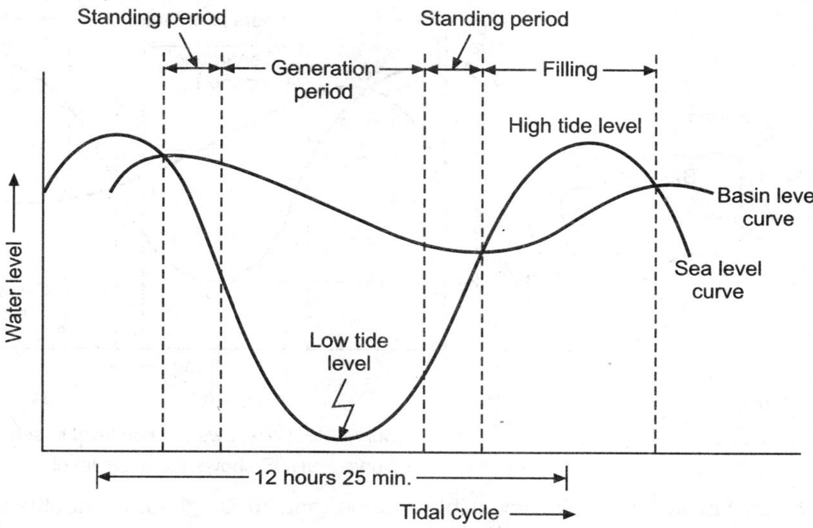

FIGURE 11.10   Operating cycle of single-basin single-effect plant.

Single basin rising tide operation

The single basin flood tide operation is similar to ebb tide operation, with the difference that power generation during rising tide is less than that at ebb tide working. In rising tide, there is rapid filling of the basin, so the turbine operates for a reduced period. In ebb tide operation, the turbine and the generator operate for a longer time giving higher output. Thus, the single basin rising tide operation, besides its lower output, also suffers from intermittent and variable output.

11.8.5   Single-basin Double-effect Plant

This arrangement makes use of the combination of the ebb tide and the flood tide working, and power is generated both during emptying and filling of the basin. With a single barrage as shown in Figures 11.11(a) and (b) the water head which produces the energy operates from the sea towards the basin during the flood tide and from the basin towards the sea during the ebb tide. The most practical method of achieving the double tide operation is by the use of the reversible turbine which can operate in both directions of flow. In the operating cycle for double-effect operation, the curve shows that the output is variable and intermittent, but to a lesser extent than that in the case of the unidirectional flow power plant. Other advantages over the one-way plant are:

(i)   The overall output from an equal turbine capacity is greater by 15%. This percentage may increase if each plant is designed to the most economic type specification.

(ii)   The period of operation is increased.

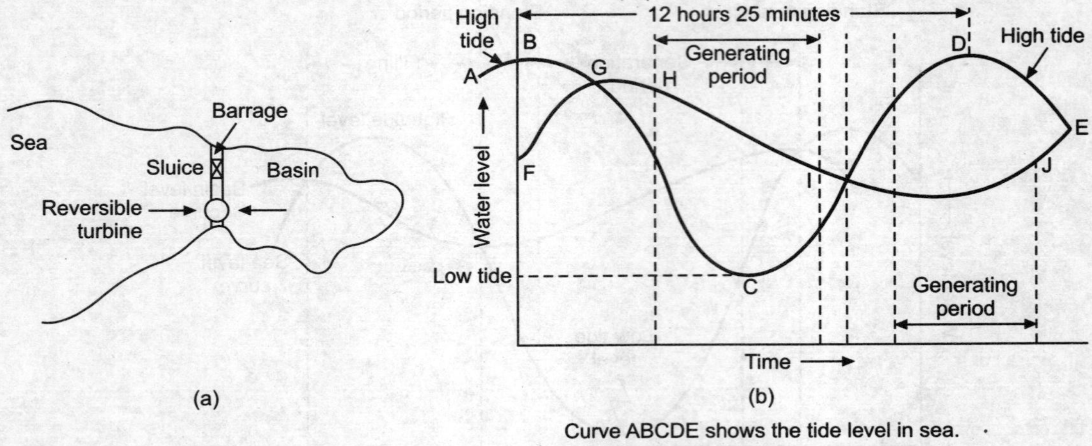

FIGURE 11.11 (a) Single basin with reversible turbine, and (b) single-basin double-effect power plant operating cycle.

The reversible turbines and other allied equipment cost a little more, this additional cost is offset by the above two advantages.

Due to its large period of operation, the operating regime of power generation can be manipulated so as to offer power to any predetermined period, either to suit the demand or to suit the tides. La Rance power plant of 240 MW in France is working on this type of operation.

11.8.6 Double-basin with Linked-basin Operation

In this arrangement a large basin is converted into two basins of suitable dimensions; one which is at higher level is called high basin and the other low basin. The scheme consists of three barrages, one separating the high basin from the sea and containing the filling gates, another separating the low basin from the sea and containing the emptying gates. The third barrage separates the high basin from the low basin and contains the powerhouse as shown in Figure 11.12.

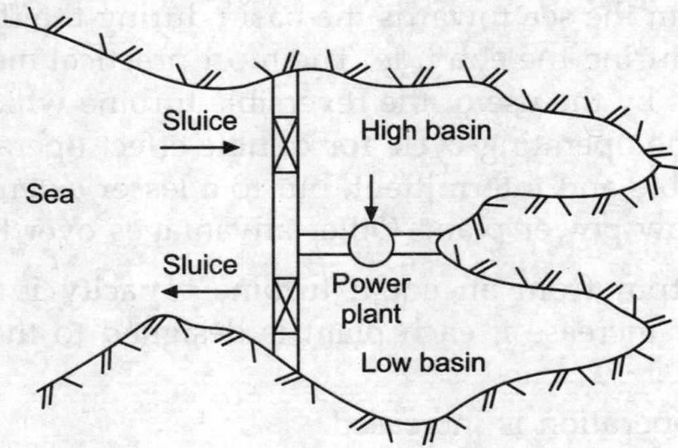

FIGURE 11.12 Double-basin with linked-basin operation.

The upper basin filling gates are opened only during the time when the sea level is higher than the upper basin. The emptying gates of the lower basin are opened only when the sea level is lower than the lower basin. The head on the turbine is the difference in elevation between the upper and lower basins.

The two-basin scheme may be economically viable where power demand is less than the guaranteed output as determined by the tide cycle. Alternatively, the two-basin system can be operated by retaining water in high basin and releasing it to meet peak demands only.

11.8.7 Double-basin with Paired-basin Operation

The paired basin scheme consists of two single-basin single-effect separate schemes located at a distance from each other. The locations are so selected that there is a difference in tidal phase between them. Both the schemes never exchange water, but are interconnected electrically. Both the basins operate in single-basin single-effect mode. One basin generates electrical energy during the 'filling' process while the other during the 'emptying' process. The scheme is shown in Figure 11.13.

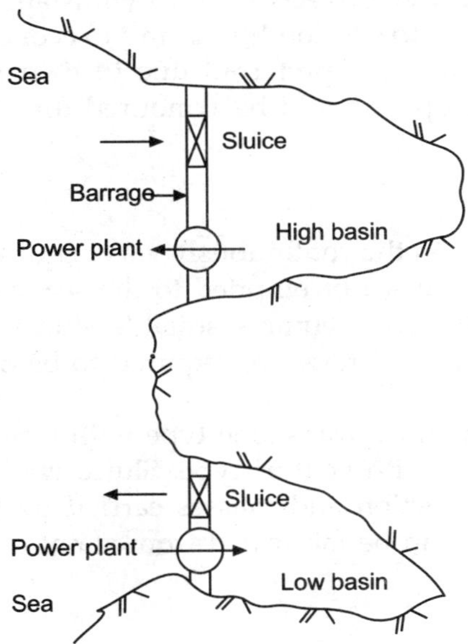

FIGURE 11.13 Double-basin with paired-basin operation.

This arrangement affords a little more flexibility in operation of the plants to meet power demands. More benefit can be derived if there is a difference in tidal phase of the sea near the two basins. In case where there is no difference in tidal phase, variations in power output can be evened out by resorting to ebb tide operation in one plant and flood tide operation in the other.

The paired-basin operation leads to a continuous output, still its power supply remains irregular and there is no solution for equalizing the great difference in output between the spring and the neap tide operation. Further, it is difficult to find two tidal sites within reasonable distance of each other having the requisite difference in time of high water.

11.9 IMPORTANT COMPONENTS OF A TIDAL POWER PLANT

There are three important components of a tidal plant:

(i) A barrage to form a basin

(ii) Sluice gates in the barrage for flow of water from the sea to the basin and vice-versa

(iii) A powerhouse equipped with turbines, each coupled to a generator along with auxiliary equipment

11.9.1 Barrage (Dam or Dyke)

The barrage should be constructed by the material available at site or from a nearby place. Barrages for tidal power projects have to withstand the force of sea waves, so the design should be suitable to site conditions and to economic aspect of development. The rockfill dams or barrages are preferred due to their stability against flows. The dyke (barrage) crest and slopes should be armoured for protection against waves.

11.9.2 Sluices

Tidal power plants operate on the continuously varying difference in level at which the basin must be filled from the sea or emptied to the sea, as required by the operating regime of the power plant. This requires suitable sluice ways equipped with gates which can be operated quickly. These are required to be operated two or more times a day.

There are two types of sluice ways, one type with crest gates and the other of the submerged gates associated with venturi type. Sluice ways with crest gates are more prone to damage by wave action and masses carried by the flow. Vertical lift gates are the natural choice and can be fabricated from stainless steel.

11.9.3 Turbines

The energy potential in tidal power development is exploited from low to very low heads, for which large size turbines are required. If the water head is more than 8 metres, a propeller type turbine is quite suitable because the angle of blades can be changed to obtain maximum efficiency while the water is falling. The main aim of the designer for a tidal power plant is to achieve as long a period of operation as possible. The turbines beginning and finishing work at the minimum head provide maximum efficiency, and this is the advantage of having turbines with variable pitch blades as shown in Figure 11.14.

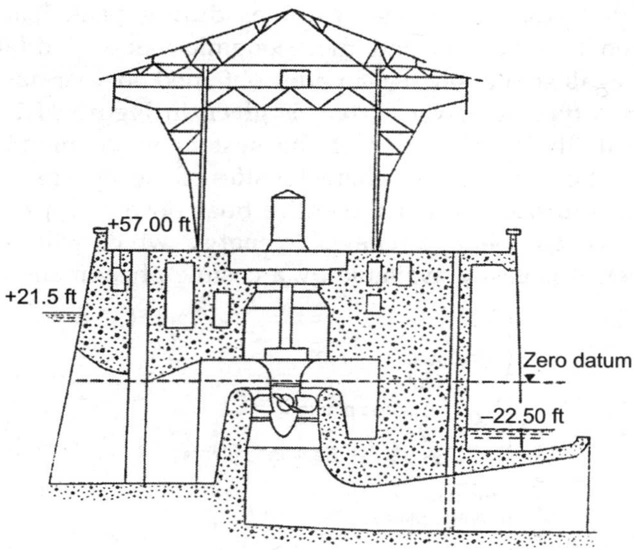

FIGURE 11.14 Kaplan turbine with adjustable blades.

For low heads three types of turbines can be used; the selection is made according to the suitability. These turbines are:

(i) Bulb turbine
(ii) Tube turbine
(iii) Straight flow rim type turbine.

11.10 GRID INTERFACING OF TIDAL POWER

Tidal power is lunar energy harnessed through sea water. The lunar period is slightly longer than the solar period by about 50 minutes of solar time. Accordingly, the time of high and low tidal levels, occurring in phase with the lunar cycle, shifts slightly forward in time each day. This phase difference changes the maximum quantities of tidal energy that can be generated, by an ebb-flow tidal electric plant, at any particular time each day. The absorption of tidal energy into power system, which have changing load demands during daily, weekly and seasonal cycles, therefore, poses certain problems.

The amount of tidal energy that a power system can absorb has a major influence on the economic viability of a tidal power plant being integrated into the system. The output of a tidal power plant is an image of tidal variations modified by the characteristics of the scheme adopted. The scheme may be single-basin single-effect or double-effect. The double-effect scheme provides marginal advantage when the size of the scheme is large compared to the system.

A power system normally adjusts supply to meet the demand by regulating production from nuclear, large thermal, hydro-plants within certain limits, by adding power from pumped storage and gas turbines during peak load hours. The tidal energy can be used to save fuel and the economic value of tidal energy absorbed can be evaluated against the savings in fuel obtained in thermal plants. The merit order of plants for a typical power system is given in Figure 11.15. The tidal energy is accommodated in the merit order of the system to be planned by taking into consideration the static and dynamic characteristics of the system constituting the load supply region. The approach is that the entire tidal power project generation should be absorbed in the system, being a renewable energy, which will be otherwise wasted if not utilised, when it is easily available by a cosmic phenomenon.

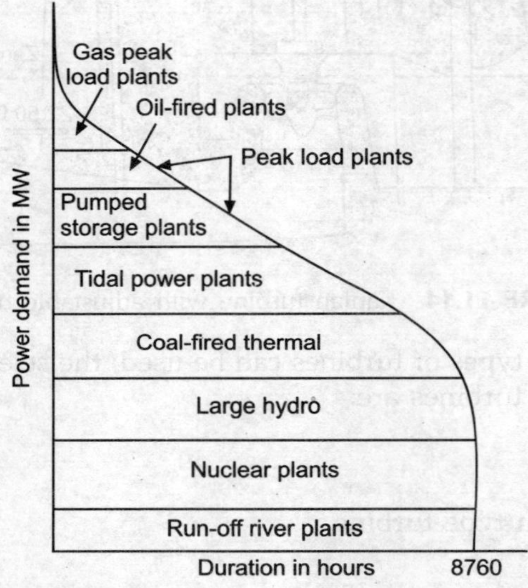

FIGURE 11.15 Plant merit order for a typical annual load duration curve.

11.11 ADVANTAGE AND DISADVANTAGE OF TIDAL POWER

Advantages

1. Tidal power is predictable.
2. Available tidal power is firm as there are no wet or dry years, no dry or wet months, nor is there any influence of summer or winter on the availability of tidal energy.
3. It is free from pollution.
4. Tidal power is inexhaustable and is a renewable source of energy.
5. Tidal power plants do not require valuable land as these are located on sea shores.
6. Tidal power when used in combination with a thermal plant can meet effectively the load demand.
7. After the capital cost of a tidal power scheme is paid off, the cost of power generated is very low.

Disadvantages

1. Tidal power plant output varies with the variation in tidal range.
2. Tidal power supply is intermittent.
3. Capital cost of a tidal plant is not economical when compared with conventional sources of energy.
4. Silting of basins is a problem with tidal power plants.

11.11.1   Global Scenario of Tidal Energy

As of today the tidal power plants in operation are detailed in Table 11.1.

TABLE 11.1   Details of tidal power plants in the world

Location	Year	Total capacity	No. of units	Tidal range (m)
La Rance Brittany, France	1966	240 MW	24	8.5
Kislaya Guba, Russia	1968	1.7 MW	1	3
Annapolis NOVA Scotia, Canada	1984	20 MW	1	5.5
China				
(i) BAISHAKOU	1978	960 kW	6 × 160 kW	3.5 to 7.8
(ii) JIANGXIA	1980	3.9 MW	6 × 500 kW	5

11.11.2   La Rance Project

The La Rance tidal electric plant was built between 1961 and 1967 and energy production started in 1966. The location and plan of the scheme are shown in Figure 11.16. The estuary of the Rance is 3 km south of St. Malo, and is 750 m wide at the project site. The barrage consists of a ship lock, a powerhouse, and a short rockfill section. There is a 115 m long sluice structure with six fixed roller gates (15 m wide × 10 m high) capable of passing the flow, at mean tide, of 9600 m^3/s under a 5 m head.

The sluiceways allow complete emptying of the basin at the end of direct generation period or complete filling at the end of reverse generation. This happens when the difference in head between the basin and the sea is smaller than the minimum working head for driving the turbine, i.e., 1.2 m outwards when emptying the basin or 1.6 m inwards when filling. The sluice section is also used for total filling of the basin in the single-effect ebb-flow mode of operation. The generating units operate as orifices to assist the sluices.

The turbo-generator units are double regulated, reversible bulb turbines capable of turbining or pumping in both directions. Twenty-four machines are installed, each with 5.35 m diameter runner, directly coupled to a 10 MW generator-motor with a rated head of 6.65 m. The maximum head is 11 m and the minimum 3 m. Sectional elevations of the powerhouse, rockfill dyke and sluiceway are shown in Figure 11.17. The project is economically attractive when compared with the average cost of nuclear kWh and thermal kWh.

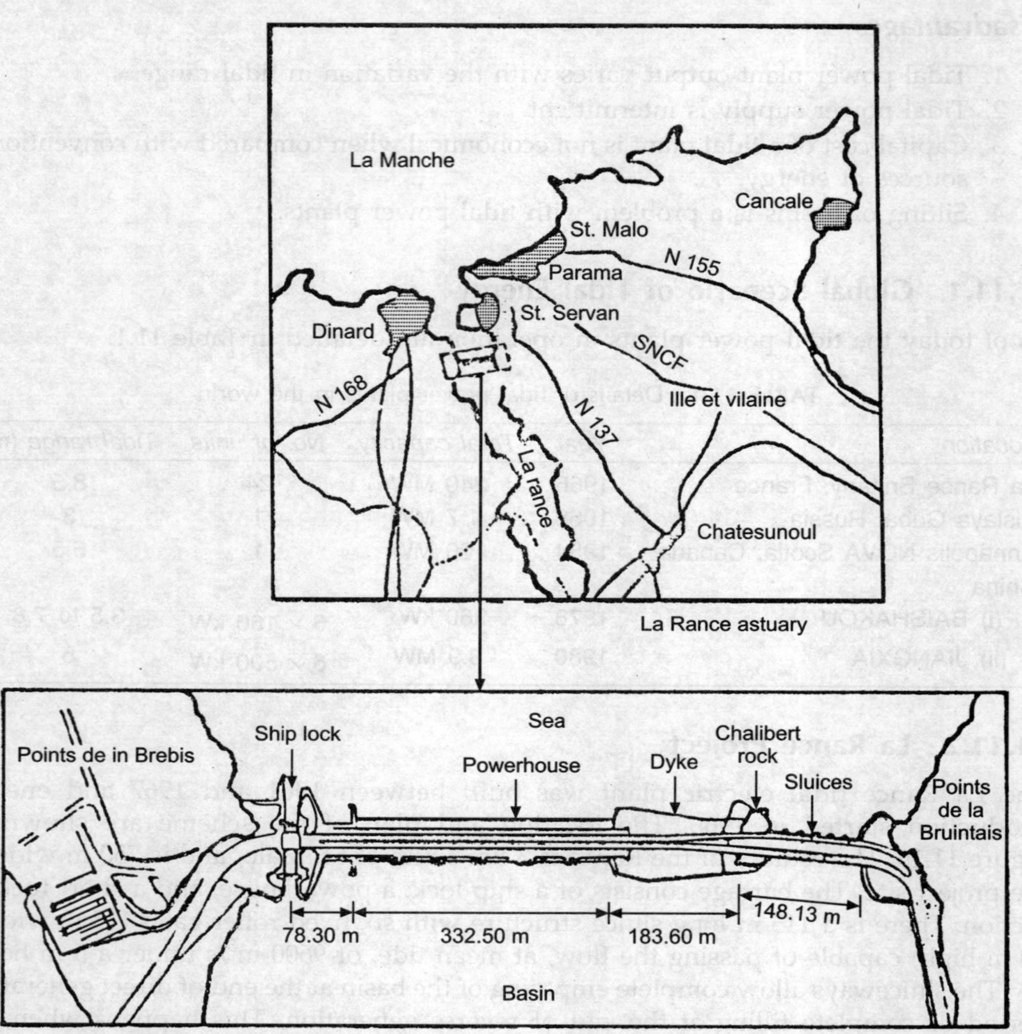

Estuary width	: 750 metres
Basin area	: 22 square kilometres
Mean tide	: 8.5 metres
Installed generating power	: 240 MW
Turbo-generators	: 24 x 10 MW bulb type Kaplan turbines with reverse flow and pumping capability
Turbine runner diameter	: 5.35 metres
rated head	: 6.65 metres
maximum head	: 11 metres
minimum head	: 3 metres

FIGURE 11.16 General layout of La Rance tidal electric power plant.

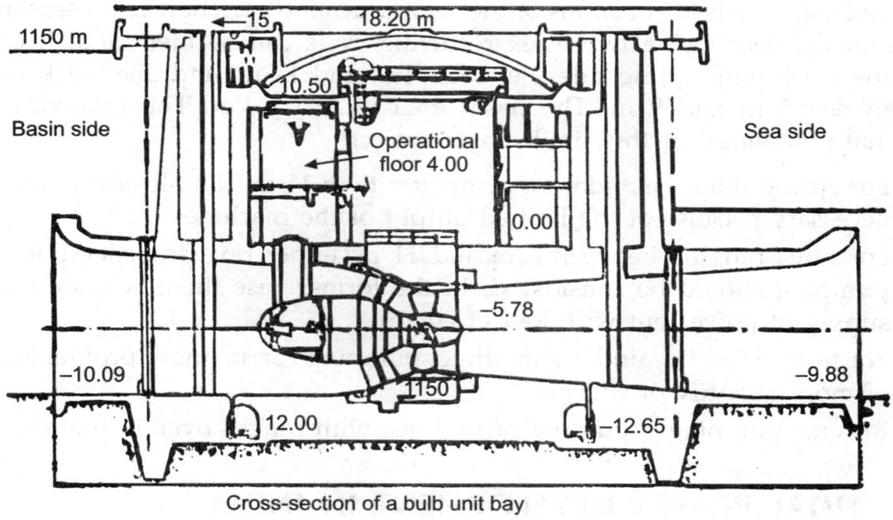

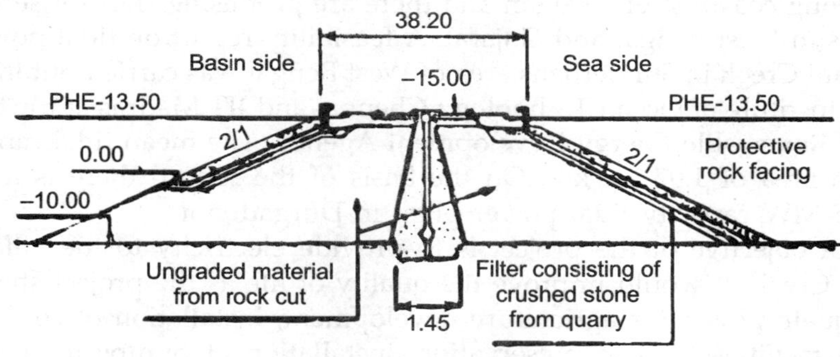

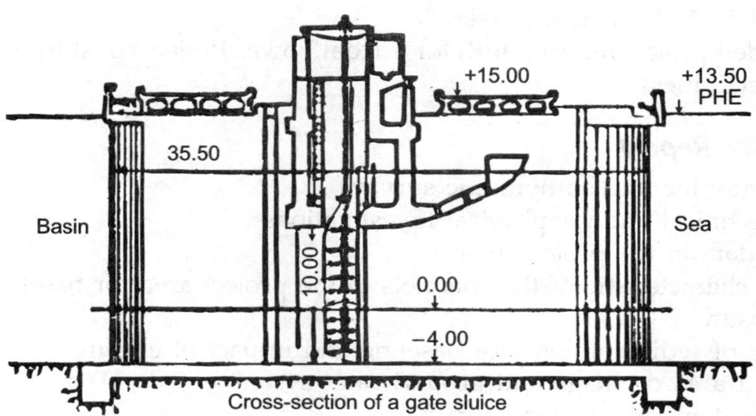

FIGURE 11.17   Sectional elevation of La Rance tidal electric power plant.

The average tide is 8.5 m and is of the semi-diurnal type. There is no seasonal effect throughout the year, but a two-week recurring cycle can be identified. It comprises nearly one week with spring tides between 9 m and 12 m, and one week with neap tides between 5 m and 9 m. The basin area is 22 sq. km. The following are the operational guidelines for this single-basin project.

  (i) For spring tides, i.e., tidal range greater than 11 m, double-effect generation is necessary because of the limited output of the machines.

 (ii) For tides ranging between 7 m and 11 m, either reverse generation or direct pumping should be considered. In the former case there is more continuous supply of power but with loss of energy.

(iii) For neap tides, i.e., under 7 m, direct pumping can be more profitable, in terms of more quantity of energy.

 (iv) Reverse pumping should be carried out with ranges over 11 metres.

11.12 TIDAL POWER DEVELOPMENT IN INDIA

India has a long coastline of 6000 km and there are promising sites for setting up tidal power plants in West Bengal and Gujarat. A feasibility report on tidal power prospect of Durgaduani Creek in Sunderbans area of West Bengal was carried out in 1995 jointly by National Institute of Ocean Technology Chennai and IIT Madras at the behest of the West Bengal Renewable Energy Development Agency. The mean tidal range is 3.54 m with a basin area of 1.07 sq. km. On the basis of the report, there is a proposal to install a 3.75 MW capacity tidal power plant in Durgaduani.

The main objective of the project is to provide electricity to the villages around Durgaduani Creek. It would improve the quality of life as the project shall be able to supply adequate power for agriculture development, installation of cold storage and refrigeration facilities for fish preservation, installation of pumps for fresh drinking water, development of industries and tourism in the area. In Sunderban area there is a potential of 50 MW of tidal power.

The detailed project report (DPR) for a Tidal Power Project constitutes the following studies and activities.

[A] *Feasibility Report*

   (i) Topographic and bathymetric surveys
  (ii) Geotechnical and geophysical investigations
 (iii) Tide data in the project area
 (iv) Flow characteristic of the channels in the project area for baseline and impact of closure
  (v) Study of sedimentology for baseline and impact of closure
 (vi) Project area drainage and groundwater system
(vii) Power demand and load survey
(viii) Selection of power generating equipment
 (ix) Power transmission and distribution system

[B] *Design Engineering*

   (i) Dam design criteria and conceptual design of closure dams
  (ii) Powerhouse conceptual design
 (iii) Design of sluice structure and gate arrangement
 (iv) Hydraulic model study
  (v) Socio-economical impact
 (vi) Environmental and ecological impact

[C] *Construction*

  (i) Organization and manpower availability
 (ii) Material availability and transportation
(iii) Cost estimate

[D] *Project Economics and Financial Analysis*

11.12.1   Importance of Ocean Resource

APJ Abdul Kalam the then President of India stressed on July 28th, 2004, to tap ocean resources on a large scale as about 37% of the Indian population is staying in the coastal areas. He asked the Department of Ocean Development to process the data related to India's claim for extended exclusive economic zone of 1.5 million sq. km for urgent submission before the UN. It will help the country to secure our claim on larger ocean resources.

11.12.2   Kalpasar Multipurpose Tidal Project in Gujarat

It is an extraordinary project; it involves building a 30 km (revised in 2017) dam across the Gulf of Khambhat (it was earlier called the Gulf of Cambay) from Ghogha in Bhavnagar district to Hansot in Bharuch district. It will turn a part of the Arabian sea into a fresh water lake. The dam will trap the water from 12 rivers that flow into the gulf, namely the Narmada, Mahi, the Sabarmati and Dhadar to create a huge fresh water lake.

The Kalpasar reservoir will be 2000 sq. km. It will store three times the water in the Sardar Sarovar reservoir.

The giant lake shall be harnessed for multipurpose activities such as:

- To generate 5880 MW of tidal power
- To provide 5.61 MCM of water annually to irrigate 10,54,500 ha of land of southern Saurashtra, where water is a scarce commodity
- To supply 900 MCM water for the industrial development of Saurashtra and Kutch
- To improve the existing ports like Ghogha and Bhavnagar due to the availability of higher water levels
- To breed fish in fresh water lake to generate extra income of about ₹70 crores
- To reclaim saline land along the coast, about 1100 sq. km, for cultivation

At present the project is on the drawing board; it will take 20 years (2035–38) to build with an estimated cost of ₹90,000 crores. When constructed, the project will lead to a quantum jump in living standard of the people in the region. It has been suggested that a multilane highway and a railway can be built across the length of the dam which would slash the distance between South Gujarat and Mumbai by about 225 km.

There are several positive factors about the execution of this project such as:

- No displacement of population from their homes
- Due to rising water levels there is a possibility to build more ports in the region
- The project is out of the threat of earthquake

It is a gigantic multipurpose project and will solve the state's water problems besides generating eco-friendly tidal power.

11.13   ECONOMICS OF TIDAL POWER

Tidal power, in its cheapest form can only be generated intermittently. To convert the intermittent low grade energy to guaranteed continuous energy, additional cost must be incurred. Another aspect is that due to the low generating heads, the cost of machinery and its supporting structure is high.

The cost economy guides that a small-scale tidal power development must be justified on its own merits, so that the unit construction cost can definitely be offset against the other consequent benefits. Planning need not be aimed at the cheapest power production, but towards the best benefit to-cost ratio of the project. The benefits can be numerous and some of them may be quite tangible.

There are some benefits other than the power benefit which can reduce the cost of energy to a competitive level. Major benefit that can accrue from tidal power are listed below:

1. It is a renewable energy source free from weather vagaries. The cost of energy produced is quite nominal, i.e., only the operational cost.
2. Performance of the plant is pollution free.
3. Tidal power combined with the pumped storage generation ensures continuous power supply.
4. Road crossing on the barrage connects the isolated areas without constructing a bridge.
5. It improves the transport and navigational facilities.
6. Creates infrastructure for regional development.
7. Recreational facilities generate tourism potential.
8. Land reclaimation of sea shore waste land is a long-term benefit.
9. Social and political benefits are quite substantial.

11.14   INTRODUCTION TO WAVE ENERGY

We are aware of the movement of ocean water in terms of waves, which become huge in height as one goes farther from the coast. This movement of large quantities of water up and down can in principle be harnessed to convert into usable forms of energy such as electricity or mechanical power. Waves are formed on the surface of water by the frictional action of the winds resulting in the radial depression of energy from the blowing winds in all directions.

The ocean is a big collector of energy transferred by wind over a large surface area which is stored as wave energy. Wave energy is more concentrated compared to wind energy, which is thinly distributed. Wave energy is available in coastal areas, islands and its potential depends upon its geographic location. Energy available in ocean waves varies in different months and seasons. Wave energy, if harnessed with improved technology, can prove to be a large dependable source of renewable energy.

11.15   FACTORS AFFECTING WAVE ENERGY

There are three major factors which govern the quantum of wave energy. The first is the wind speed, i.e., the higher the wind speed, the higher is the wave energy. The amplitude of the waves depends on the wind speed. During gusts and storms big ocean waves occur, which prove dangerous even to ships. The second factor is the 'effective fetch value', i.e., the uninterrupted distance on the ocean over which the wind can blow before reaching the point of reference. The larger the distance, the higher the wave energy. This distance may vary from 5 km to 45 km. The third factor is the depth of the sea water. The greater the depth of ocean water, the higher the wave velocity. Very large energy fluxes are available in deep ocean waves. Wave energy is abundantly available on the Indian sea coast touching the Bay of Bengal and the Arabian sea in high wind belts. All the three factors described above are available along India's long coastline of 6000 km.

11.15.1   Ocean Wave Parameters

The periodic, up and down, to and fro motion of water in seas and oceans is known as 'ocean wave' as shown in Figure 11.18.

The important wave parameters with their notations are given below:

H = wave height. It is the distance from the trough to the crest (not to the height above sea level). It mainly depends on wind speed and the fetch. The value varies from 0.2 m to 3 m.

a = amplitude of the wave = $H/2$

λ = wavelength

T = wave period which usually ranges from 4 s to 12 s

f = frequency expressed as the number of periods per second.

As a progressive wave moves, the crest line travels in a horizontal plane with a wave velocity or celerity C (wave velocity) in the direction of the x-axis, which also represents the mean sea-water level.

The frequency (f) is defined as the number of troughs or crests passing per second through a given point in the direction of wave motion.

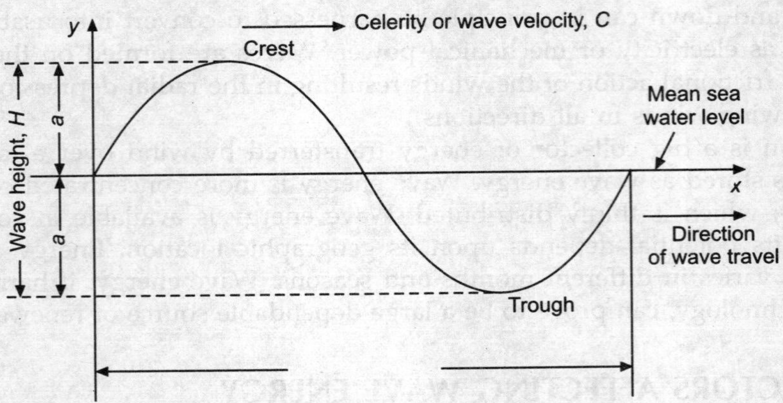

FIGURE 11.18   Representation of ocean wave.

The wavelength (λ) is the horizontal distance between two successive troughs or crests.

The wave velocity or celerity

$$C = \frac{\lambda}{T} \text{ m/s}$$

and

$$T = \frac{1}{f} \text{ s}^{-1}$$

The relation between wavelength λ and period T is given by the equation

$$\lambda = 1.56T^2 \text{ m}$$

11.15.2   Energy from High Waves

High waves are generated in deep ocean areas. As the train of waves approaches the shore, the wave period necessarily remains constant, but the wavelength, the celerity and the wave height undergo changes. As the wave approaches shallow waters (breaker zone), the decreasing depth of water gives rise to bed friction, here a part of the wave energy is dissipated in overcoming the frictional force. In this process the wave gets distorted. The water particles which have closed orbits in deeper water, now start moving forward with the wave. Consequently, when the depth of water decreases to a value nearly equal to 1.3 times the local wave height, the crest plunges forward and the wave breaks dissipating its energy on the shore.

Wave energy is defined as the rate at which it is transferred across one metre line at right angles to its direction.

The energy available in random sea is expressed as

$$P = 0.96 \ H^2T \text{ kW/m of wave crest} \tag{11.6}$$

where H is the wave height measured in metres and T the wave period in seconds.

The wave energy potential varies from place to place depending upon its geographic location. Even at a given place, the energy availability varies during the different parts of the day, for different months and from season to season.

Waves in ocean are not regular sine waves but are randam in nature. This indicates that a wave condition with a wave height of 1.5 m and a zero crossing period of 7 seconds possesses a power of about 15 kW/m of the wave crest. During a severe gale, the ocean fluxes could be as high as 1000 kW/m of wave crest. During the protracted calm or in sheltered inlets, the power could be as little as 0.001 kW/m.

11.16 WAVE ENERGY RESOURCE AND POTENTIAL OF INDIAN COAST

Maximum concentration of wave energy is available between the latitudes 25° and 55° in both the hemispheres. Latitudes of arround 30° are most suitable for harnessing wave energy. There is low pressure about the equator and high pressure near the tropics. So, trade winds blow from the sub-tropical belts towards the equator and create wave energy potential.

Winds blowing over long distances on Atlantic and Pacific Oceans generate high waves with peaks of about 50 metres between crests, displacing enormous quantity of water in each wave. The open west coasts of United States, Europe, Japan, Australia and New Zealand are the attractive sites for exploitation of wave energy.

India is also quite rich in wave energy potential. Measurements were made in respect of wave height and wave period at five different locations on Indian coasts and the wave energy calculated thereof is given in Table 11.2. South-west monsoon is between May and September while the north-east monsoon is between November and February.

TABLE 11.2 Wave energy potential on Indian coast

Location	North-east monsoon			South-west monsoon		
	Mean wave height (m)	Mean wave period (s)	Wave power (kW/m)	Mean wave height (m)	Mean wave period (s)	Wave power (kW/m)
Near Kolkata 20–25° N and 85–95° E	1.33	8.00	13.85	1.95	7.65	28.80
Near Vishakapatnam 20–25° N and 85–95° E	1.60	6.25	15.70	2.05	8.25	33.65
Near Chennai 10–15° N and 85° E	1.55	5.85	13.45	1.70	5.80	16.60
Near Cape Camorin 10–15° N and 70° E	1.20	5.35	7.80	1.80	6.30	19.55
Near Mumbai 15–25° N and 70° E	1.00	5.00	4.90	2.65	6.95	47.00

From Table 11.2 it may be seen that the wave energy potential varies from nearly 5 kW/m to 47 kW/m; the variation depends on the geographic location of the given site and the monsoon condition. During the non-monsoon period the wave power available is always less than that in monsoon months. The wave power available at a particular site also varies during each month of a year. A study was carried out and measurements were taken near a Tamil Nadu coast at a location having latitude 12°–15° N and longitude 81°–84° E. The results are tabulated in Table 11.3.

TABLE 11.3 Wave energy potential near a Tamil Nadu coast

Month	Mean wave height (m)	Maximum wave height (m)	Mean wave period (s)	Mean power (kW/m)	Maximum power (kW/m)
Jan.	1.1	3.0	3.4	2.3	16.8
Feb.	0.8	2.0	3.6	1.3	7.9
March	0.7	1.5	2.1	0.6	2.6
Apr.	0.8	2.0	2.8	1.0	6.2
May	1.1	2.5	3.8	2.5	13.1
June	1.7	4.5	4.0	6.4	44.6
July	1.3	3.0	4.4	4.1	21.8
Aug.	1.2	4.0	3.9	3.1	34.3
Sept.	1.1	2.0	3.8	2.5	8.4
Oct.	0.8	1.5	3.7	1.3	4.6
Nov.	1.1	2.5	3.4	2.3	11.7
Dec.	1.3	3.0	4.7	4.4	23.3

Another study of the Nayachara island having latitude 22° and longitude 88° 7′, near Haldia in West Bengal, was conducted to measure the wave height, the time period for shallow water during different speeds, and the results are tabulated in Table 11.4.

TABLE 11.4 Significant wave height and wave period for the Nayachara island

Wind speed (km/h)	Mean wave height at depth		Mean wave period(s) at depth	
	6 m	10 m	6 m	10 m
50	1.08	1.34	4.25	4.50
60	1.24	1.56	4.50	5.0
70	1.36	1.77	4.75	5.25
80	1.47	1.96	5.00	5.50
90	1.58	2.13	5.25	5.75
100	1.69	2.29	5.50	6.0
125	1.93	2.65	6.0	6.60

Thus, the maximum observed wave height is 2.65 m during the cyclonic weather at the location of higher water depth. The wave period also increases with the increase in wind speed and water depth.

11.17 WAVE POWER DATA

The information on wave data is built up with the aid of three main types of wave recorders, namely on the sea bed, on the surface and on ships. A 'crest' occurs when the vertical motion of water particles changes from upward to downward and it is 'trough' when the change is from downward to upward.

A simple recorder is the spark-plug recorder, consisting of a series of spark plugs each with a horizontal electrode, stationed in a vertical line with the plugs a few centimetres apart. As the sea water rises and falls, the plugs under water are shortened and the wave height is recorded.

Another method is a pressure sensing device, located on the sea bed which records water pressure. As the wave crosses the device the pressure increases and the device records the wave height. The information is recorded on the graph or on a cassette in the device itself.

A third method is a shipborne 'Wave Recorder' developed by the Institute of Oceanographic Science. A pressure sensor is mounted on a stationary ship, such as a weather or lightship, below the waterline. As the water level rises and falls outside the ship, the pressure sensor records the change. However, it should be coupled with an accelererometer to register the movement of the ship itself. The two measurements are added together and the graph prints out a picture of the waves.

Now, an electronic equipment has also been developed by the Centre for Earth Science Studies in Kerala. This instrument has already been installed in Trivandrum, Alleppey, Calicut and Tellicherry to record the wave data. These days the computer-based methods are used for the analysis of wave data. Such analysis of wave data helps to estimate the potential of maximum, minimum and mean wave power at desired locations. The following characteristics are calculated by analysing the wave data.

H — height of a significant wave

T — period of the significant wave

E/A — energy density

P/A — power density (energy per unit time)

P/W — power available per unit width

The values so computed are used for planning a wave power plant.

11.18 WAVE AREA FOR DETERMINING ENERGY

Ocean wave is a moving sheet of energy with a certain large breadth B, and a standard length equal to wavelength λ. It is represented in Figure 11.19. For determining the 'energy density' and 'power density' of the wave, it is necessary to know the area of the wave. The wave area is the projection on a horizontal plane of one full wave having width B and length equal to the wave length λ. The total content of the wave energy E is proportional to the area A as shown in Figure 11.19.

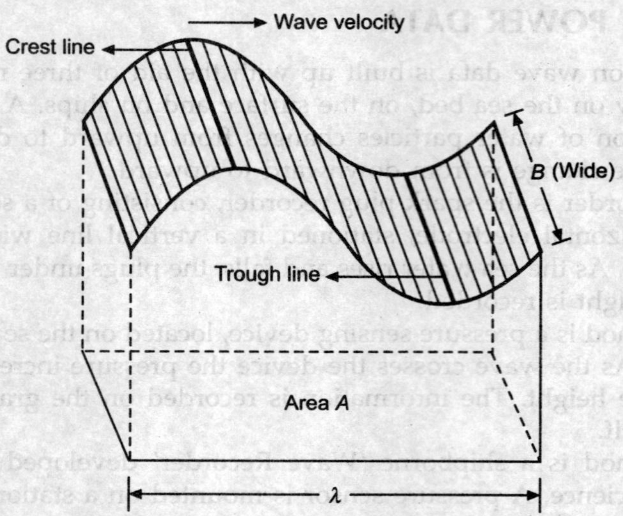

FIGURE 11.19 Water wave width B and length λ ($B > \lambda$).

EXAMPLE 11.3

A progressive sea wave has a wave width of 100 m with a period of 5 seconds. Calculate the wavelength, the wave velocity and the wave area.

Solution

$$\text{Wave Length, } \lambda = 1.56T^2$$

$$= 1.56 \times 5^2 = 39 \text{ m}$$

$$\text{Wave velocity, } C = \frac{\lambda}{T} = \frac{39}{5} = 7.8 \text{ m/s}$$

$$\text{Wave area, } A = \text{wave length} \times \text{wave breadth}$$

$$= \lambda \times B$$

$$= 39 \times 100$$

$$= 3900 \text{ m}^2$$

11.19 MATHEMATICAL ANALYSIS OF WAVE ENERGY

A progressive water wave can be represented by a sine curve as shown in Figure 11.20. Though the sea waves are highly irregular, such a wave is assumed to be of sinusoidal harmonic wave shape for the purpose of mathematical analysis. The wave is moving in the direction of the x-axis, $2a$ is the height of the wave from crest to trough in the direction of y-axis, i.e., the amplitude is a.

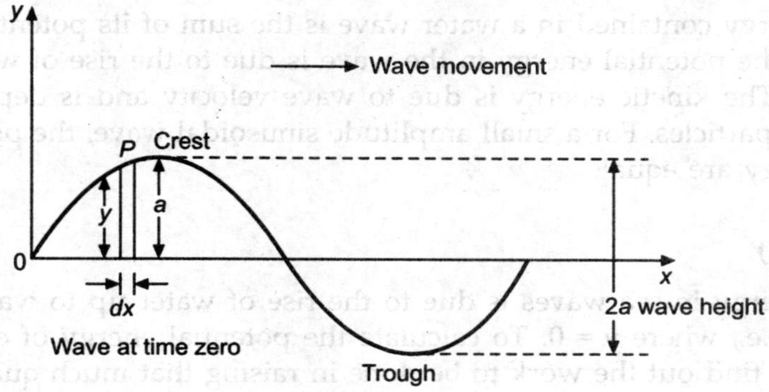

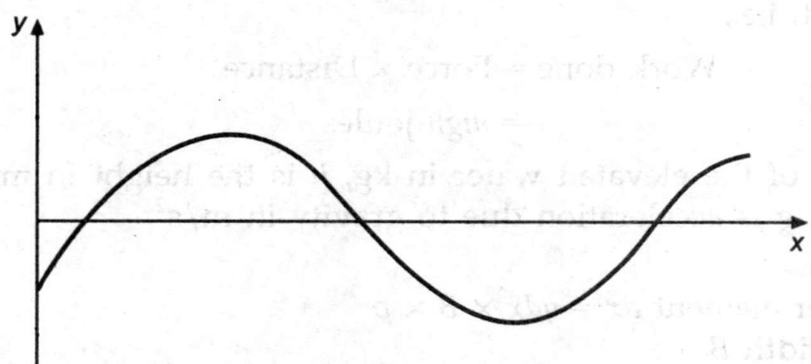

FIGURE 11.20   Two-dimensional progressive wave at time zero and time θ.

Let there be a point P on the wave surface with an element of thickness dx along the x-axis with a co-ordinate y on the y-axis. The wave being sinusoidal the value of coodinate y can be expressed with the following equation

$$y = a \sin\left(\frac{2\pi x}{\lambda} - \frac{2\pi\theta}{T}\right) \tag{11.7}$$

where the coordinates of point P are (x, y), and

  a = wave amplitude, in metre
  λ = wavelength, in metre
  θ = time, in second
  T = wave period.

For a particular wave, $\dfrac{2\pi}{\lambda}$ and $\dfrac{2\pi}{T}$ are constants.

Let $\dfrac{2\pi}{\lambda} = M$ and $\dfrac{2\pi}{T} = N$ (phase rate)

then

$$y = a \sin(Mx - N\theta) \tag{11.8}$$

$$(Mx - Nt) = 2\pi\left(\frac{x}{\lambda} - \frac{\theta}{T}\right) = \text{phase angle, dimensionless}$$

The total energy contained in a water wave is the sum of its potential energy and kinetic energy. The potential energy in the wave is due to the rise of water above the mean sea level. The kinetic energy is due to wave velocity and is dependent on the motion of water particles. For a small amplitude sinusoidal wave, the potential energy and kinetic energy are equal.

Potential energy

The potential energy in sea waves is due to the rise of water up to wave crest above the mean level, i.e., where $y = 0$. To calculate the potential energy of elevated water, it is necessary to find out the work to be done in raising that much quantity of water to the elevated height, i.e.,

$$\text{Work done} = \text{Force} \times \text{Distance}$$

$$= mgh \text{ joules}$$

where m is the mass of the elevated water in kg, h is the height in metres to which the water is elevated, g is acceleration due to gravity in m/s^2.

Here,

m = mass of water element $dx = ydx \times B \times \rho$
    with wave width B

$h = y/2$

ρ = water density, in kg/m^3

$\therefore$
$$\text{Work done} = ydx \times B \times \rho \times (y/2) \times g$$

$$\text{Potential energy} = \frac{1}{2} g\rho y^2 Bdx \tag{11.9}$$

Substituting the value of y from Eq. (11.7) and integrating from θ to λ for wave area $\lambda \times B$,

$$\text{Potential Energy (PE)} = \frac{1}{2} g\rho B \int_0^\lambda a^2 \sin^2(Mx - Nt)dx, \text{ assuming } t = 0$$

$$= \frac{g\rho Ba^2}{2} \left(\frac{1}{2}x - \frac{1}{4M} \sin 2Mx \right)_0^\lambda$$

$$= \frac{1}{4} g\rho Ba^2 \lambda \tag{11.10}$$

Wave area, $A = B\lambda$

Potential energy density per unit area $= \dfrac{\text{PE}}{A}$

Hence,
$$\frac{\text{PE}}{A} = \frac{1}{4} g\rho a^2 \tag{11.11}$$

If g_c is the conversion factor[*],

$$g_c = 1.0 \text{ kg/m/N-s}^2$$

then,

$$\frac{\text{PE}}{A} = \frac{1}{4}\rho a^2 \frac{g}{g_c} \text{ J/m}^2 \tag{11.12}$$

Kinetic energy

Due to blowing of wind on the surface of ocean, water waves moving over the ocean surface are fast. Due to their high speed, the ocean waves have a lot of kinetic energy.

When the amplitude a of the wave is small compared to its wavelength, then the potential energy and the kinetic energy are equal. The kinetic energy of the wave is same as in Eq. (11.10). It is therefore expressed as

$$\text{KE} = \frac{1}{4} g\rho B a^2 \lambda \tag{11.13}$$

The density of kinetic energy is given by

$$\frac{\text{KE}}{A} = \frac{1}{4} g\rho a^2$$

Total energy

The total energy contained in the ocean wave having wavelength λ, period T and breadth B is the arithmetic sum of potential and kinetic energies.

$$E = \text{PE} + \text{KE}, \text{ in joules}$$

$$= \frac{1}{4} g\rho B a^2 \lambda + \frac{1}{4} g\rho B a^2 \lambda$$

$$= \frac{1}{2} g\rho B a^2 \lambda$$

Energy density is the energy of the wave per unit area.

$$\text{Area, } A = \lambda B, \text{ in m}^2$$

Hence, energy density $= \dfrac{E}{A} = \dfrac{1}{2} g\rho a^2$, in J/m $\tag{11.14}$

Wave power

Power is expressed as energy per unit time. Thus,

---

[*] If m = mass of a body in kg, g is acceleration due to gravity, then the weight $W = mg$ and the unit of weight = kg-m/s^2 = N (newton).

The unit of force in SI is newton, i.e., 1 newton force can accelerate 1 kg mass to 1 metre per second^2. Force 1 kgf = 9.81 N.

$$\text{Power} = \frac{\text{Energy supplied}}{\text{Time taken}}, \text{ in J/s}$$

$$= \text{Energy} \times \text{Frequency, in W}$$

$$= \frac{1}{2} g\rho B\lambda a^2 f, \text{ in W}$$

$$\text{Power density} = \frac{P}{A} = \frac{1}{2} g\rho a^2 f, \text{ in W/m}^2 \tag{11.15}$$

where $A = \lambda B$ in sq. metres, f is the frequency in cycles per second and a is the amplitude of the wave in metres.

EXAMPLE 11.4

Ocean waves on an Indian coast had an amplitude of 1 m with a period of 5 s measured at the surface water 100 m deep. Calculate the wavelength, the wave velocity, the energy density and the power density of the wave. Take water density as 1000 kg/m^3.

Solution

$$\text{Wavelength, } \lambda = 1.56T^2$$

$$= 1.56 \times 5^2$$

$$= 39 \text{ m}$$

$$\text{Wave velocity, } C = \frac{\text{Wavelength } \lambda}{\text{Period } T}$$

$$= \frac{39}{5} = 7.8 \text{ m/s}$$

Wave frequency,
$$f = \frac{1}{5} \text{ s}^{-1}$$

Energy density,
$$\frac{E}{A} = \frac{1}{2} \times 1000 \times 1^2 \times 9.81$$

$$= 4905 \text{ J/m}^2$$

$$\text{Power density, } \frac{P}{A} = \left(\frac{E}{A}\right) f = 4905 \times \frac{1}{5}$$

$$= 981 \text{ W/m}^2$$

11.20   EMPIRICAL FORMULAE ON WAVE ENERGY

Wave energy is derived from wind energy. The higher the wind speed over the sea surface, the higher the wave height and so is the wave energy.

Scripps formula

The Scripps formula proposed by the Scripps Institution of Oceanography in La Jolla, California gives a relationship between wave height and wind velocity as

$$H = 0.085U^2$$

where H is the wave height in metres, and U the wind speed in knots (1 knot = 1.4 km/h).

Zuider Zee formula

The Zuider Zee formula combines five variables, namely the wind speed, the fetch, the rise of water level, the water depth and the angle between the wind direction and the fetch, i.e.,

$$H = \frac{KV^2 F \cos a}{D}$$

where

    H = rise in water level above the normal, in metres
    $K = 6.08 \times 10^{-3}$ (constant)
    F = fetch, i.e., unobstructed largest dimension of the lake, in metres
    V = wind speed, in km/h
    D = average water depth, in metres
    a = angle between the wind direction and the fetch

11.21 WAVE ENERGY CONVERSION

Waves with an amplitude of 2 m and period of 10 s are of considerable interest for power generation with energy fluxes averaging between 50 kW and 70 kW per metre width of the oncoming wave. Wave energy can be better concentrated than the solar energy. Devices that convert energy from waves can therefore produce much higher power densities than those produced by solar devices.

Ocean wave energy is primary energy. Our approach is to convert it into usable secondary energy. Based on the design data developed in the laboratories, a demonstration plant of 150 kW capacity for conversion of wave energy into electrical energy has been built at Vizhinjam near Trivandrum. This site was selected considering its good wave power potential, easy access to deep water, away from cyclonic zone and nearness to the available infrastructural facilities. The plant was commissioned in October 1991.

11.22 PRINCIPLE OF WAVE ENERGY PLANT

The wave energy plant utilizes an 'oscillating water column' chamber and a self-rectifying air turbine to produce power. The device works similar to the operation of a bellow. Ocean waves enter the chamber inside the caisson and cause the water mass to move up and down producing a bidirectional air flow through an opening at the top of the caisson, as detailed in Figure 11.21. The special design of the turbine makes it rotate unidirectionally even though the actuating air flow is bidirectional. The turbine drives an induction generator connected to the grid.

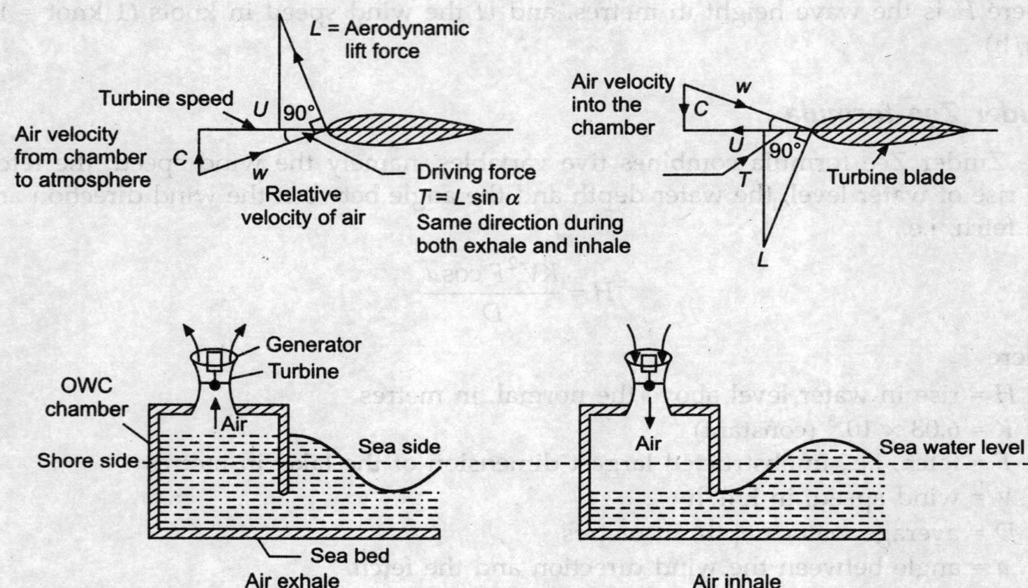

FIGURE 11.21 Principle of the oscillating water column device giving bidirectional air flow to unidirectionally rotating turbine.

11.22.1 Design Parameters

The power plant consists of a concrete caisson of size 17 m × 23 m accommodating the 'oscillating water column' chamber. The waves entering through the front of the caisson through a submerged opening below the lip wall produce a bidirectional air flow on the top of the caisson as shown in Figures 11.22(a) and (b). The caisson structure is designed for an extreme wave height of 7 m and a wave period of 10 seconds. The generating system is designed to deliver a peak power of 150 kW at a significant wave height of 1.52 m.

The turbine duct of 2 m diameter placed at the top of the caisson encloses the air turbine and a squirrel cage induction generator. The turbine blades have a special profile with a chord of 380 mm. There are eight blades on the rotor with a hub-tip ratio of 0.6. The squirrel cage induction generator with a synchronous speed of 1000 rpm operates on 440 volts, 3 phase, 50 Hz supply and is connected to the grid through a control panel and long cable to the shore transformer station.

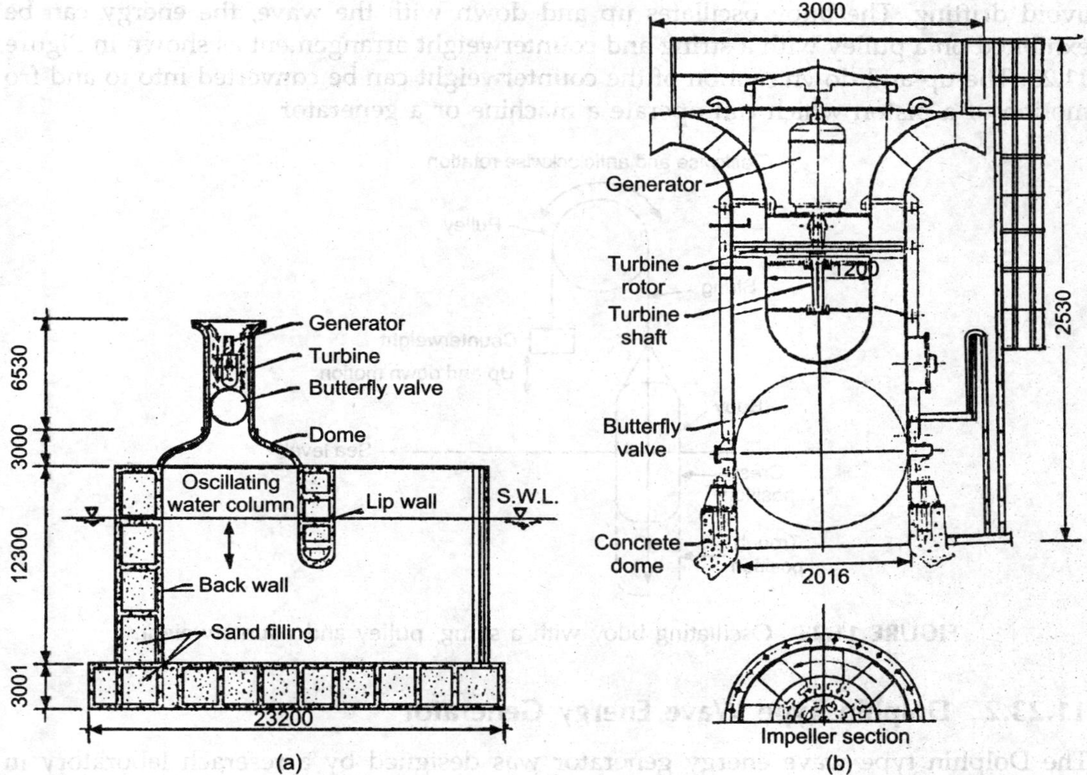

FIGURE 11.22 (a) Cross section of the caisson module at Vizhinjam, and (b) cross section of 150 kW power module.

11.23 WAVE ENERGY CONVERSION MACHINES

Wave energy is a combination of kinetic and potential energies available in sea waves. The forward motion of sea water can easily be seen on sea beaches, lashing up to 100 metres. In deep sea this forward motion of the wave strikes the ships, depicting the presence of kinetic energy. The potential energy is due to rise of sea water at the wave crest. The difference of head between the crest and the trough of sea wave is the potential energy. It can easily be experienced when a large ship in the ocean is lifted up by swell and oscillates up and down due to huge ocean waves.

Thus, if the wave advances in a horizontal plane it is due to kinetic energy; when the water moves in the vertical plane, it is the action of potential energy. Engineers of different countries have prepared several designs of wave machines to harness wave energy. Few of them, which have scope of improvement, are discribed.

11.23.1 Buoy Type Machine

The buoy is a floating part of a system which rises and falls with rise and fall of sea waves. However, the device is moored and anchored as per design methodology to avoid drifting. The buoy oscillates up and down with the wave, the energy can be exhibited on a pulley with a string and counterweight arrangement as shown in Figure 11.23. The up and down motion of the counterweight can be converted into to and fro motion of a piston which can operate a machine or a generator.

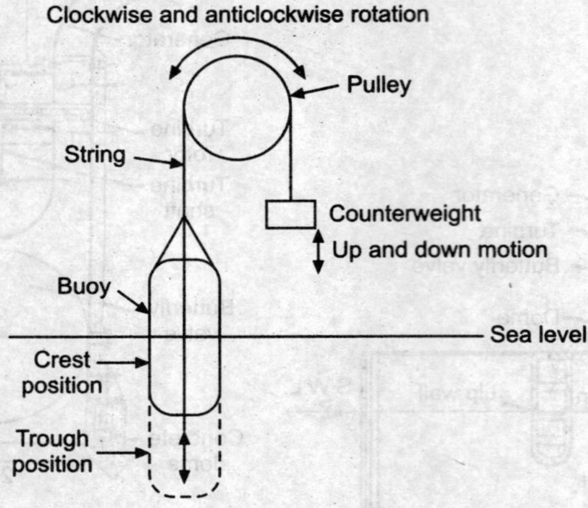

FIGURE 11.23   Oscillating buoy with a string, pulley and counterweight.

11.23.2   Dolphin Type Wave Energy Generator

The Dolphin type wave energy generator was designed by a reserach laboratory in Japan. It essentially consists of the following components as detailed in Figure 11.24.

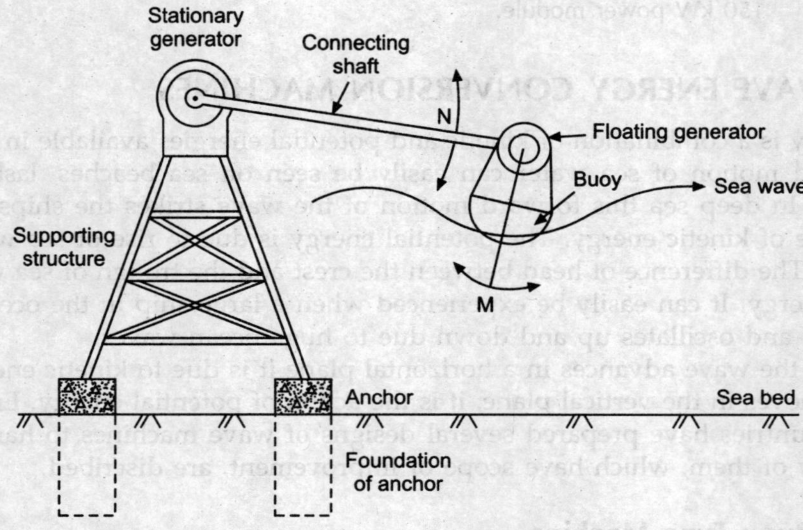

FIGURE 11.24   Dolphin type wave energy generator.

A supporting structure is built in the sea bed to provide a firm position for the equipment. The structure is erected on pile foundations. One generator is installed on the top of the structure which collects wave energy from the connecting shaft with rolling motion. The gear arrangement with the stationary generator rotates the rotor to generate electric power. The buoy is at the other end of the connecting rod floats and has two motions, namely rolling motion and oscillatory motions represented by N and M respectively. The floating generator collects wave energy from the buoy through a gear arrangement and continuously generates power.

$$\text{Power density, } \frac{P}{B} = 1740a^2T, \text{ in W/m}$$

where
   B = width of the wave, in metres
   a = amplitude of the wave, in metres
   T = wave period, in seconds

Normally one dolphine type wave energy generator is of 100 kW capacity. Several such wave energy generating systems are installed, say 50 numbers, along a width of 500 metres to have an installed capacity of 5 MW.

EXAMPLE 11.5

An array of Dolphin type wave energy generators is installed along a width of 500 m. The mean amplitude of the wave is 2 m with a period of 10 s. Calculate the installed capacity of the plant.

Solution

$$\frac{P}{B} = 1740a^2T, \text{ in W/m}$$

$$= 1740 \times 2^2 \times 10 = 69600 \text{ W/m}$$

or
$$P = 69600 \times 500$$

$$= 34800000 = 34.8 \text{ MW}$$

11.23.3 Oscillating Ducks

This wave power equipment was designed by Stephen Salter at Edinburgh university in Scotland. It is a float type wave energy conversion plant in which several duck-shaped devices (each 25 m long) are installed in a linear width-wise array along a line which is perpendicular to the direction of the wave. The system consists of a long cylindrical spine of 15 m diameter on which cam shaped ducks are installed in an array to form an assembly as shown in Figure 11.25. It responds to the incoming wave with a nodding action.

When the forward moving wavefront strikes the head on the face of the ducks, wave energy is passed on and the ducks start to oscillate. The face of the duck is designed for maximum wave energy absorption. Power is generated by the relative motion of the ducks where the wave energy is converted into mechanical energy. The cylindrical spine transfers motion through linkages and gears to the generator rotor. The overall length of the cylindrical spine varies between 100 m and 500 m.

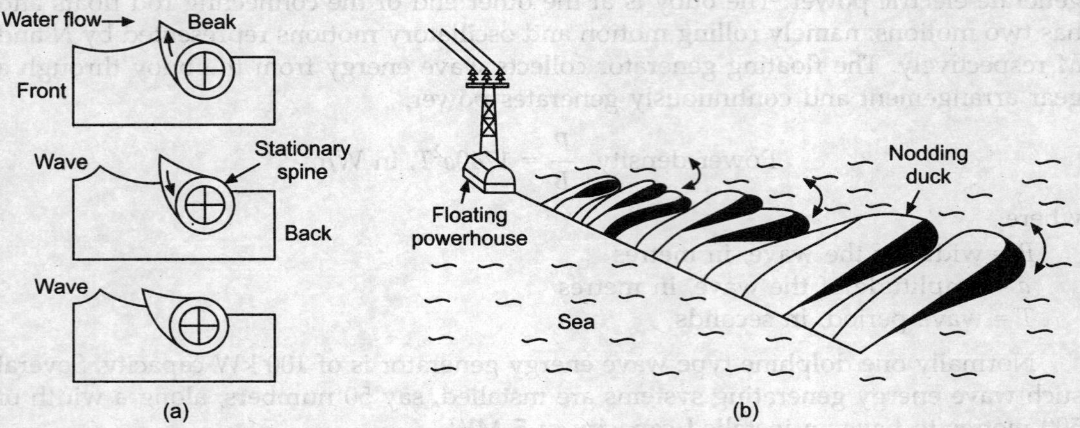

FIGURE 11.25 (a) Phases of duck motion, and (b) oscillating ducks with a floating powerhouse.

To achieve a highly efficient absorption it is necessary to mount a series of ducks on a non-movable spine. If the spine is sufficiently long (more than wavelength), the angular distribution of the waves incident on this structure will produce phase cancellations of translation force components along the spine and the spine will remain stationary.

11.24 WAVE POWER DEVELOPMENT IN INDIA

Wave power development is making a headway in India under the 'Department of Ocean Development' Government of India and Wave Energy Group, Indian Institute of Technology Madras. India has a big wave energy potential of about 90,000 MW along its 6000 km long coast. Though the wave energy is thinly distributed, its development is of prime importance in the face of energy crisis of the 21st century.

India has made a good beginning for harnessing wave energy by designing, installing and operating a 150 kW conversion system at Thiruvananthapuram. The location of the station is at the end of a breakwater wall where water depth is about 10 metres. The wave power available at the site is nearly 13 kW/m. The wave energy system consists of an oscillating water column, a chamber floating in the sea experiencing wave action through a side opening as shown in Figures 11.26(a) and (b). The chamber size is 10 × 10 m with a height of 15 m. A Wells turbine has been installed in the same direction irrespective of the direction of air flow. The turbine operates on symmetrical aerofoil concept. The turbine is 2 m in diameter and coupled to an induction generator. The wave patterns are irregular in amplitude, phase and direction, so the wave energy devices parameters match the irregular slow medium and high amplitude wave motion.

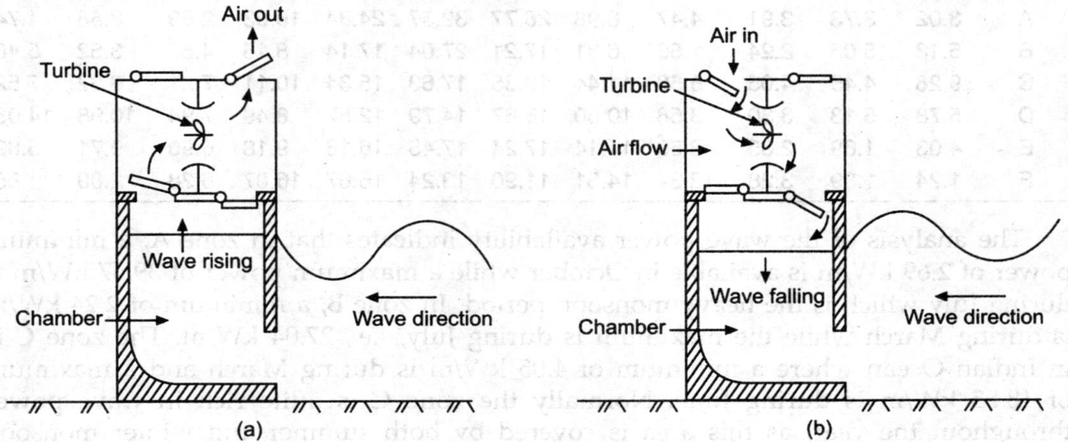

FIGURE 11.26 (a) OWC wave energy conversion system (wave rising), and (b) OWC wave energy conversion system (wave falling).

11.25 SIX ZONES OF INDIAN COASTLINE

To expedite and identify high wave energy areas suitable for power development, the National Institute of Oceanography Goa has divided the Indian coastline into six zones, namely A, B, C, D, E and F as shown in Figure 11.27.

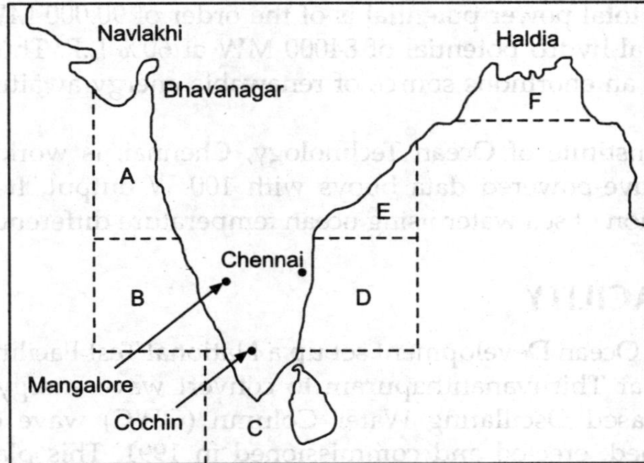

FIGURE 11.27   Map of wave power zones in India.

As there is variation in wave regime in different zones during different months the estimated wave power in kW/m is given in Table 11.5

TABLE 11.5 Analysis of wave power

Zone	Jan.	Feb.	Mar.	Apr.	May.	Jun.	Jul.	Aug.	Sept.	Oct.	Nov.	Dec.
A	3.02	3.73	3.91	4.47	6.98	26.77	39.57	24.84	10.03	2.69	3.58	4.74
B	5.13	5.05	2.24	1.56	6.31	17.21	27.04	17.14	8.15	4.55	3.52	5.40
C	9.26	4.45	4.05	5.50	11.44	18.85	17.69	15.34	10.11	7.21	6.67	7.52
D	5.78	5.13	3.30	3.58	10.60	16.67	14.79	12.57	8.49	7.94	10.98	14.05
E	4.03	1.69	2.35	3.69	11.14	17.24	17.45	16.16	9.18	6.90	9.71	5.62
F	1.24	1.39	3.28	12.34	14.31	11.90	13.24	16.67	16.07	6.28	2.80	1.85

The analysis of the wave power availability indicates that in zone A, a minimum power of 2.69 kW/m is available in October while a maximum power of 39.57 kW/m is during July which is the heavy monsoon period. In zone B, a minimum of 2.24 kW/m is during March while the maximum is during July, i.e., 27.04 kW/m. The zone C is in Indian Ocean where a minimum of 4.05 kW/m is during March and a maximum of 18.85 kW/m is during June. Normally the zone C is quite rich in wave power throughout the year, as this area is covered by both summer and winter monsoon winds. Wave power potential in zone D is a minimum of 3.30 kW/m during March and a maximum of more than 14 kW/m occurs during June, July and December. Zone E is of normal pattern where a minimum of 1.69 kW/m is during February and a maximum of 17.45 kW/m during July. Zone F clearly depicts wave power potential of more than 11.90 kW/m from April to September, while during the remaining six months it is quite low.

Thus, on the estimates of the distribution of wave energy (kW/m) of sea frontage, the potential is seen to vary from 39 kW on the West coast to 15 kW on the East coast. With an average estimated wave power potential of 15 kW/m and total coastline of about 6000 km the total power potential is of the order of 90,000 MW. It is comparable to total conventional hydro potential of 84000 MW at 60% L.F. Thus, the wave power on Indian coasts is an enormous source of renewable energy awaiting to be harnessed commercially.

The National Institute of Ocean Technology, Chennai, is working on design and development of wave-powered data buoys with 100 W output. Its activity is on the project of desalination of sea water using ocean temperature difference and wave power.

11.26 TEST FACILITY

The Department of Ocean Development set up a National Test Facility off the Vizhinjam fishing harbour near Thiruvananthapuram to convert wave energy into electricity. A concrete caisson-based Oscillating Water Column (OWC) wave energy conversion system was designed, erected and commissioned in 1991. This plant has been taken over by the National Institute of Ocean Technology (NIOT) in Chennai for R&D work. The NIOT has developed a special impulse turbine for this project.

As waves are a random and intermittent source of energy, the average power is generally low, though a peak power output of over 50 kW was obtained from the plant. The NIOT is now working on floating wave-powered buoys of smaller rating, using the energy conversion technique already developed. Special control devices with battery back-up will be provided to make up for the variations in wave energy availability.

As an offshoot, the NIOT took up another project to use this caisson-based wave energy facility to provide desalinated water to the local community. For this purpose, a commercially available reverse osmosis plant was procured and sinked to the wave energy caisson system at Vizhinjam.

11.27 ECONOMICS

A new methodology has emerged in energy analysis, which introduces the concept of 'energy ratio'. It is defined as the ratio of total energy output over the lifetime of the installation to the total energy required to build and operate the installation. Systems which have the energy ratio of 8 or above are considered 'useful' irrespective of their financial viability. In contrast, a coal-based powerhouse may be economically attractive, but its energy ratio is obviously less than one. Energy analysis thus compliments financial analysis.

A study of energy analysis of wave and tidal power was carried out by K.G. Smith and J.S. Varley in the UK. It has been analysed that the major components of the energy requirement for a wave power installation are due to (i) tools and plants about 60%, (ii) concrete about 12%, and (iii) steel about 4%.

For any assumed wave power, $P = 50$ kW/m, the energy ratio is 13 at an extraction efficiency of 23% for an RCC duck converter.

For harnessing wave energy, breakwater is constructed which checks the erosion of the coast. Thus, extraction of wave energy is of added attraction to coastal communities like Kerala, particularly vulnerable to sea erosion. Wave energy installation and anti-erosion works are complimentary and in totality economically viable.

11.28 INTRODUCTION TO THERMAL ENERGY CONVERSION

The oceans and the seas which cover about 70% of the earth are constantly receiving solar radiation and act as the largest natural solar collector. An ocean as a collector has an enormous storage capacity. Energy from the ocean is available in several forms, such as ocean thermal energy, wave energy and tidal energy.

Ocean Thermal Energy Conversion (OTEC) is a new technology, needed to be harnessed especially in India where the coastline is about 6000 km. Basically, the OTEC converts the thermal energy, available due to temperature difference between the warm surface water and the cold deep water, into electricity. Power from the OTEC is renewable and eco-friendly. An OTEC plant can operate in remote islands and sea-shore continuously. It is very low grade solar thermal energy, so the efficiency of energy recovery is quite low. However, since the ocean thermal energy is dispersed over a large ocean surface area, it has a big potential. According to MNRE, the overall potential of ocean energy in the country may be in excess of 50,000 MW. There is an enormous opportunity to tap this renewable source of energy.

11.29 WORKING PRINCIPLE—OTEC

There exists a temperature difference of about 20°C between the warm surface water of the sea (receiving and absorbing solar radiation) and the cold deep water (which flows from the Arctic regions in deep layers) in equatorial areas between latitude 30° S and 30° N. Solar heat energy is absorbed by ocean water. It can be explained by 'Lambert's law of absorption'. The law states that "each water layer of identical thickness absorbs an equal fraction of light that passes through it". Thus, the intensity of heat decreases with the increase in water depth. Due to large heat transfer at the ocean surface water, the highest temperature is attained just below the top surface. A typical temperature variation curve with distance from the surface is shown in Figure 11.28.

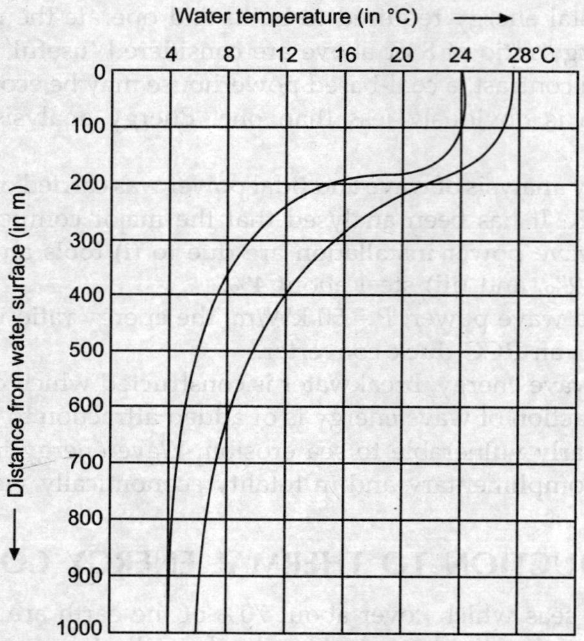

FIGURE 11.28 Ocean water temperature variation with distance from water surface.

It may be seen that the temperature at the surface changes slowly, then remains constant at a depth of about 200 m. Subsequently, the temperature decreases asymptotically and approaches a low value of about 4°C at a depth of 1000 metres. The difference in temperature between the surface and the deeper parts of the ocean is utilised to generate electrical energy. The basic process of OTEC is to bring the warm surface water and the cold water from a certain depth of the sea through pipes so as to act as 'heat source' and 'heat sink' for operating a heat engine. It will form the same system as that of conventional thermal power station with nil fuel consumption.

The OTEC plants are of three types, namely 'closed', 'open', and thermoelectric. The important broad features of these plants are as follows:

11.30 CLOSED RANKINE CYCLE OR ANDERSON CLOSED CYCLE OTEC SYSTEM

The closed cycle system using a low boiling point working fluid like ammonia or propane is shown in Figure 11.29.

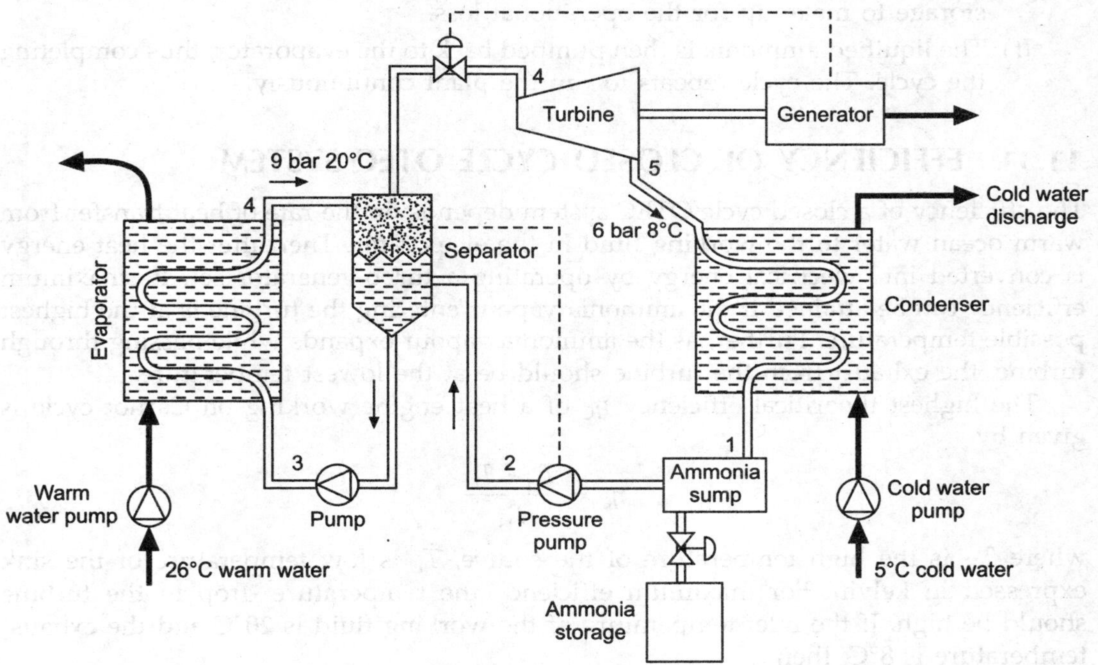

FIGURE 11.29   Schematic diagram of a closed Rankine Cycle OTEC system.

It may be seen that warm water from the surface which is at a temperature of about 26°C is brought in one pipe, and cold water at a temperature of around 5°C is brought in another pipe from a depth of about 1000 metres. In OTEC plants two water pipes are used in conjunction with a working fluid to generate electric power. Different operational activities of the plant are:

(a) The warm sea water evaporates the liquid ammonia into vapour in a unit called an evaporator. This can be done because ammonia exists in the form of gas at the temperature corresponding to the surface sea water.

(b) The liquid ammonia which is not evaporated collects in a unit known as separator, which again recirculates through the evaporator.

(c) The evaporated ammonia in the form of high pressure vapour is made to pass through a turbine where its pressure and temperature make the turbine to rotate, thus converting thermal energy into mechanical energy. The rotating turbine if coupled to an electric generator produces electric power.

(d) The ammonia vapour coming out of the turbine, which is now at the lower pressure than when it entered the turbine is condensed back into liquid ammonia by cooling it with the colder sea water brought up from the deep part.

(e) The liquified ammonia collects in an ammonia sump. After a few hours of operation, the make-up quantity of ammonia is added from the ammonia storage to make up for the operational loss.

(f) The liquified ammonia is then pumped back to the evaporator, thus completing the cycle. The cycle repeats to run the plant continuously.

11.31 EFFICIENCY OF CLOSED CYCLE OTEC SYSTEM

The efficiency of a closed cycle OTEC system depends on the rate of heat transfer from warm ocean water to the working fluid in the evaporator. Thereafter the heat energy is converted into electrical energy by operating a turbo-generator set. Its maximum efficiency can be attained if the ammonia vapour entering the turbine is at the highest possible temperature. Further, as the ammonia vapour expands while passing through turbine, the exhaust from the turbine should be at the lowest temperature.

The highest theoritical efficiency η_C of a heat engine working on Carnot cycle is given by

$$\eta_C = \frac{T_H - T_L}{T_H}$$

where T_H is the high temperature of the source, T_L is low temperature of the sink expressed in kelvin. For maximum efficiency the temperature drop in the turbine should be high. If the inlet temperature of the working fluid is 20°C and the exhaust temperature is 8°C, then

$$\eta_C = \frac{(273 + 20) - (273 + 8)}{(273 + 20)} = \frac{12}{293}$$

$$= 0.041 \text{ or } 4.1\%$$

It is a theoretical value, but in actual practice the efficiency is only 2%.

11.32 THERMOELECTRIC OTEC

The thermoelectric OTEC system was developed by Solar Energy Research Institute Colorado USA, during 1979. The OTEC system which operates on the thermoelectric principle is simple in construction and economical. Semiconductors are used to design two separate packs covered by a thin thermal conducting sheet as shown in Figure 11.30. Warm water from the surface of the ocean is circulated over one device and the cold water pumped from the depth of the ocean is allowed to flow over the other device. The temperature difference between these two water with the solid state semiconductor devices generates the electric power. The OTEC plant economy is dependent on large variation of water temperature used from the surface and the deep ocean (minimum 20°C).

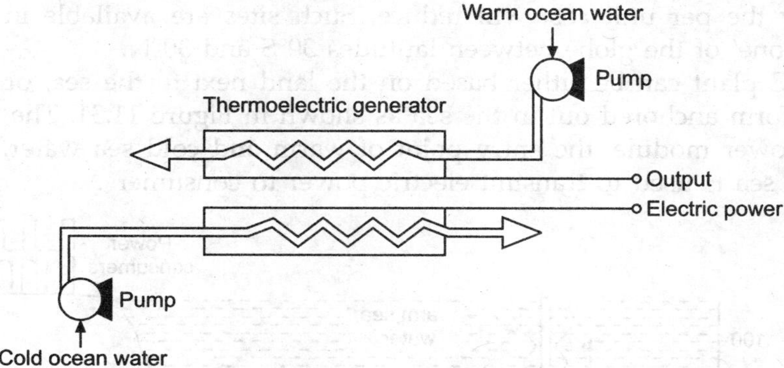

FIGURE 11.30 Thermoelectric ocean thermal energy conversion equipment.

A part of electric power generated is used to operate the pumps and other equipment. For commercial operation of OTEC plants, it is better to install land-based OTEC equipment on shores of those islands where the ocean is steep for easy access to cold water.

11.32.1 Bio-fouling

The raw ocean water which is pumped in for evaporator and condenser, contains micro-organisms which stick on the water side of both the heat exchangers. This biological impurity of sea water that deposits and grows on the evaporator and condenser metal surfaces, creating thermal resistance for heat transfer, is known as 'bio-fouling'. A thin layer of slime, i.e., a sticky substance from marine organism, also known as micro-fouling, continuously grows thicker by attaching to itself more and more biological contents from sea water. This creates a serious problem for heat transfer, consequently reducing the efficiency of the OTEC plant.

To maintain the optimum plant performance and efficiency, bio-fouling should be cleaned mechanically in addition to chemical treatment by chlorination. The problem of bio-fouling is more predominent in closed-cycle OTEC plants. In open-cycle plants the flow rate of sea water is quite large, so organisms have a less chance to stick with the heat exchanger surface, thus causing little bio-fouling. However, as per the present approach in OTEC, a closed-cycle system needs to be installed for harnessing the ocean thermal energy. Necessary design input is required in OTEC plants to off-set bio-fouling effects, especially in evaporators where warmer water is more conducive to the growth of micro-organism.

11.33 LOCATION OF OTEC PLANTS

The selection of a suitable site for an OTEC plant needs a temperature difference of about 20°C between the surface and the deep sea ocean water. If the temperature difference is higher the site becomes more suitable as it will increase the power output, consequently the per unit cost will reduce. Such sites are available in 'Torrid and Temperate Zone' of the globe between latitudes 30°S and 30°N.

An OTEC plant can be either based on the land next to the sea, or placed on a floating platform anchored out in the sea as shown in Figure 11.31. The arrangement shows the power module, the entry point of warm and cold sea water, and a cable immersed in sea is used to transmit electric power to consumers.

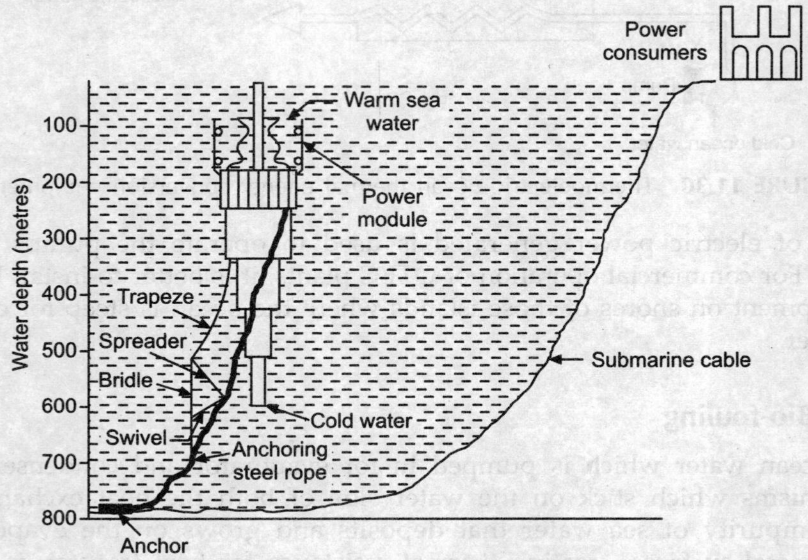

FIGURE 11.31 Artistic view of OTEC system.

For a land-based plant, the cold water pipe has to be brought in from an adequate distance so as to provide 600–800 metre depth of sea. A place where the slope falls off sharply from the land may be ideal, so that the length of the cold water pipe is kept to a minimum. The floating type plant produces electrical power on the platform, which has to be transmitted by a submarine cable to the shore. It is a costly preposition. In the plant ship variety, the plant is placed on the ship which can move on the sea to get the maximum temperature difference and the energy generated is consumed on the ship itself, where even an industrial unit may operate to manufacture a suitable product.

The most economic form of plant is the shore-based one which can be installed where considerable water depth occurs near the shore. With this type of plant there is no need to use expensive special cables for transmitting power. Further, one does not need any special floating type platform or mooring device which is needed for floating type plants. Many islands and certain continental sites are suitable for land-based plants.

11.34 APPLICATION OF OTEC

Energy from an OTEC can be converted into either electrical, chemical or protein form. Electric power generation, some forms of energy products and by-products that may result are shown in Figure 11.32. These plants may be combined with electrical energy intensive industries like ammonia, hydrogen or aluminium production. Further the OTEC plants are quite suitable for cogeneration of electricity and fresh water. In an open-cycle OTEC, sea water is used to generate steam which after operating the turbine, condenses to deliver fresh water. The turbine generator delivers the electric power.

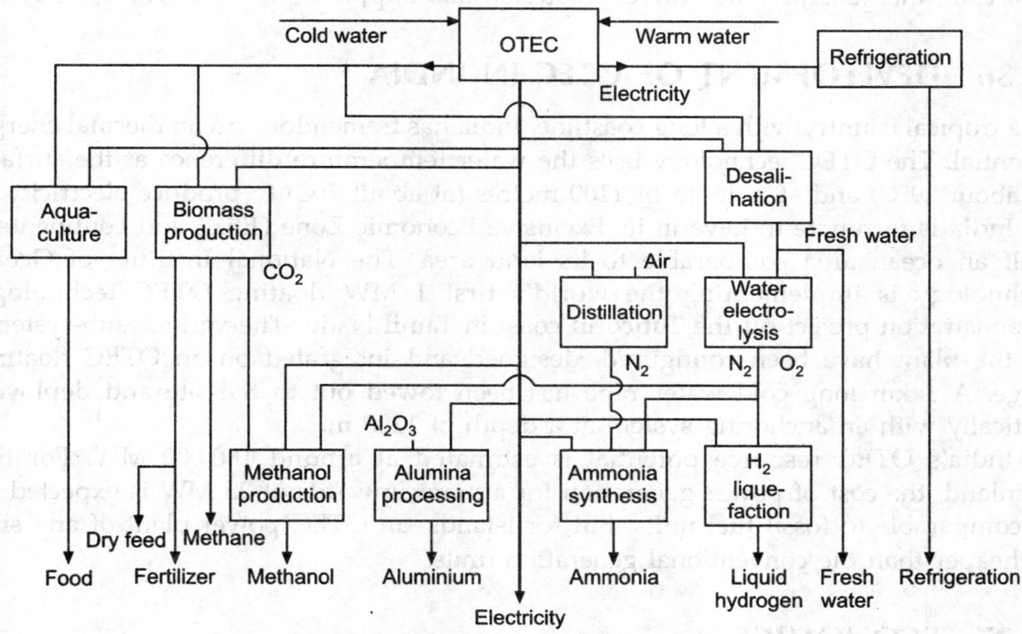

FIGURE 11.32   OTEC energy forms and by-products.

When the OTEC plants are combined with aquaculture, they provide marine food, besides valuable pearls. The cold water from the deeper sea which is rich in nutrients can be placed in a lagoon or lake, where these nutrients can help to raise fish, oysters or other type of biological life. These are rich food products and oysters can be used for raising costly pearls. Thus, the OTEC power generation is a multipurpose project producing and supplying several useful products, like the river valley multipurpose projects.

11.35   GLOBAL DEVELOPMENT OF OTEC PLANTS

The first OTEC power plant was developed in 1979 in the Hawaii state of the USA. It was a prototype 50 kW floating plant operated on closed Rankine cycle principle with ammonia as the working fluid. The plant was designed with the ocean water tempreature difference of 21°C. The available net power was only 15 kW, as 35 kW was used in pumping the warm and cold water. Its successful operation established that power generation through an OTEC system is technically and economically viable.

Another plant was installed in Nauru (Japan) in the Central Pacific Ocean during 1981. It was an on-land plant, which was economical in construction cost. But, it required a longer pipeline, 945 m for cold water pumping. The capacity of the plant was 100 kW (gross power), operated with a sea water temperature difference of 21.7°C. The net power output was 31.5 kW. The turbine used was an axial flow type with 3000 rpm, the generator was directly coupled and supplied power at 415 V, 50 Hz.

11.36 DEVELOPMENT OF OTEC IN INDIA

As a tropical country with a long coastline, India has tremendous ocean thermal energy potential. The OTEC technology uses the water temperature difference at the surface (at about 29°C) and at a depth of 1100 metres (at about 7°C) to produce electricity.

India is fortunate to have in its Exclusive Economic Zone (EEZ) and continenteal shelf an ocean area comparable to its land area. The National Institute of Ocean Technology is implementing the world's first 1 MW floating OTEC technology demonstration project off the Tuticorin coast in Tamil Nadu. The various sub-systems for the plant have been configured, designed and integrated on an OTEC floating barge. A 1 km long cold water pipe has been towed out to the site and deployed vertically with an anchoring system at a depth of 1200 m.

India's OTEC resource potential is estimated at around 180,000 MW. For the mainland, the cost of power generation for a plant upwards of 25 MW is expected to be comparable to fossil fuel units. But for islands, an OTEC power plant of any size is cheaper than the conventional generation units.

11.37 ECONOMICS

Given favourable geographical and easy conditions and a sufficient difference in temperature between the surface and the deeper sea, the OTEC plant can produce electricity at rates which are comparable with conventional oil or coal-based electric power at certain locations. However, as the price of oil and coal rise, the number of sites where the OTEC will be competitive shall increase.

It is estimated by the United Nations that 99 nations have direct access to a possible OTEC thermal resource with an average monthly temperature difference exceeding 20°C within their own exclusive economic zone of sea. The world's potential of OTEC electric power projects is about 100,000 MW. As a thumb rule, for each megawatt power generation, an OTEC plant requires a sea surface of about 1 sq. km, which acts in a way like a solar collector, since it is the sun which warms the surface water. There is a constraint on the capacity of a turbine generator unit which is limited to 25 kW only, due to limitations of small temperature difference of 20°C between the warm and cold water. It entails a large flow of warm and sea water, so the capacity of a single unit has a limit keeping in view the engineering workability. Accordingly, the number of units needed will increase to obtain higher installed capacities.

REVIEW QUESTIONS

1. Derive the expression for the average theoretical power generated from a single-basin single-effect tidal scheme during one filling. The expression to be derived is

$$P = 225\,AR^2 \text{ kW}$$

   where

   A = area of basin, in m^2
   R = tidal range, in m.

2. Write three important components of a tidal power plant. Briefly explain the Kaplan type turbine with working range of water head.

3. Discuss the relative advantages and limitations of tidal power projects. How can the tidal power supply be made continuous?

4. What are important factors needed to absorb tidal power into the grid supply? Draw a load–duration curve in support of your answer.

5. Compare the sequential operation modes of a single-effect tidal scheme with those of the double-effect tidal scheme.

6. (a) Write short notes on single-basin and double-basin power plants.

   (b) Where is the largest operational tidal power station located?

7. (a) Distinguish between a 'Tidal Power Plant' and an OTEC plant with particular reference to:

   (i) An Anderson closed-cycled OTEC plant
   (ii) A 'two pool tidal' power plant
   (iii) A 'modulated' single pool tidal system

   (b) What is meant by 'tidal range' R?

8. A 56×10^6 m^3 tidal pool surface has a level following a sinusoid during a tidal cycle of 6.2083 hours. The maximum head is 10 m. Estimate the power generated during a cycle for the case of discharge into a reservoir with a constant level and at a constant mass flow rate of water.

   What are the practical existing ways of extracting this power and with what modifications?

9. Establish the equation for the power extracted from ocean waves in terms of wavelength, amplitude, period, ocean water density, etc.

10. Name in each case two schemes or methods for utilisation of wave energy and tidal energy.

11. (a) Derive an expression for the 'potential energy' of a sea wave as:

$$\frac{PE}{A} = \frac{1}{4}\rho a^2 \frac{g}{g_c}$$

   where

   A = area = $\lambda \times L$

λ = wavelength = C_t = $1.56t^2$

t = period

g = acceleration due to gravity

$2a$ = crest-to-trough height of the wave

L = width perpendicular to wave propogation in x-direction

g_c = conversion factor = 1.0 kg/m/N-s^2

ρ = water density = 1025 kg/m^3

C = wave velocity

(b) If the kinetic energy density is also equal to the above expression, find the wavelength (for a 10 s period wave) and wave velocity. Hence for a wave height of 4 m and 2 m at the surface of water 100 m deep, calculate the energy density and power density. Comment on wave equations and wave shapes with suitable diagrams.

12. (a) Derive an expression for the power that can be extracted from a sea wave in terms of amplitude and period of wave.

   (b) Calculate the power in a deep water of wavelength 100 m and amplitude 1.5 m.

13. Describe with the help of a simple schematic diagram the process of an ocean thermal energy conversion based on steam cycle.

14. (a) Explain Carnot efficiency of an OTEC plant with the help of a thermodynamic cycle on T-s plane

   (b) Discuss the problem faced in harnessing ocean thermal energy.

   (c) Bio-fouling is a biological process mostly associated with OTEC systems. Explain this statement.

15. (a) Compare and contrast an OTEC power generation with a 'drilled geothermal' system with particular reference to its working, efficiency and cost required as closed-cycle and open-cycle, vapour-dominated and liquid-dominated systems.

   (b) What is the efficiency of OTEC?

16. The success of 'Claude Open Cycle' OTEC generation depends on improved evaporator system and cogeneration. Explain this statement with particular reference to *T-s diagram of the closed cycle. What are the improvements proposed in a closed-cycle generation?*

17. (a) Explain the basic principle of ocean thermal energy conversion.

   (b) Describe the closed cycle OTEC system and mention its advantages and limitations.

18. (a) Give the schematic diagram of an OTEC plant and point out its major differences with a conventional thermal electrical plant.

   (b) Estimate the flow rate and the pipe diameter of the cold water intake for a 100 kW OTEC plant with the usual temperature difference between hot and cold water and an assumed turbine-generator efficiency of 8%.

(c) Indicate with reasons the suitable sites available for locating OTEC plants in India.

19. Write short notes on:

  (i) Tidal energy
  (ii) Bio-fouling and OTEC
  (iii) Wave energy
  (iv) Site selection of tidal wave and OTEC plants

20. Discuss the reasons for your selection of a power plant from geothermal, OTEC, nuclear, tidal, or wind for:

  (i) Base load operation
  (ii) Peak load operation.

BIOMASS ENERGY

12.1 INTRODUCTION

Biomass refers to solid carbonaceous material derived from plants and animals. These include residues of agriculture and forestry, animal waste and discarded material from food processing plants. Biomass being organic matter from terrestrial and marine vegetation, renews naturally in a short span of time, thus, classified as a renewable source of energy. It is a derivative of solar energy as plants grow by the process of photosynthesis by absorbing CO_2 from the atmosphere to form hexose (dextrose, glucose, etc.) expressed by the reaction

$$6CO_2 + 6H_2O \xrightarrow[\text{photosynthesis}]{\text{sunlight}} C_6H_{12}O_6 + 6O_2$$

Biomass does not add CO_2 to the atmosphere as it absorbs the same amount of carbon in growing the plants as it releases when consumed as fuel. It is a superior fuel as the energy produced from biomass is 'carbon cycle neutral'.

Biomass fuel is used in over 90% of rural households and in about 15% urban dwellings. Agriculture products rich in starch and sugar like wheat, maize, sugarcane can be fermented to produce ethanol (C_2H_5OH). Methanol (CH_3OH) is also produced by distillation of biomass that contains cellulose like wood and bagasse. Both these alcohols can be used to fuel vehicles and can be mixed with diesel to make biodiesel.

12.2 BIOMASS RESOURCES

Biomass resources for energy production are widely available in forest areas, rural farms, urban refuse and organic waste from agro-industries. Biomass classification is illustrated in Figure 12.1.

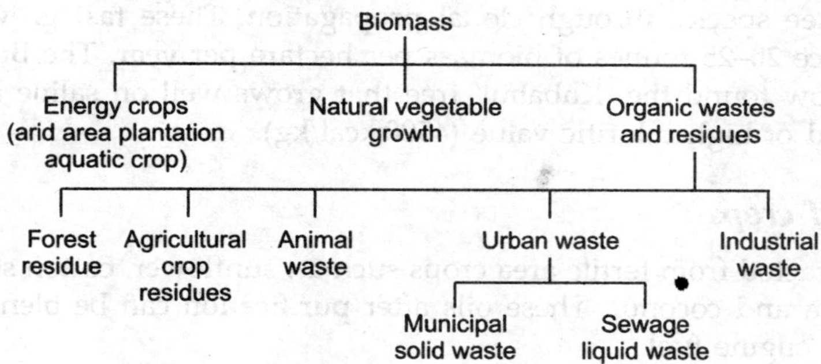

FIGURE 12.1 Biomass classification.

India produces over 550 million tonnes of agricultural and agro-industrial residues every year. Similarly, 290 million cattle population produces about 438 million tonnes of dung annually. Prime biomass sources are discussed below:

Forests

Forests, natural or cultivated are a rich source of timber, fuel wood, charcoal and raw material for paper mills and other industries. Fast growing trees like Eucalyptus, Neem, Kikar and Gulmohar are grown along canals, railway tracks and on lands of marginal quality. Wood, saw dust, and bark residue are generated in sawmills. Forests also provide foliage and logging residues. An important characteristic of forest residue is its calorific value, which is 4399 to 4977 kcal/kg for softwood foliage and 3888 to 5219 kcal/kg for hard wood species.

Agricultural crop residues

Crop residues are available in abundance as natural resource, easily collected and stored. These are, rice husk, wheat straw, corn cobs, cotton sticks, sugarcane bagasse, groundnut and coconut shells. These are converted into briquettes or pellets for use as clean fuel. These are called 'biofuels' which are high efficiency solid fuels.

Energy crops

Energy farming refers to the cultivation of fast growing plants which supply fuel wood, biomass that can be converted into gaseous and liquid fuels like biogas, vegetable oil and alcohol. To harvest biomass for power generation, energy plantation is done on degraded or wastelands which are saline, wind eroded lands in arid areas and water-logged lands.

Energy farming is promoted by MNRE in nine different agro-climate regions, namely, Garhwal (U.P.), Gwalpahar (Haryana), Udaipur (Rajasthan) and Shantiniketan (West Bengal). The other four centres are Madurai (Tamil Nadu), Calicut (Kerala), Raipur (Chhattisgarh), Bhubaneswar (Orissa). These centres produce quality seedlings of about 35 tree species through clonal propagation. These fast growing fuel wood species produce 20–25 tonnes of biomass per hectare per year. The Biomass Research Centre Lucknow found the 'Kubabul' tree that grows well on saline and rocky soils, provides wood of high calorific value (4500 kcal/kg).

Vegetable oil crops

Oil can be extracted from fertile area crops such as, sunflower, cotton seed, groundnut, rapeseed, palm and coconut. These oils after purification can be blended with diesel oil suitable as engine fuel.

There is an arid area shrub 'Jajoba', its seeds provide oil which is an important renewable source of energy. It is cultivated in Rajasthan, Gujarat and Orissa under hot-arid conditions. It is an ideal plant for areas of scanty rain with low fertility soil and produces up to 2000 kg of dry seed per hectare annually. Jajoba oil having good insulating property can be used as transformer oil. Its products are high quality lubricants and waxes, suitable for industry and transport sector. It is a good raw material for paints and varnishes.

Aquatic crop

Prominent Aquatic crops are algae, water hyacinth and sea weed. These plants grow abundantly in water bodies and provide organic matter for biogas plants.

Energy plantation programme is directed to bring sub-standard soil under cultivation. It restores the fertility of land, halts desertification, prevents soil erosion, reduces flooding and improves microclimate.

Animal waste

Animal waste, an organic material with combustible property, is a rich source of fuel. Dung cakes prepared with animal waste are used for cooking in rural and semi-urban areas. It is also a raw material for biogas plants.

Urban waste

Urban waste is of two types: (i) Municipal Solid Waste (MSW) which includes human excreta, household garbage and commercial waste. (ii) Liquid Waste from domestic sewage and effluents from institutional activities. As per MNRE estimate about 42 million tonnes of solid waste (1.15 lakh tonnes per day) and 6000 million cubic metres of liquid waste are generated every year in urban areas. At present MSW is dumped in sanitary landfills, where fuel gas is produced which is a valuable source of renewable energy. Sewage is suitably processed to produce biogas.

Industrial waste

Energy recovery from industrial waste was taken up in 1994. Projects are implemented with technical assistance of national laboratories. Projects developed under this programme are:

Pulp and Paper Industry Effluent, Starch and Glucose Industry Waste, Palm Oil Industry, Distillary Waste and Tanneries Waste. Each project is aimed to treat its waste for the production of bio-energy which can be used for power generation.

12.3 BIOFUELS

Biomass is an organic carbon-based matter obtained from plants. Biomass is a source of energy and 40% of the total energy consumed in India comes from wood, crop residues, cow dung, etc. for cooking and various domestic uses. Dry biomass gives heat energy by direct combustion.

Direct burning of firewood in traditional chulhas utilises only 10% heat. Besides inefficient burning, smoke discharge in kitchens is a health hazard. To harness fuel value, technologies are required to convert biomass into a high quality usable solid, liquid and gaseous fuels called 'biofuels'. Such fuels are discussed below.

Charcoal

Charcoal is a smokeless dry solid fuel with high energy density. Modern charcoal retorts (furnaces) operate at about 600°C to produce charcoal from 25–35% of dry biomass feed. It contains 75–80% carbon and is useful as a compact fuel. It can be burnt to provide heat for domestic, commercial and industrial applications.

Briquetting

Biomass briquetting is densification of loose biomass into a high density solid fuel. Biomass of any form such as cotton sticks, rice husk, coconut shells, saw dust and wood chips can be converted into briquetts. It reduces the volume-to-weight ratio, thus making transportation easy for efficient commercial and industrial use. The calorific value is about 3500 kcal/kg. Biomass briquettes can replace 'C' grade coal used in industrial boilers.

Vegetable oil

Vegetable oils such as rapeseed, palm, coconut and cotton seed oil can substitute diesel as engine fuel. Jajoba trees cultivated in marginal lands produce oil seeds. Jajoba oil is considered liquid gold like crude oil as it can be processed into a wide range of products like motor oil, lubricants, mono-unsaturated alcohols and oil of cosmetic value. Euphorbia species produce latex which after water removal give light hydrocarbon oil.

12.4 BIOGAS

Biogas can be produced by digestion of animal, plant and human waste. Digestion is a biological process that takes place in a digester with anaerobic organism in absence of oxygen at a temperature between 35°C and 70°C. In rural areas, household biogas plants operate from cow and buffalo dung which provide gas for cooking and lighting. Biogas is a mixture of CH_4 (55% to 65%), CO_2 (30% to 40%), H_2, H_2S and N_2 (< 10%) having a calorific value between 5000 and 5500 kcal/kg.

12.5 PRODUCER GAS

Producer gas is obtained by partial combustion of wood or any cellulose organic material of plant origin. It is a mixture of a few gases and its constituents are CO_2 (19%), CH_4 (1%), H_2 (18%), CO_2 (11%) and N_2 (45–60%). Hydrogen and methane keep heating value between 4.5 MJ/m^3 and 6 MJ/m^3 depending upon the volume of its constituents. Producer gas can be burnt in a boiler to generate steam. It is used as fuel in IC engines used for irrigation pumps, in spark ignition engines and gas turbines for power generation.

12.6 LIQUID FUEL (ETHANOL)

Ethanol (C_2H_5OH) is a flammable colourless biofuel. It can be produced by fermentation of any feedstock which contains sugar or starch and even cellulose material. Biomass containing sugar are: sugar-beets, sugarcane, sweet sorghum; starch crop covers corn, wheat, cassava and potato. Cellulose is found in all plant tissues, is available in wood, solid waste and agriculture residues. Ethanol is suitably used as a fuel additive to cut down a vehicle's carbon monoxide and other smog-causing emissions. In nine sugar producing Indian states, petrol blended with 5% ethanol is supplied.

12.7 BIOMASS CONVERSION TECHNOLOGIES

Biomass material from a variety of sources can be utilised optimally by adopting efficient and state-of-the-art conversion technologies such as:

1. Densification of biomass
2. Combustion and incineration
3. Thermo-chemical conversion
4. Bio-chemical conversion

Densification

Bulky biomass is reduced to a better volume-to-weight ratio by compressing in a die at a high temperature and pressure. It is shaped into briquettes or pellets to make a more compact source of energy, which is easier to transport and store than the natural biomass. Pellets and briquettes can be used as clean fuel in domestic chulhas, bakeries and hotels.

Combustion

Direct combustion is the main process adopted for utilising biomass energy. It is burnt to produce heat utilised for cooking, space heating, industrial processes and for electricity generation. This utilisation method is very inefficient with heat transfer losses of 30–90% of the original energy contained in the biomass. The problem is addressed through the use of more efficient cook-stove for burning solid fuels.

Incineration

Incineration is the process of burning completely the solid biomass to ashes by high temperature oxidation. The terms incineration and combustion are synonymous, but the process of combustion is applicable to all fuels, i.e., solid, liquid and gaseous. Incineration is a special process where the dry Municipal Solid Waste (MSW) is incinerated to reduce the volume of solid refuse (90%) and to produce heat, steam and electricity.

Waste incineration plants are installed in large cities to dispose off urban refuse and generate energy. It constitutes a furnace with adequate supply of air to ensure complete combustion up to a capacity of 1000 tonnes/day.

Thermo-chemical conversion

Thermo-chemical conversion is a process to decompose biomass with various combinations of temperatures and pressures. It includes 'pyrolysis' and 'gasification'.

Pyrolysis

Biomass is heated in absence of oxygen, or partially combusted in a limited oxygen supply, to produce a hydrocarbon, rich in gas mixture (H_2, CO_2, CO, CH_4 and lower hydrocarbons), an oil like liquid and a carbon rich solid residue (charcoal).

The pyrolitic or 'bio-oil' produced can easily be transported and refined into a series of products similar to refining crude oil. There is no waste product, the conversion efficiency is high (82%) depending upon the feedstock used, the process temperature in reactor and the fuel/air ratio during combustion.

Production of methanol

Liquefaction of biomass can be processed through 'fast' or 'flash' pyrolysis, called 'pyrolytic oil' which is a dark brown liquid of low viscosity and a mixture of hydrocarbons. Pyrolysis liquid is a good substitute for heating oil.

Another liquefaction method is through methanol synthesis. Gasification of biomass produces synthetic gas containing a mixture of H_2 and CO. The gas is purified by adjusting the hydrogen and carbon monoxide composition. Finally, the purified gas is subjected to liquefaction process, converted to methanol over a zinc chromium calatyst. Methanol can be used as liquid fuel. The reaction takes place at 330°C as 150 at pressure

$$4H_2 + 2CO \rightarrow 2CH_3OH$$

12.8 BIOCHEMICAL CONVERSION

There are two forms of biochemical conversions:

1. Anaerobic digestion
2. Ethanol fermentation

12.8.1 Anaerobic Digestion (Anaerobic Fermentation)

This process converts the cattle dung, human wastes and other organic waste with high moisture content into biogas (gobar gas) through anaerobic fermentation in absence of air. Fermentation occures in two stages by two different metabolic groups of bacteria. Initially the organic material is hydrolyzed into fatty acids, alcohol, sugars, H_2 and CO_2. Methane forming bacteria then converts the products of the first stage to CH_4 and CO_2, in the temperature range 30–55°C. Biogas produced can be used for heating, or for operating engine driven generators to produce electricity. Fermentation occurs in a sealed tank called 'digester' where the sludge left behind is used as enriched fertilizer.

12.8.2 Ethanol Fermentation

Ethanol can be produced by decomposition of biomass containing sugar like sugarcane, cassava sweet sorghum, beet, potato, corn, grape, etc. into sugar molecules such as glucose ($C_6H_{12}O_6$) and sucrose ($C_{12}H_{22}O_{11}$).

Ethanol fermentation involves biological conversion of sugar into ethanol and CO_2.

$$C_{12}H_{22}O_{11} + H_2O \rightarrow 2C_6H_{12}O_6$$

$$C_6H_{12}O_6 \xrightarrow{\text{Fermentation}} 2C_2H_5OH + 2CO_2$$

Ethanol has emerged as the major alcohol fuel and is blended with petrol.

12.9 BIOMASS GASIFICATION

Gasification is conversion of a solid biomass, at a high temperature with controlled air, into a gaseous fuel. The output gas is known as producer gas, a mixture of H_2 (15–20%), CO (10–20%), CH_4 (1–5%), CO_2 (9–12%) and N_2 (45–55%). The gas is more versatile than the solid biomass, it can be burnt to produce process heat and steam, or used in internal combustion engines or gas turbines to generate electricity. The gasification process renders the use of biomass which is relatively clean and acceptable in environmental terms.

12.9.1 Gasifiers

Gasifiers (fixed bed type) can be of 'updraft' or 'downdraft' type depeding upon the direction of the air flow. The working of biomass gasification can be explained by considering a typical downdraft gasifier (Figure 12.2) where fuel and air move in a co-current manner. In the updraft gasifier, fuel and air move in a countercurrent manner. However, the basic reaction zones remain the same.

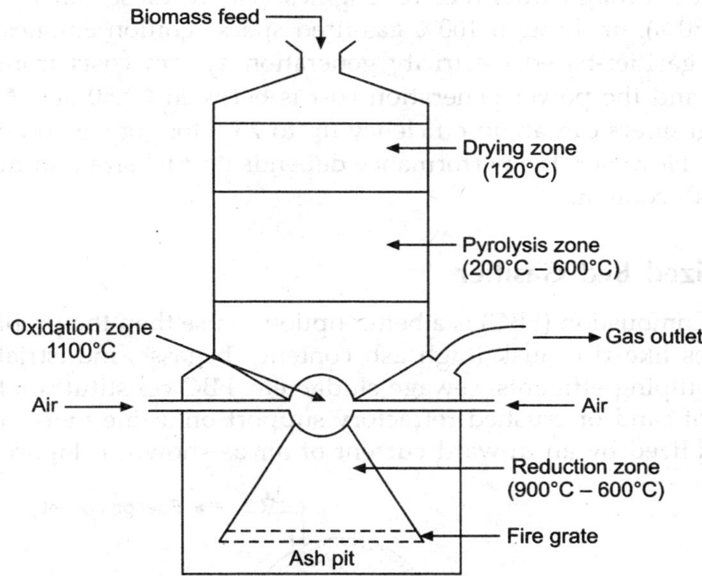

FIGURE 12.2   Downdraft gasifier.

Fuel is loaded in the reactor from the top. As the fuel moves down it is subjected to drying (120°C) and then pyrolysis (200–600°C) where solid char, acetic acid, methanol and water vapour are produced. Descending volatiles and char reach the oxidation zone where air is injected to complete the combustion. It is the reaction zone and the temperature rises to 1100°C. This helps in breaking down the heavier hydrocarbons and tars.

As these products move downwards, they enter the 'reduction zone' (900–600°C, reaction being endothermic) where producer gas is formed by the action of CO_2 and water vapour on red hot charcoal as detailed below:

$$\text{Moist biomass + Heat} \rightarrow \text{Dry biomass + Water vapour}$$
$$C + O_2 \rightarrow CO_2 + 393.8 \text{ kJ/mole (combustion)}$$
$$C + H_2O \rightarrow CO + H_2 - 131.4 \text{ kJ/mole (water gas reaction)}$$
$$CO + H_2O \rightarrow CO_2 + H_2 + 41.2 \text{ kJ/mole (water shift reaction)}$$
$$C + CO_2 \rightarrow 2CO - 172.6 \text{ kJ/mole}$$
$$C + 2H_2 \rightarrow CH_4 + 75.0 \text{ kJ/mole (hydrogenation reaction)}$$

Producer gas formed in the reduction zone contains combustible products like CO, H_2 and CH_4. Hot gas flowing out is usually polluted with soot, tar and vapour. For purifying, it is passed through coolers, tar is removed by condensation, whereas soot and ash are removed by centrifugal separation.

Clean producer gas provides the process heat to operate stoves (for cooking), boilers, driers, ovens and furnaces. The major application is in area of electric power generation either through dual-fuel IC engines (where diesel oil is replaced to an extent of 60%–80%), or through 100% gas-fired spark ignition engines.

A biomass gasifier-based electricity generation system costs from ₹4.0 crores to 4.5 crores/MW and the power generation cost is between ₹2.50 and ₹3.50 per kWh.

Fixed bed gasifiers can attain efficiency up to 75% for conversion of solid biomass to gaseous fuel. However, the performance depends on fuel size and moisture content, volatiles and ash content.

12.9.2 Fluidized Bed Gasifier

Fluidized Bed Combustion (FBC) is a better option to use than the problematic biomass of farm residues like rice husk (high ash content), bagasse, industrial waste such as saw dust and pulping effluents, sewage sludge etc. FBC constitutes a hot bed of inert solid particles of sand or crushed refractory support on a fine mesh or grid. The bed material is fluidized by an upward current of air as shown in Figure 12.3.

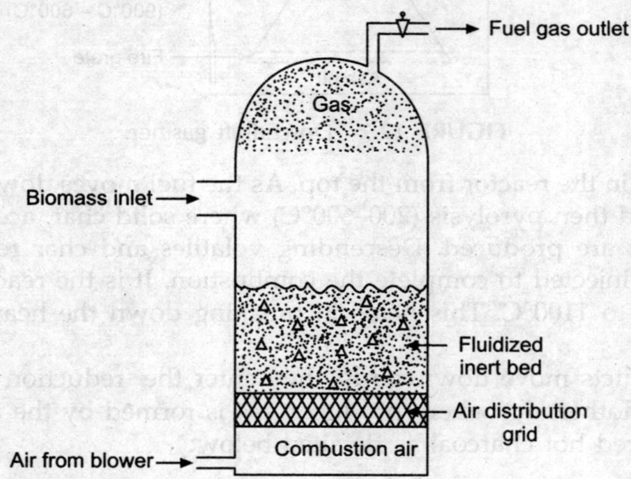

FIGURE 12.3 Fluidized bed gasifier.

Pressurized air starts bubbling through the bed and the particles attain a state of high turbulence, and the bed exhibits fluid like properties. A uniform temperature within the range of 850–1050°C is maintained. Large surface area is created in the fluidized bed and the constantly changing area per unit volume provides a higher conversion efficiency at low operating temperatures compared to the fixed beds. High heating capacity of sand and the uniform temperature of fluidized bed makes possible to gasify low-grade fuels of even non-uniform size and high moisture content.

When the gasifier is put in use, the bed material is heated to ignition temperature of the fuel, biomass is then injected causing rapid oxidation and gasification. Fuel gas so produced contains impurities, dust, char particles and tar. It needs conditioning and cleaning for utilization as an engine fuel.

12.10 BIOGAS

Biogas is a renewable energy derived from organic wastes such as cattle dung, human waste, etc. It is a safe fuel for cooking and lighting. Left-over digested slurry is used as enriched manure in agriculture lands.

12.10.1 Biogas Technology

Biogas is produced from wet biomass through a biological conversion process that involves bacterial breakdown of organic matter by micro-organisms to produce CH_4, CO_2 and H_2O. The process is known as 'anaerobic digestion' which proceeds in three steps.

1. Hydrolysis
2. Acid formation
3. Methane formation

Hydrolysis

Organic waste of animal and plants contains carbohydrates in the form of cellulose, hemicellulose and lignin. A group of anaerobic micro-organisms (celluolytic bacteria/ hydrobytic bacteria) breaks down complex organic material into simple and soluble organic components, primarily acetates. The rate of hydrolysis depends on bacterial concentration, quality of substrate, pH (between 6 and 7) and temperature (30°C–40°C) of digester contents.

Acid formation

Decomposed simple organic material is acted upon by acetogenic bacteria and converted into simple acetic acid.

Methane formation

Acetic acid so formed becomes the substrate strictly for anaerobic methanogeric bacteria, which ferment acetic acid to methane and CO_2. Gas production is stable for pH between 6.6 and 7.6.

Biogas consists of CH_4, CO_2 and traces of other gases such as H_2, CO, N_2, O_2 and H_2S. Gas mixture is saturated with water vapour. The methane content of biogas is about 60% which provides a high calorific value to find use in cooking, lighting and power generation.

12.10.2 Factors Affecting Biogas Production

There are eight major factors which affect the quality and quantity of biogas.

Solid-to-water ratio

Cattle dung (gobar) contains about 18% solid matter and the remaining 82% is water. Anaerobic fermentation proceeds at a faster rate if the slurry contains about 9% solid matter. Digester feed is prepared by mixing water in the ratio 1 : 1 by weight to reduce the solid content. To increase the solid matter, crop residues and weed plants may be mixed with the feed stock.

Volumetric loading rate

It is expressed as the quantity of organic waste fed into the digester per day per unit volume. In general, the municipal sewage treatment plants operate at a loading rate of 1.0 to 1.5 $kg/m^3/day$. Overloading and underloading reduce the biogas production with a fixed retention time. For a desired retention period of 30 days, a quantity equal to 1/30th of digester volume needs to be fed daily.

Temperature

Temperature affects bacterial activity; methane formation is optimum in the temperature range 35°–38°C. Biogas production decreases below 20°C and stops at 8°C. In cold regions a solar canopy is built over the biogas plants to maintain the desired temperature.

In hot regions, another micro-organism called 'thermophilic' is utilised for anaerobic fermentation in the temperature range 55°C–60°C. Gas production rises with the increase in average ambient air temperature. As the temperature increases, the total retention period decreases and vice-versa. However, the total gas production remains practically the same.

Seeding

Cattle dung contains both acid forming bacteria and methane forming bacteria. Acid forming bacteria multiply fast, while the methane forming bacteria grow slowly. To start and accelerate fermentation, seeding of methane forming bacteria is required. Accordingly, a small quantity of digested slurry rich in methane-forming bacteria is added to freshly charged digester.

pH value

Measure of pH value indicates the concentration of hydrogen ions. Micro-organisms are sensitive to pH of the digested slurry. For optimum biogas production, pH can be varied between 6.8 and 7.8. At pH of 6.2, acid conditions prevail which restrain the growth of methanogenic bacteria. Control on pH should be exercised by adding alkali when it drops below 6.6.

Carbon-to-nitrogen (C/N) ratio

Methanogenic bacteria needs carbon and nitrogen for its survival. Carbon is required for energy while nitrogen for building cell protein. The consumption of carbon is 30 to 35 times faster than that of nitrogen. A favourable ratio of C : N can be taken as 30 : 1. Any deviation from this ratio lowers the biogas production. A proper balance of C : N ratio is maintained either by adding saw dust having a high C : N ratio or by poultry waste having a low C : N ratio.

Retention time

The period for which the biomass slurry is retained inside the digester is called 'retention time'. It refers to the volume of digester divided by the volume of slurry added per day. Thus, a 120 litre digester which is fed at 5 litres per day would have a retention time of 24 days. It is optimized to achieve 80% complete digestion considering ambient temperature. Indian states are divided into three zones where the retention period is decided in days for cattle dung feedstock detailed in Table 12.1.

TABLE 12.1   Retention time

Zone	States	Mean ambient temp (°C)	Retention time (in days)
I	Kerala, T.N., A.P., Andaman, Karnataka, Maharashtra	>20	30
II	W.B., Bihar, Odisha, MP, UP, Rajasthan, Haryana, Punjab	15–20	40
III	North-eastern region, Sikkim, Uttarakhand, Himachal, J&K and areas with long winter	10–15	50

Stirring digester contents

Stirring the contents of the digester is necessary to mix the bacteria rich fluid in the slurry. It provides better contact between micro-organism and the substrate and uniformly distributes the volatile solids in the slurry. Gas production improves by 15% over the full cycle.

12.11   BIOGAS PLANTS

The biogas plant is a device that converts cattle dung and other organic matter into inflammable gas called biogas and into a good quality organic manure under anaerobic conditions. There are two popular designs of biogas plants: (i) Floating drum (constant pressure) type and (ii) Fixed dome (constant volume) type.

12.11.1   Floating Drum Type Biogas Plant

A popular model developed by Khadi Village Industries Commission (KVIC) was standardized in 1961. It comprises an underground cylindrical masonary digester having an inlet pipe for feeding animal dung slurry and an outlet pipe for sludge. There is a steel dome for gas collection which floats over the slurry. It moves up and down depending upon accumulation and discharge of gas guided by the dome guide shaft (Figure 12.4).

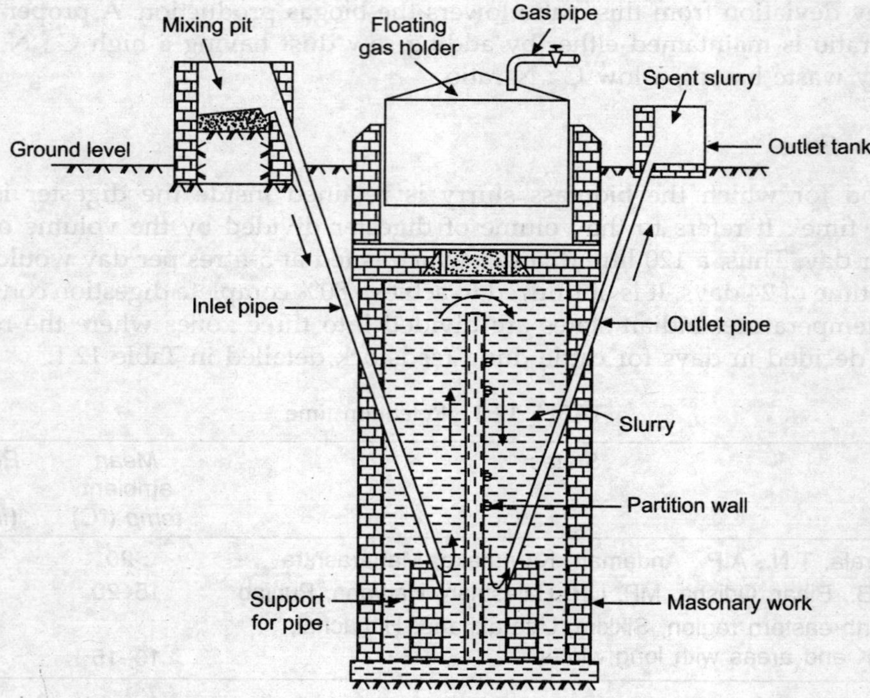

FIGURE 12.4 Floating drum biogas plant (KVIC model).

A partition wall is provided in the digester to improve circulation, necessary for fermentation. The floating gas holder builds gas pressure of about 10 cm of water column, sufficient to supply gas up to 100 metre. Gas pressure also forces out the spent slurry through a sludge pipe.

12.11.2 Fixed Dome Type Biogas Plant

It is an economical design where the digester is combined with a dome-shaped gas holder (Figure 12.5). It is known as Janata model; the composite unit is made of brick and cement masonary having no moving parts, thus ensuring no wear and tear and longer working life. When gas is produced, the pressure in the dome changes from 0 to 100 cm of water column. It regulates gas distribution and outflow of spent slurry.

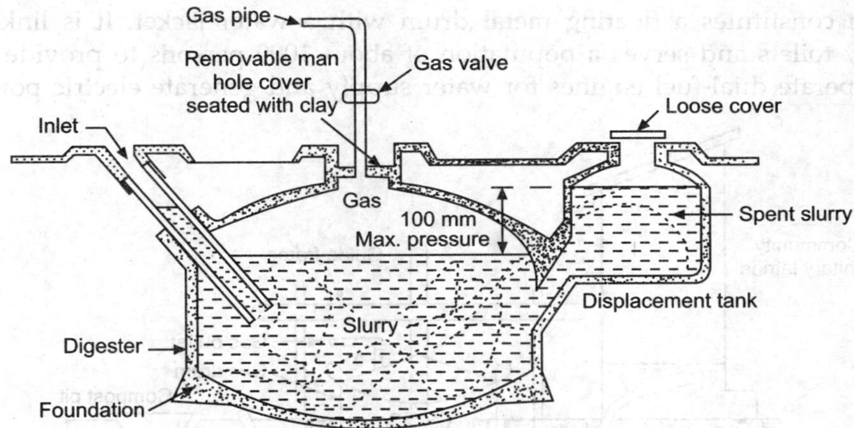

FIGURE 12.5   Fixed dome biogas plant (Janata model).

12.11.3   Deenbandhu Biogas Plant (DBP)

This plant developed by AFPRO (Action For Food Production) with the objective to extend the biogas technology to places where the availability of bricks is a limiting factor and bamboo is easily available. Its cost is reduced as the surface area is minimized by joining segments of two different diameter spheres at their bases as given in Figure 12.6. This plant requires less space being mainly undergound. It is 30% economical compared to the Janata biogas plant. After intensive trial and testing it has been approved by MNRE for family size installation.

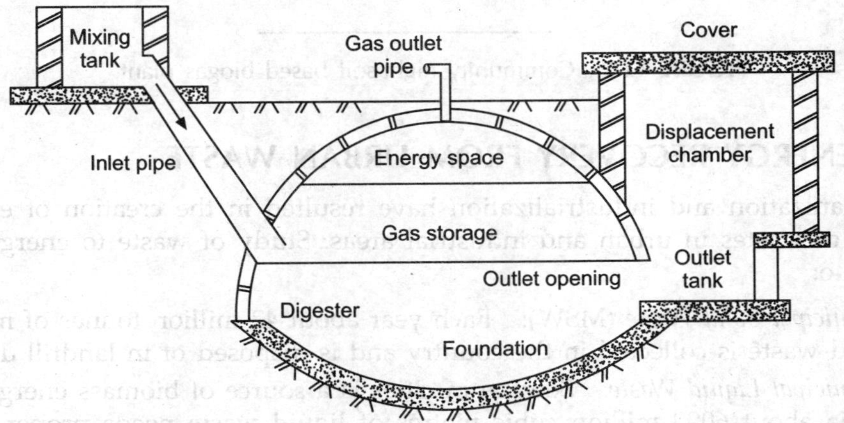

FIGURE 12.6   Deenbandhu biogas plant.

12.11.4   Community Night-soil Based Biogas Plant

Community night-soil based biogas plants have been developed to facilitate sanitary treatment of human waste at community and institutional level (Figure 12.7). This installation constitutes a floating metal drum with a water jacket. It is linked with community toilets and serves a population of about 1000 persons to provide fuel for cooking, operate dual-fuel engines for water supply and generate electric power.

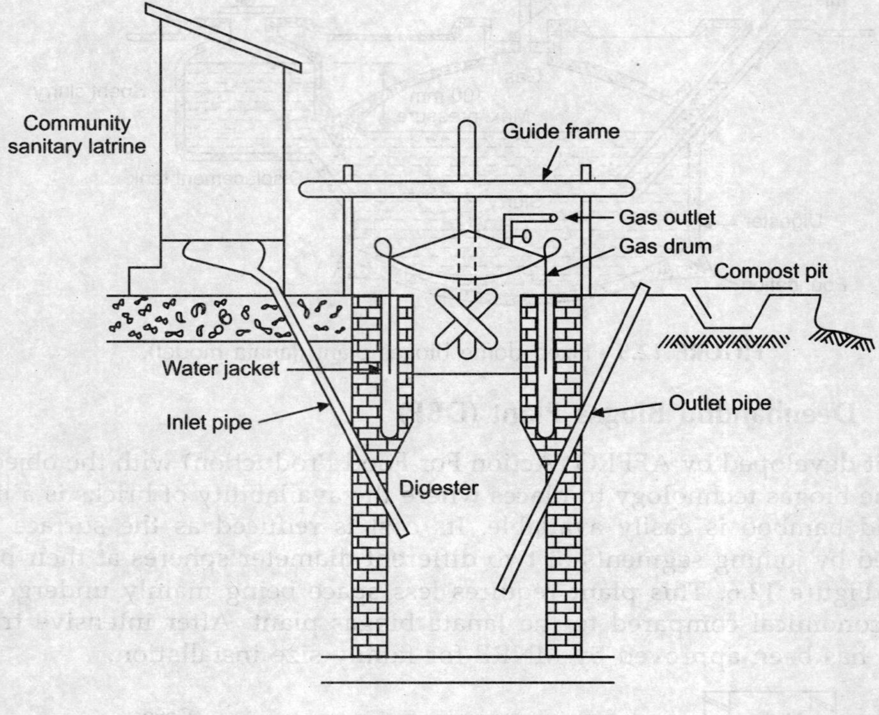

FIGURE 12.7 Community night-soil based biogas plant.

12.12 ENERGY RECOVERY FROM URBAN WASTE

Rapid urbanization and industrialization have resulted in the creation of enormous quantities of wastes in urban and industrial areas. Study of waste to energy can be divided into:

1. *Municipal Solid Waste* (MSW): Each year about 42 million tonnes of municipal solid waste is collected in the country and is disposed of in landfill dumps.

2. *Municipal Liquid Waste:* Sewage in cities is a source of biomass energy and in India about 6000 million cubic metres of liquid waste needs proper disposal every year.

3. *Urban Industrial Waste:* Industries produce a large number of residues as by-products that can be used as biomass energy sources. Food industry includes pealings and scraps from fruit and vegetables. Wine making produces distillary waste water (spent wash). Paper and pulp making effluent is 'black liquor'

which is a source for bio-oil. Starch and glucose industry wastes are maize husk, tapioca fibre and stems. Rice mills provide large volumes of rice husk. Sugar mills waste is a source of huge quantity of bagasse.

12.12.1 MSW-based Power Project (5 MW Capacity)

Urban waste represents a large source of substrate for energy production. It contains dry waste of household (waste paper etc.) mixed with kitchen scrap. It is subjected to a segregation system where inorganics (metal, glass, grit) and plastic material are sorted out, keeping items which are largely cellulosic with fats and proteins, i.e., are digestible. The major components of a power project are shown in the block diagram of Figure 12.8. It is based on high-rate biomethanation technology developed by ENTEC, Environment Technology, Austria.

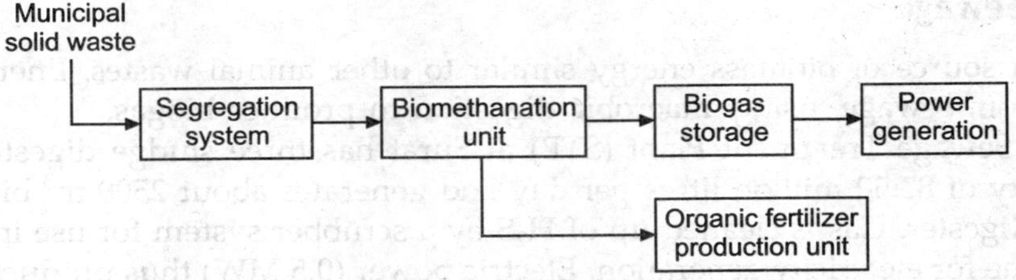

FIGURE 12.8 Block diagram of an MSW-based power project.

The project is designed to process 500–600 tonnes of MSW per day from a city. The collected MSW is converted into about 115 M.T. of dry volatile solids which produce 50,000 m^3 of biogas per day. The spent slurry in the digester (75 MT) is used as organic fertilizer. The biogas so produced is fed into five 100% biogas engines to generate 5 MW grid-quality power.

12.13 POWER GENERATION FROM LANDFILL GAS

Recycling of city garbage and MSW poses a serious problem due to its enormous quantity. Its sanitary disposal through landfill is a successful method even in the UK and USA. A large pit at the outskirt is prepared and a pipe system for gas collection is laid down before the waste is filled. For anaerobic digestion, MSW is buried, eventually the gas produced does not escape into the atmosphere. After 2–3 months, depending on the climate, landfill gas can be extracted by inserting perforated pipes into the landfill (Figure 12.9).

The gas flows through pipes under natural pressure. As the gas has calorific value of about 4500 kcal/m^3 it can be used either for direct heating/cooking applications or to generate power through IC engines. One of the largest landfill gas plants in the world is a 46 MWe plant in California.

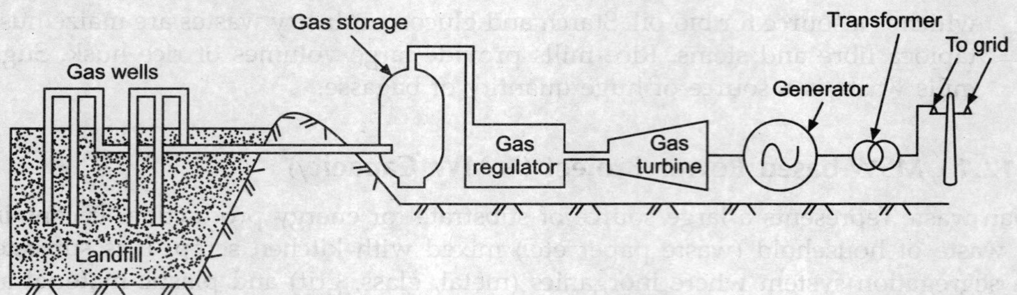

FIGURE 12.9 Landfill gas for power generation.

12.14 POWER GENERATION FROM LIQUID WASTE

12.14.1 Sewage

Sewage is a source of biomass energy similar to other animal wastes. Energy can be extracted from sewage, using anaerobic digestion to produce biogas.

Anjana Sewage Treatment Plant (STP) at Surat has three sludge digesters with a total capacity of 82.50 million litres per day and generates about 2500 m^3 biogas daily from each digester. Gas is cleaned up of H_2S by a scrubber system for use into a 100% biogas engine for electricity generation. Electric power (0.5 MW) thus produced accrues to the STP in a saving of ₹10 lakhs per month.

12.14.2 Distillary Waste

Distillary liquid wastes carry rich raw material for producing biogas. Liquid effluent from a distillary is collected in a tank where the suspended solids settle down. Decanted effluent which contains fermented molasses is pumped into a digester through a heat exchanger. Effluent is cooled to maintain the digester temperature at 36–38°C, allowed to be digested anaerobically for about 12–15 days, during which biogas is produced (Figure 12.10).

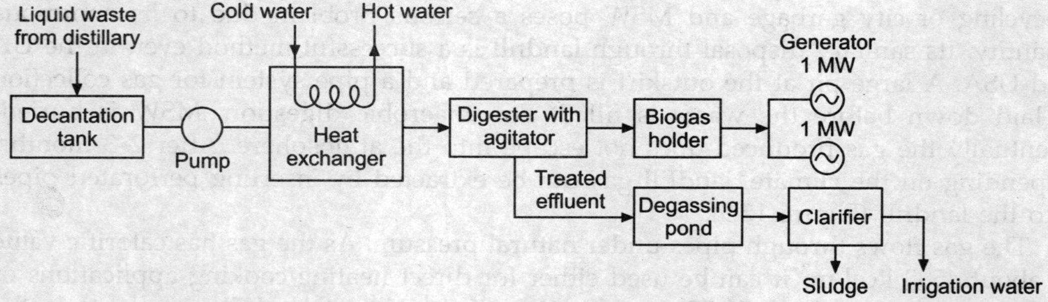

FIGURE 12.10 Power generation from distillary waste.

Biogas accumulates in a gas holder and is stored under pressure using a pressue control device. Generation of power is based on two IC engines, each coupled with

1 MW capacity generator, fuelled solely by biogas. Distillary waste is sufficient to produce about 21000 m³ of biogas per day.

The treated effluent overflows into a degassing pond to remove traces of gas. The liquid then flows into a clarifier where the sludge collects to be used as farm manure. The clarified water is utilised for irrigation after secondary treatment.

12.14.3 Pulp and Paper Mill Black Liquor Waste

The pulp and paper industry consumes a large amount of energy and water in its various unit operations. The waste discharged water contains compounds from wood and raw material, useful for recovery of energy.

A plant for biomethanation of bagasse wash effluent is installed at Karur in Tamil Nadu, based on UASB technology. Biochemical Oxygen Demand (BOD) and Chemical Oxygen Demand (COD) removal is 94% and 89% respectively with gas production of 0.37 cu metre per kg.

Presently about 15000 m³ of gas is generated every day, which is used in a lime-mud re-burning kiln. The gas output from the plant meets 50% heat load of the kiln, equivalent to 12–13 kilolitre of furnace oil.

12.15 BIOMASS COGENERATION

Cogeneration is defined as the sequential generation of two different forms of useful energy from a single primary energy source, typically mechanical energy and thermal energy. Mechanical energy is used to drive an alternator for producing electricity. Thermal energy can be used either for direct process applications or for producing steam.

Sugar industry in India uses bagasse-based cogeneration for achieving self sufficiency in steam and electricity. Cogeneration cleans up the environment, generates power for in-house consumption and earns additional revenue from the sale of surplus electricity.

The main equipment required for bagasse-based cogeneration projects comprises high-temperature/high-pressure bagasse-fired boilers, a steam turbine and a grid-interfacing system. Experience shows that when steam generation temperature/pressure is increased from 400°C/32 bar to 485°C/66 bar, more than 80 kWh of additional electricity is generated from each tonne of cane crushed. Additional power generation with increase in pressure and temperature of a typical 2500 TCD sugar mill is tabulated in Table 12.2.

TABLE 12.2 Biomass cogeneration as function of steam pressure/temperature

Steam pressure/ temperature	Electricity generation (MW)	In-house consumption (MW)	Surplus electricity (MW)
21 bar/300°C	2.0	2.5	−0.5
33 bar/380°C	3.5	3.5	0
45 bar/440°C	6.0	4.0	2.0
64 bar/480°C	13.5	4.5	9.0
85 bar/510°C	17.0	6.0	11.0

Case study

A progressive sugar mill in UP crushing 11000 tonnes of cane per day is deployed at 87 bar/525°C steam configuration to cogenerate over 18 MW of surplus electricity.

Case study No. 1: Biomass and Bagasse cogeneration programme is promoted by Ministry of New and Renewable Energy (MNRE). The potential of power generation from agricultural and agro-industrial residues is estimate at about 18000 MW, while the potential of surplus power generation through bagasse cogeneration in sugar mills is estimated at 7000 MW.

Over 300 biomass power and cogeneration projects aggregating to about 4000 MW capacity have been installed in the country up to December 2014 for feeding power to the grid.

Case study No. 2: 15 MW Grid connected cogeneration plant at a sugar factory in District Wardha Maharashtra Commissioned.

Case study No. 3: MNRE proposed a National Bioenergy Mission in 2011 to ramp up biomass power generation of 20,000 MW by 2022. It plans to increase grid connected biomass power by raising feed stock in plantation. The ministry reckons that 5000 MW of power can be generated by using dedicated plantation on 2 million hectares (ha) of forest and non-forest land. This would mean 400 hectare per MW. This is on higher side as a US based company acquired 9000 acres in Valliyur and Tuticorin in Tamil Nadu to support a 32 MW biomass power plant. This would mean 113 ha of land per MW. For a successful plantation, India needs to increase supply water for irrigation in the area of plantation.

12.15.1 Cogeneration Plant in Rice Mill

Rice production from paddy has undergone changes from traditional soaking and drying to modern method of parboiling at higher temperature. It results in increased productivity of husk and quality rice. Husk produced is effectively utilised for steam production, which is used for both process and power generation. The characteristics of rice husk as fuel are given in Table 12.3 where the figures show percentage by weight.

TABLE 12.3 Ultimate analysis of rice husk-fuel

Carbon	36.14
Hydrogen	3.70
Sulphur	0.08
Nitrogen	0.46
Oxygen	29.34
Water	8.92
Ash	21.36

The calorific value of rice husk varies from 2637 to 3355 kcal/kg depending on variety. Rice husk is difficult to handle because of its silica-cellulose structure.

Considering this aspect the 'Fluidized Bed Combustion' (FBC) boiler is used to ensure complete combustion as firing is balanced between buoyancy and gravitational force. Steam from the boiler (6000 kg/h) is fed to the back pressure turbine coupled with a 350 kW electric generator as shown in Figure 12.11. Input steam conditions are: 6000 kg/h having pressure 32 atm at 400°C.

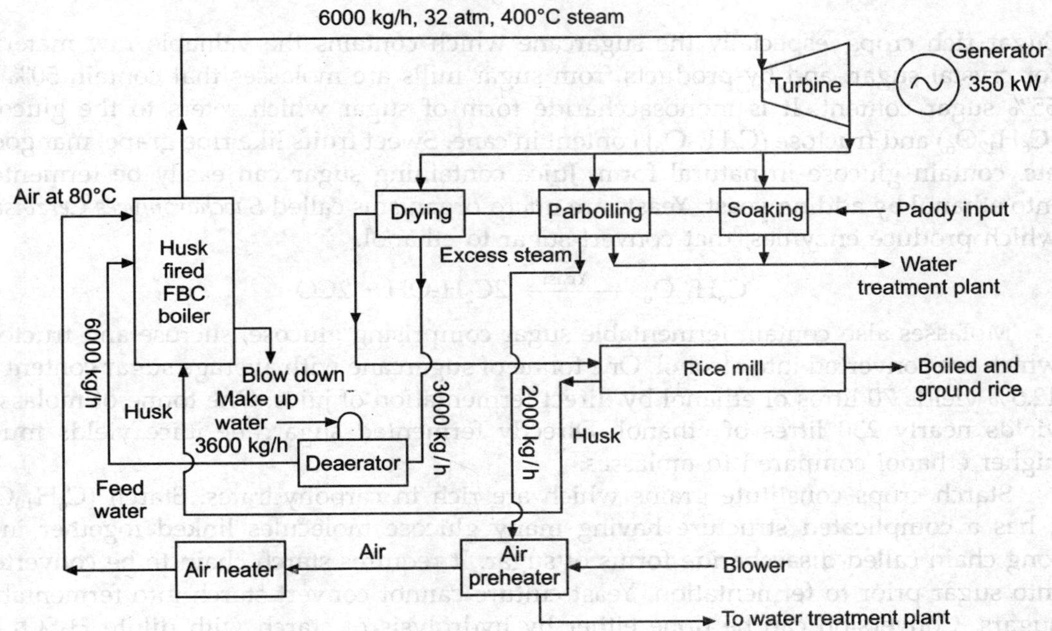

FIGURE 12.11 Cogeneration plant in a rice mill (50 tonnes capacity).

Steam pressure after expanding in turbine 3.5 kg/cm^2 is utilised for three processes (i) Soaking (ii) Parboiling and (iii) Drying. Pressurized deaerator supplies water at 6000 kg/h to the boiler. Make-up water 3600 kg/h at 30°C is added into deaerator besides 3000 kg/h of hot water received from turbine exhaust after the drying process. Excess steam from parboiling process is fed into an air preheater and the hot air is further heated to maintain 80°C before feeding into the boiler.

Thus, a 50 tonne/batch capacity rice mill generates 350 kW from rice husk which is normally dumped and wasted. At present 47% rice husk is used as energy source.

Cogeneration in India excluding sugar industry has a potential of 10,000 MW from rice mills, distillaries, paper mills, petrochemicals and fertilizer plants.

12.16 ETHANOL FROM BIOMASS

Ethanol is ethyl alcohol (C_2H_5OH), a colourless flammable liquid. It is a renewable energy source which can substitute petroleum products. Ethanol can be produced from a variety of biomass materials, containing sugar, starch and cellulose. The best-known feedstock under three categories are:

Sugars: sugarcane, sugar beet, sweet sorghum, grapes, molasses
Starches: maize, wheat, barley, potatoes, cassava, rice
Cellulose: wood, straw, stems of grasses, bamboo, sugarcane bagasse.

Production process

Sugar rich crops, especially the sugarcane which contains the valuable raw material for crystal sugar, and by-products from sugar mills are molasses that contain 50% to 55% sugar content. It is monosaccharide form of sugar which refers to the glucose ($C_6H_{12}O_6$) and fructose ($C_6H_{12}O_6$) content in cane. Sweet fruits like ripe grape, mangoes, etc. contain glucose in natural form. Juice containing sugar can easily be fermented into ethanol by adding yeast. Yeasts are micro-organisms called *Saccharomyees Cerevisiae* which produce enzymes, that convert sugar to ethanol.

$$C_6H_{12}O_6 \xrightarrow{\text{Yeast}} 2C_2H_5OH + 2CO_2$$

Molasses also contain fermentable sugar comprising glucose, sucrose and fructose which are converted into alcohol. One tonne of sugarcane with average sugar content of 12.5% yields 70 litres of ethanol by direct fermentation of juice. One tonne of molasses yields nearly 230 litres of ethanol. Directly fermented sugarcane juice yields much higher ethanol compared to molasses.

Starch crops constitute grains which are rich in carbohydrates. Starch $(C_6H_{10}O_5)_n$ has a complicated structure having many glucose molecules linked together in a long chain called disaccharide forms of sugar. It requires starch chain to be converted into sugar prior to fermentation. Yeast culture cannot convert starch into fermentable sugars. Conversion can be done either by hydrolysis of starch with dilute H_2SO_4 or through enzymatic method. Starch is converted into maltose and glucose prior to initiating ethanol production.

$$2(C_6H_{10}O_5)_n + nH_2O \xrightarrow{\text{Hydrolysis}} nC_{12}H_{22}O_{11}$$
$$\text{(Starch)} \qquad\qquad\qquad\qquad\qquad \text{Maltose}$$
$$C_{12}H_{22}O_{11} + H_2O \longrightarrow 2C_6H_{12}O_6$$
$$C_6H_{12}O_6 \xrightarrow{\text{Fermentation}} 2C_2H_5OH + 2CO_2$$

Cellulosic material comprises dry biomass abundantly available, but difficult to utilise carbohydrate in cellulose. Cellulose contained in wood, grasses and crop residue contain a long chain of sugars and lignin available in plants which hinders hydrolysis to sugars. This complex material is called 'polysaccharides' in which breaking the chemical bond of cellulose is not as easy as that of a starch to simple sugars.

The conversion of cellulosic material is carried out by special hydrolysis with dilute H_2SO_4 at high temperature 180°–200°C, which causes the product sugar to decompose into glucose.

$$(C_6H_{10}O_5)_n + nH_2O \longrightarrow nC_6H_{12}O_6$$
$$\text{(Cellulose)} \qquad\qquad\qquad\qquad \text{(Glucose)}$$

Optimum glucose production is achieved by adjusting three variables, i.e., acid concentration, operating temperature and reaction time. Finally, ethanol is obtained by fermentation of glucose sugars.

$$C_6H_{12}O_6 \xrightarrow{\text{Fermentation 32°C}} 2C_2H_5OH + 2CO_2$$

(Glucose)   (Ethanol)

Ethanol production from various biomass crops is given in Table 12.4.

TABLE 12.4 Ethanol production from biomass crops

Raw material	Ethanol (litre) *per tonne of crop* (l/t)	Ethanol (litre) *per hectare per year* (l/ha)
Sugar beet	90–100	3800–4800
Sugarcane	60–80	3500–7000
Sweet sorghum	80–90	2500–3500
Potato	100–120	2200–3300
Maize	360–400	1500–3000
Cassava	175–190	2200–2300
Wheat	370–420	800–2000
Barley	310–350	700–1300
Soft wood (hydraulic agent dilute acid)	190–220	1800–3100
Hard wood (dilute acid)	160–180	1500–2500
Straw (dilute acid)	140–160	200–500

Source: Internet Alternate Energy Development Board (Biomass Energy Systems)

Microbial growth and conversion of sugars to ethanol is best at its 10% concentration as the fermentation process drops down (micro-organism in the yeast is poisoned) with increase in alcohol concentration. Concentration of ethanol can be increased to 95% by volume by successive fractional distillation. The product is called hydrated ethanol and used as fuel in modified IC engines.

Removal of balance 5% water from 95% ethanol concentration is not possible by simple distillation as a constant boiling mixture (azeotrope) is formed which prevents further separation due to the absence of differential vaporization. An hydrous ethanol is produced with azeotropic removal of water by co-distillation using benzene as solvent. Production of ethanol from three biomass resources is given in Figure 12.12.

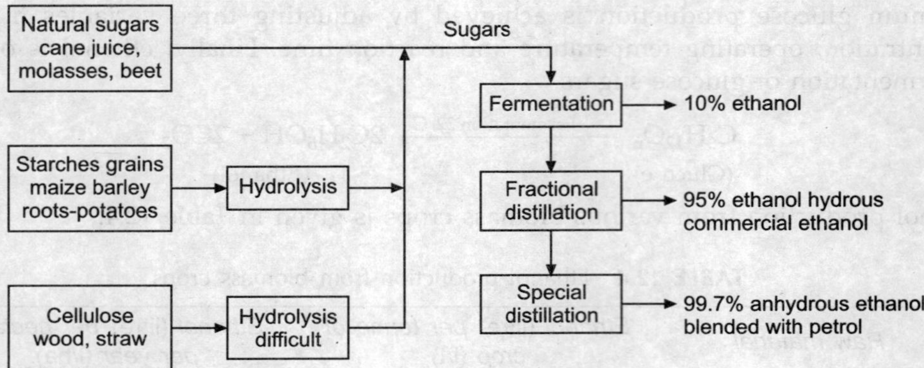

FIGURE 12.12 Ethanol production from biomass.

12.16.1 Ethanol as Fuel & Blending with Petrol

In the USA, anhydrous ethanol (10%) is blended with petrol (90%) to produce 'gasohol' a good substitute for petrol in automobiles without any engine modifications. Ethanol being a high octane fuel raises the octane rating of the mixture. Octane rating is explained as the fuel's quality to increase its antiknock property. Considering the advantage, Canada and Sweden also utilise 10% ethanol blended petrol. Brazil, the leader in ethanol production enhanced ethanol doping to 25–26% with petrol to tide over the soaring oil prices.

The level of sugar production in India is 31.18 million metric tonnes in 2020–2021 per year, ensuring ethanol production to 1700 million litres. It is assessed that the requirement for potable purpose and chemical sector shall consume 1200 million litres—leaving a clear balance of 500 million litres, sufficient for 5% blending with petrol in the country.

Molasses is a residue of sugar factory from which balance 40–47% sugar cannot be obtained by conventional methods. But molasses are fermented with a yeast (*saccharomyces cerevisiac*) and alcohol is separated in a distillation column. In Indian conditions, alcohol recovery from molasses is about 230 litres from one tonne of molasses.

Ethanol yield is 6 times higher if the sugarcane juice is directly fermented instead of molasses. One tonne of sugarcane with sugar content of 13% yields about 70 litres of ethanol through direct fermentation of juice. Sugar content in molasses is only 2%. The Reliance group of industries is venturing for such a project in Maharashtra to reduce crude oil import. It is established fact that blending would lead to major saving on fuel and foreign exchange.

12.17 BIODIESEL

Biodiesel is a liquid fuel produced from non-edible oil seeds such as Jatropha, Pongamia pinnata (Karanja), etc. which can be grown on wasteland. However, the oil extracted from these seeds has high viscosity (20 times that of diesel) which causes serious lubrication, oil contimination and injector choking problems. These problems are solved through trans-esterification, a process where the raw vegetables oils are treated with alcohol (methanol or ethanol with a catalyst) to form methyl or ethyl esters. The monoesters produced by trans-esterifying vegetable oil are called 'biodiesel' having low fuel viscosity with high octane number and heating value. Endurance tests show that biodiesel can be adopted as an alternative fuel for existing diesel engines without modifications.

In EU and USA, edible vegetable oil like sunflower, groundnut, soyabean and cotton seed, etc. are used to produce biodiesel. India is endowed with a number of non-edible vegetable oil producing trees which thrives in inhospitable conditions of heat, low water, rocky and sandy soils, a renewable resource of economic significance (Jojoba in Rajasthan).

Biodiesel is the name of diesel fuel made from vegetable oil or animal fats. The concept dates back to 1885, when Dr. Rudolf Diesel developed the first diesel engine to run on vegetable oil. In recent past the use of bio oil as an alternative renewable fuel to compete with petroleum was proposed during 1980.

The advantages of biodiesel as engine fuel are: (i) biodegradable and produces 80% less CO_2 and 100% less SO_2 emissions, (ii) renewable, (iii) higher octane number, (iv) can be used as neat fuel (100% biodiesel) or mixed in any ratio with petro-diesel, and (iv) has a higher flash point making it safe to transport. Selected fuel properties of biodiesel and petrodiesel are given in Table 12.5.

TABLE 12.5 Properties of biodiesel and petrodiesel

Properties	Petrodiesel	Biodiesel
Boiling point, °C	188–343	182–338
Viscosity at 40°C	1.3–4.1	1.9–6.0
Carbon, wt%	87	77
Hydrogen, wt%	13	12
Oxygen, wt%	0	11
Sulphur, wt%	0.05 max	0.0–0.0024
Heating value, kcal/litre	7278	6491

12.17.1 Production of Biodiesel from Jatropha

Jatropha curcas drought resistant perennial shrub with 4–5 metre height is ideally suited to green up the wastelands in arid areas. Commercial seed production commences from the 6th year onwards with yield of 6000 kg/ha under rain-fed conditions and 12000 kg/ha in irrigated areas. The average oil production is 0.25 kg oil/kg seed. The oil cake is used as organic fertilizer.

Scientists of Central Salt & Marine Chemical Research Institure (CSMCRI) Bhavnagar (Gujarat) have confirmed the use of Jatropha curcas and Jojoba seed oil as promising substitutes for diesel. The yield of Jojoba seed is 0.5 kg per plant after

10 years of plantation, Jojoba seed costs ₹200/kg, so presently it is uneconomical as feedstock for engine oil.

The characteristics of four biodiesels obtained from vegetable oils of peanut, soyabean, sunflower Jatropha and diesel are given in Table 12.6

TABLE 12.6 Characteristics of four biodiesels

Name	Flash point (°C)	Density at 20/40°C	Viscosity	Octane *number*	Heating value (MJ/litre)
Diesel	32	0.82–0.86	2.0–7.5	42	34.5–36.0
Biodiesel (Jatropha)	161	0.878	4.54	65	33.7
Biodiesel (Sunflower)	183	0.880	4.60	49	33.5
Biodiesel Soyabean	178	0.885	4.50	45	33.5
Biodiesel Peanut	176	0.883	4.90	54	33.6

The heat of combustion for biodiesel is up to 95% by volume of conventional diesel, but biodiesel being oxygenated provides the same fuel value as the diesel. The parameters in Table 12.5 justify Jatropha seed (cost ₹5.0/kg) as an economically favourable feedstock to produce biodiesel.

Oil is extracted from Jatropha seeds in an oil press. It is treated with methanol (CH_3OH) to produce three methyl ester molecules and one glycerol molecule. Alkalis like NaOH or KOH are used to catalyze the reaction having the following constituents: 1000 litre Jatropha oil + 400 litre (CH_3OH) + 10 litre catalyst. The reaction process is completed rapidly, glycerol is separated and methyl ester is obtained as biodiesel.

The Ministry of Petroleum and Natural Gas has opened a biofuel centre in Delhi to build awareness of importance of Jatropha curcas cultivation and manufacture of biodiesel. The Indian Oil Corporation (IOC) has already established a biodiesel plant at Faridabad and another one being established in Panipat refinery to prepare 30,000 litres of biodiesel daily by crushing 100,000 kg Jatropha seeds. Biodiesel shall be blended with diesel to the extent of 5% in different Indian climatic conditions. Approximately, 40 million tonnes of HSD is consumed annually in India, thus, only 5% replacement of petroleum fuel by biodiesel would save the country approximately ₹4000 crores in foreign exchange yearly.

12.18 BIOFUEL PETROL

Shell oil company started selling petrol containing 10% cellulosic ethanol in Ottawa. Biofuel is produced from wheat straw. Logen's process converts biomass into cellulostic ethanol using a combination of thermal, chemical and biochemical techniques. Yield of cellulosic ethanol is 340 litres per tonne of fibre. Lignin is the plant fibre is used to drive the process by generating stream and electricity, thus, eliminating the need of coal or natural gas. Cellulosic ethanol is identical to ethanol, but produces up to 90% less CO_2 than petrol.

12.19 BIOMASS RESOURCE DEVELOPMENT IN INDIA

The energy scenario in India indicates that 'biomass' is a promising form of renewable energy matching with the agricultural base in rural areas and industrial development in urban set-ups. The estimated potential and physical achievement of biopower are given in Table 12.7.

TABLE 12.7

Resource	Potential in MW	Total achievement MW up to 31/12/2016
1. Bio energy		
(a) Biomass power	17536	
(b) Bagasse cogeneration	5000	7907.34
(c) Waste to energy	2554	
2. No of Family Biogas Plants (in Lakhs)	—	49.40

Advantages of Biomass Power

1. India has an agriculture based economy and thus bio-energy potential is quite high.
2. U.P. has utilized a large part of its biomass potential. This is largely due to sugar-cane industry with cogeneration power plants.
3. Government policy to support biomass power can play a big role in enhancing the viability of biomass power plants in states with potential.
4. 10MW Biomass based power plant provides employment to nearly 100 works for 18 months during construction, full time employment to 35 for maintenance and operation and 35 persons for collection.
5. Water requirement is six time less than coal power thermal power stations—a great advantage in a water stressed country like India.
6. It gives birth to a chain of entrepreneurs among the locals for secured and sustained fuel supply.
7. Biochar the byproduct can also be sold back to farmers either for use as kitchen fuel or for enriching the soil before the next sowing.
8. Biomass based energy will help in meeting the goal of "electricity to all".
9. Unlike solar and wind, biomass is a more reliable source of renewable energy and is relatively free of constraints.
10. Capital cost associated with biomass plants ranges from ₹4.5–5 crores per MW. So, the investment for biomass electricity generation is 18 times less than same units of solar energy generation.
11. National biomass cooking stoves are used in Anganwadis to prepare Mid-day meal provided in schools, Tribal hostels, etc. Projects taken up under Unnat Chulha Abhiyan are now eligible for carbon credits under the CDM mechanism with Sardar Swaran Singh National Institute of Renewable Energy.

Commissioned Projects

(i) A grid connected power plant of capacity 7.5 MW based on poultry waste was setup by M/s Redan Infrastructure Pvt. Ltd. Chittoor, Andhra Pradesh.

(ii) Three projects based on biogas for power generation/thermal application from industrial effluents/urban waste were set up in Andhra Pradesh, Telangana and Kerala.

(iii) A plant for production of 1000 kg/day of Bio-CNG from 2500 m^3/day of biogas generated from cattle manure was set up by M/s NRB Bio-energy in Hanumangarh, Rajasthan.

(iv) A plant for production of 1847 kg/day from 5000 m^3/day biogas from urban waste was set up by M/s Arc-Bio Fuel Ltd. Barnala, Punjab.

12.20 FUTURE OF BIOMASS ENERGY IN INDIA

Use of Biomass is growing globally. Modern biomass has potential to penetrate in four segments.

- Process heat applications in industries generating biomass waste.
- Cooking energy in domestic and commercial sectors (through charcoal and briquetts).
- Electricity generation and
- Transportation sector with liquid fuels.

Future of biomass energy lies in its use with modern technologies.

Case study: A Model Biogas Plant in Pune (Anaerobic process)

In Pune, twenty biogas plant are operational with the capacity to take 5tonnes of wet waste every day. Detail is as under:

1. Waste comes in daily by trucks arranged by Pune Municipal Corporation (PMC). The trucks empty the waste at collection unit. The waste is passed on to the shredder with help of 2 employees, who pull out any waste that is unfit for the bio-methanation process. Only organic waste enters the shredder.

2. The shredder turns the solid waste into a semisolid state. At this stage the waste is passed on to the scum remover where particles that are unfit for processing such as oil are separated. The waste settled at the bottom is sent to the anaerobic digester for building the culture. A culture is prepared in the digester before injecting waste for bio-methanation. It is built by adding cowdung and poultry droppings.

3. The process of anaerobic digestion produces biogas, largely containing methane. After this process, waste in the digester, the slurry material is dried. It is high in nutrients and is used as excellent manure which is sold in the market as compost. Cabinet has approved promotion of city composting as role model for other.

4. The biogas is transferred to a scrubber to remove hydrogen sulphide. The clean biogas is then transferred to a gas balloon, which pumps the gas to the biogas engine to generate 375 units of electricity.

Advantages of anaerobic digestion

1. It helps to produce energy from waste matter, which would have been disposed otherwise.
2. It allows two main benefits, i.e. generation of biogas fuel and production of fertilizer in the form of sludge.
3. This process conserves complete nitrogen content of the biomass. Nitrogen is recycled, produces fertilizer that boosts agriculture production.
4. It leads to good management of Municipal solid waste that improves sanitation and hygiene of towns and cities.

Factors affecting the performance of a digester

1. *Temperature:* Anaerobic bacteria optimally grow and operate in the temperature range of 20–65°C.
2. *Feeding rate:* A uniform feeding rate is maintained for continuous operation of the digester. Faster feed will retard digestion process.
3. *Seeding:* To start and accelerate digestion, a small amount of digested slurry containing methane forming bacteria is added in the digester. This process is called seeding of bacteria.
4. *Retention time:* Normally the retention period range from 30 to 50 days depending upon region temperature and type of biomass.
5. *Type of biomass:* The biomass may be, cow dung, poultry dropping, night soil, rice husk, algae and water hyacinth depending upon local availability.

After dealing with anaerobic system for solid waste management, let use find out what happens with aerobic (presence of oxygen) solidwaste treatment.

Two common methods of processing include composting and vermi-composting. Both produce rich manure for improving agricultural production.

Composting: Composting is a biological process in which organic matter present in waste is converted into enriched inorganic nutrients. The manure obtained has high nitrogen, phosphorus and potassium content. The quality of compost depends upon the waste being composted. The presence of high nitrogen, phosphorus and potassium contents in organic waste, facilitates production of high quality manure after compositing as detailed in Table 12.8.

Table 12.8 Characteristic of compost

Content	Typical value %
Total nitrogen	1.3
Total phosphorus	0.2–0.5
Total potassium	0.5
Organic phosphorus	0.054
PH	8.6
Moisture	40–50
Organic matter	30–70

Vermi-composting: This is a process whereby food materials, kitchen waste including vegetables and fruit peelings, papers and so on can be converted into compost by the action of earthworms. An aerobic condition is created due to the exposure of organic waste to air. Major species of earthworms, Eisenia fetida and Eisenia andrei are effective in decomposition of organic waste. Certain biochemical changes in the earthworm intestine result in excretion of cocoons and undigested food known as vermi castings, these are an excellent manure due to the presence of rich nutrients—vitamins and enzymes, nitrates,phosphates and potash. The enzymes produced also facilitate the degradation of different molecules present in solid waste into simple compounds for utilization by microorganisms. Different stages of vermi composting are shown in Figure.

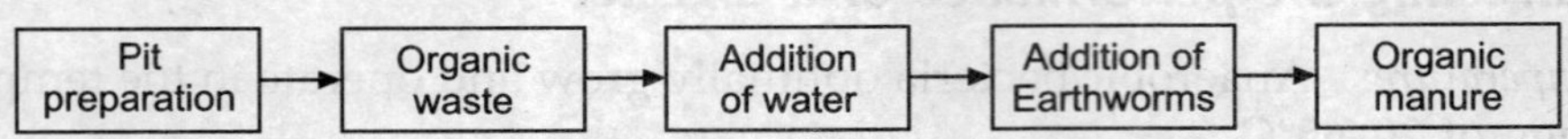

FIGURE 12.13 Different stages of vermi-composting.

National Biogas and Manure Management Programme (NBMMP)

Biogas is obtained from biodegradable organic waste through anaerobic digestion and enriched organic bio-manure is produced as byproduct from this process. NBMMP provides for setting up of single family stand alone Biogas Plants in rural and semi-urban areas with following objectives.

 (i) To provide clean gaseous fuel mainly for cooking purposes and organic manure to rural and semi-urban households.

 (ii) To mitigate drudgery of rural women, reduce pressure on forests and accentuate social benefits.

 (iii) To improve sanitation in villages by linking sanitary toilets with cattle dung based biogas plants.

 (iv) To provide bio digested slurry as a source of organic enriched bio manure and wormy composting to reduce use of chemical fertilizers.

 (v) To help in mitigation and combating climate change by preventing emissions of Green House Gases.

Two 32 kW off grid biomass gasification units in Sahebganj vilalge in Bihar Muzzaffarpur district, supply electricity in the evening from 6 pm to 12 pm. There plants convert solid biomass into gas which is then used to generate electricity.

Each plant uses 300 kg of rice husk per day to power 40% of the household is Sahabganj. Husk Power System (HPS) charges RS.50 per household a month for a 15 watt CFL connection for 6 hours a day to illuminate houses for children studies and to avoid use of kerosene lamp which was more expansive.

More than 100 such plants of varying capacities have been setup across Bihar. This has become possible because the Electricity Act exempts developers from taking any license to provide power in rural area. Off-grid energy solutions create acceptance among people as it helps to generate employment. HPS has employed women to make incense sticks out of char, a by-product of biomass gasification most of these gasifier are running on basic technology with 65% efficiency.

12.21 GLOBAL SCENE

In US 71.4 billion kWh (in 2016) of electricity is from biomass and 12.63 billion gallon of ethanol is used in vehicles. Biomass supports 13,000 jobs in the US. In Sweden, biomass and peat contribute 12% of total energy while in Austria this-figures is 13%. World wide biomass contributes 10% of total energy and it is 38% in developing countries especially in rural sector.

12.22 ENVIRONMENTAL BENEFITS

Biomass energy brings several environmental benefits–reduces air and water pollution, increases soil quantity and reduces erosion, and improve wildlife habitat.

Biomass reduces air pollution by being a part of carbon cycle. Actually, carbon cycle is nature's way of moving carbon around to support life on the earth. CO_2 is common vehicle for carbon. Plant photosynthesis breaks CO_2 in two, keeping carbon to form carbohydrates that make the plant, releasing oxygen into air. When plant is burnt, it gives its carbon back to air, which is reabsorbed by other plants.

On the other hand, when fossil fuels are burnt, there is no extra plant to absorb that carbon, so, the cycle becomes out of balance. There are two different carbon cycles in operation now; the natural one between plants and air, which is in balance, and man made cycle, where carbon is pulled from the earth (fossil fuel) and emitted into the atmosphere. Thus, biomass use for energy reduces CO_2 emissions by 90% compared with fossil fuel.

Water pollution is reduced, as little fertilizers and pesticides are used to grow energy crops. Planting poplar trees in buffers along water ways, runoff from corn field is captured, making streams cleaner.

High-yield food crops pull nutrients from the soil, while energy crops improve soil quality. Prairie grasses, with their deep roots, build up top soil, putting nitrogen and other nutrients into the ground.

Finally, biomass crops create better wildlife habitat than food crops, being native plants which attract birds and small mammals.

In addition to above, biomass offers economic and energy security benefits. By growing our fuel at home, oil import is reduced and farmers get money for their products.

Keeping in view above, Ministry of Environment and Forest (MOEF) in December 2009, exempted biomass and nonhazardous municipal water power plants up to 15 MW from environmental clearance.

If planned well, our capital requirement for pellitser machines, biomass fuel fired stoves and agroprocessing industries can be financed by earning large carbon credits.

REVIEW QUESTIONS

1. What is biomass? What are the different resources used to extract biomass energy?

2. What you mean by biogas? Write a note on producer gas and liquid fuel.

3. Discuss the different technologies to produce biogas. Also, discuss the factors affecting the production of biogas.

4. With a neat diagram, discuss the working of Deenbandhu biogas plant.

5. Discuss the state-of-the-art conversion technologies that help to use biomass material efficiently.

6. Discuss the process of production of ethanol from biomass.

7. With a neat diagram discuss the biomass gasification method.

8. Discuss the method of power generation from liquid waste.

9. Explain the production process of ethanol from biomass.

10. What is biodiesel? Discuss the production of biodiesel from Jatropha.

FUEL CELLS

13.1 INTRODUCTION

A fuel cell is an electrochemical device that converts the chemical energy of a fuel into electricity without involving a combustion cycle. The first fuel cell was developed in 1839 in England by Sir William Grove. However, the application of fuel cell was first demonstrated by Francis T. Bacon in 1959 when his model generated 5 kW at 24 V. Its practical application began during the 1960s when the US space programme chose fuel cells over nuclear power and solar energy. Fuel cells provided power to the Gemini, Apollo and Skylab spacecraft, and continue to be used to provide electricity and water to space shuttles.

In India, the fuel cell laboratory of Bharat Heavy Electricals Ltd. (BHEL), Hyderabad is developing 'Phosphoric Acid Fuel Cells' (PAFC) since 1987. PAFC power packs of 1, 5, 10 and 50 kW ratings have been successfully developed and demonstrated by BHEL, Hyderabad.

13.2 PRINCIPLE OF OPERATION OF AN ACIDIC FUEL CELL

It is known that electrolysis of water produces hydrogen and oxygen. In a fuel cell the process is reversed where these gases combine in an electrochemical cell to generate electricity and water.

A basic hydrogen-oxygen fuel cell with phosphoric acid as electrolyte is shown in Figure 13.1. Fuel cell electrodes are made porous in order to provide a large number of pockets where the gas, the electrolyte and the electrode are in contact for chemical reaction. A fuel cell like any other battery consists of two electrodes and an electrolyte. However, a fuel cell differs from a battery in the sense that both the reactants (i.e., hydrogen and oxygen) are not permanently contained in the electrochemical cell, but are fed into it from an external supply, when electric power is required. In fuel cells, platinum coated special graphite plates are used as the electrodes, separated by an electrolyte. The fuel is hydrogen gas which is supplied at the anode side where the hydrogen molecules are effectively reduced to hydrogen ions which move on into the electrolyte.

$$H_2 \rightarrow 2H^+ + 2e^-$$
$$\text{(gas)} \qquad \text{(ion)}$$

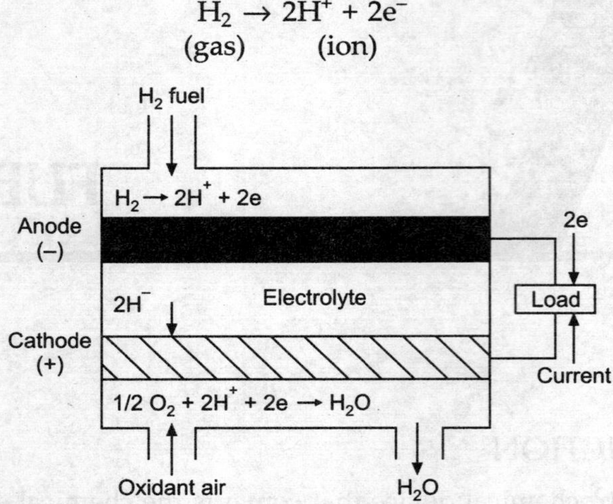

FIGURE 13.1 Fuel cell operation.

Electrons so liberated at the anode build up a negative potential and travel towards the cathode through an externally connected circuit. Oxygen gas is supplied at the cathode where it is reduced by hydrogen ions to produce water.

$$4H^+ + O_2 + 4e^- \rightarrow 2H_2O$$

Electrochemical reactions coupled with movement of hydrogen ions through the electrolyte generate an electric potential, which causes electric current to flow through the load.

$$2H_2 + O_2 \xrightarrow{\text{Fuel cell}} 2H_2O + \text{Electric energy generated} + \text{Heat energy released}$$

This reaction is exothermic, which results in heating up the cell. A stream of air is circulated on the cathode side of the cell which absorbs enough heat to maintain outlet air and steam at 180°C which is optimum for best performance of the cell.

13.3 TECHNICAL PARAMETER OF A FUEL CELL

An individual cell produces a small voltage, in the range of 0.55 to 0.75 V. A number of cells are therefore, arranged in 'stacks' to provide the required level of voltage. Electrode size (area) determines the total amperage, and current density in the range of 100–500 mA/cm^2 can be achieved. For any given power rating, an appropriate electrode size and the requisite number of cells are selected.

In a typical fuel cell 'stack', a number of individual cells are stacked together using bipolar plates. A bipolar plate has two functions, namely to facilitate the distribution of fuel and oxidents to the cells, and to provide continuity in electrical circuit between the cells. To meet this requirement, the plate should be impervious to gas diffusion and possess low electrical resistance. A set of such repeating elements is shown in Figure 13.2.

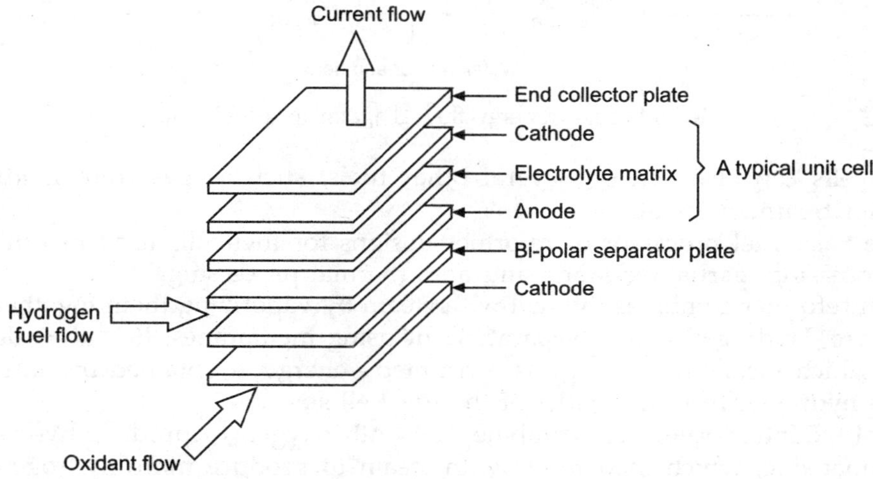

FIGURE 13.2 Repeating elements used in a cell stack.

Apart from the cell stack assembly, a fuel cell power plant consists of a number of other sub-systems, namely a fuel processing system, an inverter system, a control and instrumentation system, and a water and heat recovery system. These sub-systems are schematically shown in Figure 13.3.

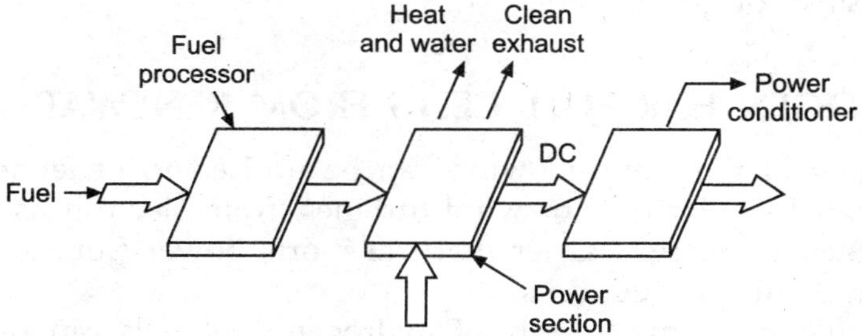

FIGURE 13.3 Major sub-systems in a fuel cell power plant.

13.4 FUEL PROCESSOR

Fuel cells generally run on hydrogen, but any hydrogen-rich material can also serve as a fuel source. This includes fossil fuels—methanol, ethanol, natural gas, petroleum distillates, liquid propane and gasified coal. Fuels containing hydrogen require a 'fuel processor' that extracts hydrogen gas as shown in Figure 13.4.

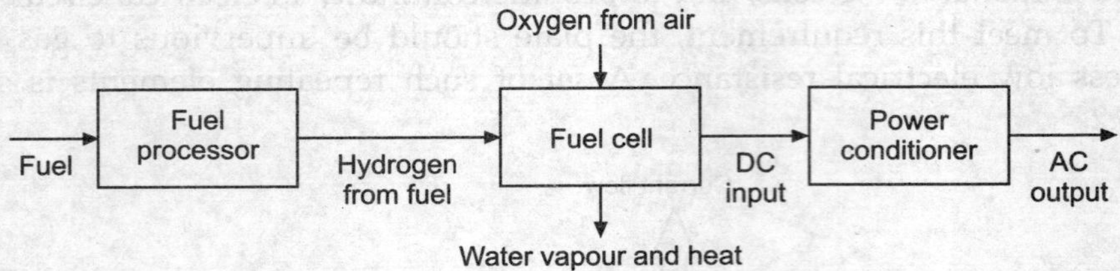

FIGURE 13.4 Energy flow diagram in a fuel cell.

Fuel cells can also run on several other fuels, such as gas from landfills and wastewater treatment plants.

Three basic fuel processor or reformer designs for fuel cells used in vehicles are: steam processing, partial oxidation and auto-thermal processing.

Steam reformer combines the fuel with steam by vaporizing them together at high temperature. Hydrogen is then separated out using membranes. It is an endothermic process, which means that energy is consumed—energy is obtained by burning fuel or excess hydrogen from the outlet of the fuel cell stack.

Partial oxidation reformers combine fuel with oxygen to produce hydrogen and carbon monoxide, which then reacts with steam to produce more hydrogen. Partial oxidation releases heat which is utilised elsewhere in the system.

Auto-thermal reformers combine the fuel with steam and oxygen, thus, the reaction remains in heat balance. In general, both methanol and gasoline can be used in any of the three reformer designs. Differences in the chemical nature of the fuels, however, can favour one design over another.

Fuel cells are ideal for power generation, particularly for on-site service in areas that are inaccessible for grid supply.

13.5 HYDROGEN FOR FUEL CELLS FROM RENEWABLE SOURCES

Renewable energy of the sun and wind can be utilised to generate hydrogen, by using power from PV solar cells or wind turbines, from electrolysis of water. Thus, hydrogen becomes an energy carrier that transports power from generation site to another location for use in fuel cells.

If it is difficult to arrange supply of hydrogen, fuel cells can easily operate on methanol.

13.6 METHANOL (METHYL ALCOHOL) FUEL CELL

In places where hydrogen is not readily available, methanol (CH_3OH) can be used as fuel. To generate one kilowatt of power the fuel cell uses about 12 litres of hydrogen gas, obtained from 0.6 litre of methanol that costs about ₹45 per litre. The overall efficiency of the cell, from fuel to electricity, is about 60%. The efficiency of the fuel cell directly depends upon its use, as in a chlor-alkali cell; besides electricity generation the waste heat, as a by-product, is also utilised usefully in industry.

Methanol is a liquid hydrocarbon fuel. It can directly be introduced from the anode side of the fuel cell without converting to hydrogen. Such an arrangement is used in mobile applications of fuel cells, such as in buses and in remote military sites where noise and smoke discharge, such as in diesel generators, are prohibited. Methanol can directly be oxidized to operate as fuel.

$$2CH_3OH + 3O_2 \rightarrow 2CO_2 + 4H_2O$$

In another method, methanol is converted to hydrogen and carbon oxides. Hydrogen rich fuel gases so produced are fed into the fuel cells to generate electrical energy.

For clarity between Fuel cell and primary battery, similarities and differences between them are presented point wise:

1. Fuel cell and primary battery both have two electrodes and electrolytes.

2. In fuel cell, electrodes acts as a catalyst, hence are nearly stable while in primary battery electrodes takes part in electro-chemical reaction hence they are consumed.

3. In fuel cell, electrodes are not replaced while in primary battery electrodes are replaced periodically.

4. In fuel cell, fuel and oxidant are stored outside the cell while in primary battery they are stored in cell.

5. Primary battery is storage device that are charged and discharged.

13.7   FUEL CELL TYPES

Fuel cells are identified by their most important component, i.e., the electrolyte. The electrolyte determines the operating temperature, the catalysts to be applied to the electrodes and the requirement of the process gas.

There are five types of 'fuel cells' based on the electrolyte used in each of them.

- Alkaline Fuel Cells (AFC)
- Proton Exchange Membrane Fuel Cells (PEMFC)
- Phosphoric Acid Fuel Cells (PAFC)
- Molten Carbonate Fuel Cells (MCFC)
- Solid Oxide Fuel Cells (SOFC)

Various fuel cells, their operating temperatures, unit sizes and applications are given in Table 13.1.

TABLE 13.1 Fuel cell characteristics

Type	Operating temp. (°C)	Unit size (kW)	Applications
Alkaline Fuel Cell (AFC)	70–100	1–100	Space and military
Proton Exchange Membrane Fuel Cell (PEMFC)	50–100	0.1–100	Residential, portable laptops, cellular phones, video cameras, buses, cars, railway locomotives
Phosphoric Acid Fuel Cell (PAFC)	160–210	5–200 (also MW sized plants)	Dedicated power (+heat), railways
Molten Carbonate Fuel Cell (MCFC)	650	100–2000	Dispersed power and utility power
Solid Oxide Fuel Cell (SOFC)	800–1000	2.5–250 (plants up to 100 MW)	Domestic and commercial utility power, mobile applications for railways

13.7.1 Alkaline Fuel Cells (AFCs)

Alkaline fuel cells use KOH as electrolyte with porous electrodes of carbon having nickel as the electro catalyst. Hydrogen is used as fuel and oxygen as oxidant, as shown in Figure 13.5. Its operating temperature is about 80°C. At anode, the hydrogen gas reacts with hydroxide ions present in the electrolyte solution to form water, and electrons are released.

$$H_2 + 2(OH)^- \rightarrow 2H_2O + 2e$$

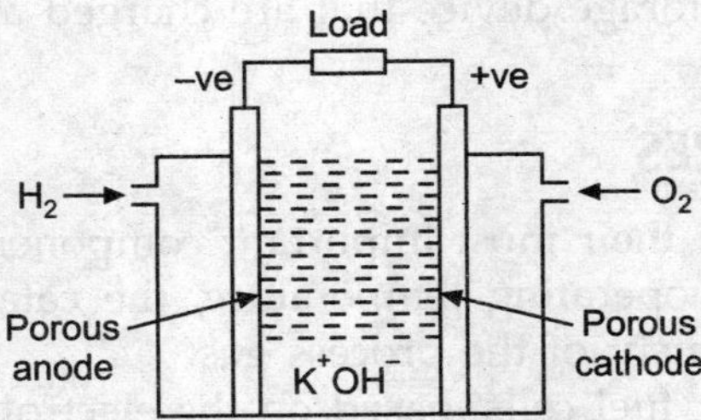

FIGURE 13.5 Alkaline electrolyte fuel cell.

Electrons so produced build up a negative potential and move towards the cathode through an externally connected circuit.

At cathode the electrons are picked up by the oxygen atoms available there, react with water present in the electrolyte to form hydroxide $(OH)^-$ ions.

$$\frac{1}{2}O_2 + H_2O + 2e^- \rightarrow 2(OH)^-$$

which combine with hydrogen ions to form water.

$$H^+ + OH^- \rightarrow H_2O$$

Thus, with hydrogen and oxygen continuously supplied, the fuel will steadily be oxidized by the ions produced in the process to generate electric power, causing current to flow in the external circuit. The voltage across the terminals of the cell is about 1 volt. For greater outputs a number of single cells can be connected in series. The efficiency of fuel cells is high, about 73%. M/s Apollo, have developed AFC-based power packs in 2 kW to 25 kW range.

As the KOH electrolyte used in AFCs readily reacts with the CO_2 to form K_2CO_3, so this cell is not considered suitable for terrestrial applications. Even traces of CO_2 present in the ambient air limits the life of fuel cells.

However, where pure H_2 and O_2 reactants are available, as in rockets and spacecraft, there is no other fuel cell that can compete with the high power densities offered by AFCs.

13.7.2 Polymer Electrolyte Membrane Fuel Cells (PEMFC)

A polymer electrolyte fuel cell, also known as Protone Exchange Fuel Cell (PEFC), consists of a solid electrolyte which is an ion exchange membrane as shown in Figure 13.6. The fuel used is hydrogen with air as the oxidant. The membrane is impermeable to gases but allows the hydrogen ions to move across it which makes the current to flow in the circuit. The cell operates at low temperatures (80–100°C).

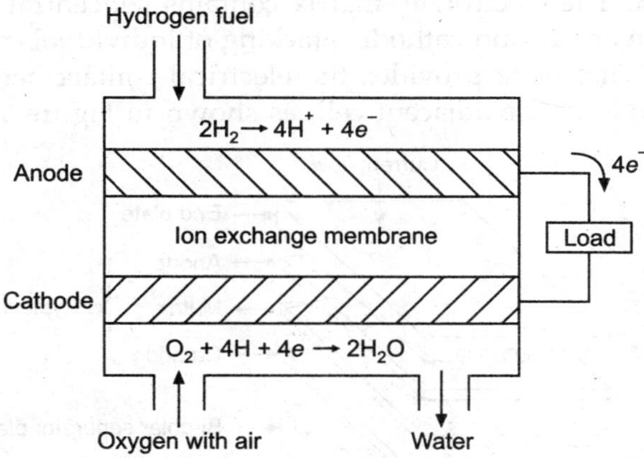

FIGURE 13.6 Polymer electrolyte fuel cell.

A polymer electrolyte membrane fuel cell developed in the laboratories of the Siemens Power Generation Group in Erlangen (Germany), has a maximum power output of 50 kW. The primary objective to develop this PEMFC has been to create a noiseless, pollution free power pack. Siemens is also working on a low-cost fuel cell design and simplified manufacturing methods which are suitable for mass production.

Polymer electrolyte fuel cells are the electrochemical energy converters with an efficiency level of more than 60%. No gaseous environmental pollutants are produced like in internal combustion engines. This fuel cell operates at low temperature, so its start-up and shutdown times are short. This type of fuel is especially suitable for small stationary applications in the power range below 1 MW.

Each PEMFC is a thin wafer with a catalyst-coated membrane resembling a plastic film that is enclosed in graphite or ceramic plates. One side of the membrane acts as an anode and is fed with hydrogen gas. The other side of the membrane serves as the cathode and is in air to provide oxygen.

At the anode, a catalytic reaction takes place and the hydrogen atoms release their electrons to become hydrogen ions (protons). The protons then diffuse through the membrane to reach the cathode. The electrons take a different path around the membrane, creating an electric current as they travel to the cathode. At the cathode another catalytic reaction takes place as they recombined hydrogen atoms with oxygen to produce water.

An undisputed world leader in the PEM area is M/s Ballard Power Systems Canada. Ford Motor Company is designing Zero Emission Vehicles (ZEVs) to be operated by PEM stacks. In India, SPIC Science Foundation is associated with the development of PEM technology with the support from MNRE.

13.7.3 Phosphoric Acid Fuel Cell (PAFC)

The Phosphoric Acid Fuel Cell (PAFC) consists of an anode of porous graphite substrate with platinum alloy as the catalyst. The cathode in similar to the anode but made with a noble-metal catalyst. The electrolyte* matrix contains concentrated phosphoric acid and is located between anode and cathode. Stacking of individual cells is accomplished with a bipolar plate. This plate provides the electrical contact between the anode of one cell and the cathode of the adjacent cell, as shown in Figure 13.7.

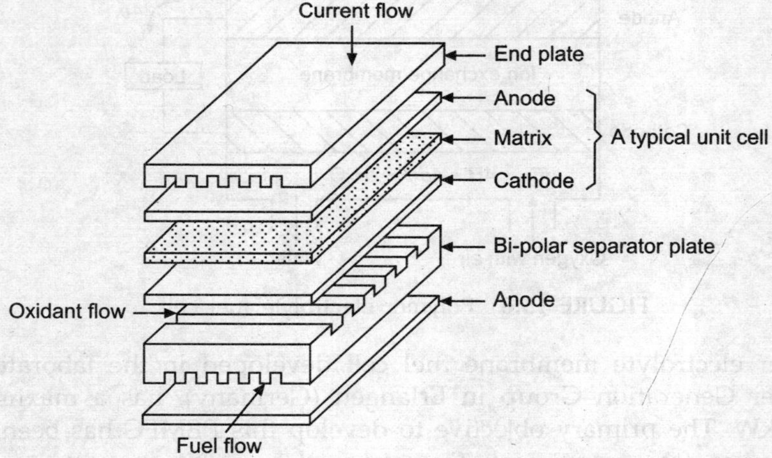

FIGURE 13.7 Schematic of repeating elements in a PAFC stack.

---

* *Matrix:* The matrix serves the dual purpose of holding the electrolyte and acting as a separator between the cathode and the anode. The matrix must have high electrical resistance, good acid wicking capability and should be chemically stable.

It is a low temperature fuel cell and so requires high purity hydrogen. The low-temperature fuel cell systems comprises a second pre-processing step called the 'water gas-shift reaction' in which CO reacts with steam and is converted into hydrogen and CO_2. Such pre-processing steps are performed in separate reactors, and the fuel cell system is referred to as an 'external reforming system'.

This cell has an advantage that it can be used without any danger of poisoning the electrolyte, but CO has to be converted to avoid platinum poisoning.

PAFC technology is presently in the initial phase of commercialization at the level of 200 kWe. Due to low operating temperature ($<200°C$), PAFC is not attractive for large-scale power production, limiting its applications to uninterrupted power source for computer centres, airports, etc. BHEL R&D Hyderabad operated these cells using LPG in both modes i.e., grid independent as well as grid connected. During 2000, BHEL developed 50 kW PAFC power pack, i.e., two stacks of 25 kW each.

M/s International Fuel Cell Corporation, a joint venture of M/s United Technology & M/s Toshiba are the sole world leaders in this field. More than 200 units of their 200 kW Fuel Cell Power Plants (FCPPs) have already been sold. Recently FCPPs have been integrated with digester gas from sewage treatment plants in large cities. Another niche area is their use in Combined Heat & Power (CHP) mode. CHP application of fuel cells can be more than 80% efficient, i.e., 40% electrical efficiency with another 40% energy made available as heat.

13.7.4 Molten Carbonate Fuel Cells (MCFCs)

A molten carbonate fuel cell needs a molten mixture of alkali carbonates as an electrolyte. Its operating temperature is around 650°C which allows the use of catalysts like nickel in the electrodes. High temperature keeps the carbonate electrolyte in liquid phase. The electrolyte is retained between two porous nickel electrodes as shown in Figure 13.8.

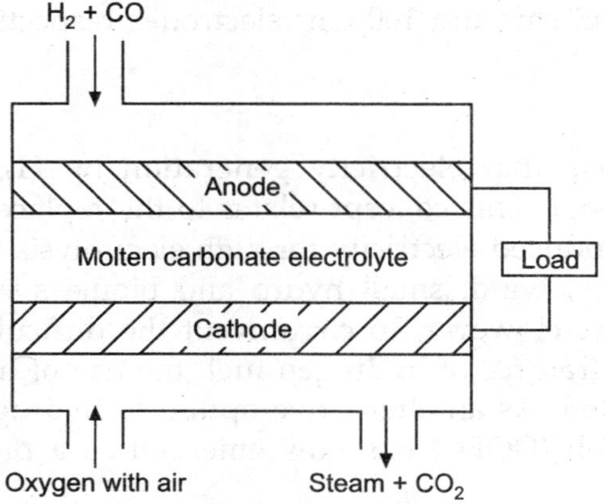

FIGURE 13.8 Molten carbonate fuel cell.

A fuel gas derived from fossil fuels contains CO_2 and CO. The MCFC is insensitive to CO_2 and with nickel/nickel oxide electrodes it is also immune to poisoning by CO.

Being a high temperature fuel cell, there is an internal reforming system which takes place almost simultaneously with the electro-chemical reactions. With the operating conditions in this cell, CO is oxidized, via the water-gas shift reaction, to CO_2 with the production of hydrogen.

The oxidizing agents for hydrogen are carbonate ions which are formed at the cathode. Thus, the oxidant gas must contain CO_2. In practice, CO_2 is provided by recycling the anode off-gas to the cathode.

At anode
$$H_2 + CO_3^{--} \rightarrow H_2O + CO_2 + 2e$$
$$CO + CO_3^{--} \rightarrow CO_2 + 2e$$

At cathode
$$\frac{1}{2}O_2 + CO_2 + 2e \rightarrow CO_3^{--}$$

The by-products of this cell are, steam and carbon dioxide at a high temperature of 545°C and a source of cogeneration. Thus the MCFC, in addition to electricity, also provides industrial process heat. Waste heat can generate steam in a boiler which can drive a generator to supply additional electric power, thus improving the total efficiency of the system.

At international level the world's largest MCFC-based power plant of 2 MW rating with internal reforming system was commissioned at Santa Clara in 1996. Currently, M/s Fuel Cell Energy Research Corporation is engaged in commercialization of 300 kW, 1.5 MW and 3 MW modular, "Direct Fuel Cell" power plants for stationary, distributed power applications.

In India, M/s Central Electro-Chemical Research Institute, Karaikudi is spearheading the MCFC R&D in association with the Tata Energy Research Institute Delhi. There is an MNRE funded project with CECRI for making single cells of 45 and 100 cm^2 area. They are successful in establishing current densities of 168 mA/cm^2@0.62 V and 105 mA/cm^2@0.67 V for the 45 cm^2 and 100 cm^2 electrodes respectively.

Hydrogen Economy

It is a research prospects that electricity generation by H_2/O_2 Fuelcells leads to Hydrogen economy concept. This concept relates to the replacement of our fossil fuel economy to hydrogen produced electricity through electrolysis from water. Renewable energies like solar energy, wind, small hydro and biomass would play the role of oil gas and coal of today. However, to circumvent the difficulty associated with the production, storage and transfer of hydrogen fuel, the use of liquid fuels in fuel cells has been widely advocated. As an alternative option to hydrogen based fuel cell, the Direct Oxidation Fuel Cell (DOFC) has now emerged as a promising technology in the fuel cell arena.

In contrast to the use of hydrogen, DOFCs can use organic and inorganic fuels such as alcohols, carbohydrates, hydrazine (N_2H_4) and borohydrides which are directly fed in liquid phase into the anode compartment. Such fuel cells commonly use Polymer

Electrolyte Membrane (PEM) in acid environment. Use of liquid fuel can provide considerable weight and volume advantages over PEMFC.

Among all the liquid fuels methanol is a good choice due to its relatively higher electro-chemicals activity compared to the other liquid fuels. However,methanol is volatile and relatively toxic. So, ethanol (C_2H_5OH) has emerged as a green fuel since it is eco-friendly and has more energy density compared to methanol (8.01 kWh/kg for ethanol, 6.9 kWh/kg for methanol as against 10.3 kWh/kg for gasoline).

Moreover, bio-ethanol is a renewable energy source and can be easily produced in great quantity by the fermentation of sugar containing raw materials from agriculture and even from organic fraction of municipal solid wastes. Also, use of bioethanol in fuel cell has the significant advantage of being nearly CO_2 neutral since the CO_2 produced in the process is consumed in biomass growth.

The liquid methanol CH_3OH is oxidized at anode in presence of water, producing CO_2, hydrogen ions and the electrons. The hydrogen ions move through the electrolyte and react with oxygen to form water at anode and electrons flow in the external circuit, thus generating electricity as detailed in chemical equations.

Anode Reaction $\qquad CH_3OH + H_2O \rightarrow CO_2 + 6H^+ + 6e^-$

Cathode Reaction $\qquad 3O_2 + 12H^+ + 12e^- \rightarrow 6H_2O$

13.7.5 Solid Oxide Fuel Cell (SOFC)

An SOFC is based on a solid metal oxide electrolyte (zirconium dioxide) called zirconia. It allows ionic conductivity of oxgyen ions from cathode to anode. It is a high temperature fuel cell which operates in the temperature range of 800°C to 1000°C. The electrodes are electric conductors with a high porosity. The operating temperature is high enough for internal reforming of natural gas in the anode chamber. The water gas shift reaction takes place at the anode, thus enabling H_2 and CO mixtures to be used as fuel feedstock. Configuration of a single SOFC in tubular shape is shown in Figure 13.9. The construction materials used are metal oxides and ceramics. The central hollow space is for air flow that acts as an oxidant. Its operation is efficient at 1000°C and 1 atmospheric pressure. Fuel gas flows through the outermost layer of the fuel electrode. Next to it is the electrolyte layer. Fuel gas permeates through the porous electrodes and is oxidized by air containing oxygen. The air electrode is next to electrolyte and air flows axially through the central hollow space. Both fuel gas and oxidant are fed into the cell continuously which get consumed and the SOFC delivers electrical energy.

For large-scale power generation the SOFC can also be fuelled with coal gas and gases derived from biomass. Due to high operating temperature, integration with gas and steam turbines is possible. Natural gas is clean, efficient and economically versatile than fossil fuels. An SOFC fuelled by natural gas can attain a high electrical efficiency rating up to 55%. Each cell delivers 25 A current at 0.7 V and a pack of 50 cells gives an output of 1000 W.

A world leader in the Solid Oxide Fuel Cells is M/s Siemens Westinghouse Power Corporation of USA (SWPC). The company markets 25 kW multi-tube stacks. They installed a power plant that feeds 109 kW into the grid besides 64 kW worth of hot water/steam to a local heating system. Electrical efficiency of 46% has been achieved.

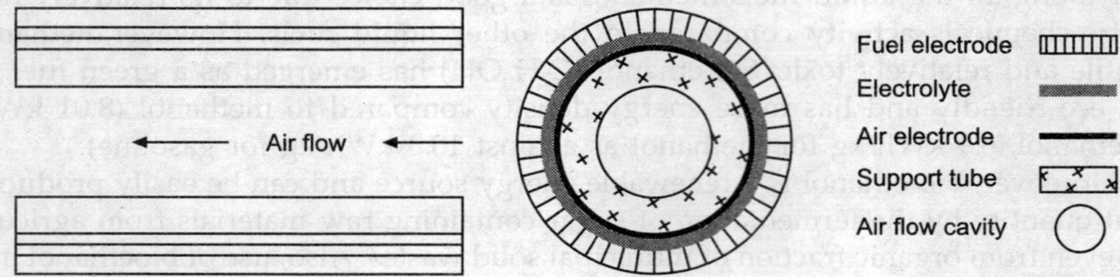

FIGURE 13.9 Solid oxide fuel cell (tubular shape).

In Asia, NKK of Japan introduced 5 kW and 50 kW units as distributed power generators for residential and commercial consumers. NKK also plans to introduce 250 kW and 550 kW SOFC units for use in offices, shoping malls, multistoried buildings, hospitals and hotels.

In India, SOFC research is being carried at the Central Glass & Ceramic Research Institute Kolkata, Corporate R&D, BHEL Hyderabad where various materials and electrode processing technologies have been developed. Both have assembled and tested SOFC single cells.

Solid oxide fuel cells promise a vast potential in utilisation of low grade, high ash, graded coals through Fluidized Bed Gasification. BHEL R&D Hyderabad developed a fuel cell system control as shown in Figure 13.10.

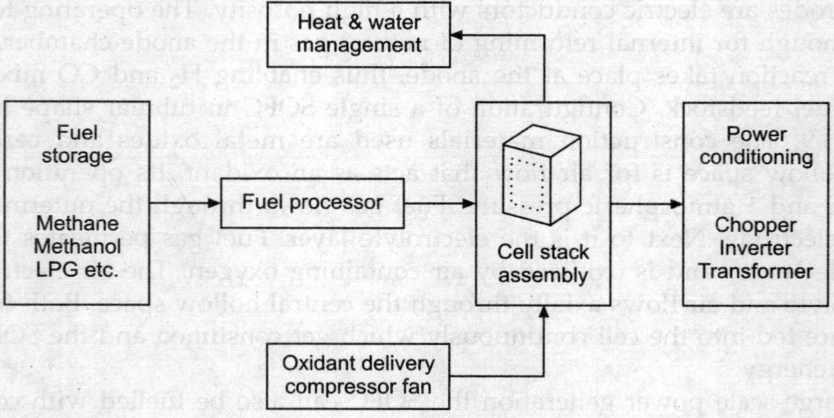

FIGURE 13.10 Fuel cell system control.

Fuels—for Fuel Cells

Fuel is important to operate the Fuel cells, especially for their use in distributed generation, to ensure the local availability of the same. Basically all Fuel cells operate with Hydrogen and Oxygen. Air is generally used as a source for oxygen.

Hydrogen is generally supplied to the low-temperature Fuel cells by the process of external reformation of the hydrocarbon-rich fuels, whereas internal reformation is possible in the high temperature Fuel cell stocks.

Possible option for Urban Areas are Hydrogen or Natural gas (wherever available), Methanol, LPG, Diesel and Petrol. Only suitable reformers should be developed and made available. Economics needs to be worked out.

Hydrocarbon gases or liquid are first-steam reformed to get H_2 and CO. Carbon monoxide is then removed by the reaction.

$$CH_4, C_3H_8 \xrightarrow[\text{Ni catlyst}]{\text{steam reforming at 90°C}} H_2 + CO$$

For rural areas ethanol can be used, which can be produced from agro-waste. Also Gobar gas (actually CH_4) is easily available for supply of hydrogen for Fuel cells.

Where no other fuel is available, but plenty of sunshine is available, hydrogen can be generated using solar photovoltaic panels and the stored hydrogen can be used after the sunset for generating power through Fuel cells.

13.8 ADVANTAGES OF FUEL CELL POWER PLANTS

Some of the important advantages of fuel cell power plants are:

1. Fuel cell power plants are eco-friendly, noiseless, carry no rotating components. In contrast, in coal-based stations, ash slurry, discharge of smoke through chimney adversely affect the environment.

2. It is a decentralized plant, can be operated in isolation for military installations and hospitals where noise and smoke are prohibited. Besides, no power is wasted in transmission and distribution.

3. Fuel cell power sources attain a high efficiency up to 55% whereas conventional thermal plants operate at 30% efficiency.

4. A large degree of modularity is available, with capacity ranging from 5 kW to 2 MW. The number of fuel cells can be increased as per the requirement.

5. There is a wide choice of fuels for fuel cells. These can be operated with natural gas, ethanol, methanol, LPG and biogas supplied from local biomass. All these are hydrogen rich materials and hydrogen gas can be produced by using fuel reformers.

6. Fuel cells can operate at landfills and wastewater treatment plants from the methane gas they produce. Fuel cells operate on waste gases at breweries, also on gas from sewage sludge proving to be the cleanest and most cost-effective energy conversion technology.

7. In addition to electric power, fuel cell plants also supply hot water, space heat and steam. Fuel cells have cogeneration capabilities.

8. Potential areas of cogeneration systems where fuel cells can be effectively installed are sugar, paper, cotton, textile, caustic soda, iron and steel mills and refineries which will enhance system efficiency and reduce demand on grid.

13.9 FUEL CELL BATTERY–POWERED BUS SYSTEM

In recognition of several benefits which could accrue with large-scale introduction of fuel cell technology in transport sector, the US Department of Energy has taken up a R&D programme on fuel cell propulsion technologies for potential applications.

Immediate efforts are directed only at the phosphoric acid fuel cell (PAFC) as it is a suitably developed technology for the transport sector. The programme is to build PAFC fuel cell/battery-powered test buses with methanol as fuel.

A schematic of a typical fuel cell/battery-powered bus system is shown in Figure 13.11.

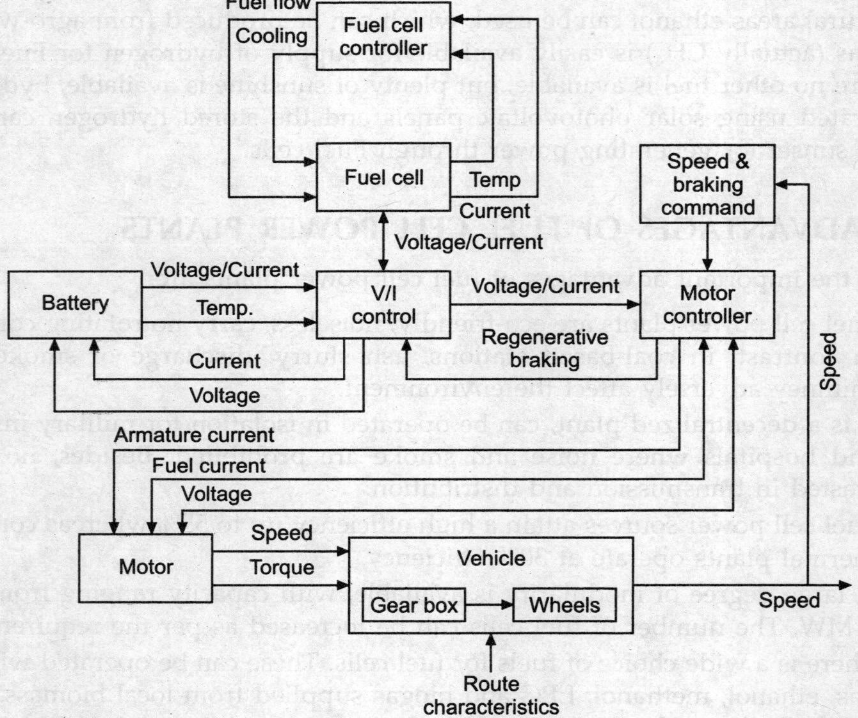

FIGURE 13.11 Typical fuel cell battery-powered bus system.

These buses are expected to offer performance equivalent to, or better than diesel buses with tail pipe emissions reduced by more than 99%.

13.10 COMPARISON BETWEEN ACIDIC AND ALKALINE HYDROGEN–OXYGEN FUEL CELLS

Comparison between acidic and alkaline H_2–O_2 fuel cells may be made with respect to their contents and the reaction that takes place to produce electrical energy.

Electrodes: In both the cells, electrodes are of carbon, porous in construction with nickel as the electro catalyst.

Electrolyte: In an acidic fuel cell the electrolyte is phosphoric acid, while in an alkaline cell it is aqueous alkaline like KOH.

Reaction: An electrochemical difference between the two cells is, that in an acidic electrolyte the migration is by hydrogen ions, while with an alkaline electrolyte the migration is with hydroxyl ions (OH).

However, the total reaction in both the fuel cells is the same, i.e., hydrogen and oxygen supplied to either acidic or alkaline fuel cell produce electrical energy, water and heat.

$$2H_2 + O_2 \xrightarrow{\text{Fuel cell}} 2H_2O + \text{Electrical energy generated} + \text{Heat energy released}$$

13.11   STATE-OF-THE-ART FUEL CELLS

13.11.1   Microbial Fuel Cell

Researchers from the US-based Pennsylvania State university have developed an electricity generator that is fuelled by human waste. The device, the microbial fuel cell (MFC), is useful for countries where large-scale waste-processing plants are needed, but are prohibitively expensive, because of high power requirement. Offsetting this cost by producing electricity, while treating waste, the programme is economically viable.

Operating process

Human waste contains undigested food comprising organic matter such as carbohydrates, proteins and lipids. Bacteria use enzymes to oxidize the organic matter. In this process, electrons are released. Normally the electrons power the respiratory reactions of the bacteria cells, and eventually combine with oxygen molecules. However, by depriving the bacteria of oxygen on one side of MFC, the electrons are used to power an external circuit.

An MFC comprises a 15 cm long cylindrical metal container with a central cathode rod that is surrounded by a proton exchange membrane (PEM) as shown in Figure 13.12. Eight anodes (long, slender graphite rods) are arranged around the cathode. Bacteria clustering around the anodes breakdown the organic waste as it is pumped, releasing electrons and protons. With no oxygen to help mop up the electrons, the bacteria enzymes transfer them to the anodes, while the protons migrate to the central cathode. Molecules on the PEM encourage the protons to pass through the cathode. There, they combine with oxygen from the air and electrons from the anodes to produce water. During the transfer of the electrons from the cathode, a voltage is created, enabling the MFC to power an external circuit. The equipment is designed for a laboratory test. It will be put in commercial use and the system may produce about 51 kW of power from the waste of 100,000 people.

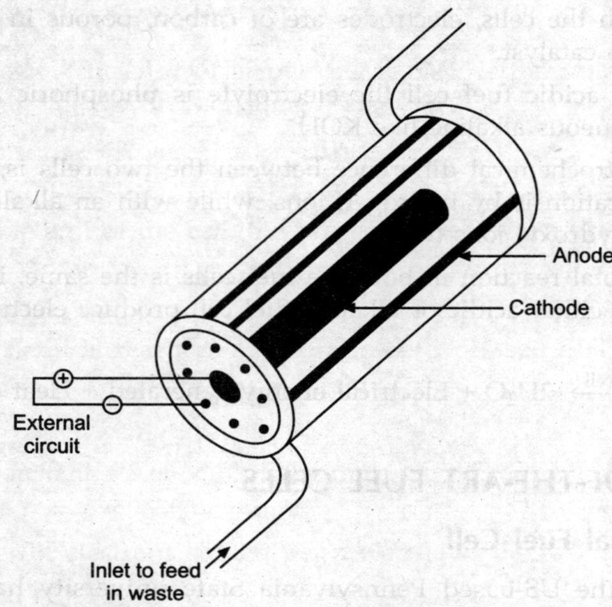

FIGURE 13.12 Microbial fuel cell.

13.11.2 World's First Fuel Cell–Gas Turbine

The fuel cell–turbine hybrid system combines a Siemens–Westinghouse solid oxide fuel cell with an Ingersoll Rand microturbine. Solid oxide fuel cells are electro-chemical devices that convert energy in a fuel into electricity, while microturbines are small high-speed gas turbines. The synergy realized by fuel cell–turbine hybrids is derived primarily from using the rejected thermal energy and combustion of residual fuel from a fuel cell in driving the gas turbine. This leveraging of the thermal energy makes the high temperature molten carbonate and solid oxide fuel cell most suitable for the hybrid system. Use of a recuperator contributes to improving the thermal efficiency by transferring heat from the gas turbine exhaust to the fuel and air used in the system.

Linked together in a mini-power plant, the advanced generator is being tested at the National Fuel Cell Reserach Centre in University of California–Irvine. The two technologies combined can produce 190 kW of electricity. The combination has pushed power efficiency to 60% for smaller systems and 70% or higher for larger systems, which is much higher than that of the fuel cell system on natural gas (about 53%). Its emissions of nitrogen oxide are nearly 50 times less than an average natural gas turbine. It is environmental friendly and can be sited anywhere, even in the most environmentally sensitive regions.

13.12 ENERGY OUTPUT OF A FUEL CELL

Different forms of energy differ in quality. A few forms of energy are of high quality, while others are of low quality. Work, heat and electrical energy differ in quality. From the second law of thermodynamics, "heat cannot be completely converted to work" and

$$Q > W$$

where Q represents the heat and W the work. However, the work can be fully converted to heat, i.e.,

$$W = Q$$

Thus, the work is high grade energy. Similarly, electrical energy is a higher form of energy as it is convertible into heat or work.

In a fuel cell a chemical reaction takes place as fuel and air are continuously fed into it at normal atmospheric temperature and pressure (NTP). The products after reaction are electric power and water. There is a steady flow of fuel and air at the inlet of the fuel cell and so is the steady flow at the outlet, of course, maintaining pressure and temperature equilibrium with the atmosphere.

In a steady-flow condition, the equation from the first law of thermodynamics is

$$\Delta Q = \Delta W + \Delta H \tag{13.1}$$

where

ΔQ = net heat transferred to the steady flow stream from the surrounding atmosphere

ΔW = net work done by the steady stream on the surrounding

ΔH = change in enthalpy[†] of the flow stream from inlet to exit.

(In Eq. (13.1) the changes in potential energy, kinetic energy and internal energy are not mentioned, being negligible.)

Entropy is expressed as a quantity of a 'system's thermal energy' not available for conversion into mechanical work. Entropy like heat is a characteristic of the system. When a system receives heat from surroundings, the entropy of the system increases. Also, if a system rejects heat to surroundings the entropy of the system decreases. Entropy is an indicator of heat per kelvin temperature (T). For a system of high absolute temperature the entropy will also be high. For a reversible process, considering the second law of thermodynamics, the change in entropy can be expressed as

$$\Delta Q = \int T ds \tag{13.2}$$

At any given instant, the surrounding atmosphere maintains a constant temperature T. Thus, reversible heat transfer takes place at temperature T, which is the prevailing temperature at inlet and exit.

Thus, $$\Delta Q_{rev} = T\Delta s \tag{13.3}$$

From Eqs. (13.1) and (13.3),

$$\Delta W = T\Delta s - \Delta H$$

$$= -(\Delta H - T\Delta s) \tag{13.4}$$

[†] *Enthalpy (H) of a substance is represented by the equation ($U + pV$) where U is the internal energy, p is the pressure and V is the volume of the system. The unit of enthalpy is joule (J). Enthalpy is a thermodynamic quantity equivalent to the total heat content of a system.*

13.13 EFFICIENCY AND EMF OF A FUEL CELL

The energy conversion efficiency of a fuel cell at NTP is the ratio of the net useful work done to the heat of the fuel consumed, i.e.,

$$\eta = \frac{\Delta W}{-\Delta H} \tag{13.5}$$

To find the reversible emf of the cell, the reversible electrical work is expressed as

$$\Delta W_{rev} = E\Delta q \tag{13.6}$$

where Δq is the charge shifted.

For a fuel cell chemical reaction, Δq can also be expressed as

$$\Delta q = NF \tag{13.7}$$

where

$$F = \text{Faraday's constant} = 9.65 \times 10^7 \text{ C/kg-mole}$$
$$= 96500 \text{ c/g-mole}$$

N = total number of electrons shifted per molecule of the reactant.

From Eqs. (13.6) and (13.7),

$$\Delta W_{rev} = NFE \tag{13.8}$$

The emf of the cell can be expressed as

$$E = \frac{W_{rev}}{NF} \tag{13.9}$$

13.14 GIBBS–HELMHOLTZ EQUATION

Chemical reactions can be made to do work, or to serve as pathways for the conversion of heat into work. Energy available to perform useful work is called 'Gibbs Free Energy'. It is represented by the symbol 'G' in honour of Josiah Willard Gibbs, a professor of chemistry at Harvard University who developed both the concept and the quantitative equation which describe it.

$$G = H - Ts$$

where H is the enthalpy, s is the entropy and T is the absolute temperature.

A change in free energy ΔG is expressed as

$$\Delta G = \Delta H - \Delta(Ts)$$

As the temperature is a state variable and the entropy is a state function, they can be expressed as

$$\Delta G = \Delta H - T\Delta s - s\Delta T \tag{13.10}$$

However, at constant temperture, the change in temperature ΔT is zero, so the term $s\Delta T$ is zero. For any constant-temperature (isothermal) change, Eq. (13.10) is expressed as

$$\Delta G = \Delta H - T\Delta s$$

Rearrangement of this equation, for larger changes we get the Gibbs–Helmholtz equation

$$dH = dG + Tds \qquad (13.11)$$

where Tds represents the change in entropy or the energy not available for doing work. Thus, Eq. (13.11) has the physical meaning.

Total energy available as heat = dG + energy not available for doing work[‡]

Therefore, dG is the part of energy which is available for doing work.

The convention of heat flow 'q' is also that of enthalpy change. When dH is negative, heat is given off in the exothermic reaction, whereas when dH is positive, heat is absorbed in the endothermic reaction from the surroundings.

Free energy change follows the same conventions. When dG is negative, energy is given up in the reaction and the change can do useful work. This is a spontaneous process, called an exergonic (energy-giving) process. When dG is positive, energy is absorbed in the reaction and useful work must be done on the materials so that reaction can occur. This is non-spontaneous process and is called an endergonic (energy-requiring) process.

13.14.1 Free Energy Changes in Chemical Reactions

Under the standard conditions of temperature and pressure (STP), the free energy change of an isothermal reaction is:

$$dG^0 = dH^0 - Tds^0 \qquad (13.12)$$

This is true for any reaction, and for the particular case of formation reactions:

$$dG^0_f = dH^0_f - Tds^0_f$$

The entropy change in a chemical reaction can either be calculated from the enthalpy change and the free energy change or as the difference in entropies of products and reactants. Both dH^0_f and dG^0_f are zero, by definition, for the elements in their standard state at 25°C and one atmosphere (one bar) pressure.

For many chemical reactions, the enthalpy and the free energy are nearly the same and thus from the values of enthalpy change alone the change in free energy can be calculated.

If the free energy change in a chemical reaction is negative, the reaction can occur spontaneously. When the free energy change in a chemical reaction is positive, the reaction cannot occur spontaneously. It is obvious that free energy changes correspond to the work being done. When dG is negative, the system will do work for us. When dG is positive, one has to do work on it.

For example, the reaction $C(s) - O_2(g) \rightarrow CO_2(g)$ which takes place spontaneously, is opposite to the direction of the reaction $CO_2(g) \rightarrow C(s) + O_2(g)$ for which dG is positive and this reaction does not take place spontaneously. The value for dG, can be expressed as

$$dG = \text{The sum of } dG_f \text{ (Products)} - \text{The sum of } dG_f \text{ (Reactants)}$$

[‡] Unavailable energy is a product of the lowest temperature of heat rejection and change in entropy.

The value of the free energy in a chemical reaction changes only in sign when the direction of the reaction is reversed.

13.14.2 Helmholtz Free Energy

The Helmholtz free energy F is expressed by the equation

$$F \underset{\text{Helmholtz free energy}}{\downarrow} = U \underset{\text{Internal energy}}{\downarrow} - \overset{\text{Final entropy}}{\overset{\uparrow}{Ts}} \underset{\text{Absolute temperature}}{\downarrow} \tag{13.13}$$

Here Ts represents the energy which one can get from the system's environment by heating. The internal energy U is the energy required to create a system in the absence of changes in temperature or volume. In case, the system is created in an environment of temperature T, then some of the energy is obtained by spontaneous heat transfer from the environment to the system. The measure of this spontaneous energy transfer is Ts where s is the final entropy of the system. In such a case one does not need to put much energy. However, if a more disordered (higher entropy) final state is created, less work is required to create the system.

Thus, the Helmholtz free energy is defined as "the amount of energy one has to put in to create a system once the spontaneous energy transfer from the environment is accounted for".

The four thermodynamic potentials are related by offsets of the "energy from the environment" term Ts and the "expansion work" term pV. A mnemonic (Figure 13.13) as suggested by Schroeder shows the relationship between the four thermodynamic potentials.

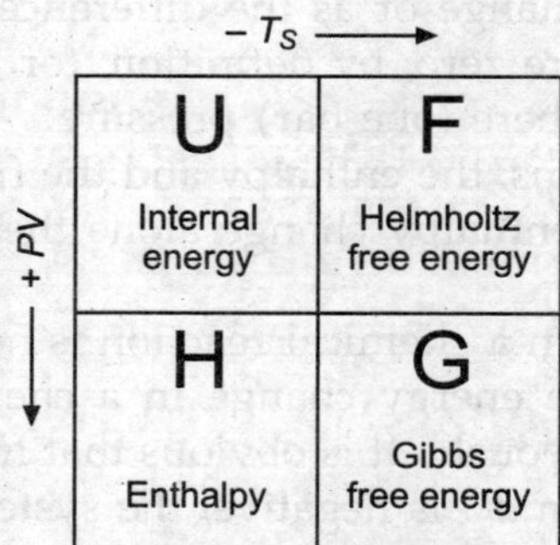

FIGURE 13.13 Mnemonic diagram showing the relation between Ts and pV.

13.14.3 Gibbs Free Energy

In chemical reactions of non-cyclic processes, four quantities are called "thermodynamic potentials". These are internal energy, the enthalpy, the Helmholtz free energy and the Gibbs free energy.

Gibbs free energy G as already expressed is

$$G = H - Ts$$

where H is enthalpy, T is absolute temperature, and s is final entropy. However,

$$H = U + pV$$

where U is internal energy, p is absolute pressure and V is final volume.
So,

$$G = U - Ts + pV \tag{13.14}$$

Also,

Ts = energy that one can get from system's environment by heating

pV = work to give the system the final volume V at constant pressure p

The internal energy U might be thought of as the energy required to create a system when there are no changes in temperature or volume. But in defining enthalpy, an additional amount of work pV must be done if the system is created from a very small volume in order to "create room" for the system. Also while defining the Helmholtz free energy, an environment at constant temperature T will contribute an amount Ts to the system, reducing the overall energy investment necessary for creating the system. This net energy contribution for a system created in environment temperature T from a negligible initial volume is the 'Gibbs free energy'.

The change in Gibbs free energy, ΔG, in a reaction is a useful parameter. It provides a clear picture of the maximum amount of work obtainable from a reaction. For example, in the oxidation of glucose, the change in Gibbs free energy is $\Delta G = 686$ kcal = 2870 kJ. This reaction is the main energy reaction in living cells.

13.15 HYDROGEN FUEL CELL ANALYSIS WITH THERMODYNAMIC POTENTIALS

Hydrogen and oxygen when combined in a fuel cell produce electrical energy. A fuel cell uses a chemical reaction to provide an external voltage like a battery. However, it differs from a battery in the sense that the fuel is continuously supplied in the form of hydrogen and oxygen gas. It produces electrical energy at a higher efficiency than just burning the hydrogen to produce heat to drive a generator. Its only by-product is water, so it is pollution free, as detailed in Figure 13.14.

One mole of hydrogen gas combines with a half-mole of oxygen gas from their normal diatomic forms to produce a mole of water. This process takes place at 298 K and one atmosphere pressure. The relevant values are taken from a table of thermodynamic properties and detailed in Table 13.2.

TABLE 13.2 Thermodynamic properties

Quantity	H_2	$0.5O_2$	H_2O	Change
Enthalpy	0	0	−285.83 kJ	$\Delta H = -285.83$ kJ
Entropy	130.68 J/K	0.5×205.14 J/K	69.91 J/K	$T\Delta s = -48.7$ kJ

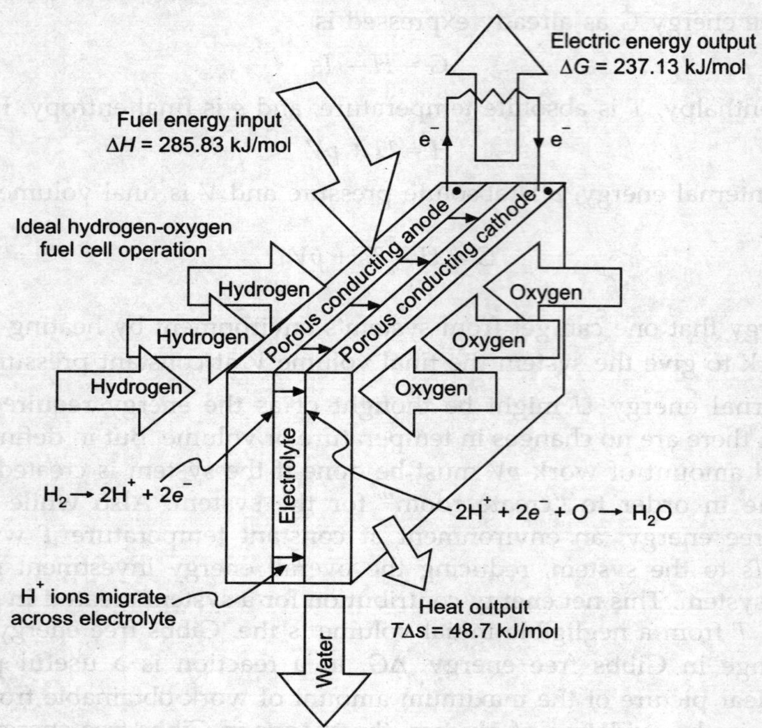

FIGURE 13.14 Hydrogen fuel cell showing input and output energy.

Energy is produced by the combination of the atoms and also from the decrease in volume of both the gases. At temperature 298 K and one atmosphere pressure, the system work is:

$$W = p\Delta V = (101.3 \times 10^3 \text{ Pa}) (1.5 \text{ moles})(-22.4 \times 10^{-3} \text{ m}^3/\text{mol})(298 \text{ K}/273 \text{ K})$$

$$= -3715 \text{ J}$$

As the enthalpy $H = U + pV$, the change in internal energy U is given by

$$\Delta U = \Delta H - p\Delta V = -285.83 \text{ kJ} - (-3.72 \text{ kJ})$$

$$= -282.1 \text{ kJ}$$

The entropy of the gases decreases by 48.7 kJ in the process of combination as the total number of water molecules is less than the total molecules of hydrogen and oxygen. The total entropy will not decrease in the reaction, the excess entropy in the amount $T\Delta s$ must be expelled to the environment as heat at temperature T. The amount of energy per mole of hydrogen that can be provided as electrical energy is the change in the Gibbs free energy:

$$\Delta G = \Delta H - T\Delta s = -285.83 \text{ kJ} + 48.7 \text{ kJ} = -237.1 \text{ kJ}$$

For this ideal case, the fuel energy is converted to electrical energy at an efficiency of $\frac{237.1}{285.8} \times 100\% = 83\%$.

This is far greater than the ideal efficiency of a generating facility which burns the hydrogen and uses the heat to power a generator. Actually the real fuel cells do not approach that ideal efficiency, yet they are more efficient than any electric power plant which burns a fuel.

13.16 COMPARISON OF ELECTROLYSIS AND THE FUEL CELL PROCESS

Fuel cell process is a reverse reaction to that of electrolysis of water. It can be explained by enthalpy change as the overall energy change. The Gibbs free energy is that which one has to supply to drive a reaction, or the amount of energy that one can get if the reaction is working for release of energy. So, with electrolysis the enthalpy change is 285.8 kJ, and it is necessary to put in 237 kJ of energy to drive electrolysis, and the heat from the envioronment will contribute $T\Delta s = 48.7$ kJ to complete the reaction.

Contrary to the above reaction, when a fuel cell operates, one gets 237 kJ as electric energy, and the energy $T\Delta s = 48.7$ kJ is dumped into the environment.

13.17 OPERATING CHARACTERISTICS OF FUEL CELLS

A fuel cell consists of two electrodes surrounded by an electrolyte. Hydrogen fuel is fed into the 'anode' of the fuel cell, while oxygen enters through the cathode. Excited by a catalyst, the hydrogen atom splits into a proton and an electron, which take different paths to the cathode. The proton passes through the electrolyte while the electrons create a current which is utilised before they return to the cathode, to be reunited with the hydrogen ions and oxygen to form water.

It has already been explained that the amount of energy per mole of hydrogen which can be supplied as electrical energy is the change in the Gibbs free energy, i.e., 237.1 kJ at 25°C. The heat energy (or enthalpy) of the reaction is 285.83 kJ under the same conditions. Thus, the maximum efficiency of a hydrogen–oxygen fuel cell is 83%.

However, when a fuel cell operates, it can harness the fuel's energy via a chemical reaction into electricity to the extent of 65% only. The performance of a fuel cell can be evaluated from a curve of the cell voltage V_C drawn against current density I_D at electrode surface at a given temperature, as shown in Figure 13.15.

The difference in open-circuit voltage (V_o) and closed-circuit voltage (V_c) is due to the polarization effect within the cell. The drop in voltage V_p is expressed as

$$V_o = V_c + V_p$$

The 'polarization loss' at the electrodes denotes the difference in the open-circuit voltage and the closed-circuit voltage. Electrode losses can be divided into three categories: (1) chemical polarization, (2) internal resistance polarization, and (3) electrolyte concentration polarization.

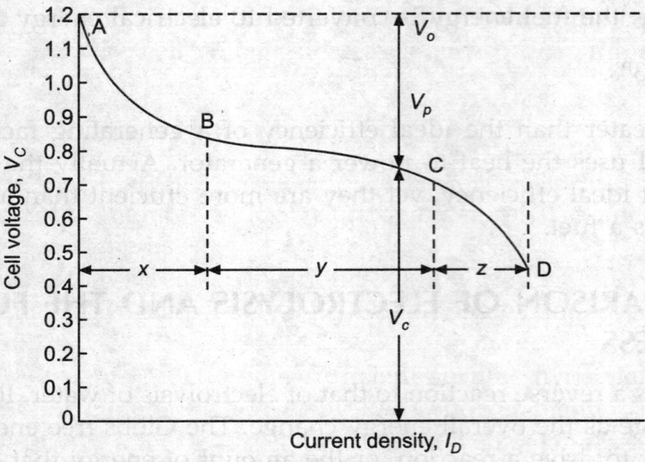

x represents chemical polarization,
y the internal resistance polarization, and
z shows the electrolyte concentration polarization.

FIGURE 13.15 Voltage–current density curve (polarization curve).

Chemical polarization

In fuel cells, elctrons are released and their movement causes the current to flow in the external circuit. At low current density, less number of electrons are liberated. To increase the current, the process needs acceleration for which the energy is supplied by the fuel cell represented by the curve part AB. It causes voltage drop, the output loss is due to chemical polarization. The effect can be reduced by using a superior electrode catalyst and operating the cell at a higher temperature.

Internal resistance polarization

When a fuel cell operates, the ions liberated at one electrode move to the other electrode through the electrolyte causing flow of current in the external circuit. The internal resistance is the total of electrode resistance, the contact resistance between electrode and electrolyte and the electrolyte resistance. It is represented by the curve part BC.

The internal resistance polarization can be reduced by decreasing the electrode size, coating the electrodes with a good electric conductor, increasing the electrolyte concentration and reducing the distance between the two electrodes.

Electrolyte concentration polarization

When a fuel cell operates, there is a reduction in voltage due to concentration loss of the electrolyte which causes slow ionic movement. It is represented by the CD part of the curve. A remedy for this polarization is to increase the electrolyte concentration by continuous stirring and operate the cell at a higher temperature. The optimal operating temperature for an acidic cell is 180°C.

Voltage efficiency of a fuel cell

Polarization in a fuel cell reduces the output voltage. All the losses in a fuel cell are converted to waste heat that is released to the atmosphere. The efficiency of a fuel cell in terms of voltage is expressed as

$$\eta_V = \frac{\text{closed circuit voltage}}{\text{open circuit voltage}} = \frac{V_o - V_p}{V_o}$$

Putting values from the curve, $\eta_V = \dfrac{0.78}{1.20} \times 100 = 65\%$

13.18 THERMAL EFFICIENCY OF A FUEL CELL

The voltage efficiency of a fuel cell has already been considered which is 65% as calculated from the curve. However, a large number of energy converters convert heat energy into electricity. To compare the efficiency of electrochemical energy converters, i.e., 'fuel cells' the heat input may be taken as the enthalpy change of the reaction (ΔH).

It is now assumed that the only useful work done by a fuel cell system is electrical work which is represented by the quantity of charge that flows from the cell, multiplied by the potential difference of the cell. Thus, the quantity of charge can be calculated by multiplying the number of moles[§] of electrons operative in the cell reaction by the number of coulombs per mole of electrons.

$$\text{Electrical work} = W_{el} = -nFV_o \tag{13.15}$$

where

V_o = open-circuit voltage of the fuel cell
F = Faraday constant
n = number of moles of electrons.

The work output in a fuel cell which operates at the thermodynamic reversible voltage (which drives the electrons liberated at the anode through the external load) is equal to Gibbs free energy change (ΔG). Thus,

$$W_{el} = \Delta G = -nFV_o$$

The thermal efficiency of a fuel cell (not considering the losses associated with the accessories) is the work output of the system divided by the heat reaction [see Eq. (13.10)], i.e.,

$$\eta_{\text{th}} = \frac{\Delta G}{\Delta H} = \frac{-nFV_o}{\Delta H} \tag{13.16}$$

$$= -\frac{ItV_o}{\Delta H} \tag{13.17}$$

---

§ One gram mole of electron = 6.02×10^{23} electrons = total charge of 96,493 coulombs. A charge equal to 96493 coulombs is called a 'Faraday' constant F. If a wire passes a current of one ampere for 96493 seconds, it shall pass one mole of electrons or a charge of 96493 coulombs.

where I is the current and t is the time in second for which the current flows. When a fuel cell feeds a load the cell voltage decreases and so does the efficiency.

13.19 FUTURE POTENTIAL OF FUEL CELLS

Market studies predict that large-scale application of fuel cells is expected to enter into the niche market of high technology electronic devices like cellular phones, laptop computers and video cameras.

Casio computers have released a notebook PC powered by a fuel cell using methanol reforming. MTI Micro Fuel Cells, USA, has developed an external fuel cell for use in cell phones.

With chloralkali plants functioning, sufficient surplus hydrogen is available. This can partly be used for generating power by PAFC power plants.

The real revolution is expected to come from the vehicular application of fuel cells. It will help the developed countries to propel their compliance of Kyoto protocol, i.e., cut down the "greenhouse gas emissions" to levels that are 5% below the 1990 level.

With all-round applications, the generation of electricity from fuel cells is expected to jump to 15000 MW in 2011 from a mere 75 MW in 2001. It will cover all the three market segments, i.e., portable, stationary and vehicular.

REVIEW QUESTIONS

1. (a) Derive the expression for the power output and efficiency of a fuel cell.
    (b) Give a complete description of the working and constructional features of a hydrogen–oxygen fuel cell.
    (c) Find the maximum possible value of the internally generated voltage of the hydrogen–oxygen cell at STP (1 atmosphere and 298 K) where the product is in its liquid state. Find also the internally generated voltage of the hydrogen–oxygen cell at a temperature of 32°C when the air is used as the oxidant and the product is water. The air is supplied at a pressure of 1.2 atmosphere and the hydrogen is at a pressure of 1.1 atmosphere. The partial pressure of oxygen in air is 0.21 Gibbs free energy at STP for water is 237×10^6 coulomb for $H^+ = 0$, $R = 8314$.

2. (a) Explain the basic theory of electrochemistry applied to fuel cells.
    (b) Compare the H_2–O_2 acidic fuel cell with the H_2–O_2 alkaline fuel cell.
    (c) Discuss the performance characteristic and applications of fuel cell.

3. Fuel cells operate at a fairly high efficiency and are in demand as auxiliary supplies in certain places. This trend will continue in the future. Explain this with examples of different types of 'electrolytes' and 'electrodes' and their construction and working bringing out the relative merits and demerits.

4. (a) In connection with a fuel cell, explain the meaning of $WE \leq \Delta G$ and use this for an isothermal thermodynamic all gas chemical reaction in a fuel

cell to find an expression for the maximum value of internal cell voltage and the conversion efficiency.

(b) Draw a neat sketch of a 'matrix type' hydrogen, oxygen alkaline cell with facilities for electrolyte recirculation and water removal. How is this accomplished in other types?

5. Briefly explain the principle of a fuel cell by taking a hydrogen–oxygen fuel cell. Which factors are responsible for limiting the efficiency of such a cell?

6. (a) Describe a solid electrolyte fuel cell with a suitable diagram.

(b) Explain with the help of current–voltage characteristic the normal performance of a fuel cell with particular reference to an acid electrolyte hydrogen cell and its 'cell voltage' and its 'maximum conversion efficiency'.

7. Distinguish between battery and fuel cell.

8. Discuss the operating characteristics of the fuel cell.

9. Discuss and differentiate between "electrical efficiency" and "thermal efficiency" of the fuel cell.

10. Write notes on:

(a) Fuel processor
(b) Fuel cell battery powered system
(c) Microbial fuel cell
(d) Polarization in a fuel cell.

HYDROGEN ENERGY SYSTEM

14.1 INTRODUCTION

At present most of the world's energy demand (about 70%) is met by fossil fuels because of their availability and convenient use. However, fossil fuels (i.e., petroleum, natural gas and coal) are being depleted rapidly. Also, their combustion products cause global problems, such as the greenhouse effect, ozone layer depletion, acid rains and pollution which pose danger to our environment. Now, engineers and scientists agree that the solution to all these global problems is to replace the existing, fossil fuel system with the 'Hydrogen Energy System'. Hydrogen is an efficient and clean fuel. It has minimum carbon context compared to other fuels as shown in Table 14.1.

TABLE 14.1 Carbon contents of fuels

Fuel	*Carbon : Hydrogen*
Wood	9 : 1
Coal	1.5 : 1
Oil	1 : 2
Natural gas	1 : 4

A carbon-rich fuel produces more CO_2 which contributes to global warming. By adopting a leaner carbon and richer hydrogen content, it is a step towards better environment-friendly sources of fuel.

14.2 EMERGENCE OF HYDROGEN

Having understood the importance of hydrogen, energy planners can foresee a carbon-free, all-hydrogen energy scenario for future. However, it is made clear that hydrogen is not a source of energy, it is a currency or an energy carrier like elelctricity. This is clarified by the details presented in Table 14.2 which shows the energy chain from the source of energy, transformation technology to the end use.

TABLE 14.2 Hydrogen as an energy carrier

Energy source	Transformation technology	Energy currencies	Service technologies	Energy use
Crude oil	Oil refineries	Petrol, Diesel, Natural Gas	Internal combustion engine (ICE) Automobiles motors	Transport Irrigation
Oil, Natural gas, Coal	Generators Electrolyzers Reformers	Hydrogen	Fuel cell vehicles ICE Automobiles, Motors	Industry, Transport Heating System
Nuclear, Wind, Solar, Geothermal (carbon free)	Generators Electrolyzers	Hydrogen	Fuel cells for electricity, ICE Automobiles, Fuel Cell Vehicles	Water supply Transport Defence industry

The above details in Table 14.2 clarify the role of hydrogen in the energy chain. Though hydrogen has the highest energy content, its volumetric density is just 0.0899 kg/Nm^3 which makes it difficult for packing, storage and public distribution. It requires a high percentage of its energy content to be consumed in either pressurization or liquification in order to enable its commercial utilisation.

The hydrogen makes up 90% of all the matter in the universe. However, it does not exist in free state, therefore, it is not a primary energy source. It occurs in water and all organic compounds such as fossil fuels (natural gas, coal, methanol and petrol) and biomass.

Hydrogen – An Energy Carrier

Energy carriers move energy in a usable form from one place to another place. We use electricity as most common form of energy carrier which is obtained by moving energy of coal, uranium, potential energy as head of water in dams and other energy sources from power plants to residential houses, commercial establishment, industries etc. It is much easier to use electricity than the energy sources themselves.

Like electricity, hydrogen is an energy carrier and it is produced from another substance like water, fossil fuels and biomass. It can be obtained as a by-product of other chemical processes, unlike electricity hydrogen can be stored in large quantities and can be used in future, moved to a place wherever it is needed.

14.3 HYDROGEN PRODUCTION

The world's annual production of hydrogen gas from all sources is nearly 500 billion m^3 (plus 125 billion m^3 as a co-product) weighing 87 million tonnes with an energy content of 6×10^{18} joules. A large part of this hydrogen is produced from steam reformation of natural gas and some form of fossil fuel. A major portion of this hydrogen is utilised for the production of ammonia and methanol. It is also used in semiconductor industries and rockets.

Hydrogen is produced from different raw materials and from different processes. The method adopted is dictated by economic viability. Some of the methods used are:

1. Steam reformation
2. Electrolysis of water
3. Thermal decomposition of water through solar energy or nuclear energy
4. Thermo-chemical methods
5. Biological production
6. Sunflower oil, air and water vapour
7. Solar electrolytic Hydrogen produce

14.3.1 Steam Reformation

Steam is passed over hot sponge iron sheets at a suitable tempeature (550°C to 800°C) where hot iron and steam react to produce ferric oxide, hydrogen, CO_2 and CO in small quantities. The gases are passed through a scrubber where dilute NaOH absorbs CO_2 and CO.

$$3H_2O + 2Fe \rightarrow Fe_2O_3 + 3H_2$$

Another method is the steam reformation of natural gas or crude oil depending upon the availability.

14.3.2 Hydrogen Production by Electrolysis

Water is abundantly available which can be split into hydrogen and oxygen by using dc power. Hydrogen evolves at the cathode and oxygen at the anode. As an experiment, two pairs of stainless steel electrodes, each having 500 cm^2 surface area were used with different electrolyte solutions. Readings with current ratings 25, 50, 75 and 100 ampere were taken and the average power consumption was recorded as given in Table 14.3. The quantity of hydrogen generated was measured by a flow meter. The power required for the generation of one Nm^3 hydrogen was computed.

TABLE 14.3 Energy requirement to liberate 1 Nm^3 of hydrogen

Electrolyte (water + salt)	Energy consumed (kWh/Nm^3)
5% NaCl	11.3
10% NaCl	10.3
15% NaCl	8.3
5% KOH	8.8
10% KOH	7.8

It can be seen that 10% KOH is an optimum electrolyte solution. It was observed that with an increase in the electrode surface area the power consumption further reduced. The results with different surface area of electrode are given in Table 14.4.

TABLE 14.4 Energy consumed with different pairs of electrodes

Electrode surface area (cm^2)	Energy consumed (kWh/Nm^3)
Two pairs (4 electrodes)	7.8
Three pairs (6 electrodes)	7.6
Five pairs (10 electrodes)	3.2

The results show that the most economical value is 3.2 kWh/Nm^3.

The temperature of the cell is maintained between 65°C and 80°C. Hydrogen thus produced is of 99% purity.

$$2H_2O \text{ (liquid)} \xrightarrow{\text{Electrolysis}} 2H_2 \text{ (gas)} + O_2 \text{ (gas)} + (\Delta \text{ Heat)} \qquad (14.1)$$
$$\text{(At cathode)} \quad \text{(At anode)}$$

As the rate of hydrogen production is proportional to the current, a high operating current density is required to be maintained for economic viability. The net efficiency of generating electricity from fossil fuel plants is 35%. Considering the electrolysis efficiency of 80%, the overall efficiency of producing hydrogen would be 28% only. With hydroelectric power plants which operate at 90% efficiency, hydrogen production shall have an efficiency of 70%.

The efficiency of electrolysis E is expressed by the equation

$$E = \frac{\text{Hydrogen produced in m}^3 \times K}{\text{Power input in kWh}} \times 100 \qquad (14.2)$$

From experimental studies the value of constant K can be taken as 3.3 kWh/m^3. The average power input in kWh from Table 14.3 is 4 kWh/m^3. The value of E is about 82%.

Nangal (Punjab) fertilizer plant used hydropower for a low-cost energy efficient water electrolysis plant to produce hydrogen for manufacturing synthetic ammonia by Haber's process.

14.3.3 Thermal Decomposition of Water

A better conversion efficiency may be obtained if the heat produced by the primary fuel is directly applied to decompose water, without utilising an intermediate electrical energy. In thermal decomposition, heat energy is required to be used at a very high temperature of 2500°C or more to dissociate water from hydrogen and oxygen. Considering the requirement of such a high temperature conversion process equipment, single-step water decomposition is not commercially viable.

14.3.4 Thermo-Chemical Method

This method involves thermal chemical reactions between primary energy, water and specific chemicals to produce hydrogen at temperature range of 700°C to 1000°C, which is lower than temperature required for thermal decomposition. As an illustration,

Natural gas which is a mixture of CH_4 and CO is reformed with steam at 900°C to produce a mixture of H_2 and CO_2. Then CO_2 is separated by scrubbing process to obtain hydrogen. Reaction of reforming of natural gas with steam is given:

$$CH_4 + 2H_2O \xrightarrow{\text{(steam) 900°C}} CO_2 + H_2$$

$$CO + H_2O \longrightarrow CO_2 + H_2$$

Besides natural gas other fuels such as LPG, biogas landfill gas and coal gas can commercially be used for production of Hydrogen.

14.3.5 Biological Production of Hydrogen

A project on biological hydrogen production is being undertaken at Shri AMM Murugappa Chettiar Research Centre, Chennai. Under this project a 12.5 m^3 capacity bioreactor has been installed to demonstrate the biological production of hydrogen from distillery waste. Another reactor of 125 m^3 capacity has been designed with the support of MNRE, New Delhi.

14.3.6 Production of Hydrogen Fuel from Sunflower Oil

Scientists in the University of Leads England have developed an experimental hydrogen generator which needs only sunflower, air and water vapour along with two highly specialized nickel-based and carbon-based catalysts. These are alternatively used to store and then release oxygen or CO_2 while producing hydrogen intermittently. This process does not involve burning of fossil fuel, hence the hydrogen fuel becomes renewable.

First, the nickel-based unit catalyst absorbs oxygen from the air and this interaction heats up the reactor bed of the device. Simultaneously, in the presence of heat, another catalyst (a carbon-based adsorbent) releases CO_2 previously trapped in the device.

Once the reactor bed is hot and all the CO_2 has been released from the reactor, the mixture of vaporized oil and water is fed into the reactor chamber. The heat from the reactor bed breaks down the carbon–hydrogen bonds in the vaporized oil. Water (steam) binds its oxygen to the carbon, releasing its hydrogen and yielding carbon monoxide. Water vapour and carbon monoxide tend to form carbon dioxide and hydrogen in the presence of each other. This overall process results in a cyclical production of hydrogen.

Hydrogen fuel thus produced is of 90% purity which is more efficient than other hydrogen producers which produce hydrogen fuel of about 70% purity. Methane and CO_2 are the by-products of sunflower oil transformation which are generated in equal proportions.

At present the generator is heated electrically, but in the near future all the heat necessary to carry out the reaction of steam with oil vapour will come from the intake of oxygen on the nickel catalyst.

14.3.7 Solar Electrolytic Hydrogen Production

The German government started several projects to produce hydrogen from renewable sources of energy like solar and wind. One of these projects utilises solar energy collected from 50 m^2 of solar cells mounted on the roof. This energy is used to break water molecules into hydrogen and oxygen. Hydrogen thus produced is utilised to power a specially converted Toyota, a hybrid vehicle that can run on hydrogen, gasoline or biogas with no loss in vehicle performance. Hydrogen has been used in this vehicle to provide more than 90% of the total fuel required.

Electrolytically produced hydrogen is useful for sintering and melting in the glass and jewellery industries.

From an economic point of view the larger the electrolyzer project the lower is the hydrogen cost. Hydrogen produced electrolytically from wind or solar power is used in fuel cell vehicles which provide zero emission during transportation. The production of electrolytic hydrogen is attractive in Europe, North Africa and United States where the availability of biomass-derived fuels is limited.

R&D project for Hydrogen Production

A New R&D project on hydrogen production by transformation of green house gases using low temperature plasma catalysis is in progress at Indian Institute of Technology Hyderabad.

A 5 Nm^3/h hydrogen production facility through electrolysis of water and powered by a solar photovoltaic array of 120 kWp capacity along with associated storage and dispensing facilities is established in the campus of National Institute of Solar Energy Gwalpahari Gurgaon (Haryana).

14.4 COST ANALYSIS OF HYDROGEN PRODUCTION

Hydrogen can be produced by several methods, but currently it is an expensive option in comparison with other available fuels. The approximate cost of hydrogen production by various methods as listed in the literature is given in Table 14.5.

It is evident that 'Green hydrogen' (produced from carbon-free sources) is in the higher cost slab compared to the 'Black hydrogen' (produced from carbon-rich fuels). However, the production of hydrogen from renewable sources of energy is the ultimate answer to energy security and environmental friendliness.

The higher cost of hydrogen is neutralized by its special property of being versatile and more convenient to utilise. A fuel is said to be more versatile if it can be converted through more than one process to various forms of energy at the user end. All other fuels can be converted through one process only, i.e., combustion, while hydrogen can be converted in five different ways in addition to flame combustion, namely to steam, to heat through catalytic combustion, heat sink through chemical reactions and to electricity through electrochemical processes.

TABLE 14.5 Cost of producing hydrogen

Method	Cost (₹/GJ)
Reformation of natural gas	225
Coal gasification	495
Hydroelectric electrolysis	540
Wind electrolysis	1440
Solar thermal electrolysis	1755–2340
Solar photovoltaic electrolysis	2115–4680
Biomass gasification	585
Partial oxidation of oil	405

14.5 CHARACTERISTICS AND APPLICATIONS OF HYDROGEN

Hydrogen as a fuel has a huge potential for use by individual consumers and in industrial sectors. It is a clean burning fuel superior to other fuels, i.e., petrol, methane, methanol and ethanol. It carries a high specific energy per unit weight, so is very useful for surface vehicles and aeroplanes which carry their fuel over a long distance before replenishing. It is most suitable for space vehicles which carry their fuel and oxidant for their long scheduled trips.

Its heat of combustion is twice compared to fossil fuels, six times of methanol and four times more than ethanol. Its utilization efficiency is high as hydrogen can be converted to the desired energy form more efficiently than other fuels. The utilisation efficiency factor is defined as the fossil fuel utilisation efficiency divided by the hydrogen utlilisation efficiency. For various applications the utilisation efficiency factors 'ϕ_u' are given in Table 14.6.

TABLE 14.6 Comparison of utilisation efficiency of fossil fuels and hydrogen

Application	Utilisation efficiency factor, $\phi_u = \eta_f/\eta_h$
Thermal energy	
(i) Flame combustion	1.00
(ii) Steam generation	0.80
Electric power, Fuel cells	0.54
Internal combustion engines	0.82
Subsonic jet transportation	0.84
Supersonic jet transportation	0.72
Hydrogen utilisation efficiency factor	1.00
Fossil fuel utilisation efficiency factor	0.72

14.6 HYDROGEN STORAGE

For developing hydrogen-based energy systems, there is a bottleneck in that it has a molecular weight of only two. For a given mass of the fuel, the molar quantity of hydrogen calculates to be larger compared to the number of moles for a fuel of higher molecular weight. Thus, to store a given mass of hydrogen, the container needs to be designed to withstand higher pressures than those for greater molecular weight fuels. As a typical example, to store 2 kg of hydrogen in a 40-litre (0.4 cubic metre) petrol tank of a car, it would require a chamber pressure of nearly 6.5 MPa, i.e., 65 atmospheres. Normally a petrol tank stores fuel at 0.1 MPa (1 atmosphere).

For utilisation of hydrogen as an energy source there is a need to store it. Several possible modes of storage are:

- Gaseous storage
- Liquid storage
- Solid state storage

14.6.1 Compressed Gas Storage

Hydrogen can be stored in compressed gaseous state in underground reservoirs similar to natural gas or can be stored in high pressure cylinders. The method of storage is costly as a larger quantity of steel is required to store a small amount of hydrogen. For industrial use of hydrogen as a fuel, gaseous storage of hydrogen is economically not viable.

14.6.2 Liquid Storage

Liquid storage of hydrogen is economically feasible for stationary and mobile applications. Liquid hydrogen fuel is used as a rocket propellent in space vehicles as it has the highest energy density (energy per unit mass), almost three times more than the conventional fuels. To store liquid hydrogen [boiling point (–253°C)] it is necessary to use vacuum insulated cylinders to avoid air condensation over its surface. Concentration of liquified air* (rich in oxygen) around liquid hydrogen cylinder is a fire hazard. Liquification of hydrogen gas requires 25–30% of the calorific value of this fuel to attain cryogenic storage for space programmes.

14.6.3 Solid State Storage

Solid storage in the form of metallic hydrides is the most attractive method of storing hydrogen. The metal hydride system is based on the principle that a few metals absorb hydrogen in an exothermic reaction when treated with the gas and the absorbed gas is released when the metal hydride is heated. The chemical equations are:

---

* Air liquifies at a higher temperature than hydrogen.

$$\text{(Gas is stored) } H_2 + \text{Metal} \xrightarrow{\text{Charge}} \text{Hydride} + \text{heat} \qquad (14.3)$$

$$\text{(Gas is released) Metal hydride} + \text{heat} \xrightarrow{\text{Discharge}} \text{Metal} + H_2 \qquad (14.4)$$

In this technique hydrogen gas is reacted with powdered metallic alloy in a closed evacuated pressure vessel. As hydride formation is accompanied by a negative enthalpy change, the excess heat of formation is removed during charging. On completion of the charging cycle, the cylinder is maintained at room temperature using cylinder pressure at maximum equilibrium vapour pressure of the hydride. When hydrogen gas is required, the cylinder is heated to a suitable temperature corresponding to the discharge pressure in order to maintain the required gas flow rate.

There are a few metallic alloys such as magnesium–copper and iron–magnesium–titanium with high storage capacities of hydrogen. The reaction is reversible as hydrogen is released when metallic hydride is heated. Reactions are represented by the equations:

$$H_2 + FeTi \rightarrow FeTiH_{1(1.7)} + \text{heat} \qquad (14.5)$$

$$Fe.Ti.H_{1(1.7)} + \text{heat} \rightarrow Fe.Ti.H_{0.1} + 0.8H_2 \qquad (14.6)$$

A project on the development of a metal hydride reactor is in progress at IIT Madras. Under this project carbon nano-materials have been developed and characterized as reported by MNRE.

The metal–hydride storage system competes well with the conventional pressure gas storage system particularly in volume-specific storage capacity. For hydrogen-powered vehicles, hydride storage tanks are compact and the rejected heat from the engine is utilised for dehydriding. Hydrogen fuel-operated buses and cars are suitable for short distances up to 160 km. Such vehicles are similar to electric vehicles which run on fuel cells or batteries and are environmentally benign.

14.7 HYDROGEN STORAGE USING NANO-CRYSTALLINE MAGNESIUM-BASED NICKEL HYDRIDE

Magnesium has a storage capacity of nearly 7.6 wt%, however, its kinetics is low and enthalpy of formation is high. These two bottlenecks are solved by alloy additions and decreasing the particulate size in nano range (10^{-9} m) to generate a large surface area for absorption. Pure Mg and Ni powders of mean particle size of 400 microns and 100 microns respectively, with 99% purity were selected as the base material. The micron-size particles were reduced to nano-crystalline size in a high energy ball mill. An R&D study revealed that hydrogen storage in the 20 hours milled Mg–Ni sample can be charged with hydrogen at 300°C. The maximum absorption (formation of MgH_2) was measured as 2.5 wt% in about 50 minutes time. This value is nearly 65% of the theoritical absorption capacity of the 2-part Mg and 1-part Ni alloy used for this work. This reveals that about 35% of the Mg particles could be potentially deactivated due to surface reaction with the atmosphere.

14.8 DEVELOPMENT OF HYDROGEN CARTRIDGE

Researchers have reported the development of metal–organic 'nano-cubes' which can store and release hydrogen to power fuel cells for laptop computers and other portable electronic devices. The nano-cubes, made from terephthalic acid and zinc oxide, absorb hydrogen under moderate pressure, and could be used in cartridges to precisely control the release of hydrogen to the fuel cell. The hydrogen in the cartridge would be subject to ten times the atmospheric pressure.

14.9 NATIONAL HYDROGEN ENERGY BOARD

To coordinate national effort on developing hydrogen energy systems and to prepare National Hydrogen road map, National Hydrogen Energy Board (NHEB) has been set up under MNRE. Besides, preparing codes and standards for hydrogen, it clears projects on production, storage, delivery and application of hydrogen energy.

A steering group of NHEB conducted demonstration of hydrogen-powered systems including motor cycle, three wheelers, generators, cooking system and lighting lamps.

Indian Oil Corporation assigned a project to produce "Hyathane", a blend of hydrogen and CNG, in Delhi, Faridabad and Mathura. Initially, 10% hydrogen would be mixed with CNG, and this ratio would go up to the extent of 30%. Members of NHEB expressed "Hydrogen has the potential to emerge as an alternative fuel to India's growing energy needs".

14.10 APPLICATION OF HYDROGEN IN INDUSTRY

Hydrogen produced from biomass is the cheapest and costs only 50% to hydrogen produced electrolytically. Hydrogen can be converted into several useful form as detailed

1. Fuel in internal combustion engines, similar to gasoline to produce mechanical energy or electric power.
2. For electrochemical conversion in Fuel cells to generate electrical energy.
3. Hydrogen and oxygen combustion to generate steam.
4. Hydrogen is used for catalytic combustion and can replace LPG for domestic cooking.
5. For running vehicles using metal hydride technology.
6. In gas turbines, hydrogen can be used instead of conventional fuels. Hydrogen turbines have higher efficiencies, lesser pollution problems and create no difficulty in corrosion of blades.
7. Hydrogen has high energy density hance used as aviation fuel. It is superior fuel for turbojet aircraft due to greater fuel economy, low noise level and little pollution.

8. Hydrogen is utilized for coolling of large electric generators in thermal and nuclear plants.

9. Hydrogen is used for hydrogenation of vegetable oil.

10. Hydrogen is widely used to manufacture ammonia and urea (ammonium nitrate).

Global corporate sector use of hydrogen

- In 1999, shell Hydrogen and Daimler-Chrysler executed an agreement with the Government of Iceland to make the world's first hydrogen- powered economy. In August 2003, the hydrogen-powered buses started playing in Reykjavik (Capital of Iceland).

- Nissan Motors have started manufacturing a gas-electric hybrid version of its Altima sedan for the US market. It suits with the strict fuel economy and emission standards in California state.

- Hybrid vehicles made by Toyota and Honda are not any more. During 2021 about 801,550 hybrid light vehicles were sold in the US. They make up 2.4% of all cars sold in USA.

- BMW, a leading automobile company of Germany in association with Messer–Griesheim a cryogenic gas firm took up a project of developing hydrogen powered cars.

- The Tatas are working on the modification of internal combustion engines in the present vehicle that can run on hydrogen fuel.

- The National Aeronautics and Space Administration (NASA) is using hydrogen as an energy fuel in its space programme. Liquid hydrogen fuel lifts the space shuttle into orbit. Hydrogen batteries – known as Fuel cells – power the shuttle's electrical systems. It produces pure water as a by-product, which the crew uses as drinking water.

- Hyrogen fed fuel cells (batteries) generate electricity as emergency power in hospitals and military applications in remote locations.

14.11 LIQUID HYDROGEN TURBOPUMP FOR ROCKET ENGINE

Recently, the US Air Force, NASA and two aerospace companies have tested a liquid-hydrogen turbopump and an oxidizer preburner, under the Next Generation Launch Technology programme. It is a development activity of a hydrogen fuelled, full flow, staged-combustion rocket engine in the 25,000 pound thrust class. The engine uses dual preburners that provide both oxygen-rich and hydrogen-rich staged combustion, which will cool down engines during flight providing higher engine efficiency and low exhaust emissions.

The turbopump is designed to provide high pressure hydrogen to rocket engine thrust chamber which initiates the combustion process and generates the thrust. The turbopump extracts energy from hot gases that are generated by the oxidizer preburner and flow through a turbine.

The oxidizer preburner which initiates the combustion process, is designed to generate oxygen-rich steam for use by the oxygen turbopump's turbine. The preburner burns a large quantity of liquid oxygen with a small quantity of hydrogen to produce this steam, which then mixes with additional hydrogen fuel to be burned in the main combustion chamber.

14.12   GAS HYDRATE

Gas hydrate is one of the non-conventional sources of energy identified recently by ONGC and to be studied for exploration to ensure energy security for the country.

Gas hydrates are naturally occurring ice-like compounds of methane and are water formed under low temperature and pressure conditions. These ice formations consist of water molecules that have trapped gas molecules in a cage-like structure, found at varying depths in areas of low temperature. The first naturally occurring gas hydrate deposit was found in Messoyakha field in Russian Permafrost region. Subsequently, gas hydrates were also found in shallow marine sediments of arctic region and in tropical deep-water areas where water depth exceeds 650–750 metres.

Methane trapped in marine sediments as a hydrate represents such an immense carbon reservoir that it is considered as a dominant factor in estimating the unconventional energy resources. In a conservative estimate the world-wide amounts of carbon bound in gas hydrates are equal to twice the amount of carbon that can currently be found in all known fossil fuels on earth. The distribution of organic carbon in earth reservoirs as per the estimate prepared by the United States Geological Survey (USGC) is given in Figure 14.1.

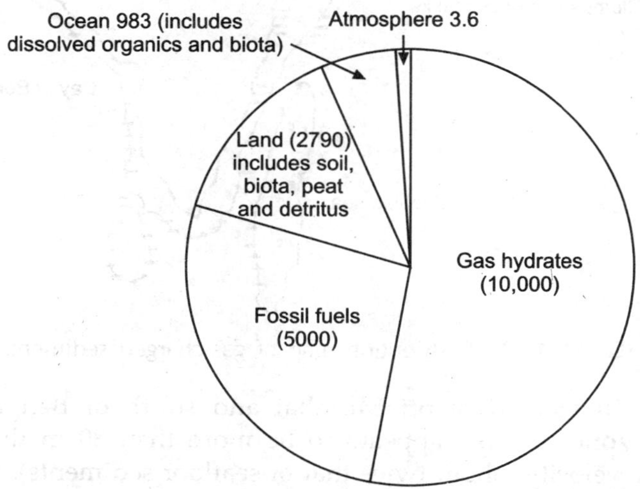

FIGURE 14.1   Global resource estimate of organic carbon in billions of tonnes.

14.12.1 Gas Hydrate Occurrence

Hydrates tend to form along margins of continental slopes, where the seabed drops from the relatively shallow shelf, about 150 m below the surface towards the ocean abyss several kilometres deep. It is difficult, so far, to bring the methane hydrates from the seabed to the sea floor due to technical difficulties and for which R&D work is in progress.

Methane hydrates are stable only at near-freezing temperature and under high pressure of nearly 500 metres of overlaying water. These ice-like hydrate deposits become unstable when the surrounding temperature rises above the freezing point and the pressure becomes less. Gas hydrates need to be dealt with advanced technology, as methane escaping from the disturbed undersea hydrates may be an ecological threat.

14.12.2 Indian Resource of Hydrates

Preliminary studies carried out by the National Institute of Oceanography, India, indicated the presence of gas hydrate deposites in the subcontinent within EEZ. The fuel stored in these hydrates is enough to meet India's energy requirement for several centuries. India has carried out detailed analysis of high-resolution seismic data from the continental shelf and slopes to detect acoustic indicators of the presence of gas hydrates. A map of the gas-charged sediments along the Indian coast is shown in Figure 14.2.

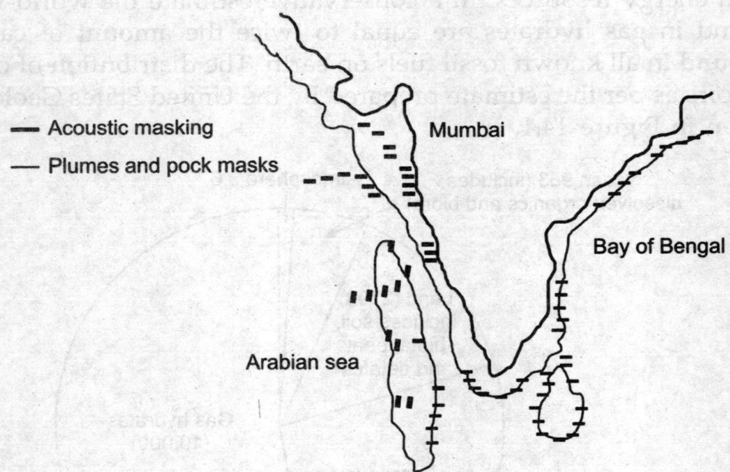

FIGURE 14.2   Distribution map of gas-charged sediments.

In two areas, in southwest off Mumbai and south of Bengal Delta, the gas hydrate stability zone (GHSZ) appears to be more than 80 m thick. Gas hydrates have high acoustic velocity (about twice that of seafloor sediments). They show strong reflection called the Bottom Simulating Reflection (BSR) which indicates a promising gas hydrate region. The largest number of BSR occurrences are traced in Western offshore area with concentration in West of Mangalore coast. The shape of Laccadive ridge seem to be characterized by the occurrence of BSR in large numbers.

BSR occurrences are also observed in Northern Andaman sea, while in Southern region they are noticed in Nicobar islands.

India's estimated gas hydrates resources are to the tune of 1894 trillion cubic metres which is over 1700 times the proven gas reserves with the country (1.08 trillion cubic metres).

14.12.3   Importance of Gas Hydrates

Gas hydrates are important due to several reasons such as:

- They contain a great volume of methane and are a source of cleaner fuel.
- One volume of gas-hydrate produces 164 volume of gas at standard temperature and pressure.
- The global energy contained in the hydrates is estimated to be twice the amount of the total fossil fuel.
- Methane made available by drilling around these gas hydrates can be captured, stored and fed into pipelines that feed our nation with natural gas.
- Methane thus obtained can be used to extract hydrogen and use it to power fuel cells, one of the most promising 'energy alternatives'.

India, Russia to develop gas hydrate technology

Indo-Russian Joint Council for the Integrated Long Term Programme (ILTP) for cooperation in science and technology is working to develop a technology for deep sea mining of 'gas-hydrate'. This project set-up has started in Indo-Russian Centre at the Indian Institute of Oceanography, Chennai.

14.13   SUMMARY

Hydrogen is highly reactive with metals and non-metals, thus, well suited for a large number of applications. It is used in petroleum refining, is convenient for ore-reduction in metallurgical processes and is widely used in the manufacture of vanaspati, fertilizers and alcohols. In household cooking and space-heating, hydrogen fuel offers a special advantage of flameless catalytic combustion. Hydrogen generates nearly three times energy compared to per unit mass of petrol.

National Hydrogen Energy Board (NHEB) has prepared a workable plan to make hydrogen as a commercial fuel. In the near future, large amounts of hydrogen could be produced in remote wind farms, solar stations and ocean power plants and stored underground. Homes could produce some of their own hydrogen using rooftop solar cells, storing it in basements.

REVIEW QUESTIONS

1. Why is hydrogen called a secondary energy source? Name the various methods of hydrogen production.
2. Explain the characteristics of hydrogen as fuel. How can hydrogen compete with fossil fuel?
3. Explain how hydrogen is a more versatile fuel? What are various methods of hydrogen storage?
4. Write short notes on:
   (a) Metal hydride
   (b) Hydrogen storage using nano-crystalline Mg-based Ni-hydride
   (c) Production of hydrogen from sunflower oil
   (d) Application of hydrogen as fuel
5. What are gas hydrates? Explain their importance for energy security.
6. Briefly describe the distribution of organic carbon in the earth reservoirs as identified by USGC.
7. Explain the importance of gas hydrates.
8. Indian resources of gas hydrates shall meet the future energy needs. Explain.
9. Areas of gas hydrate in Indian subcontinent need extensive R&D effort to extract methane as fuel. Explain.
10. Find out the technological advancements that have taken place to extract hydrogen and gas hydrate as an alternative resource of fuel.

15

HYBRID ENERGY SYSTEMS

15.1 INTRODUCTION

Renewable energy sources dealt in various chapters are distributed systems of energy widely spread in the country that is most suitable for dispersed population located not reachable by state grid. It is inherent with renewable energy systems that energy supply is not continuous. Reason for this shortcoming is to be understood and solution searched.

15.2 NEED FOR HYBRID SYSTEMS

Solar water heaters, air heaters, solar distillation and wax melters, PV arrays, PV pumps, operate at optimal efficiency for the months of April to September when solar radiation contain high energy flux. To meet the load demand during night and cloudy days, battery bank is provided. During winter, load demand shoots up and solar energy reduces, so designer is compelled to select large size equipment, PV arrays and battery bank. Similar situation is faced for a stand alone wind power generating system, when wind speed drops below cut-in speed and Wind Turbine Generator (WTG) stops. For emergency, loads of hospitals, defense installations and communication services, a back up source (1) diesel generator, (2) gas turbine generator, (3) biogas, (4) small hydro, and (5) fuel cell is required. Two different energy systems installed at a location to ensure continuity of electrical supply is known as *hybrid energy system*. Thus, hybrid energy system provides an edge over the *stand-alone* and even *grid interactive systems* for reliability of energy supply and lower capital cost. However, engineer's selection of the back up source is done by maximum capacity of the prime energy source at peak energy demand period.

15.3 TYPES OF HYBRID SYSTEMS

Few hybrid energy systems that are operative in prevailing Indian conditions in various states are given:

It is assumed that a battery bank of a suitable size is installed as the storage tank for the period of low wind speed, during 'No Sun' cloudy day and night period. Correct choice for an option will include the parameters

   (i) available solar insolation at optimum array tilt,
   (ii) free wind velocity at 10 m or 20 m height,
   (iii) number of cattle available in a village or a cluster community.

(A) PV – Diesel	(B) Wind – Diesel	(C) Biomass – Diesel
(D) Wind – PV	(E) Micro hydel – PV	(F) Biogas – Solar Thermal
(G) Solar – Biomass	(H) Electric and electric hybrid vehicles	

15.3.1 PV Hybrid with Diesel Generator

Renewable energy technologies are possible for electrification of remote villages including small hydro, wind, biomass and solar energy, yet solar PV lighting remains the most preferred. Such systems are used in Orissa, Assam, Sikkim, Jammu and Kashmir, and Uttarakhand. This power plant contains one PV array with a Diesel electric generator and a battery bank. Energy generated from PV array feeds load demand and then charges the battery bank. Diesel generator keeps the battery fully charged and some time supplies load demand when PV output is not sufficient and battery charge is low to supplement. Figure 15.1 is a block diagram of such a power plant where.

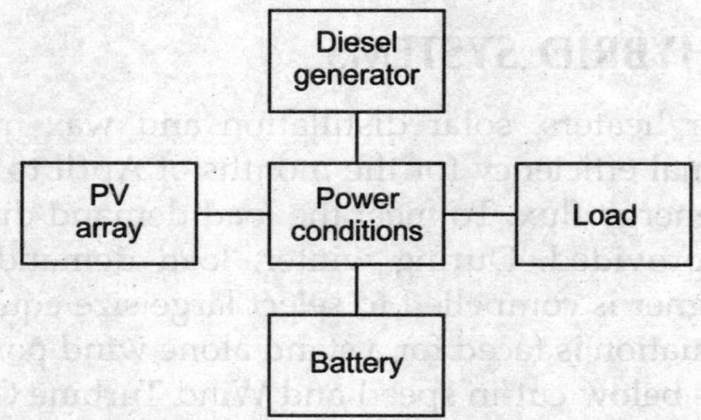

FIGURE 15.1 Block diagram of PV–Diesel hybrid power plant.

Power conditioner perform three functions:

   (i) To convert alternating current (ac) diesel generated output into direct current (dc) for charging battery bank.
   (ii) To invert direct current (dc) from PV array and battery bank into ac for feeding load.
   (iii) To regulate battery current and voltage for input from generator and output for load.

Several experiments have been carried out to find where 10 per cent diesel fuel would be required with a given solar PV array area to replace 90 per cent of diesel fuel that would be consumed for a diesel system only. Experimental values have been used to draw a graph. Figure 15.2 shows 'life cycle cost' versus array area ($10^3 \times m^2$).

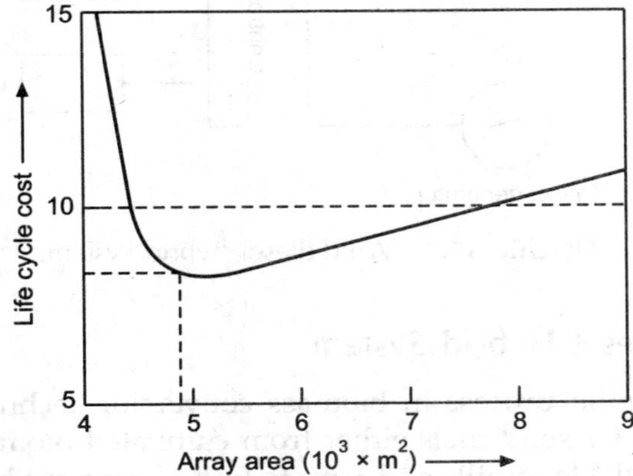

FIGURE 15.2 Graph of photo voltaic-diesel system i.e., life-cycle cost and array area.

Graph indicates a minimum cost point corresponding to a cost effective design for a PV-diesel hybrid power plant where PV has replaced 90 per cent of the diesel fuel; had it been a diesel system only. Thus, a PV-diesel hybrid power plant ensures continuous power supply and is more cost effective as compared to stand alone PV system or stand alone diesel.

15.3.2 Wind-Diesel Hybrid System

Remote coastal areas where wind speed is sufficient to operate a wind turbine but there is no state grid supply, wind generators are installed to electrify the area. Wind energy being intermittent a back up of diesel generator is required to maintain 24 hour power supply. Thus, wind-diesel hybrid system is installed to supply electric power to emergency load of hospitals, communication services, defence installations, commercial and domestic load.

Wind-Diesel hybrid system constitutes components, wind turbine, diesel generator, Controller battery and the load, detailed in Figure 15.3.

During favourable wind 400 V ac is delivered to the controller. The controller converts AC voltage to 120 V dc for charging the battery and it also controls the current required for its charging. Controller also ensures continuous power supply to the load.

As the wind speed drops the lower limit, WTG stops and the diesel generator automatically starts to supply energy to the load and also for battery charging. Thus, wind-diesel hybrid system ensures maximum utilisation of free wind energy and continuity of power supply in remote inaccessible areas.

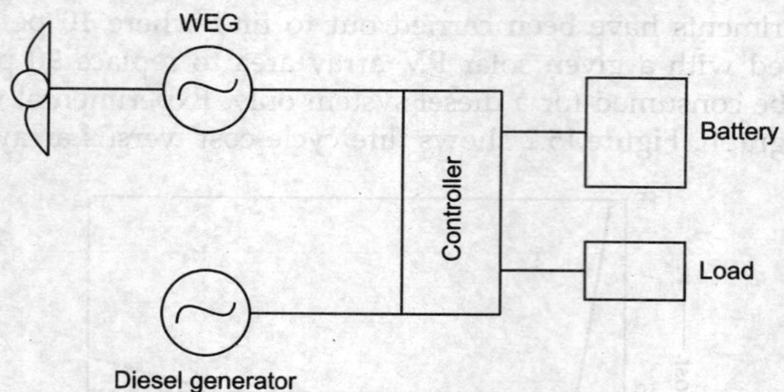

FIGURE 15.3 Wind-diesel hybrid system.

15.3.3 Biomass-Diesel Hybrid System

Combustion is a common process in biomass conversion technology. Application of combustion process is for solid fuels either from cultivated biomass or waste biomass. Biomass is widely available in hills and remote forest areas but becomes scarce during snowy winter. When its supply stops and stock dwindles, energy route of biomass to electrical energy by incineration suffers a set back. This system needs a back up by diesel power electric generator to meet the known lighting and plug loads of residences, commercial establishments, hospitals and other life sustaining loads. Essential components of this hybrid configuration are:

(a) 25 kW biomass generator
(b) Battery bank of 1000 Ah capacity
(c) 15 kVA diesel generator.

A biomass-fired steam power plant is made hybrid with a diesel generator along with a controller, battery bank and load is shown in Figure 15.4.

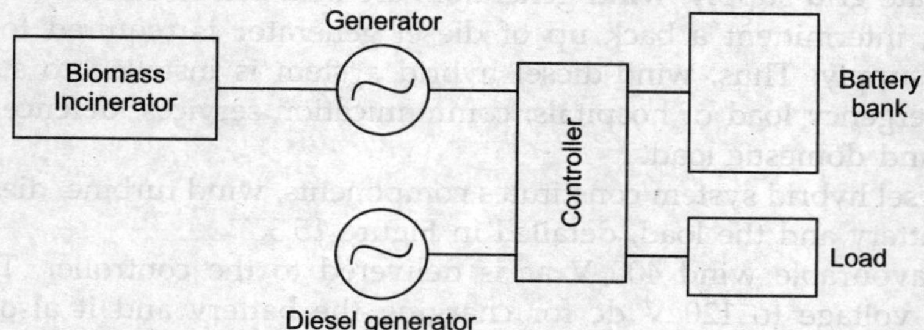

FIGURE 15.4 Biomass-diesel hybrid system.

To operate this system, economic viability is necessary by utilising biomass generator to the full capacity and minimum use of diesel generator, for essential and life saving load during crisis period of biomass availability.

15.3.4 Wind-PV Hybrid System

Wind and solar hybrid energy systems are located in open terrains away from multistorey buildings and forests. Locations are selected in those areas where the sunshine and wind are favourable for more than 8 months during a year.

A schematic wind-PV hybrid system is shown in Figure 15.5. During the day when sun shines, the solar photovoltaic plant generate dc electric energy conditioner provided, converts dc to ac and supplies power to the load. During favourable wind speed, wind turbine generator produce ac electrical power. It supplies power to the load and excess energy after conversion to dc is stored by the battery bank. The plant may operate as stand alone load or may be connected to the state grid.

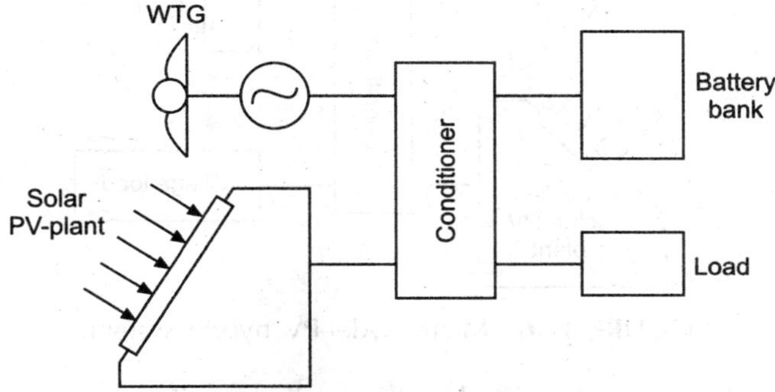

FIGURE 15.5 Wind-PV hybrid system.

Case study: Asia's biggest solar-wind hybrid park in village CHARANKA, Rann of Kutch, Gujarat

Rows of photovoltaic cells will spread over nearly 49,600 hectares in a small village CHARANKA in RANN OF KUTCH, GUJARAT, will generate 30 GW of electricity every year. It is ASIA's largest solar park. The second largest is of 2.2 GW at GOLMUND Solar Park in China's Qingzai province.

If it goes according to plan, the park will be able to generate 15 GW of energy by 2024. Upon achieving planned plant capacity CHARANKA will be the world's biggest solar wind hybrid park.

15.3.5 Micro Hydel-PV Hybrid System

Micro hydel (up to 100 kW) power stations are low head (less than 3 m) installations and provide decentralised power in mountain regions, also in plains on canal falls. In remote areas of J & K, boarder districts of Arunachal Pradesh micro hydro power plants are the only source of energy. With the help of micro hydro power, rural electrification can be achieved besides providing power for pumped irrigation and grinding mills.

In Arunachal Pradesh, 425 villages are being electrified by completing 46 small/micro hydro power projects.

However, there are 1058 villages which cannot be illuminated by micro hydel projects as at several locations, head is very low, while at other, quantity of water is small. Solution is to provide micro hydel-PV hybrid system as sunshine is available practically at all locations.

Portable micro hydel sets of 15 kW capacity are installed with solar PV panels to compliment each other as given in Figure 15.6.

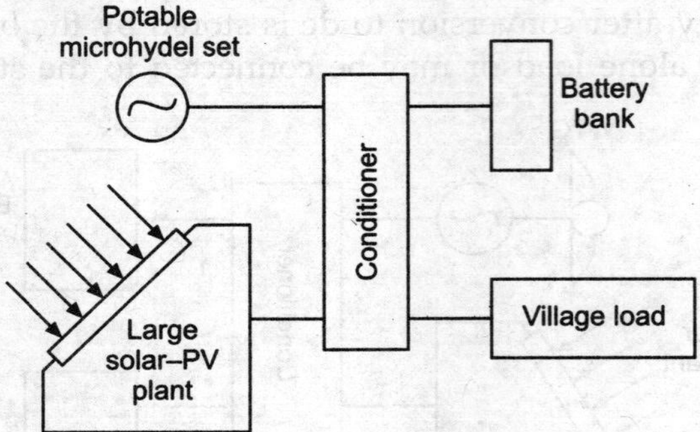

FIGURE 15.6 Micro hydel-PV hybrid system.

Micro hydel systems are provided with small dam store water to be used during night when solar PV panels stops power supply. A battery bank may be provided for emergency power supply. A battery bank may be provided for emergency power supply wherever required. Load management is carried out to maintain continuity of supply for 24 hours matching with the capacity of generating equipment.

15.3.6 Biogas-Solar Thermal Hybrid System (A Case Study)

It is a case study of a milk chilling centre located in a remote village Vasna Margia in Kheda district of Gujarat. Major components of this system are:

 (i) Biogas plant to be operated with an input of 300 kg cow dung daily.

 (ii) One 5 H.P. dual fuel engine. Initially it starts with 100 per cent diesel fuel. Subsequently, engine switches to dual fuel mode with fuel ratio 80 per cent biogas and 20 per cent diesel.

 (iii) Flat plate solar collector is installed on the roof of a building for the supply of hot water (100 litres per day at 60°C) required for cleaning the cans and milk chilling equipment parts.

 (iv) An insulated water storage tank placed over the building connected to the solar collector. Biogas is generated in a KVIC type floating dome vertical design plant with a capacity of 12 cu m/day gas production. Biogas from the plant is taken to milk chilling centre (Figure 15.7) through a G.I. pipeline.

To begin operation, duel fuel engine is started where power transmission to the chilling plant is obtained with a common shaft coupled to the engine. This shaft further operates refrigeration compressor, chilled water circulating pump and air blower detailed in Figure 15.7.

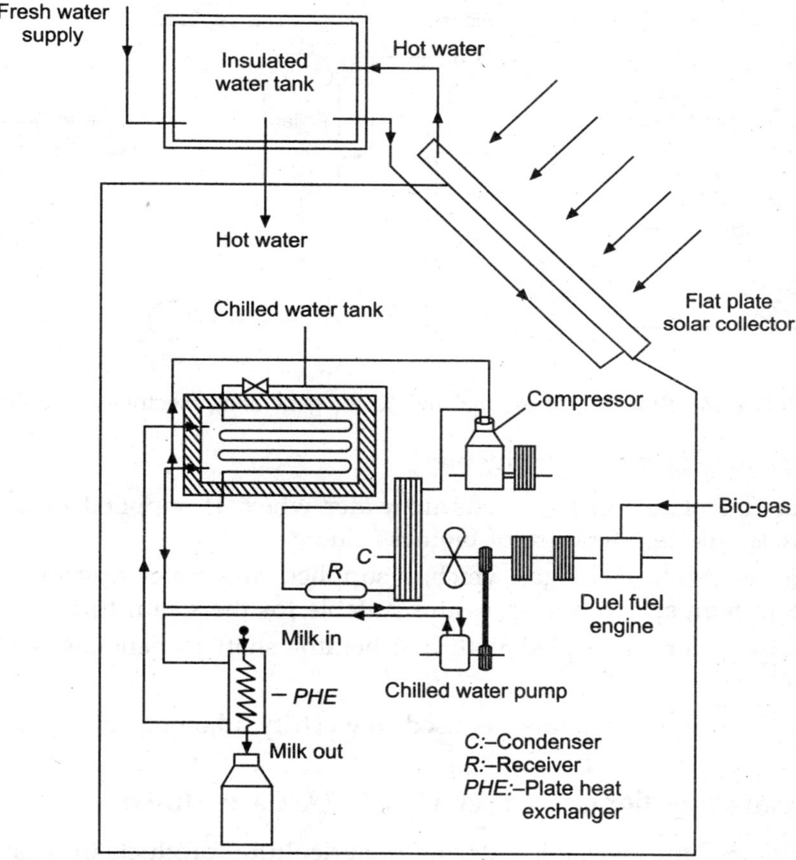

FIGURE 15.7 Schematic of milk chilling centre powered by Biogas and solar energy.

With the successful commissioning of this plant, milk cooperative society has become economically viable and milk producing farmers of the area are earning well. Really a combination of two renewable source of energy, i.e., biogas and solar thermal has proved a boon for remote villages.

Biogas-Solar Thermal Hybrid Plant for Electricity Generation in Hilly Area

In India, hilly regions produce a huge quantity of biomass round the year; but are deficient in electric supply. Biogas-solar thermal Hybrid plants are planned for village electrification on a large scale.

In hilly areas, the average temperature hovers around 13°C. The anaerobic bacteria grow and digest biomass in the temperature range 20–65°C. Due to low temperature of biomass cannot be converted into biogas in cold areas. This problems can be solved by utilizing solar thermal integration as detailed in Figure 15.8.

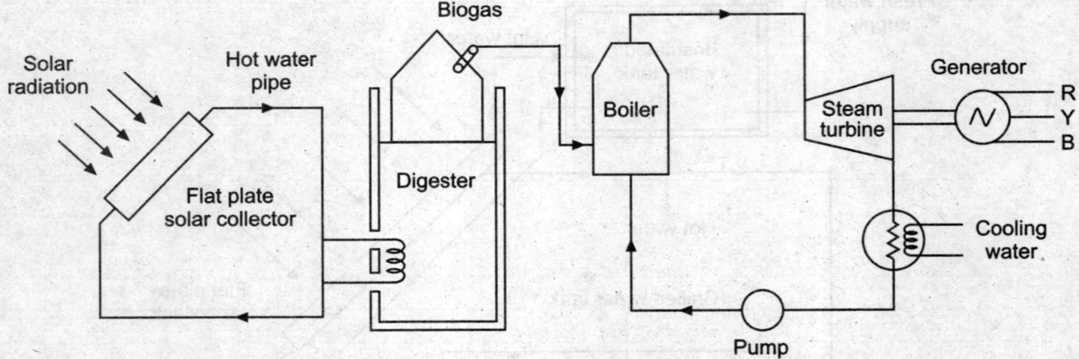

FIGURE 15.8   Biogas-solar thermal hybrid for generating electricity in hilly areas.

Various steps to generate electricity are:

 (i) Solar flat plate collector heats up water which is supplied to the digester to increase the temperature of biomass slurry.

 (ii) Digester produces biogas which is supplied to a boiler to generate steam at a certain temperature and pressure suitable for the steam turbine.

 (iii) Turbine shaft is coupled with a generator shaft to generate 3 phase electric power.

Electric power thus generated is used to electrify adjoining villages.

15.3.7   Solar-cum-Biomass Dryer Hybrid (A Case Study)

Energy is required for mechanical drying of agricultural products in rural areas where grid electric supply is scarce. Solar and biomass are two main renewable sources of energy that may be used for drying of spices, herbs, and agricultural products for commercial production at low cost. Analysis showed that traditional drying, i.e., open sun drying, dried the turmeric rhizomes in 12 days while solar cum biomass dryer took only 1.5 days and produced better quality products in terms of colour, taste and texture.

Selecting right drying technique is necessary in tropical regions where herbs and spices are harvested during winter or rainy season. The solar-cum-biomass dryer was developed at I.I.T. Delhi for 15–18 kg capacity of turmeric rhizomes and other such products. The dryer has two parts: (i) solar dryer (ii) biomass burner as shown in Figure 15.9.

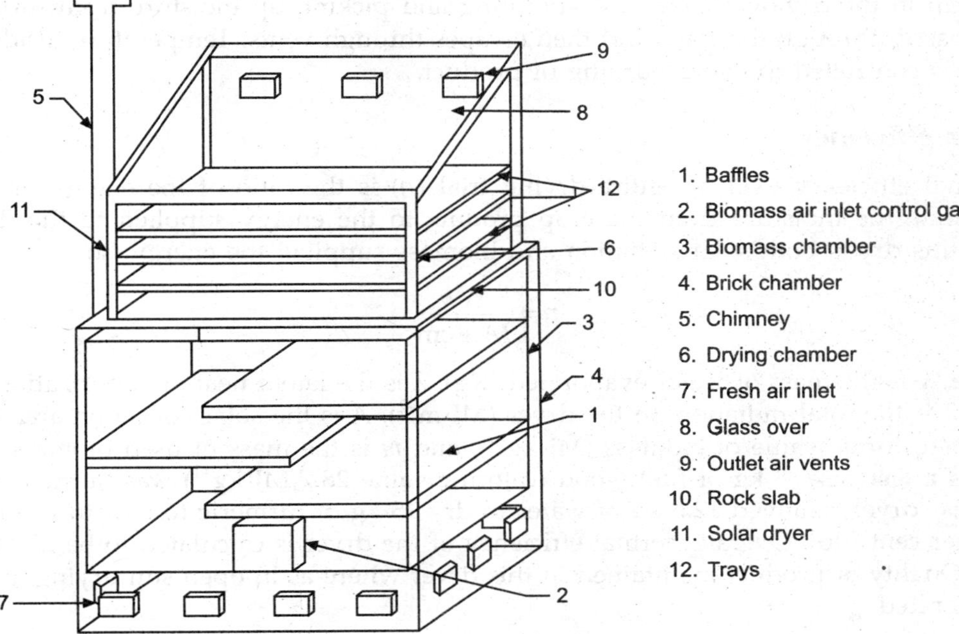

Figure 15.9 Schematic diagram of solar-cum-biomass dryer.

Solar dryer

It consists single glazed (2 mm thick) solar dryer mounted on a rock slab supported on a brick wall chamber. Top glass surface is inclined at an angle of 28.5° to capture maximum solar radiation at Delhi (lattitude 28°32′ N during winter). There are three drying trays of wire mesh with a drying area of 0.94 m^2 each. Three adjustable vents are at the top of the dryer. Two doors are at the front for unloading the products.

Biomass burner

Biomass burner is a rectangular box. A door at the bottom is to feed the biomass and the control airflow for combustion. There is a iron grate for burning biomass. Exhaust gases exit through a chimney. Three metal baffle plates are above the iron grate to lengthen the flow path of combustion gases. A brick chamber encloses the burner which is covered with a rock slab to maintain correct air temperature.

Dryer uses solar energy when solar radiations are more than 100 W/m^2. When solar radiations fall on glass surface, these are absorbed resulting in increase of dryer temperature. Heated air inside the cabinet goes upward; picks up moisture from the product and goes out from the vents. It reduces pressure inside the cabinet an ambient air is drawn into the dryer thorugh inlet holes. A continuous flow of air is thus established.

During period of low or zero solar radiation, biomass burner is used for back up heating. Combustion gases warm the air as it moves over the outer surface. Warm air rises up in the drying chamber; evaporating and picking up moisture from turmeric as it passes through the trays and then escapes through vents. Temperature inside the dryer is controlled to avoid burning of product.

Dryer efficiency

Thermal efficiency over an entire drying trial (η) is the ratio of the energy used to evaporate the moisture from the crop product to the energy supplied to the dryer. With this dryer, both solar radiation and biomass supplied the energy, so

$$\eta = \frac{W\lambda}{IA + cm}$$

where W is the mass of water evaporated (Kg), λ is the latent heat of vaporization (MJ kg^{-1}), I is the total radiation on the dryer (MJ m^{-2}), A is the solar collection area (m^2), c is the calorific value of biomass (MJ kg^{-1}), and m is the mass of used biomass (kg).

As a test case, 8 kg of fuel wood (calorific value 28.7 MJ kg^{-1}) was burned. Solar biomass dryer removed 12.6 kg of water to dry 15 kg of turmeric to moisture content of 9 per cent (db). Overall thermal efficiency of the dryer is calculated to be 28.11 per cent. Quality of product maintained in this dryer where as in open sun drying, it gets deteriorated.

15.3.8 Fuel Cell-PV Solar Hybrid System

Power after sunset remains a challenge for the solar home (Figure 15.10). However, a self-renewing catalyst discovered by MIT (US) researcher Daniel Nocera allows water to act as a storage medium by keeping the lights on at night . Solar panel home hybrid with fuel cell resolves issues faced after sunset.

15.4 ELECTRIC AND HYBRID ELECTRIC VEHICLES

Electric vehicles are propelled by an electric motor powered by rechargeable battery packs. These vehicles need not have Internal Combustion Engines (IEC) system, the drive train and fuel tank. Electric motor replaces the engine and it gets power from rechargeable batteries through a controller. The electronic motor controller provides electric power to the motor based on inputs from accelerator. Electric power is delivered from battery pack, which is like the fuel tank of an electric (e) vehicle. However, they are slow in speed and move only up to 80 km on a charge. Full battery recharge takes nearly four hours.

A hybrid electric vehicle combines a conventional internal combustion engine with an electric propulsion system. Presence of electric power train is intended to achieve better fuel economy than conventioanl vehicle or better performance. Most common of HEV is the hybrid electric car.

Hybrid vehicles use both petrol and electric propulsion systems. In such vehicles, the electric motor provides a boost during starting and is recharged during vehicle operations. This cuts emissions significantly and improves fuel economy.

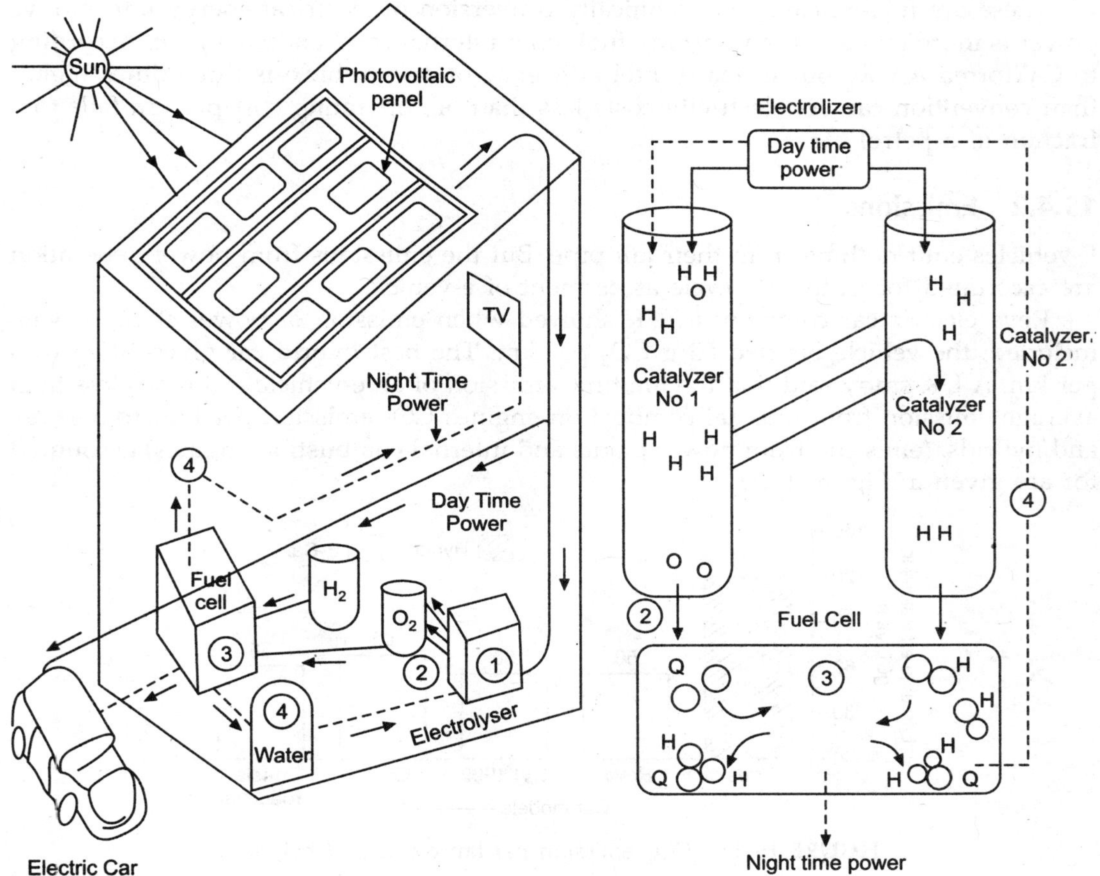

1. Excess day time electricity from solar panels goes to an electrolyser
2. Aided by the new catalyst the solar electricity splits water into hydrogen and oxygen which are stored.
3. When it is too dark for solar power, the stored elements are recombined in a fuel call, generating electricity.
4. Power flows to household appliances and recharges electric car batteries. The only by-product from the fuel cell is water that is recycled.

FIGURE 15.10   Photovoltaic solar panels over a roof of a house Hybrid with FUEL CELL for 24 hour electric power.

15.4.1   E-Vehicle Need

E-vehicle are gaining popularity concerning to:

   (i) High oil prices
   (ii) Green house gas emissions
   (iii) Ambient air quality

Concern over high oil prices and stringency in pollution and climate regulations have spurred new interest in e-vehicles.

These are fuel-efficient, as, technically conversion of electrical energy into motive power is more efficient than burning fuel in an internal combustion engine. According to California Air Resource Board, fuel efficiency of an e-vehicle is three times higher than convention car. As electricity costs less than oil, operating cost per km falls to a fraction of a petrol car.

15.4.2 Emissions

E-vehicles emit nothing from their tail pipe. But the emissions from power generation are accounted for in the life cycle assessment of e-vehicle.

Reva electric car company in UK showed when emission of power stations were included, the vehicle emitted 63 g CO_2 per km. The best hybrid car gives 104 g CO_2 per km. A UK study said that the life time emission of an e-vehicle is 3 times less than average emission from internal combustion engines CO_2 emission per km from e-cars and hybrids. (emission from power plants and internal combustion engines) accounted for are given in Figure 15.11.

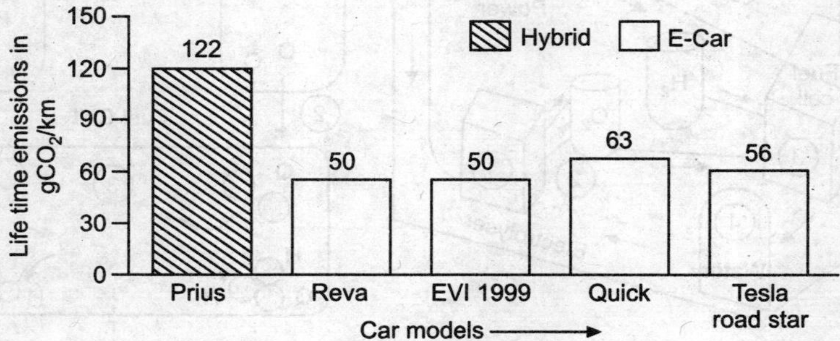

FIGURE 15.11 CO_2 emission per km e-cars and hybrids.

The other advantage of an e-vehicle is that there are no oil filters, air filters, spark plugs and radiators which need maintenance.

15.4.3 Limitations

Widespread use of battery operated vehicles is constrained by high prices, limited driving range, bar on maximum speed and battery efficiency.

In India, most e-vehicles run on lead-acid batteries which provides short bursts of power to starter motors in cars. Also, lead is a known environmental hazard with serious health consequences.

Lead-acid batteries will have to give way to lithium-ion batteries that improve performance four times over. Lithium ion is currently a dominant battery technology in portable applications. It provides the highest energy density of all rechargeable systems. REVA NXR is a new lithium-ion powered e-car claims that a first charge for 90 minute would offer a range of 320 km a day.

The biggest challenge of e-vehicle industry is to produce batteries that can store large amount of energy that can be released and recharged quickly.

15.5 HYDROGEN-POWERED-ELECTRIC VEHICLES

Hydrogen is a clean energy carrier that can replace liquid and gaseous fossil fuels. Hydrogen can be used to power electric vehicles for longer distance, better speed, acceleration and cost comparable with fossil fuel driven vehicles. Conversion of hydrogen to electricity is achieved through fuel cells. A fuel cells is an electrochemical device that converts chemical energy of hydrogen directly into electricity and heat without combustion. Fuel cell systems operates on pure hydrogen and air/oxygen to produce electricity with water and heat as by-products.

A phosphoric acid fuel cell (PAFC) is suitable for hydrogen powered electric vehicles. In India, BHEL has field tested them and commercially available in 40 kW, 200 kW and more sizes. Around Ajanta caves (Maharashtra), fuel cell operated bus ply within 7 km route to avoid pollution. Fuel cell operated vehicles are eco-friendly as it eliminates noise, SO_2 and nitric oxide emissions.

15.5.1 Clean Mobility Options

Low speed e-vehicles with power less than 250 W and speed less than 25 kW/h are exempt from Central Motor Vehicle Act and Rules. These are not categorised as motor vehicles, so driving licenses are not required.

E-vehicles are part of the solution to air pollution and climate change. The e-vehicles will truly become zero emitters when these are charged with electricity from renewable sources such as solar and wind.

REVIEW QUESTIONS

1. Define hybrid energy systems. What was the need for hybrid systems?
2. Discuss different types of hybrid systems.
3. Differentiate between wind-diesel hybrid system and wind-PV hybrid system.
4. Discuss a case study of installed hybrid energy systems in your city/state.
5. Discuss in brief, how with the help of hybrid system vehicle emission can be controlled?

ENVIRONMENT, ENERGY AND GLOBAL CLIMATE CHANGE

16.1 INTRODUCTION

As planet earth revolves around the sun, it regulates CO_2 concentration in the atmosphere which goes up sharply and then down once each year. It happens as the Northern hemisphere is tilted towards the sun during spring and summer. At this time, green leaves come out on trees, they breathe in CO_2, thus causing reduction in its quantity world wide. During winter, the Northern hemisphere is tilted away from the sun. At this time, the leaves fall and discharge CO_2 which goes back into the atmosphere. It is, as if the entire earth takes a big breadth in and out once each year.

The earth receives energy from the sun in the form of light radiation. About 70% incoming solar radiation are absorbed to warm the land, the atmosphere and the oceans, while remaining 30% is reflected back to space. Trapping of infrared radiation by atmospheric CO_2 and other gases is good to maintain earth's temperature suitable for the survival of life. Consequently, average surface temperaure is about 15°C. This is about 33°C higher than it would be in the absence of trapping of solar radiations by atmosphere (Green house effect). Without such gases in the atmosphere, the earth's surface would be frozen with a mean air temperature of −18°C.

After World War II, fast economic developments along with explosive population growth forced large scale use of fossil fuels for energy which causes sharp increase of CO_2 in the earth's atmosphere. Concentration of CO_2 increased from pre-industrial value of 280 ppm to 418 ppm in January, 2022. Estimation of atmospheric CO_2 gives an insight to cooling and warming cycles of planet earth.

It is believed that maximum 350 PPM of CO_2 is required to sustain and preserve the earth's climate. With higher concentration there is a possibility of irreversible catastrophic effects. Hence immediate action is required to bring down CO_2 concentration.

Total green house gases like CO_2, CH_4, chlorofluoro carbons (CFC) and nitrous oxide are 460 ppm in atmosphere that trap infrared rays coming from the sun and block them from being reflected back again in space by the earth's surface.

The increase in the average temperature of the earth's atmosphere and the oceans is known as *global warming*. Resultant consequences would be dire—ranging from melting of ice caps, rise in sea level, glacier retreat, increased intensity of storms and cyclones, changes in precipitation patterns, lowering agricultural yields, increase in disease vectors to species extinction.

It is envisaged that by 2050 nearly 52% of world's population may face serious drinking water shortage besides rise in sea level causing submergence of vast coastal areas and islands. It will be colossal environmental and social problem if the world fails to act boldly and quickly to mitigate global warming.

16.2 ENVIRONMENTAL STUDIES—A MULTIDISCIPLINARY APPROACH

Environmental studies encompasses the inputs from a wide range of disciplines, and this gives environment studies, a multidisciplinary nature (Figure 16.1).

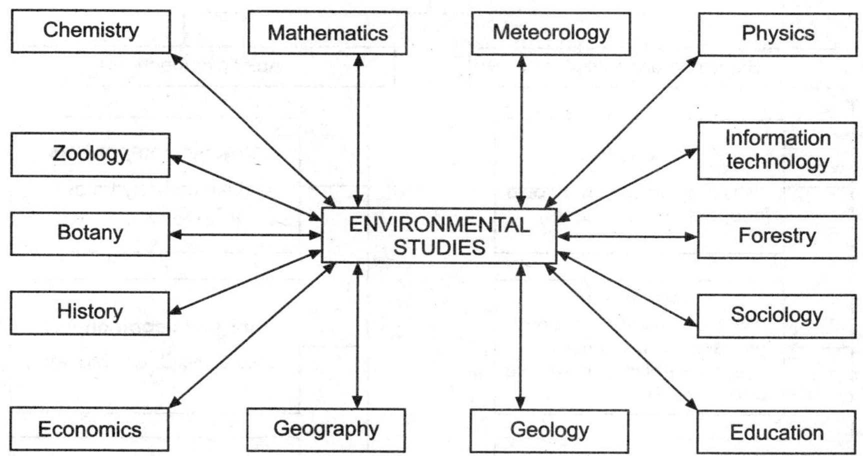

FIGURE 16.1 Environmental studies—a multidisciplinary approach.

As an illustration, *life sciences* subjects (like botany, zoology, microbiology, genetics, biochemistry, biotechnology, medicine, etc.) help, understand the biotic components and their interactions in the environment. *Natural science* subjects (like chemistry, physics, mathematics, etc.) aid in comprehending the physical and chemical properties of environmental components and help in forecasting, modelling and calculations. *Social sience* subjects (like law, economics, sociology, history, etc.) help in describing the socio-economic aspects of environment and development. *Technical* subjects (like engineering, meteorology, etc.) provide the solutions to environmental problems.

For example, to understand the effects of global warming and climate change, we first need to understand the components of the environment likely or getting affected. Life sciences would come in handy to understand this aspect. Then one would need to study natural science subjects to understand atmospheric chemistry and climate change predictions (chemistry and mathematics). The social science subjects would explain the socio, economic and political aspects of climate change. The technical subjects would offer mitigation measures like CO_2 sequestration, clean technologies, renewable options, etc.

16.3 ENVIRONMENT

Everything around us is our environment. It covers both living or 'biotic' and non-living or 'abiotic' components (Figure 15.2). The biotic part of environment consists of plants and animals which need their own habitat to grow and survive. Forests, grasslands, mountains, deserts, rivers, lakes, coastal swamps and marine enviroment all form habitats for different plants and animals.

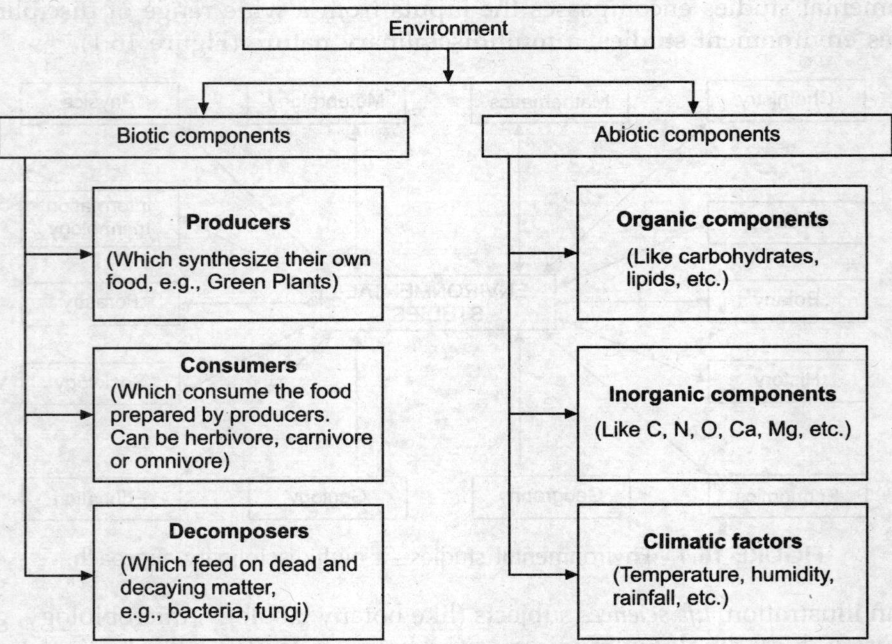

FIGURE 16.2 Components of environment.

Plants and animals are interdependent and their interaction with the physical environment is called '*Ecosystem*'. A smaller ecosystem may be a pond while largest ecosystem is the biosphere. An ecosystem consists of the following biotic components.

16.3.1 Producers

Green plants and algae are also known as autotrophic (self nourishing) because they produce their own food with sunlight. All these store energy and liberate oxygen.

16.3.2 Consumers

All animals including both herbivores and carnivores consume producers to sustain life and are thus called consumers.

16.3.3 Decomposers

Micro-organisms like bacteria and fungi which start decaying of waste material, dead producers and consumers to release inorganic elements back to the soil are called decomposers. They utilise the organic carbon and generate CO_2 which is used by algae. Thus, decomposers play a significant role in material (mineral) recycling in the eco-system.

16.4 BIOGEOCHEMICAL CYCLES

Each ecosystem maintains a dynamic equilibrium between the organisms and the environment through pathways of natural cycles. These are water cycle, carbon cycle, oxygen cycle, nitrogen cycle and the energy cycle.

All the functions of an ecosystem are related to growth and regeneration of its plants and animal species. Various processes are linked through different cycles which operate on energy from sunlight. During photosynthesis, CO_2 is used by the plants and oxygen is released. Humans and animals consume oxygen for respiration. Water cycle depends on rainfall which is a necessity for plants and animals to live. The energy cycle helps in recycling nutrients into the soil for the growth of plant life. Our own life sustains with proper functioning of these cycles.

16.4.1 Water Cycle

The earth is a blue planet as 70% of globe area is covered by oceans which contain 97% of water available in the biosphere. Remaining 3% is available on the continents. More than 70% of this portion (i.e., 3%) is locked in glaciers, ice caps of north and south poles. Remaining less than 1% water is available in lakes, streams and as ground water for our use.

Water cycle works on the reciprocity of evaporation and precipitation. Water evaporates from oceans, lakes to form clouds. Cooling of clouds cause precipitation which falls as rain, sleet, hail or snow (Figure 16.3) water is transpired from the leaves as water vapour in the atmosphere, which being lighter than air rises and forms cloud wind blows the clouds for long distances. As the clouds rise higher, vapour condenses to form droplets which fall on the land in the form of rain and hail while on hills as flakes and snow. Estimated quantity of water discharged into the oceans is about 37000 km^3/year, which is in circulation and is available for human needs. Rapid economic growth, rise in population, five-year plans for better life demand, optimum utilisation of available water for domestic, industrial, irrigation and power generation purposes.

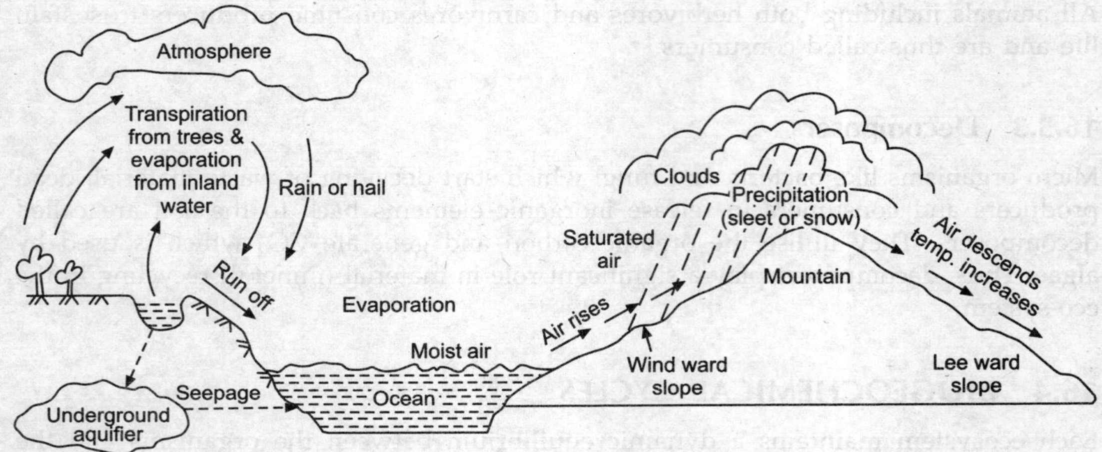

FIGURE 16.3 Water cycle.

16.4.2 Carbon Cycle

Carbon occurs in organic compounds ($C_6H_{12}O_6$, CH_4) and is present in both the abiotic and biotic parts of the ecosystem. Carbon is a building block for both plant and animal tissues. Carbon is present in the atmosphere as CO_2 which is absorbed by trees through leaves. Plants combine CO_2 with water (which is absorbed by roots from the soil) and transported to leaves. In the presence of sunlight process of photosynthesis occurs and during the process plants release oxygen into the atmosphere required by human and animals for respiration. Life on the planet depends upon the oxygen generated through this cycle. Trees and plants are paramount to sustain life on the planet.

Herbivorous animals feed on plants for their growth and return fixed carbon to the soil as their excrete. When plants and animals die they return their carbon to the soil. Buried carbon under the earth over a millions of years ago had transformed to present day major source of energy coal and oil. Coal and oil are burnt to generate mechanical energy and electricity. These processes complete the carbon cycle. Important features of this cycle are that carbon is utilised for a number of uses through self-regulating feedback mechanisms and then restored to the atmosphere and the oceans thus making it a stable system (Figure 16.4). Oceans store carbon more than 50 times as much as the atmosphere with a dynamic equilibrium.

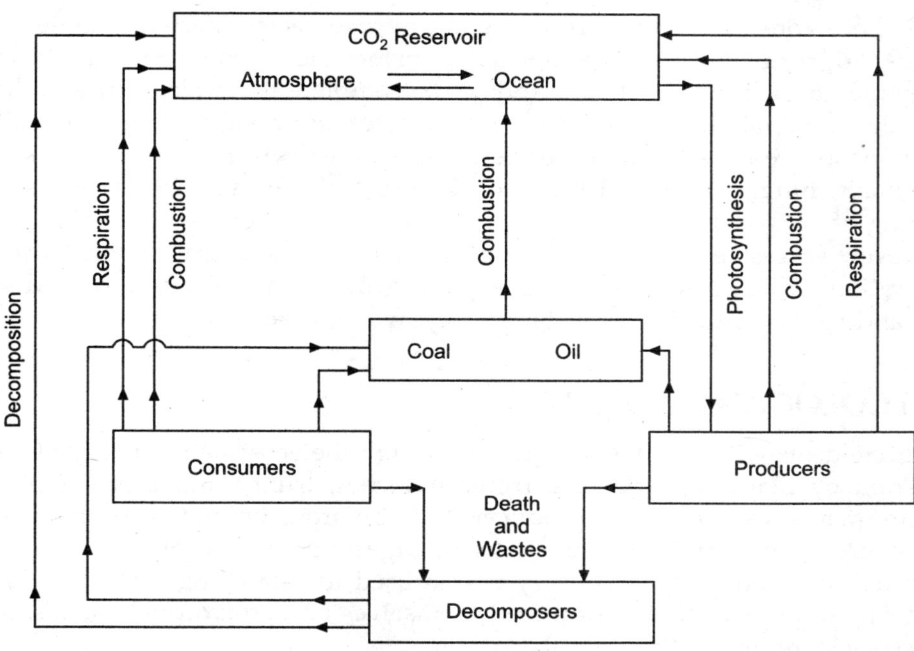

FIGURE 16.4 Carbon cycle.

16.4.3 Oxygen Cycle

Oxygen is taken up by human and animals from the air for respiration. Plants release oxygen to the atmosphere during photsynthesis. This links the oxygen cycle to carbon cycle. Oxygen is also continuously exchanged between the atmosphere and all water bodies on the earth for survival of aquatic and marine life. A part of atmospheric oxygen reaches the stratosphere and is converted to ozone which saves human, animal and plant life from ultra violet radiation of sunlight.

Oxygen is generated by plants, so deforestation reduces oxygen level in our atmosphere. To replenish vital oxygen, it is exigency to undertake afforestation programmes on a large scale.

16.4.4 Nitrogen Cycle

Nitrogen is abundant as gas and constitutes 78% of the atmosphere. However, it cannot be used directly by differnt forms of life. It requires 'fixation' before it can be used by plants and animals. For fixation, energy input is required through biological process. Plants achieve fixation of nitrogen through bacteria living in their root nodules. Another group of bacteria in soil and water convert gaseous nitrogen into nitrates. Plants obtain this nitrate from the soil necessary for their growth.

When herbivores eat plants, part of organic nitrogen is transferred to these animals, and finally to carnivores. When animals and plants die, decomposers act to convert their organic nitrogen into ammonia (NH_3). To complete the cycle, nitrifying bacteria converts the ammonia into nitrites, then into nitrates and finally into gaseous nitrogen back to atmosphere by denitrifying bacteria. Thus, our own lives are interlinked with small animals, fungi, plants and microscopic forms of life that are important for the functioning of ecosystem.

However, excessive nitrogen compounds in the aquatic environment can cause eutrophication, which leads to excessive plant growth and decay, lack of dissolved oxygen, and severe reduction in water quality, fish and aquatic life.

16.5 ECOLOGICAL PYRAMIDS

Energy cycle deals with the flow of energy through the ecosystem. Energy of sunlight is converted by plants into flowers, fruits, branches, trunks and roots of plants. In ecosystem, plants are known as 'producers'. Herbivorous animals use plants as food to gain energy, for performing functions of digesting food, growth of tissues and maintaining body temperature. Energy is also used for searching food, breeding and bringing up young ones. Carnivores feed themselves on herbivorous animals, as such animal species are linked through 'food chain'.

Energy in the ecosystem can be studied in the form of 'food pyramid'. This pyramid has a large base of plants called 'producers'. The pyramid has a narrower middle section that depicts the number and biomass of 'herbivorous animals' called 'first order consumers'. Upper middle section depicts small biomass of carnivorous animals called 'second order consumers'. Man is at the apex of the pyramid (Figure 16.5).

Similarly, several other ecological pyramids are depicted in Figure 16.6 to 16.8. Thus, to support mankind a large base of herbivorous animals and greater quantity of plant material is the necessity. These pyramids were first proposed by Charles Elton in 1927. They are sometimes also called as Eltonian Pyramids.

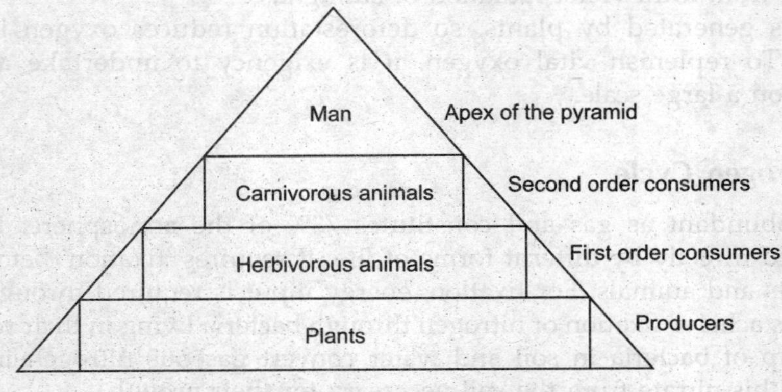

FIGURE 16.5 Ecological pyramid.

16.5.1 Pyramids of Numbers

1. Upright pyramid

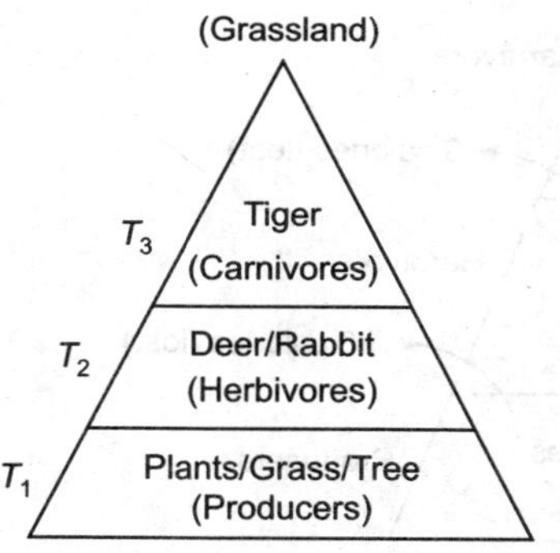

FIGURE 16.6(A)   Upright pyramid (Grassland).

2. Inverted pyramid

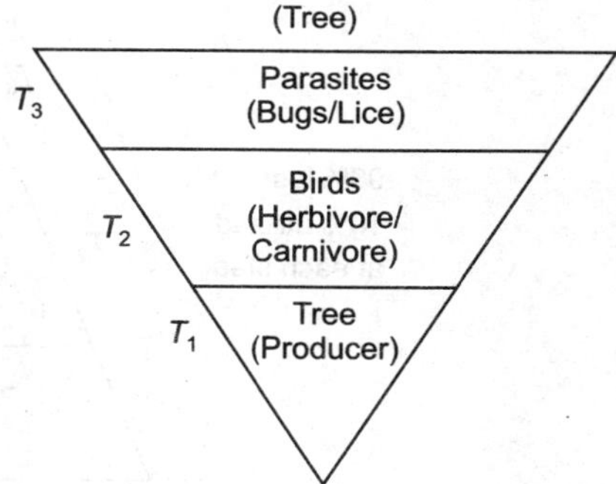

FIGURE 16.6(B)   Inverted pyramid (Tree).

T_1, T_2, T_3—Trophic levels.

16.5.2 Pyramids of Biomass

1. Upright pyramid

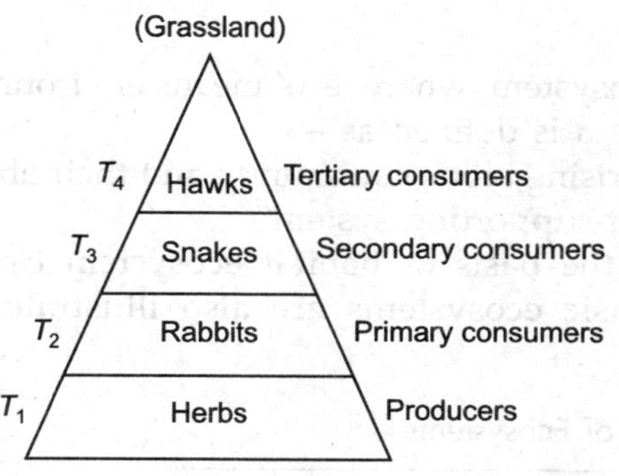

FIGURE 16.7(A)   Upright pyramid (Biomass).

2. Inverted pyramid

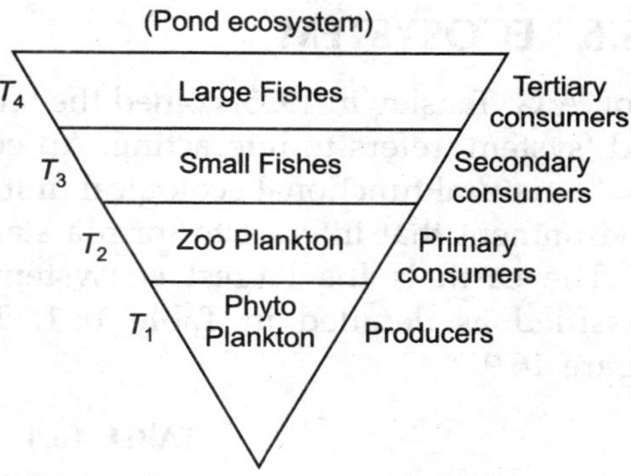

FIGURE 16.7(B)   Inverted pyramid (Biomass).

16.5.3 Pyramid of Energy

When plants and animals die, these material returns back to the soil after being broken down to simpler substances by decomposers, such as insects, worms, bacteria and fungi so that plants may absorb nutrients through their roots. Animals excrete after digesting food goes back to the soil. It links the 'Energy Cycle' to the other material cycles (carbon, nitrogen etc.)

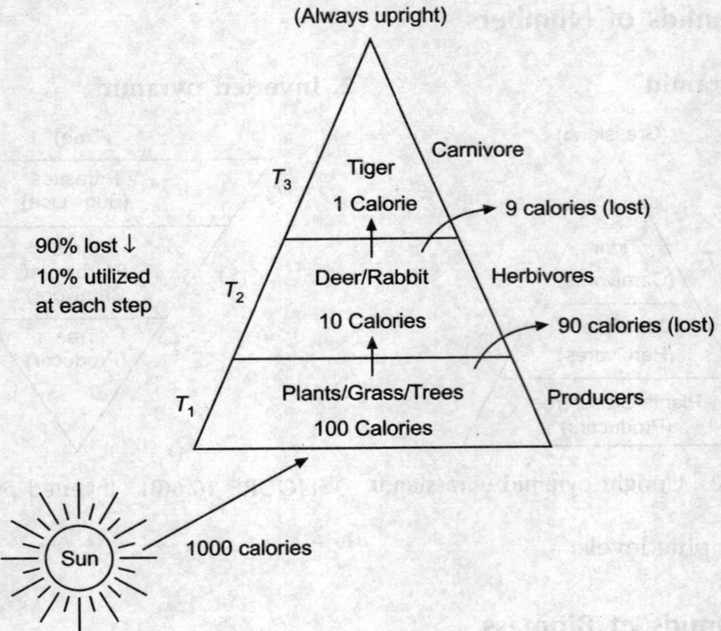

FIGURE 16.8 Pyramid of energy.

16.6 ECOSYSTEM

Prof. A.G. Tansley in 1935, coined the term 'ecosystem' where 'eco' means environment and 'system' refers to interacting. An ecosystem is defined as—

"A natural functional ecological unit comprising biotic community and their abiotic environment that interact to form a stable self-supporting system".

The earth is the largest ecosystem. On the basis of habitat ecosystem can be classified as detailed in Table 16.1. The basic ecosystems are also illustrated in Figure 16.9.

TABLE 16.1 Types of Ecosystems

Terrestrial ecosystem	Aquatic ecosystem
Forest	Pond
Grassland	Lake
Semi arid areas	Webland
Deserts	River
Mountains	Delta
Islands	Marine
	Glaciers
	Antarctica

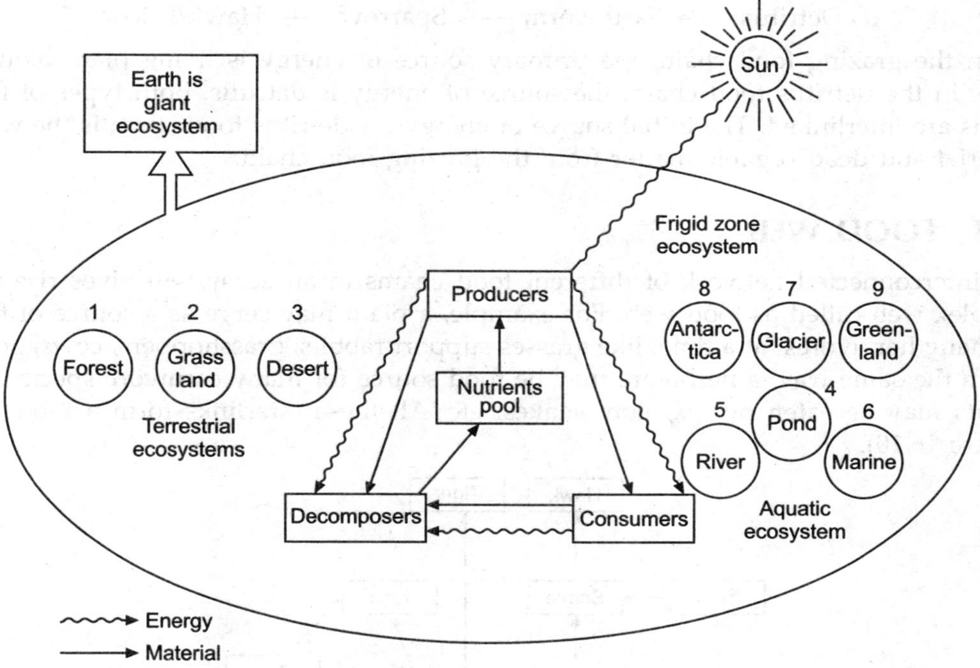

FIGURE 16.9 Basic ecosystem.

16.7 FOOD CHAIN

The transfer of food energy through a sequence of organisms by feeding and being fed upon is termed as food chain.

16.7.1 Types of Food Chains

Grazing food chain

This food chain starts from the producers (green plants) and proceeds to herbivores, and then to carnivores.

e.g., Grass ⟶ Grasshopper ⟶ Birds ⟶ Hawks/Falcon

Detritus food chain

The dead organic matter of plants or animals is known as detritus. A part of this detritus remains on surface as litter and the other part is acted upon by three kinds of organisms — scavengers, detrivores and decomposers. The scavengers eat up dead bodies, e.g., vultures. The detrivores directly feed on organic part, e.g., termites, worms, etc. The decomposers decompose the organic part through enzymes which is then absorbed e.g., fungi, bacteria.

e.g.,   Detritus ⟶ Earthworm ⟶ Sparrow ⟶ Hawk/Falcon

In the grazing food chain, the primary source of energy is living plant biomass while in the detritus food chain, the source of energy is detritus. Both types of food chains are interlinked. The initial source of energy for detritus food chain is the waste material and dead organic matter from the grazing food chain.

16.8   FOOD WEB

The interconnected network of different food chains in an ecosystem gives rise to a complex web called as food web. For example, a plant may serve as a source of food for many herbivores at a time, like grasses support rabbits, grasshoppers, cows, goats, etc. In the same way, a herbivore may be food source for many carnivore species like rabbits may be eaten by fox, lion, snakes, etc. All these interlinks form a food web (Figure 16.10).

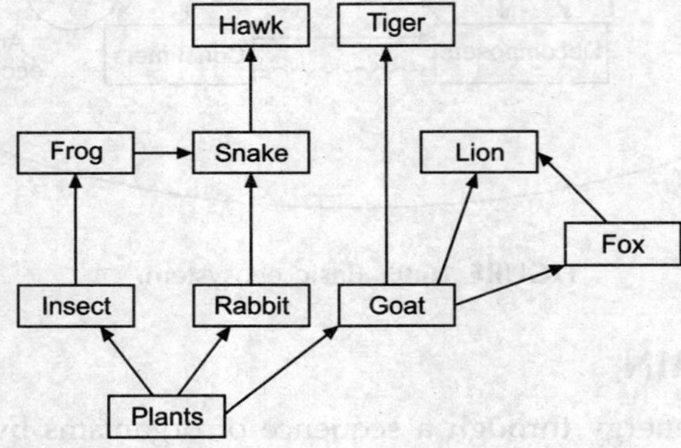

FIGURE 16.10   Food web.

16.9   TEN PER CENT LAW

Lindemann in 1942, proposed the Ten Per cent Law for the transfer of energy from one trophic level to the next. According to this law—

"During the transfer of organic food from one trophic level to the next, only about ten per cent of the organic matter is stored and the remaining about ninety per cent is lost" (Figure 16.11).

As an illustration, it is interesting to compare the vegetarian and non-vegetarian paths of food habits.

16.9.1   Path-I—Vegetarian-Route

Assuming that the sun emits 1000 J of energy, plants utilise 10% of this (i.e., 100 J) as net production available to a herbivore (say a vegetarian). When the plant is eaten by the herbivore it utilises 10% of 100 J, i.e., 10 J. This energy is available to the herbivore (vegetarian).

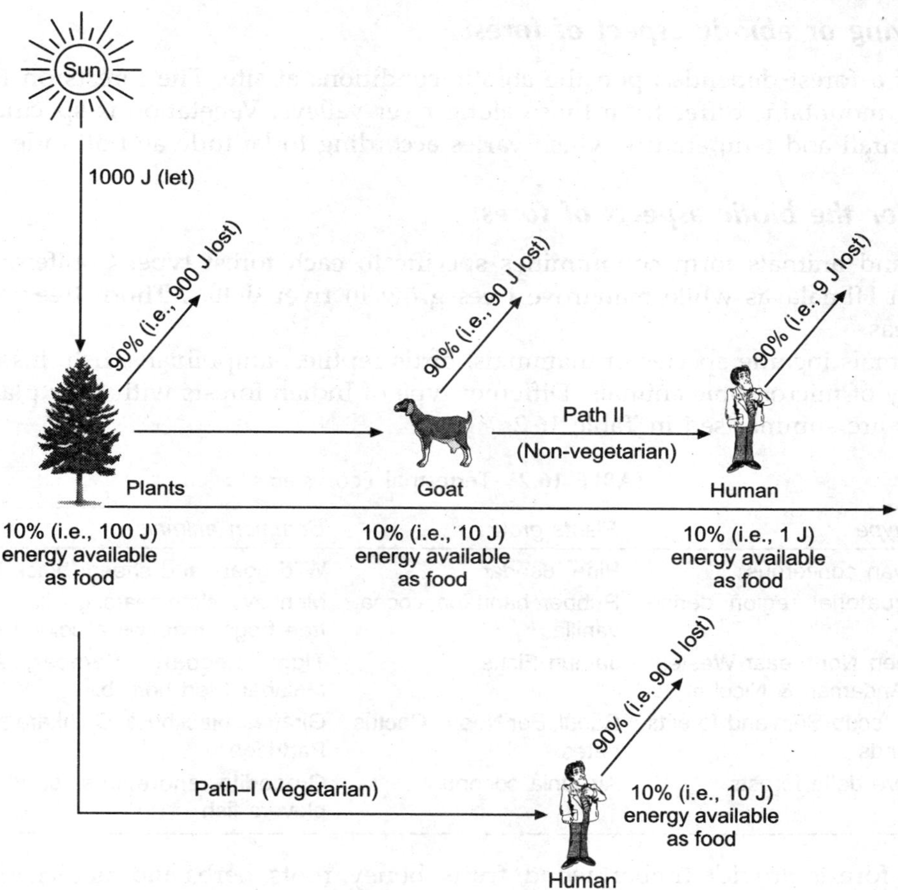

FIGURE 16.11 Law of ten per cent.

16.9.2 Path-II—Non-Vegetarian-Route

Assuming the same situation, the sun emits 1000 J of energy, plants utilise 10% of this (i.e., 100 J) as net production available to a herbivore (say, a goat). This means the goat utilises 10% of 100 J i.e., 10 J. When the goat is eaten by a carnivore (say a non-vegetarian) it utilises 10% of 10 J i.e., 1 J of energy.

Hence, through the vegetarian-route, we get 10 J energy, whereas non-vegetarian route gives 1 J energy.

16.10   TERRESTRIAL ECOSYSTEMS

16.10.1   Forest Ecosystem

Forests are formed by number of plants structurally defined by its trees, shrubs, climbers and ground cover. Forest ecosystem has two parts:

Non-living or abiotic aspect of forest

Type of a forest depends upon the abiotic conditions at site. The forests on the hills and the mountains differ from those along river valleys. Vegetation is specific to the area rainfall and temperature which varies according to latitude and altitude.

Living or the biotic aspects of forest

Plants and animals form communities specific to each forest type. Coniferous trees occur in Himalayas while mangrove trees grow in river deltas. Thorn trees occur in arid areas.

Animals include species of mammals, birds reptiles, amphibians, fish, insects and a variety of microscopic animals. Different type of Indian forests with their plants and animals are summarised in Table 16.2.

TABLE 16.2 Terrestrial Ecosystems

Forest type	Plants grow	Common animals
Himalayan coniferous	Pine, deodar	Wild goats and sheep, black bear
Wet equatorial region dense forests	Rubber, bannana, cocoa, vanilla	Monkeys, sloth bears, gorillas, snakes, tree frogs, macaws a colourful bird
Evergreen North-east Western Ghats Andaman & Nicobar	Jamun Ficus	Tiger, Leopard, Sambar, Zebras, Malabar Pied horn bill
Thorn & scub Semiarid forests Grasslands	Babul, Ber Neem, Cactus dates	Giraffes, blackbuck, Chinkara antilope, Partridge
Mangrove delta forests	Avicenia coconut	Crocodile shorebirds sand pipes, plovers fish

The forests provide timber, wood, fruits, honey, roots, herbs and medicianl plants. Their economic benefit goes to local people, sub-urban and city dwellers.

The forest ecosystem plays an important role in controlling climate and water regimes. Forest produces surface runoff of rainwater and allows ground water to be stored. Soil erosion is prevented by the forests. They also absorb CO_2 and release oxygen that we breathe.

Conserve forest ecosystem

The forests can be conserved if we use its resources carefully. Alternate sources of energy should be used instead of fuel wood. Afforestation needs to be practised vigorously.

Forests with all their diverse species must be protected as National Parks and Wild Life Sanctuaries, where plants and animals can be preserved.

16.10.2 Grassland Ecosystem

Grassland cover those areas where rainfall is low and soil quality is poor. In India, grasslands form a variety of ecosystems such as shola grasslands which consist of patches on hillslopes, Western Ghats, Nilgiri and Aamalai ranges.

Semi-arid planes of Western India and Deccan are covered by grassland tracts with patches of thorn forests. Mammals like wolf, blackbuck, Chinkara; birds such as bustard and floricans have adapted to these arid conditions.

Himalayan pasture belt extends up to snowline. Animals migrate up into the high altitude grassland in summer and move down into the forest during winter when snow covers the grassland.

Terai consists of tall elephant grass (5 metres) which are located in low-lying waterlogged areas. This ecosystem extends to the south of the Himalayan foothills.

Conserve grassland ecosystem

Grasslands should not be overgrazed. Man has made significant alterations that have caused grassland species to extinct. Cheetah, Wolf, Chinkara and great Indian bustard are vanishing. Grassland species should be protected otherwise, they will vanish from their shrinking habitat.

16.10.3   Desert Ecosystem

Deserts and semi-arid areas are located in Western India and Deccan Plateau. The climate is dry as seen in Thar Desert of Rajasthan, where rainfall is scanty. In adjoining semi-arid tract, vegetation consists thorny trees like Kher and Babul.

The great and little Rann of Kutch has special arid ecosystems. In summers, there is a desert landscape. However, these are low lying areas near the sea, so get converted into marshy salt during monsoons. During this period, aquatic birds like ducks, geese cranes and storks visit the area. Little Rann of Kutch is the only home of wild ass in India.

Desert and semi-arid regions inhabit rare animals like Indian wolf, desert cat, desert fox and birds as great Indian bustard and florican. Being an area of scanty vegetation, crops like Jowar and Bajra are grown here.

Extensive irrigation system like Indira Gandhi canal in Rajasthan has changed the natural arid ecosystem to a region of intensive agriculture.

Conservation of desert ecosystem

Desert ecosystem is sensitive, so local people protect their meagre water resources. Bishnois of Rajasthan have protected their Khejdi trees, black buck antilope for several generations. We need a sustainable development that takes care of the desert ecosystem.

16.10.4   Aquatic Ecosystems

The aquatic ecosystems constitute the fresh water systems in ponds, tanks, lakes, rivers, and wetlands. In aquatic ecosystems, plants and animals live in water. Special abiotic features are physical, like quality of water, clarity, salinity, oxygen content and rate of flow. Aquatic ecosystems are classified as 'fresh water ecosystem' and 'marine ecosystem' as detailed in Table 16.3.

TABLE 16.3 Types of aquatic ecosystems

Fresh water ecosystems				Marine ecosystems	
Flowing water		Still water	Brackish water	Saline water	
(LOTIC) Streams	Rivers	(LENTIC) Ponds wetlands lakes	Deltas	Coastal shallows coral reefs	Deep oceans

Pond ecosystem

The simplest aquatic ecosystem is pond of Indian villages. Rain water is stored in ponds, where abiotic substances are water, nutrients, oxygen and CO_2. Producers of the system are, floating algae and large rooted plants. Consumers are insects, frogs, animals and human beings. Decomposers are bacteria and fungi. They utilise organic carbon to generate CO_2 which is used by algae. Algae is eaten by microscopic organisms which are further eaten by small fish. Fish are consumed by prey birds like Kingfishers and herons. Aquatic insects, worms and snails feed upon waste, excreted by animals and decaying plants, so a dynamic equilibrium is maintained in the ecosystem. A lake ecosystem functions like a giant permanent pond.

River ecosystem

Rivers are flowing water ecosystems, where living forms adapt to different rates of flow. Snails and other burrowing animals can survive in rapid flow in hill areas. Water beetles and skaters can live in slow moving water. Mahseer fish need clear water to breed.

Community of flora and fauna of rivers depends on the clarity, flow and oxygen contents besides nature of their beds. Rivers can have a sandy, rocks or muddy bed, each type have its own species of plants and animals.

Marine ecosystem

The Indian ocean, Arabian sea and the Bay of Bengal constitute the marine ecosystems around peninsular India. In the coastal area, the sea is shallow while farther away, it is deep. Both these are different ecosystems. The producers in this ecosystem vary from microscopic algae to large seaweeds. Zooplankton and invertebrates are eatables for fish, turtles and marine mammals.

Shallow areas near kutch and around the Andaman and Nicobar islands are home to rare coral reefs of the world. Coral reefs are only second to tropical evergreen forests in richness of species. Fish, crustacea starfish, jellyfish and the polyps deposite the coral.

Ecologically rich coastal areas of Gujarat

Gulf of Kutch, the largest coral habitat encompasses 1000 km long shoreline having an area of 7350 sq kms. Average depth is 30 m while minimum depth is 5 m around Lushington Island. The Marine National Park and Marine Sanctuary are situated along the southern shore of Gulf from Okha and extends eastwards to Khijadia. This includes 42 islands and a complex of fringing reefs backed by musflats, sandflats, coastal salt marsh, mangrove forests, sand and rocky beaches which support a great diversity of fauna and flora. Detail of site is given in Table 16.4.

TABLE 16.4 Sites of Flora and Fauna

Site	Ecology	Area in km²	Coastal length (km)
Gulf of Kutch	Mangrove	1307 Mangrove	131 Mangrove
	Coral reef	406 Coral	95 Coral main
			75 Coral island
Gulf of Khambat	Estuary	6.4 Mangorve	2.6 Mangrove

Availability of deep channel and proximity to hinterland, the coast line between Jamnagar and Salaya of the Gulf of Kutch is centre of industrial development. Mega industries like Integrated Petrochemical Refinery at Sikka (RPL) and Vadinar (ESSAR, IOC), thermal power project, cement factory and fertilizer factory at Sikka are located.

Operation of these industries pose a severe threat to fragile marine ecosystem of the Gulf of Kutch. Industries are not catering protection measures of the sensitive environment. Now World Bank has stepped in to fund for the development of Integrated Coastal Zone Management in the Gulf of Kutch. It is important to obtain a baseline data on the status of the marine fauna and the ecosystem present today, before it is too late.

16.10.5 Glaciers Ecosystem

Glaciers are masses of ice which move downhill with gravity. They are located in high mountain ranges like Himalayas and Alps. A number of long valley glaciers, located in Indian sub-continent are, Siachen (72 km), Gangotri (26 km), Kanchanjunga (16 km) and Kedarnath (14.5 km). Glaciers shape landscape like valleys, ridges.

Glaciers in Himalayas are melting faster than natural rate. This affects hydrology, i.e., flow of rivers in mountains and our water security.

Institute of Himalayan Geology found that annual rate of recession of Dokriani Glacier in Gangotri Valley was 16.5 m up to 1990, which increased to 17.4 m by 1995. National Institute of Hydrology Roorkee studied Chhatru in Chenab basin receded 54 m a year. The area of two Beaskund glaciers shrunk to half in past 26 years. Conclusion is that small glaciers are receding faster. Factors in the hydrology of streams are, quantum of monsoon rain, sub-surface flows and snowmelt. Glacier water provides flow during post monsoon lean period, prime source of usable water essential for forests, animals, agriculture and human population.

16.10.6 Antarctica Ecosystem

Antarctica is the coldest, the windiest and the driest continent near south pole. Antarctica holds 70% of the earth's freshwater, and 91% of earth's ice. This continent influences the global ocean. Cold, dense and oxygen rich water originate in Antarctica and replenish the ocean's bottom water supply which helps to drive ocean circulation. The sea surrounding Antarctica supports marine life from tiny ice dwelling algae to the great whales.

The north and south poles maintain the heat budget of the world. Heat transported through the atmosphere and the oceans to the poles is dissipated in space in the form of long-wave radiation. Cold air moving from Antarctica when meets warm air in atmosphere of lower latitude, converts into moisture bearing clouds. Thus, Antarctica regulates the global climate, particularly in southern hemisphere.

The Indian, Atlantic and the Pacific oceans meet around Antarctica forming a distinct body of water which girdles the earth. Mixing process between cold and warm water in this body of water demarcates the area of Antarctic convergences which has its own physical, chemical and biological characteristics.

Antarctica provides an unpolluted and stable environment for carrying out scientific observations with global warming and climate change becoming contemporary issues, the Defence Research and Development Organisation (DRDO) has initiated a project to study and correlate weather conditions over Antarctica and the Himalayas.

Over next 5 years, scientists from the Snow and Avalanche Study Establishment (SASE), a DRDO laboratory would proceed to the frozen continent to conduct scientific experiments. IIT Roorkee would also collaborate to develop algorithms and models to retrieve information of snow cover and glaciers from satellite data of varying resolution.

During the project, titled "impact of climate variation on the cryosphere at Antarctica and Western Himalayas", observational data of different satellites will be analysed to estimate physical changes in the snow cover over Antarctica and Himalayas.

Cryosphere refers to the earth's surface where water is in solid form, including sea ice, lake ice, river ice cap and frozen ground. It is an integral part of the global climate system with important linkages, with feedbacks generated through its influence on surface energy, clouds, precipitation, hydrology, and atmospheric and oceanic circulation. Through these, the cryosphere plays a significant role in global climate change.

SASE has already commissioned two Automatic Weather Stations (AWS) on the ice sheet of Antarctica in a radius of 10 km from Indian research station, Madurai. For large spatial coverage of the study area, SASE will establish three more AWS in different snow ice media in Antarctica. For analysis, online data would be transmitted to SASE head quarters Chandigarh, through satellite.

Pacific and Atlantic oceans communicate to both North pole (Arctic) and South pole (Antarctic), while the Indian ocean has its northern boundaries, closed with landmass. It, therefore, communicates only with Antarctic ocean, from where it derives its fertility and energy. Important oceanographic features of Indian ocean are governed by Antarctic ocean as both join together. Antarctic ocean is the richest biological province on the earth. Important organism regulating the simple food chain in the Antarctic water is 'krill' (like shrimp).

Antarctic governance

Antarctica is governed by 'Antarctic Treaty System' (ATS) formed in 1959. India was admitted to ATS in 1983, and started a permanent station, called Dakshin Gangotri for scientific experiments. India built its second station 'Maitri' at Schirmucher Oasis during 1988–89, to accommodate scientists round the year for research. India ratified the environment protocol to the Antarctica Treaty in 1997, to uphold its commitment to preserve the pristine continent.

Antarctic environment is highly prone to impacts of human activities, and has less natural ability to recover the damage. To protect the environmental values, awareness is essential to manage the pollution. This area is under strict surveillance of protocol, on environmental protection to 'Antarctica Treaty' which monitors all changes due to visitors.

16.10.7 Greenland

Greenland ice cap and the swirling seas have posed a puzzle by global warming. Greenland glaciers are speeding their discharge of icebergs into the sea. This influx of fresh water could block North Atlantic current that help to, moderate the weather of Northern Hemisphere.

The most fragile component of the global climate system is in North Atlantic, where the 'Gulf stream' encounters the cold winds coming off the Arctic and across Greenland. As the two collide, heat evaporates from Gulf stream and is swept as steam by prevailing winds as earth's rotation eastward to Western Europe. The heat drawn from the Gulf stream is carried to Europe makes cities like London and Paris much warmer than Montreal, North Dokota, though they are in same latitude. Madrid is quite warmer than New York, though both are located geographically over the same latitude.

16.11 POLLUTION

Pollution is the effect of inconvenient changes in our surroundings that impinge harmful effects on plants, animals and human beings. This occurs when short-term economic gains are made at the cost of long-term ecological benefits. During last few decades, contamination of air, water and soil has increased beyond recovery levels. Concentration of pollutants determines severity of detrimental effects on human health. An average human requires 12 kg of air each day, which is nearly 13 times greater than food we eat. Thus, pollutants in air are more damaging than similar levels present in food. Pollutants that enter water spread fast with its flow on land and marine ecosystem.

16.12 AIR POLLUTION

Air pollution occurs due to the presence of harmful solid or gaseous particles in air, that causes damage to human health and environment. There are five primary pollutants, which contribute about 90% of emission. The earth receives energy from the sun in the form of radiation. It absorbs 70% to warm the land, atmosphere and oceans, remaining 30% is reflected back. This helps life forms to exist and flourish on earth. But gases like CO_2, N_2O, CH_4 absorb infrared radiation going out from earth, thereby increasing temperature across the world. It is known as green house effect (name comes from garden green houses that trap, heat helping plants to grow). Emission of carbon oxides, nitrogen oxide, hydrocarbon are known as green house gases, causing global air pollution. Emission which absorb heat radiation are called "green house gases".

- Carbon oxide (CO and CO_2)
- Nitrogen oxides (N_2O, NO, NO_2, N_2O_3 and N_2O_5)
- Oxides of sulphur (SO_2 and SO_3)
- Volatile organic compounds [CH_4 and chlorofluoro carbons (CFC)]
- Suspended particulate matters
  (size 0.0002 μ to 500 μ, where 1 μ = 10^{-6} m)

16.12.1 Sectoral Contribution to Greenhouse Gases

Energy sector

Electrical energy is the prime necessity for development. According to 2021 estimates of Central Electricity Authority (CEA), 63% of India's installed capacity of thermal powers is 53% coal, 10% gas, 1% oil, hydropower is 12.3% and nuclear power is only 3%, Renewable energy including hydropower contributes 37%. According to plans, the present installed capacity, 388,134 MW is to be increased to 8,00,000 MW by 2031–32. Coal is inevitable with mature technology (like IGCC) options with increased phase of nuclear and renewable technologies.

In industrial sector, iron and steel, cement and aluminium are energy intensive sectors. Need is to assess current technologies and future emission reduction strategies.

Transport sector

Transportation sector is growing and world is losing the battle to control transport emissions, opportunity is to leapfrog from private vehicles to mass transport providing mobility with less pollution.

Agriculture sector

Daily needs of rural India for food, fuel, fodder and fertilizer require biomass products. More than 50% of fuel consumption in India is for cooking. It is met by biomass combustion—firewood, cowdung, leaves and twigs, used desperately by poor to cook and light their homes. These are survival emissions. An action plan is required to reduce poverty and emissions.

Solution is to grow biomass for energy use, biodiesel for buses which reduces emissions and provides biomass based energy for poor households. This will require an increase in our forest cover above 23% to provide livelihood and energy option for the poorest.

Second option is new coal technology of Integrated Gasification Combine Cycle (IGCC). Coal can also be converted into gas or recover gas trapped in coal seams underground. These gases can be burnt cleanly.

16.12.2 Effects of Air Pollution

1. Living organisms are seriously affected as detailed in Table 16.5.

TABLE 16.5 Damage caused by gases

Gases	Damage to human system
CO	Respiratory system breaks, causing bronchitis asthma and lung cancer. Expose of CO for hours can cause collapse, coma and even death as it remain attached to haemoglobin, reducing oxygen carrying capacity of blood.
Nitrogen oxides NO_2	Irritate lungs, aggravate asthma, increases respiratory infections like influenza.
Volatile organic com-pounds like benzene toxic particulates like lead, cadmium and silica	Can cause mutations, reproductive problems or cancer. Inhaling ozone, a part of photochemical smog causes chest pain, irritation to eyes, nose and throat. Inhalation of silica dust causes silicosis.

2. Gaseous pollutants damage the leaves of crop plants causing excessive water loss. It interferes with photosynthesis and plant growth, causes leaves to turn yellow and drop off. Flower buds become stiff, unable to turn in flowers and fall. Food production grossly reduces.
   The other damages caused are as follows:

   - SO_2 bleaches the leaf surface and causes chlorosis (i.e., loss of chlorophyll and yellowing of the leaf) especially in leafy vegetables.
   - NO_2 causes premature leaf fall (abscission) and suppressed growth of plants, resulting in reduced yields of crop plants.
   - Ozone causes necrosis (dead areas on a leaf structure) and damages leaves.
   - Ozone also causes Epinasty (downward curvature of leaf).
   - PAN (Peroxy Acyl Nitrate) damages leafy vegetables causing premature fall, discolouration and curling of sepals.

3. Air pollutants can affect materials by:

   - Corrosion
   - Abrasion
   - Deposition and removal of materials
   - Change chemical properties.

4. Some prominent damages caused to various materials by air pollutants are:

   - Acid rains, due to air pollutants damage the building materials and the historical places like discolouration of white marble of Taj Mahal.
   - Paints are discoloured by SO_2, H_2S and particulates.

- Metals undergo corrosion and tarnishing by SO_2 and acid gases.
- Paper becomes brittle and leather gets disintegrated by SO_2 and acid gases.
- Ozone, SO_2, NO_2 and acid gases deteriorate and reduce the tensile strength of textiles.

16.12.3 Control Measures for Air Pollution

Air pollution can be controlled by preventive techniques. Installing equipments for removal of pollutants from flue gases in power houses and industries.

1. Use dry and wet collectors, filters and electrostatic precipitators (ESP).
2. Provide greater heights to stacks (Chimney) to facilitate discharge of pollutants far away from ground and effective dispersion in atmosphere.
3. Locate industries in such places so as to minimise the effects of pollution after considering topography and wind direction.
4. Substitute raw material that causes more pollution with those that causes less pollution.

A typical diagrammatical representation of air pollution control method is shown in Figure 16.12.

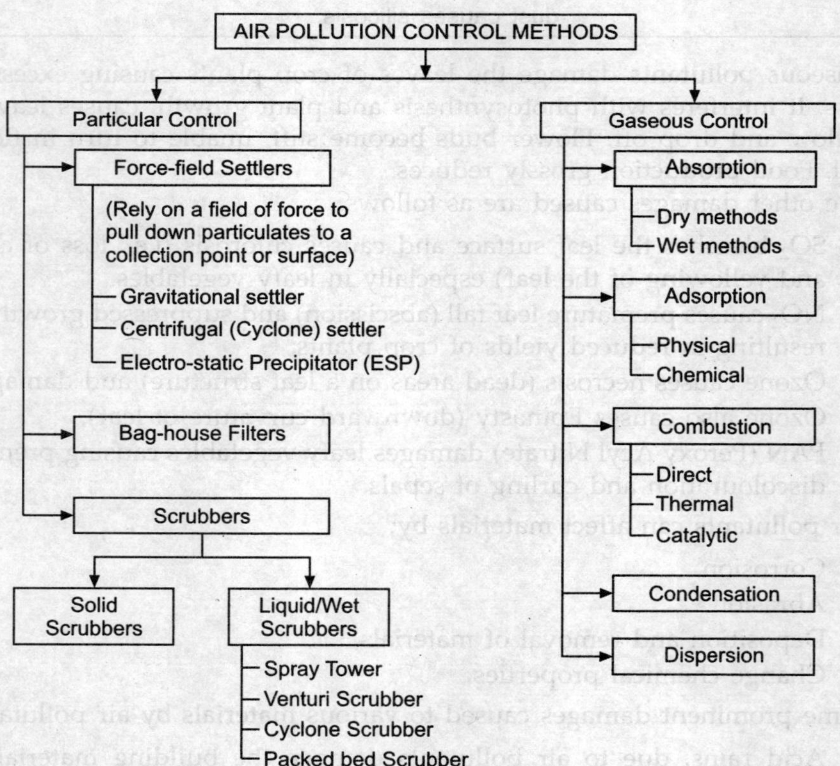

FIGURE 16.12 Typical air pollution control methods.

16.12.4 Indian Approach to Air Pollution

The Air (Provention and Control of Pollution) Act was legislated in 1981. The act provides 'no industrial pollution causing activity can come up without the permission of the 'State Pollution Control Board'. After the Bhopal disaster, a more comprehensive Environmental Protection Act (EPA) was passed in 1986. It conferred enforcement agencies necessary punitive powers to restrict any activity that can harm the environment.

To regulate vehicular pollution, the Central Motor Vehicle Act of 1939 was amended in 1989.

Gravitational Setting Chamber

These are used to remove particles greater than 50 µm in size from polluted gas streams. It consists of huge rectangular chambers, where polluted gas stream with particulates is allowed to enter from one end with very low velocity. The particulates having higher density settle at the bottom of the chamber due to gravitation and are removed (Figure 16.13).

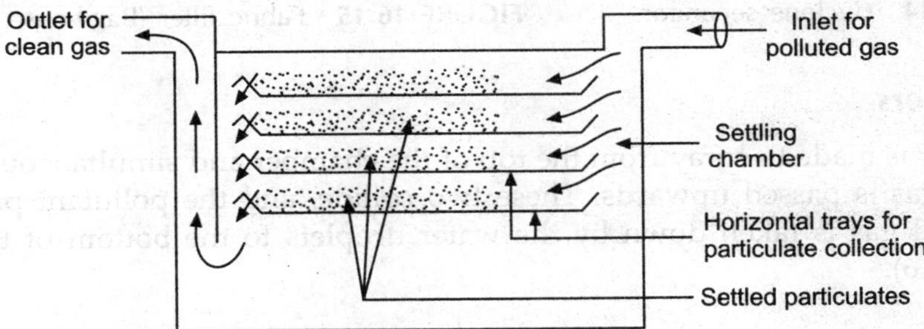

FIGURE 16.13 Gravitational setting chamber.

Cyclone separators

These are based on centrifugal force for the separation of particulate matter from the polluted gas. A simple cyclone separator consists of a cylinder with a conical base. An inlet introduces the polluted gas tangentially which spins inside the chamber. Using the centrifugal force so generated, the particulate matter is separated from the gas, and these slide down the walls of the cone and are discharged from the outlet. Smaller sized particles are also separated (Figure 16.14).

Bag-house filters/Fabric filters

A stream of polluted gas is passed through a fabric, which filters out the particulate pollutant and allows the clear gas to pass through. The particulate are removed from the filter afterwards. A bag-house filter consists of many such filter bags. Its efficiency is more and particulates of 0.5 µm size are also removed (Figure 16.15).

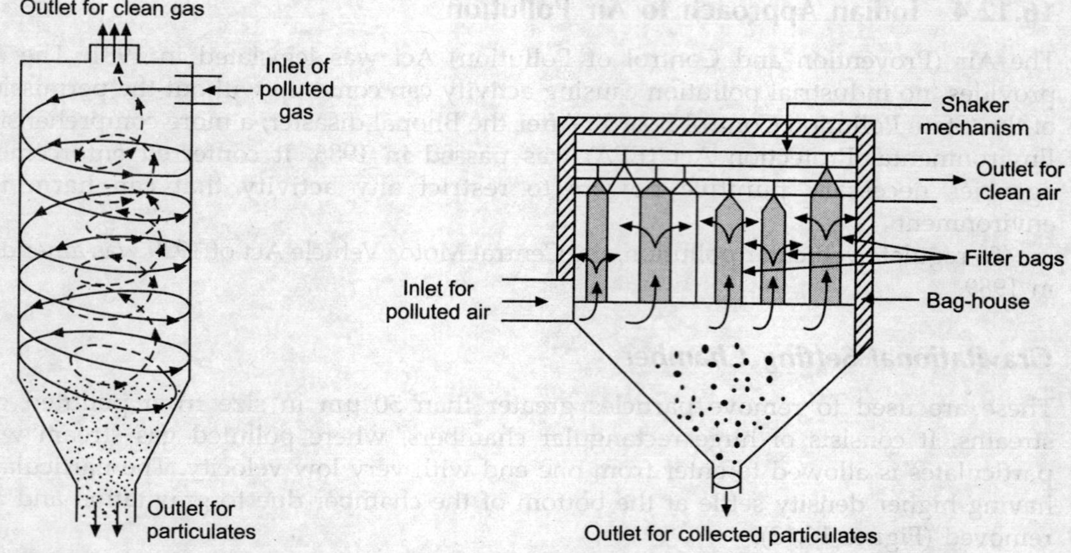

FIGURE 16.14 Cyclone separator.

FIGURE 16.15 Fabric filter (Bag-house filter).

Spray towers

Clean water is made to spray from the top of the chamber and simultaneously polluted stream of gas is passed upwards. These two collide, and the pollutant particulate in the polluted gas is taken down by the water droplets to the bottom of the chamber (Figure 16.16).

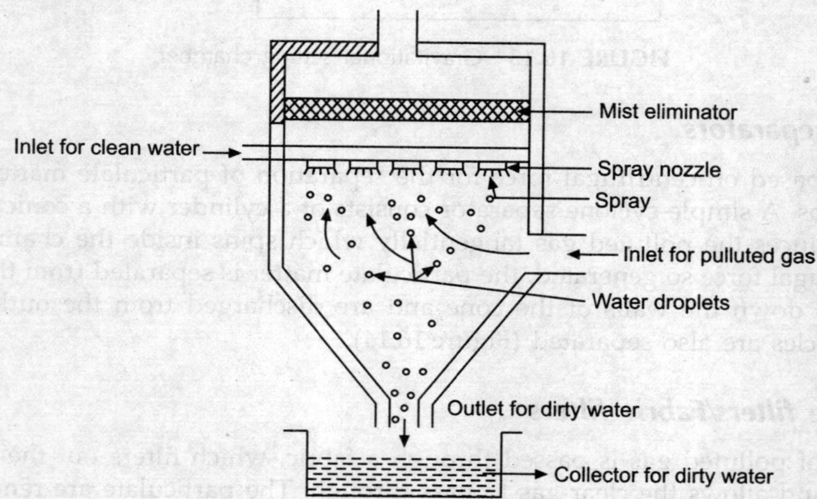

FIGURE 16.16 Spray tower.

Electro-Static Precipitators (ESP)

An ESP works on the principle of electro-static precipitation. The electrically charged particulates present in the polluted gas are separated from the gas stream under the influence of an electrical field.

The polluted gas enters from the bottom, and flows upwards (i.e., between the high voltage wire and grounded collecting surface). The high voltage (50 kV) in the wire ionises the gas. The negative ions migrate towards the positively charged collecting rounded surface, and pass on their negative charge to the dust particles also. These negatively charged dust particles are electro-statically drawn towards the positively charged collector surface, where they finally get deposited. The collecting surface is vibrated periodically to remove the collected dust particles, so that the thickness of the dust layer deposited does not exceed 6 mm, otherwise the electrical attraction becomes weak and efficiency of the electro-static precipitator gets reduced.

The ESP provides advantage of high efficiency (99%). It can be operated at high temperature (600°C) and pressure at less power requirement. Hence, it is simple to operate and an economical option for controlling air pollution. They are widely used in thermal power plants (Figure 16.17).

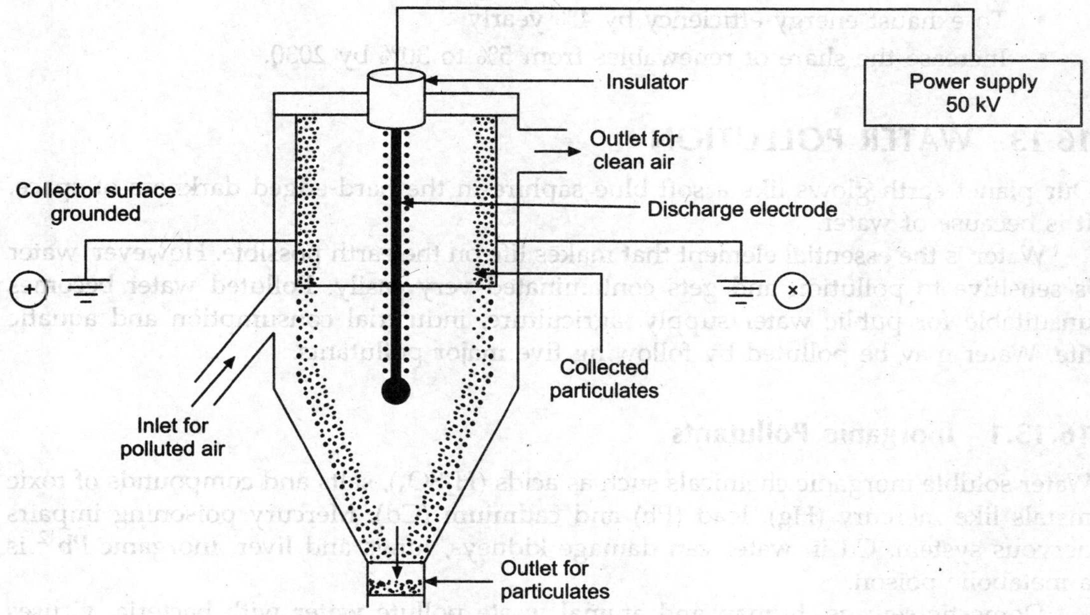

FIGURE 16.17 Electro-static precipitator.

The Central Pollution Control Board (CPCB) initiated its own National Ambient Air Quality Monitoring (NAAQM) programme in 2009 as given in Table 16.6.

TABLE 16.6 National ambient air quality standards

Area category	Concentration* of pollutant in ambient air				
	SO_2 µg/m³	NO_2 µg/m³	CO µg/m³	Small particulate matter (SPM) µg/m³	Lead (Pb) µg/m³
Industrial, Residential, Rural and Other areas	50.8	40.8	02.04	60.1	0.501
Ecologically sensitive area	20.8	30.8	02.04	60.1	0.501

*Data provided above are annual arithmetic mean of minimum 104 measurements in a year.

Data to NAAQM is supplied by the respective state pollution control boards. Air quality management needs an integrated approach as:

- Putting emphasis on pollution prevention rather than controlling by low carbon technologies.
- Reducing use of fossil fuels.
- Improve quality of vehicular fuel with ethanol and biodiesel from 5% to 10%.
- To exhaust energy efficiency by 4% yearly.
- Increase the share of renewables from 5% to 30% by 2030.

16.13 WATER POLLUTION

Our planet earth glows like a soft blue saphire in the hard-edged darkness of space. It is because of water.

Water is the essential element that makes life on the earth possible. However, water is sensitive to pollution and gets contaminated very easily. Polluted water becomes unsuitable for public water supply, agriculture, industrial consumption and aquatic life. Water may be polluted by following five major pollutants.

16.13.1 Inorganic Pollutants

Water soluble inorganic chemicals such as acids (H_2SO_4), salts and compounds of toxic metals like mercury (Hg), lead (Pb) and cadmium (Cd). Mercury poisoning impairs nervous system. Cd in water can damage kidneys, lungs and liver. Inorganic Pb^{+2} is a metabolic poison.

Domestic sewage, human and animal waste pollute water with bacteria, viruses and parasitic worms, causing gastro-intestinal diseases.

High level of these pollutants make water unfit to drink, harm fish and other aquatic life, reduce crop yield and accelerate corrosion of equipments that use this water.

16.13.2   Organic Pollutants

Organic chemicals include oil, gasoline, plastics, pesticides, detergents and fertilizers. When they enter the water bodies, water system degrades due to bacterial activity which consumes dissolved oxygen. The amount of oxygen required by microbes to breakdown a certain amount of organic matter is called '*Biochemical Oxygen Demand*' (BOD). Too much organic matter consumes entire oxygen of water causing fish and aquatic life to die. A layer of oil on water surface blocks sunlight which reduces rate of photosynthesis, causing harm to marine plants. Pesticides like DDT can cause tumour in liver and cancer in other soft organs.

16.13.3   Sediment of Suspended Matter

These are insoluble particles of soil and other solids that remain suspended in water. Soil erosion by fast flowing water in hills and due to deforestation, and suspended particles of organic and inorganic matter in water. This reduces photosynthesis activity of aquatic plants and algae, thus disrupting ecological balance of water bodies. When water velocity in the rivers decreases, suspended particles settle down at the bottom as sediments. Excessive sediments destroy feeding and spawning grounds of fish and clogs, also causes undesirable filling of lakes and man made reservoirs.

16.13.4   Thermal Pollutants (Heat)

Power plant and industries that use large volumes of water to cool the plant dispose warm left over water into water bodies resulting in rise in temperature of local water bodies. Such warm water reduces solubility of oxygen in water, causing adverse effects on aquatic flora and fauna.

16.13.5   Water Soluble Radio Isotopes

Nuclear power plants discharge traces of radioactive substances in water bodies. Other sources of pollution are, radioactive isotopes in medicine, agriculture and industry. Mining and processing of uranium besides testing of nuclear weapons add to radioactive pollution. Radioactive pollution passes through food chains and food webs. Ionising radiation emitted by such isotopes can cause both defects, cancer and genetic damage.

16.14   GROUND WATER DEPLETION

Besides water pollution, there is a serious ground water shortage in northern India. The US space agency NASA's findings are based on 'Gravity Recovery And Climate Experiment' (GRACE) with twin satellite, which can sense tiny changes in earth's gravitational field and associated mass distribution of water masses, stored above and below the earth's surface. NASA has warned that in Punjab, Haryana, Delhi and Rajasthan, water is being pumped and consumed by human activities, mainly to irrigate cropland, faster than the aquifiers could replenish by natural process. Ground water levels have been declining by an average of one metre every three years. More than 109 cubic km (26 cubic miles) of ground water had been disappeared between 2002 and 2008.

Ground water comes from natural percolation of precipitation and other surface water down through earth's soil and rock, accumulating in aquifers, cavities and layers of porous rock, gravel, sand or clay.

Ground water levels do not respond to changes in weather as rapidly as lakes, streams and rivers do. So, when ground water is pumped for irrigation or other uses, recharge to the original levels can take months or years. However, changes in underground masses affect gravity enough to provide a signal, which can be translated into a measurement of an equivalent change in water.

16.14.1 Case Study of Ground Water in Haryana

Ground water engineers conducted study in the districts of Haryana during 2009, and found that average depth of ground water is 15.66 m. Main reason is large rice sowing area, which requires 4500 litre water to produce one kg of rice. Ground water depth for few districts is in Table 16.7.

TABLE 16.7 Ground water depth of few district in Haryana

District	Depth in metre
Kurukshetra	29.66
Karnal	16.25
Panipat	15.17
Faridabad	13.10
Gurgaon	23.61
Rewari	21.90
Mohendragarh	44.0
Bhiwani	21.78
Fatehabad	17.50

Exigency of the situation dictates that sustainable ground water usage be taken on priority.

16.15 SOIL POLLUTION

Soil formation takes place from rocks by climatic factors like rain, wind, temperature and chemical weathering activities. Soil contains clay, sand silt, animal waste, fungi and decomposed organic matter. For one inch of top soil to be formed it requires 200–1000 years depending upon climate and soil type.

A useful property of soil is its fertility, which is important for agriculture and forestry. Soil can absorb plant and animal wastes, industrial and municipal discharges. Soil is a renewable resource, but gets polluted when amount of wastes is too large, and loses its fertility. Soil is damaged due to several activities, discussed with remedial measures as:

- Municipal Solid Waste (MSW) is disposed of in plastic bags. Dumping and burning of waste is not an acceptable practice, as it pollutes air and soil. Modern methods of MSW disposal are incineration and development of sanitary landfills.

- Industrial waste (including bio-medical waste) i.e., hospital waste is a big health hazard as it pollutes soil. Fly-ash from thermal power plants, is a major industrial waste, which pollutes soil adversely. Now, industrial waste is treated by integrated waste management facilities comprising a secured land facility, waste stabilization facility, incinerator and leachate treatment facility. Hospital waste treatment constitutes an incinerator autoclave; shredder and landfill. Fly-ash is utilised to produce cement and bricks as building material.

- Pesticides are chemicals and are used to boost agricultural output. India's per capita consumption of pesticides is 0.5 kg. Pesticides are absorbed by the soil and pollute it. Besides killing pests, these chemicals damage other living things including humans. Pesticides reach to human body through cereals, vegetables, fruit and milk. Solution is to keep soil healthy, by adding plenty of compost. Grow plants in soil rich in humus. Humus is the organic matter obtained from decaying plant and animal remains. It binds the soil, that checks soil erosion.

- Excessive use of nitrogenous fertilizers damage the environment. Emission of nitrous oxide from fertilizers has 296 times more global warming potential than CO_2. Nitrogen oxides are an important component of acid rain. Nitrate in drinking water affects reproduction and development of human beings. Need is to hammer out rules for right combination and more varieties of fertilizers.

16.16 GLOBAL CLIMATE CHANGE

Inter-governmental Panel on Climate Change (IPCC) in report of 2007, indicated that Green House Gas (GHG) emissions caused climate change. In 21st century, term 'global' means changes in earth's atmosphere, CO_2 increased by 31% CH_4 by 151%, N_2O by 13% and ozone by 31%. Though percentage of these gases in terms of nitrogen and oxygen is only 1%, it is the increase in these which causes drastic changes in earth's ecology. There has been temperature change of about 0.8°C per 100 years but during last 40 years, change witnessed double the rate.

16.17 CLIMATE CHANGE

In its article 1, UNFCC defines climate change as, "A change of climate which is attributed directly or indirectly to human activities, that alters the composition of the global atmosphere and which is in addition to natural climate variability observed over comparable time periods".

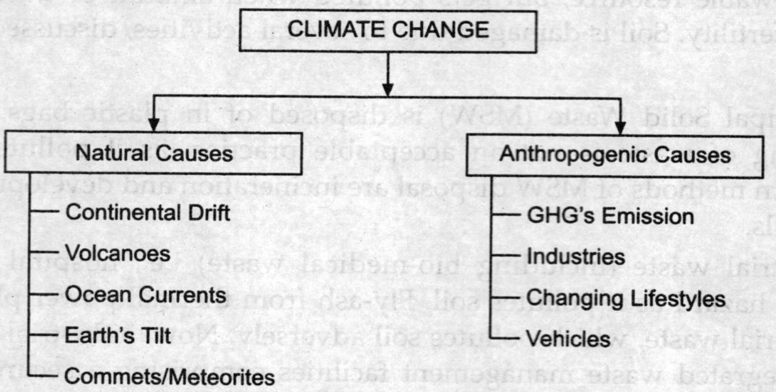

FIGURE 16.18 Causes of climate change.

Sources: Activities that Generate Green House Gases.
Sinks: Activities that remove Green House Gases. Some natural sinks
are: Forests, Oceans and Soil.

16.17.1   Climate Change vs. Global Warming

Many people use the term 'global warming' and 'climate change' interchangeable. These are two distinct concepts. Climate change is a broader term that refers to long term change in climate i.e., average of 3 decade 'weather'. Latest report of IPCC guides to keep CO_2 concentration between 350–400 ppm and total concentration of GHG in the range of 450–500 ppm to keep global mean temperature increase between 2°C and 2.4°C.

Global warming refers to the increase of earth's average surface temperature and lower atmosphere due to increase in concentration of water vapour, CO_2 (55%), CH_4 (15%), NO_2 (6%) and CFCs. This is known as 'green house effect'. Global warming is one of the main causes of changes in climate, diagrammatic representation is in Figure 16.19.

If the threat of global warming persists, the entire climate pattern of the planet will change.

Emission intensity

It is evident that the less the fuel used, the lower will be the green house gas emissions per unit of GDP. This is known as 'Emission Intensity'. Industrialized countries attained efficiency with economic growth. All climate change mitigation scenarios, bank on increasing energy efficiency.

Transport sector contributed 29% of world's green house gas emission during the year 2019. In 2019, nearly 35% of oil consumption consumed in running about 1.4 billion automobiles in the world.

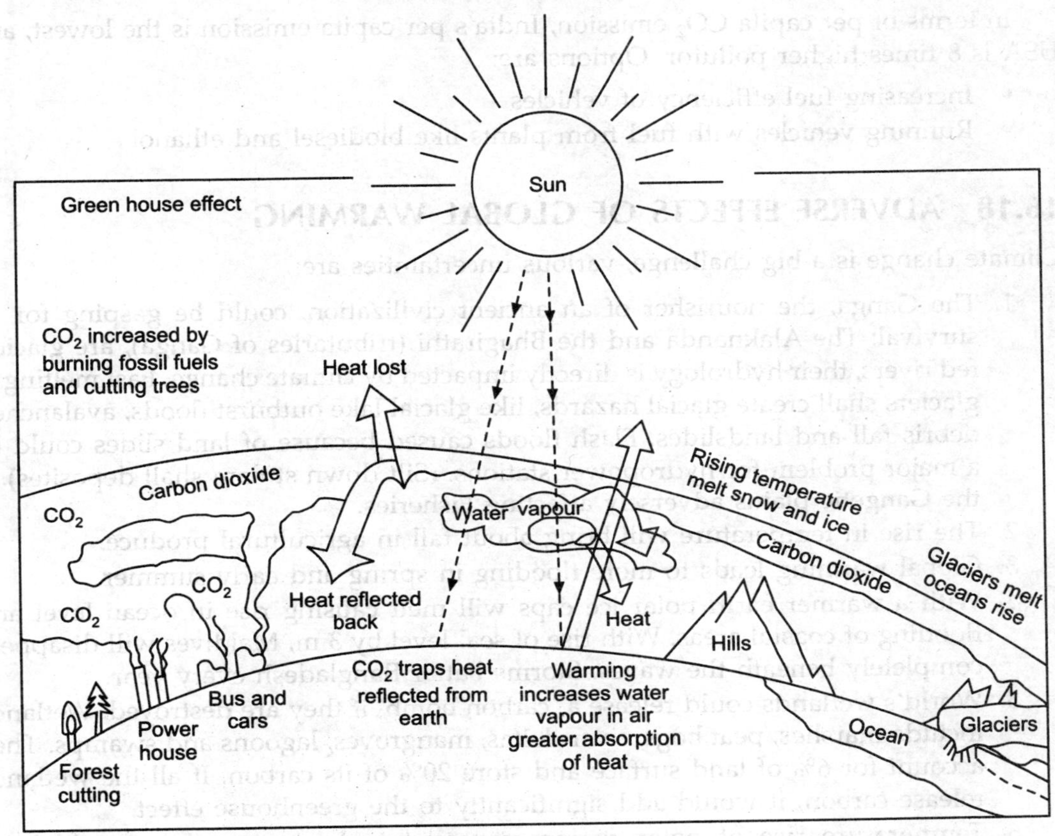

FIGURE 16.19 Global warming.

In 2019, over 29% of the world's transport related emission wave from the US, which is the largest car market, while vehicles are made in India, China, Brazil and Indonesia. CO_2 emissions of six countries during 2019 as per UNFCC report are in Table 16.8.

TABLE 16.8 CO_2 Emissions–2004

No.	Country	Total CO_2 Emissions (Billion Metric Tons in 2004)
1	USA	5.107
2	Japan	1.153
3	Germany	0.702
4	Russia	1.792
5	China	11.535
6	India	2.597

In terms of per capita CO_2 emission, India's per capita emission is the lowest, and USA is 8 times higher pollutor. Options are:

- Increasing fuel efficiency of vehicles
- Running vehicles with fuel from plants like biodiesel and ethanol

16.18 ADVERSE EFFECTS OF GLOBAL WARMING

Climate change is a big challenge, various uncertainties are:

1. The Ganga, the nourisher of an ancient civilization, could be gasping for its survival: The Alaknanda and the Bhagirathi (tributaries of Ganga), are glacier-fed rivers, their hydrology is directly impacted by climate change. Fast melting of glaciers shall create glacial hazards, like glacial lake outburst floods, avalanches, debris fall and landslides. Flash floods caused because of land slides could be a major problem for hydropower stations. (Silt down stream shall deposites) in the Gangetic plains adversely affecting Fisheries.

2. The rise in temperature will bring about fall in agricultural produce.

3. Global warming leads to more flooding in spring and early summer.

4. With a warmer earth polar ice caps will melt causing rise in ocean level and flooding of coastal areas. With rise of seal level by 3 m, Maldives will disappear completely beneath the waves. Storms batter Bangladesh every year.

5. World's wetlands could release a 'carbon bomb' if they are destroyed. Wetlands include marshes, peat bogs, river deltas, mangroves, lagoons and swamps. They account for 6% of land surface and store 20% of its carbon. If all the wetlands release carbon, it would add significantly to the greenhouse effect.

6. Temperature rise of polar region, caused by global warming would have disastrous effects. Vast quantities of CH_4 are trapped under the frozen soil of Alaska. When the permafrost[*] melts, the methane that will be released can accelerate the process of global warming.

7. A paradoxical effect of global warming is that, it produces more evaporation from oceans to fill warmer atmosphere with increased moisture, further it also sucks more moisture out of the soil. Consequently, desertification is increasing in the world. Table 16.9 shows the global impact measured in km^2 per year. The latest figures are significantly worse.

TABLE 16.9 World's desertification

Year	Desertification in world in km^2
1970	1597
1980	2150
1990	3517

[*] Permafrost is a thick layer of soil beneath the surface that remains frozen throughout the year.

    Global warming is critical to crop outcomes, shall reduce wheat production and yields of corn, soyabeans and cotton.

8. The National Oceanic and Atmospheric Administration summarised research studies. As water temperature goes up, so does storm moisture condensation causing disturbances in oceans and generate hurricans.[†] In 2006, one-third of the Guff's oil producing facilities remained crippled due to devasting effect of hurricans.

9. Global warming is linked to great increase in intensity, duration and frequency of hurricans.

10. An estimated 40 million people have been driven out of their homes by rising seas, falling rain, desertification and other climate driven factors in 2020. It is a displacement crisis to new heights.

16.19 SENSITIVITY, ADAPTABILITY AND VULNERABILITY

16.19.1 Sensitivity

The degree to which a system will respond to a change in climate conditions is called its sensitivity.

16.19.2 Adaptability

The degree to which adjustments are possible in practices, processes, or structures of systems to projected, or actual changes of climate is termed as its adaptability. Adaptation can be spontaneous or planned, and can be in response to or in anticipation of changes.

16.19.3 Vulnerability

The extent to which climate may damage or harm a system is called its vulnerability. It depends upon a system's sensitivity and ability to adapt new climatic conditions.

16.20 PROMINENT CLIMATE CHANGE, VULNERABILITY AND IMPACTS IN INDIA

16.20.1 Agriculture

Among the cereals, wheat production potential in the sub-tropics is expected to be affected the most, with significant declines anticipated in several regions including South Asia (IIASA, 2002). For e.g., wheat yields in Central India may drop by 2% in a pessimistic climate change scenario (GOI 2004). Districts in Western Rajasthan,

---

[†] Hurricans, Cyclone and Typhoons are all the same weather phenomena, depending on the ocean in which they originate.

Southern Gujarat, Madhya Pradesh, Maharashtra, Northern Karnataka, Northern Andhra Pradesh, and Southern Bihar are highly vulnerable to climate change in the context of economic globalization. Numerous physical (e.g., cropping patterns, crop diversification, and shifts to drought/salt resistant varieties) and socio-economic (e.g., ownership of assets, access to services, and infrastructural support) factors come into play in enhancing or constraining the current capacity of farmers to cope with adverse changes (TERI 2003).

Temperature rise of 1.5°C and 2 mm increase in precipitation could result in decline in rice yields by 3% to 15%. Sorghum yields are predicted to vary from +18% to –22%, depending on a rise of 2°C to 4°C in temperatures, and increase by 20% to 40% of precipitation (IPCC 2001).

16.20.2 Water Resources

Increased glacial melt due to warming is predicted to affect river flows. Increased warming might result in increased flows initially with reduced flows later as the glacier disappears. Available records suggest that Ganga glacier is retreating by about 22 metre/year. Warming is likely to increase melting far more rapidly than accumulation (IPCC 1998).

Climate change could impact the Indus river basin. The total annual run-off from the copper basin is likely to increase by 11% to 16%. It is estimated that although increased run-off could be advantageous for water supply and hydropower production, it could aggravate problems of flooding, water logging, a salinity in the upper basin.

According to UN projections, India is estimated to experience water stress by 2025 and is likely to cross the 'water scarce' benchmark by the year 2050, under high growth scenario.

(NB: Water stress and scarcity are defined as situations, where per capita annual water available is less than 1700 m^3 and 1000 m^3 respectively.)

16.20.3 Human Health

Changes in climate may alter the distribution of important vector species (e.g., mosquitoes), and may increase the spread of disease to new areas that lack a strong public health infrastructure. High altitude, populations that fall outside areas of stable endemic malaria transmission may be particular vulnerable to increase in malaria, due to warming climate. The seasonal transmission and distribution of many other diseases transmitted by mosquito (dengue, yellow fever), and by ticks (Lyme disease tick-born encephalitis), may also be affected by climate change (GOI 2004).

Climate change due to global warming is the biggest economic and political issue. CO_2 emissions are directly linked to economic growth. Control of climate change needs international cooperation. If rich world pumped excessive quantities of CO_2 yesterday, the emerging rich will do it today. Solution is to build controls for fairness and equity for the biggest cooperative enterprise.

1. **Consume less energy:** The Bureau of Energy Efficiency (BEE) rates building on a scale of 1 to 5, depending on how efficiently they use energy. Lower the energy consumed, higher the star rating. For example, RBI headquarter in Delhi got 4 starts, as it uses 90–115 kWh/sq m energy in a year. To get 5 star it must lower consumption to less than 90 kWh/sq m. In Australia energy efficiency rating is mandatory for all buildings. For Europe buildings of 1000 sqm or more need building energy rating certificates.

2. **Shrink your carbon foot prints:** An individual can reduce climate change by shrinking his own carbon footprint. We make emissions by our daily life style. Average annual carbon footprint of an Indian is only 0.56 tonnes while for a person living in industrial nations is 11 tonnes, though world wide average is 4 tonnes.

   Emissions depend on annual consumption of energy, like electricity, heating oil, coal, gas, wood and LPG in household. It covers travelling in kms by road, rail or aircraft. A house of two persons if consumes 500 units of electricity, 2 litre of heating oil, one cylinder of LPG per month will have carbon footprints of 6.3 tonnes. Driving 2000 km by a small car add 3.9 tonnes, 1000 km by journey, 5000 km rail journey, 3000 km taxi journey will add 0.93 tonnes. Two trips by air from Delhi to Bangalore will add 0.36 tonnes. This makes a total of 11.5 tonnes.

   Most homes or business establishments can reduce energy use by 10% which will result 10% reduction in GHG emissions.

3. **Transition to renewables:** If low-level polluters can trade their unused emission rights with high-level polluters, an incentive is provided to keep their emissions as low as possible. Additionally, emission trading shall promote transition to renewable energy technologies which are zero carbon energy.

4. **Urban areas use large quantities of energy:** Urban housing use materials like burnt bricks cement, steel, marble, glass, aluminium and iron which are high energy intensive. The process of extraction, refinement, fabrication and delivery are all energy consuming, which add to pollution of the earth, air and water.

   Urban areas in hot climate need energy for cooling. Prevalent system of fans changing into air-conditioning, which consumes lot of energy. We Indians are initiating western style of covering large areas by glass sheets. It is good in cold climates, where glass uses the green house effect to trap warmth of the sun which adds several degrees temperature inside. In India large areas covered by glass need central air conditioning making the house energy intensive. Buildings need to be made energy efficient to reduce CO_2 emissions.

5. **Use low energy consuming kitchen appliances:** Kitchen appliances are gas stove, oven, toaster mixer, refrigerator, water purifier lights and fans. Divide the appliances and products into survival and luxury items, because every person is entitled to a basic minimum level of GHG emissions to survive. Fuel used to cook contributes to survival emissions. But heaters, air-conditioners and imported glassware are luxury emissions. As a member of a family, reduce contribution of CO_2 load in the atmosphere.

6. **Design a cheap biomass-based chulha:** According to 2006 International Energy Agency data, nearly 13% of world's primary energy supply is renewable. It constitutes solar, wind, geothermal and cogeneration just 4%, while hydro-electricity 16%. The bulk 80% of renewable comes from biomass burning, from the chulhas of poor-families. These families live on margins of survival, already vulnerable to climate change impacts, that are in the renewable energy net. Now, exigency of climate change is to invent a cheap biomass-based chulha, that can be sold to millions of diverse households across the world.

7. The principles of contraction and convergence—the rich to reduce while the poor to grow and equal per capita entitlement—living within limits would be the basis of future global agreement.

16.21 GLOBAL WARMING POTENTIAL

Global Warming Potential (GWP) is defined as the cumulative radiative forcing, both direct and indirect effects over a specified time horizon, resulting from the emission of a unit mass of gas related to a reference gas.

GWP provides a quantified measure of the relative forcing impacts of a particular greenhouse gas.

Direct radiative forcing: Direct radiative forcing occurs when the gas itself is a greenhouse gas.

Indirect radiative forcing: Indirect radiative forcing occurs when chemical transformations involving the original gas or gases that are greenhouse gases, or when a gas influences the atmospheric lifetimes of other gases.

16.21.1 Representation of Greenhouse Gas Estimates

The estimates of greenhouse gases are represented in units of Million Metric Tons of Carbon Equivalents (MMTCE), which weighs each gas by its GWP value, or Global Warming Potential.

Carbon comprises 12/44th of CO_2 by weight. The following equation is used to convert emissions reported in teragrams (T_g) of a gas to MMTCE:

$$\text{MMTEC} = (T_g) \text{ of a gas} \times \text{GWP} \times \frac{12}{44}$$

where

T_g = Teragram (equivalent million metric tons)

$\frac{12}{44}$ = Carbon to CO_2 molecular weight ratio

GWP = Global warming potential

16.21.2 Calculation of Global Warming Potential

The GWP of a greenhouse gas is mathematically calculated as:

$$GWP_g = \frac{\displaystyle\int_0^T F_g \cdot T_g(t)\, dt}{\displaystyle\int_0^T F_{CO_2} \cdot R_{CO_2}(t)\, dt}$$

where

F_g = Radiative forcing efficiency of the concerned gas $(W/m^2)/kg$.

F_{CO_2} = Radiative forcing efficiency of CO_2 $(W/m^2)/kg$

$R_g(t)$ = Fraction of 1 kg of the concerned gas remaining in the atmosphere at time t

R_{CO_2} = Fraction of 1 kg of CO_2 remaining at time t

T = The time period for cumulative effects (in year)

16.21.3 Calculation of Total Amount of Carbon in the Atmosphere

A very important relationship between ppm of CO_2 and Giga tonne of carbon (GtC) is:

$$1 \text{ ppm } CO_2 = 2.12 \text{ GtC}$$

As an illustration, the global average CO_2 concentration in 1990 was 354 ppm. This means, the total amount of carbon in the atmosphere at that time was,

$$354 \text{ ppm} \times 2.12 \text{ GtC/ppm} = 750.48 \text{ GtC}$$

Similarly, in 2006, the global average CO_2 concentration rose to 380 ppm. This converts to 380 ppm × 2.12 GtC/ppm i.e., 805.6 GtC.

16.22 FOREST RESOURCES OF INDIA

Forests in India are of two types namely:

1. **Coniferous forests** that grow in Himalayan and mountain region, where temperature is low. Plants are like pine and decodar are found here.

2. **Broad leaved forests**, which are of four types:
   (i) Evergreen forest grows in high rainfall areas. Trees are pine, decoder which form continuous canopy.
   (ii) Deciduous forests found in moderate rainfall areas. Trees are teak and sal.
   (iii) Thorn forests found in semi-arid regions. Trees are babul, ber, neem which are sparsely distributed.
   (iv) Mangrove forests grow along coast and in river deltas. Trees grow in muddy area like avicenia.

Since the Forest Conservation Act (1980) came into existence, all projects need prior approval of the 'Ministry of Environment and Forests' (MoEF) through the state forest department. The forest land diverted to various sectors during 1981–2007 is in Table 16.10.

TABLE 16.10 Forest land diverted to various sectors in 2019

Department	Area (in Ha)
Defence	1208.72
Hydel	1107.30
Irrigation	4287.50
Mining	3846.09
Rehabilitation	170.58
Road	1487.83
Thermal	49.83
Transmission line	578.67
Wind power	18
Total	11467.83

Source: Union Ministry of Environment and Forest.

The forest area has been reducing since mid-twentieth century, and is now 24.62% (in 2021) of India's geographical area (640134 km as per TERI 1991). National Forest Policy 1988 depicts forest cover 33% of total geographical area.

16.22.1 Forest Functions

(i) Watershed protection i.e., to reduce rate of run-off water, prevent flash flood and soil erosion.

(ii) Atmospheric regulation i.e., absorption of solar heat, maintaining CO_2 level to control climate.

(iii) Consumptive use of forest produce i.e., fodder, fuel wood, timber, fruits, honey and medicinal plants.

(iv) Market use of products for industrial use i.e., paper, timber for furniture and construction, fibre for baskets, poles or buildings, silk for clothing.

16.22.2 Forest Conservation Act

Forest Conservation Act (1980) was amended in 1988, to empower government to create (a) Reserved forests, (b) Protected forests and (c) Village forest. Penalties for effluences in reserved and protected forests are imprisonment up to six month, or a fine of ₹ 500 or both.

To check deforestation, Environmental Protection Act of 1986 was passed to create an instrument to deal the Environment Impact Assessment (EIA) for all development projects. The EIA is done by a competent organisation who look into physical, biological and social parameters. Projects of industries, roads, railways and dams affects the lives of local people. This must be adversed in the EIA.

16.22.3 Biodiversity Conservation

Most of us complain about deteriorating environmental situation in our country. Our ecofriendly activities can make an improvement. A famous dictum is to think globally and act locally to improve your own environment. Remember, we are a part of a complex web of life, and our existence depends on the integrity of 1.8 million species of plants and animals on the earth that live in several ecosystems. Following are few activities that one can do to ensure ecological security and biodiversity conservation.

- Plant more trees of local species around your home and work place.
- If urban garden is small, plant local shurbs and creepers to support bird and insect life.
- Prevent cutting of trees: Participate in preservation of greenary by planting, watering and caring for plants.
- Participate in events that highlight need for creating sanctuaries and national parks, nature trails, and saving forests.
- Learn and identify birds that are common in your area. Understand their food requirement and support their survival.
- Enrich your garden by compost and manure from poultry and cows. Healthy soil grows healthy plants.

16.22.4 Conserve Energy

Crores of rupees are being spent to extract, process and distribute coal, petroleum and electricity. At an individual level, every one of us should try to conserve energy. Following are few things that one can do to conserve energy and save the environment.

- Turn off light fans and air conditioning when not necessary.
- All incandescent lamps should be replaced with compact fluorescent lamps (CFLs). It reduces electricity bill with higher lumen.
- Use a pressure cooker to save energy.
- All electric geysers should be converted to solar gysers. They are cost efficient and also save energy.
- Minimal use of cars should be promoted. Smaller distances be travelled by public transport. For 8 km distance it will same 1.5 kg CO_2. For a short trip, use a bicycle which is pollution free.
- Select a light shade of paint for walls and ceilings, as it will reflect more light and reduce electrical consumption.

TABLE 16.11 Major UN conferences on environment

Year	Location	Theme/Title	Issues
1972 (June 5–16)	Stockholm	UN conference on human environment	Environmental degradation that affects quality of human life
1985	Vienna	Vienna convention	Protection of ozone layer
1989 (January 1)	Montreal	Montreal protocol	Total elimination of Ozone Depleting Substances (ODS)
1989 (March)	Basal	Basal convention on control of transboundary movements of hazardous wastes and their disposal	Minimise generation of hazardous wastes, protection of human health, disposal as close as possible.
1990 (November)	Geneva	Geneva convention	Access to technology and financial resources to help developing countries.
1992 (May)	New York	UN convention on climate change	Economic developments and environmental protection
1992 (May)	Nairobi	Biodiversity convention	Preservation of the earth's biodiversity
1992 (June 3–14)	Rio de Jaineiro	UN Conference on Environment and Development (UNCED)	Environment and sustainable development
1997	New York	Earth Summit (+5)	Sustainable development at local, national and global levels
1997 (Dec. 1–11)	Kyoto	Kyoto protocol	Stabilization of green house gases
2002 (26 Aug.–4 Sep.)	Johannesburg	World summit on sustainable development (+10)	Accelerate implementation of agenda 21, development of north/south partnership, international solidarity
2007 (Dec. 3–15)	Bali (Indonesia)	UN climate change conference (COP 13)	Agreement on a timeline and structured negotiation on post 2012 framework.
2009 (Dec. 7–18)	Copenhagen (Denmark)	UN climate change conference (COP 15)	Establishment of global climate agreement for the period from 2012, when the first commitment period under Kyoto protocol expires.
2012 (June)	Rio de Janeiro	UN conference on sustainable development	"The future we want"
2015 (September)	New York	UN sustainable development Summit	"Transforming our world: the 2030 Agenda for sustainable development"
2015 (December)	Paris	UN Framework Convention on climate change	Paris Agreement was adopted
2019 (September)	New York	UN climate action summit	

16.23 WATER MANAGEMENT IN INDIA

For India, out of total annual precipitation of about 4000 Billion Cubic Metres (BCM), annual availability of water is nearly 1120 BCM. This provides for an annual per capita availability of less than 1000 cubic meter, which is considered a slots of scarcity. With this annual quantity, domestic needs water for food production, industrial activities and cultural requirements are to be met. Irrigation engineers in India have exhausted many rivers and aquifers, leaving little water for satisfying needs of environmental flows for the aquatic ecosystems.

Ownership of ground water being treated as private, totally unsustainable extraction of water for industry and irrigation has deprived large areas of Peninsular India of water from accessible aquifers. No state can survive in absence of water, lack of it will lead to water mass.

16.23.1 Remedial Action

To ensure that vital ecosystem processes and related services may continue, attention to ecosystem research and ecosystem management of water system is needed.

Case Study I

Immediate action is to revive age old technique of making 'Johads', sort of a catchment area that holds rainwater. In Rajasthan, people participation created several Johads, the rainwater stored in those helped the arid land to get back its moisture, which ultimately filled up wells and rivers of the place. It rejuvenates nature and now people fetch all their water from village wells.

Case Study II

Another remedy used in Punjab 'Preservation of Sub Soil Water Act' has led to rise in water table in north Punjab and also reduced electricity consumption. This Act, prohibits sowing of paddy nursery before May 10 and transplantation before June 10. Discipline imposed by this Act preserved ground water, essential for sustainable agriculture, also saved power required for pumping ground water. This activity has increased the production, yield and procurement of rice.

Case Study III

Third activity is 'Yamuna Campaign' of Delhi which started in the year 2000. It has infused, more public participation and ownership on environmental issues, as large number of youth groups at college and schools have taken initiatives to green their surroundings.

Situation in Delhi is alarming; as 70% of Delhi's drinking water comes from Yamuna, while 3500 million litres of city's sewage flows untreated into the river. Yamuna's stretch in Delhi is 22 kms only, but city contributes more than 80% of total pollution in the river. Immediate steps needed are:

- Establish community sewage treatment plant to ensure less polluted water into Yamuna.

- Stop dumping of religious waste in it.
- Ensure that drains in each locality are connected to other larger drains.
- Conserve rain water.

16.23.2 Joint Action of Corporate Sectors

Soldiers of the earth: The 'Soldiers of the Earth' programme, a joint initiative of ONGC, TERI and NDTV is a movement that aims to inspire, sensitive and educate students for sowing seeds for a greener planet:

- Promoting sustained growth and development by harnessing clean, natural, renewable energy sources, as a 50 MW wind power plant in Gujarat Commissioned.
- Endeavouring towards carbon neutral company as 4 Carbon Development Mechanism (CDM) projects registered with UNFCC.
- Cover green coastal soil erosion control project, as plantation of 5 lakh mangroves along the Gulf of Cambay.
- Afforestation project in Uttarakhand, i.e., to plant 3 lakh Ringal Bamboo plants.

16.23.3 Clarion Call to Address Climate Change

Two resources, air and water are life sustaining. Human activities have polluted both, that has threatened life. We need to clean them on war footing. About 3,00,000 people die from climate related changes annually. Worst hit are those, who live on the margins. To face the situation, we need the following:

- Green energy investment in developing countries to be increased.
- As per United Nation Environment Programme (UNEP), a minimum of \$750 billion is needed to finance a sustainable economic recovery in 5 key sectors – buildings, energy, transport, agriculture and water of the global economy. Cost should be borne by those directly responsible for the crisis.
- Industrialised nations to make available expertise in technology and aid the developing countries to grow their economies.
- Reduce emissions from present 450 ppm heat trapping CO_2 and other GHGs blanket, surrounding the planet to 350 ppm to contain 2°C temperature, increase from 1990 levels. It will require huge conversion to renewable energy system.
- Use of recycled water should be made mandatory. Reduce effluents to 30 mg/l BOD for irrigation, 5 mg/l BOD for human use.
- Increase forest cover to 33% of the land area. At present, it is 19% only.

Global climate change may appear gradual in the context of a single life time, but on the scale of the earth's history it is happening at considerable speed and it is accelerating so fast that we are looking at the bubbles of a boiling pot. Survival is at stake, but we have ability to rescue ourselves.

16.24 ECOLOGICAL SUCCESSION

Definition: An orderly and progressive replacement of one community by another, till the development of a stable community in that area, is termed ecological succession.

Characteristics:

1. It is a gradual and orderly progress towards a stable community.
2. The seral stages in ecological succession are so regular and directional that by studying a particular seral community, the future communities and their sequence can be predicted.
3. Both the number and species of organisms change at each successive seral stage.
4. The succession of flora and fauna communities occurs simultaneously.

16.24.1 Stages in Succession

- The entire series of communities is known as sere and the individual transitional communities are termed as seral communities or seral stages.
- The first community to inhibit an area is called Pioneer community. The succeeding communities are called transitional communities, and the last stable community is called the climax community.

16.24.2 Kinds of Succession

1. **Primary succession:** When succession begins on a previously unoccupied area by a community, it is known as Primary Succession. E.g., an exposed rock area, new islands, etc.
2. **Secondary succession:** When community development is proceeding in an area from which a community was removed and where conditions for existence are already favourable, it leads to secondary succession. This kind of succession occurs rapidly, since some species are already present. Moreover, a previously occupied territory is more condusive to community development than a previously unoccupied territory, e.g., a cutover forest.
3. **Autogenic succession:** When ecological succession is the product of the organisms themselves, it is called Autogenic Succession.
4. **Allogenic succession:** When succession occurs due to external forces like fire or flood, affect changes is termed allogenic succession.

16.24.3 Significance of Ecological Succession

The basic principles of ecological succession are of immense help in aforestation and range management.

16.25 BIODIVERSITY

Definition: The totality of genus, species and ecosystem of a region is termed as biodiversity.

16.25.1 Levels of Biodiversity

Genetic diversity

Definition: The diversity in the genetic organisation of a species is termed as genetic diversity.

Causes:

- Difference in alleles (different variants of same genes)
- Difference in entire genes.
- Difference in chromosomal structure.

Advantages:

- Better adaptation to environment.
- Better response to natural selection.

Result of lower diversity:

- Monoculture of genetically similar plants.
- Eradication of entire crop, when an insect or fungal disease attacks.

Species diversity

The richness of a species in an ecosystem is called species diversity. In nature, the number and kind of species and number of individuals per species differ. This leads to more species diversity.

Community and ecosystem diversity

It relates to variety of habitats, biotic communities and ecological processes in the biosphere.

16.25.2 Aspects of Ecosystem Diversity

1. **α-Diversity (Within–community diversity):** It includes the diversity of organisms sharing the same habitat or community.

2. **β-Diversity (Between–community diversity):** It is the replacement of species, while moving from one habitat to another within a given geographical area. Communities differ in species composition along environmental gradients of altitudinal gradients, moisture gradient, etc. Greater the dissimilarity between communities, higher is the β-diversity.

3. *v*-**Diversity:** It refers to replacement of species between similar habitats in different geographical areas.

16.25.3 Value/Importance of Biodiversity

1. It maintains ecosystem services like fixation of solar energy, protection of soil, regulation of water flow, recycling of nutrients.
2. It is a source of food.
3. It is a source of medicines.
4. It provides aesthetic value.
5. It is economically important in ecotourism.
6. Biodiversity is linked with human population through cultural and religious beliefs.
7. It has many social uses like firewood, mining, etc.

16.25.4 Categories of Species

1. **Extinct:** A particular species is considered extinct when the last surviving member dies, and has not been seen in wild from last 50 years. Example dodo.
2. **Endangered:** Those species which are under the threat of extinction, and whose survival is unlikely if the causal factors continue to operate. Their number has reduced beyond critical level and their habitat reduced severely, e.g., Great Indian bustard, one horned Rhinoceros, Salamander, Cycas Beddomei, Ghariyal, Pelicans.
3. **Vulnerable:** Species whose population has still not reduced, but still face the threat of extinction as the casual factors especially reduction in habitat, e.g., Asiatic wild ass, golden langur.
4. **Rare:** These have small populations in the world, and are confined to limited areas or thinly distributed over a wider area, e.g., Asiatic Pheasants.

Other kinds of species

1. **Endemic species:** These are specific to a certain region or place. This makes them unique, e.g., Penguin, Panda.
2. **Keystone speices:** Those species which influence the ability of many other species to survive in a community are keystone species.
3. **Exotic species:** Any species, which is not a natural inhabitant of the locality, but accidentally or deliberately introduced into a geographical region. Their introduction can cause disappearance of native species through changed biotic interactions.

16.25.5 Biodiversity Conservation Measures

I. Ex-situ conservation

It refers to conservation of species in suitable locations outside their natural habitat. Ex-situ conservation needs arise, when the population of a species is so fragile/fragmented that its survival in the wild may no longer be possible.

Some popular ex-situ conservation measures are:

- seed banks, gene banks, germ plasm reserves
- long term captive breeding
- short term propagation and release
- animal translocations and reintroduction
- artificial insemination
- cryo preservation of gametes and embryos
- botanical gardens and zoos.
- tissues culture banks
- pollen storage.

II. In-Situ conservation

The preservation of species in its natural eco-system is termed In-situ conservation. These can be done by declaring certain areas as protected, like national parks, sanctuaries and biosphere reserves.

National parks: It is an area which is strictly reserved for the betterment and conservation of wild life, and where activities like forestry, grazing and cultivation are not permitted. Its financial maintenance is done by the central government. Its boundaries are well marked, circumscribed and a lot of warning placards imposed. It spreads over 100–500 sq. km Tourist activity and limited biotic interference on the outskirts (buffer zone) may be permitted. Any interference in core area is not permitted in any case (Figure 15.20).

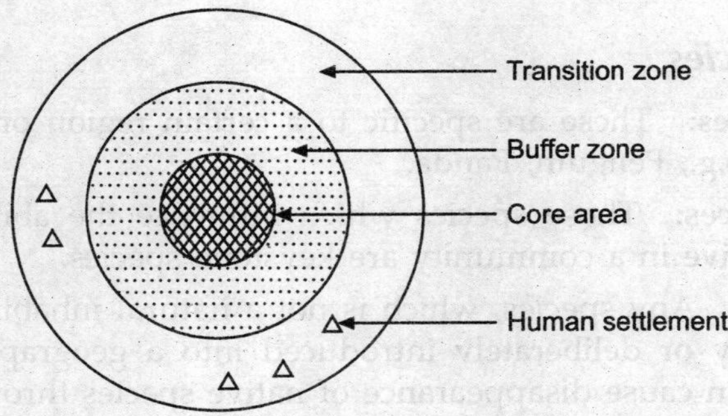

FIGURE 16.20 Zone in protected areas.

Sanctuaries: A sanctuary is a protected area, reserved for the conservation of wild life and is maintained by state government with assistance from central government. The boundaries are not wll defined and controlled biotic interference like tourism is allowed. Some human activities by native people and private ownership rights are allowed by taking permission from the forest authority on the condition that these activities would not affect the wild life.

Biosphere reserve: These are large protected areas spanning more than 5000 sq. km. The boundaries are governed by legislation. Multiple land use is permitted by dividing it into certain zones, each specific for a particular activity. The concept of biosphere reserves was launched in 1975, under UNESCO's man and biosphere programme with the aim to preserve the representative sample of entire biotic spectrum of the locality. The three main zones of biosphere reserve are:

1. *Outer zone (Manipulation zone):* It is the outermost periphery of the reserve, where human activities are allowed.
2. *Buffer zone (Transition zone):* It is the middle region where limited human activity including research and educational activities are permitted without affecting the wild life.
3. *Core zone (Central zone):* It is the inner most, undisturbed and legally protected area, where no human activity is allowed.

16.25.6 Hot Spots of Biodiversity

Norman Myers in a paper in 1988, first identified ten tropical forest "hotspots" characterized by both exceptional levels of plant endemism and a serious loss of habitat. In 1990, Myers added eight more hotspots. Conservation International in 1989, adopted Myer's hotspots as its institutional blue print and in 1999, an extensive global review was carried out which resulted in the adoption of quantitative limits for the designation of a region as a hotspot of biodiversity.

Criteria to be designated as a biodiversity hotspot

(i) The region must have at least 1,500 species of vascular plants (i.e., > 0.5% of the world's total) as endemics.
(ii) The region must have lost at least 70% of its original habitat.

Key statistics

- The hotspots combined together host about 1,50,000 plant species as endemics, which is about 50% of the world's total.
- More than half of the planet's species are endemic to only 16 per cent of its land area.
- About 11,980 terrestrial vertebrates i.e., 42% of total are endemic to hotspots.
- Nearly 22,022 terrestrial vertebrate species, accounting for 77% of the world's total are endemic to these hotspots.

16.25.7 List of Hotspots of Biodiversity

So far 34 hotspots of biodiversity have been identified as listed below by region:

(I) North and Central America

1. California Floristic Province
2. Caribbean Islands
3. Madrean Pine-Oak Woodlands
4. Mesoamerica

(II) South America

5. Atlantic Forest
6. Cerrado
7. Chilean Winter Rainfall Valdivian Forests
8. Tumbes Choco-Magdalena
9. Tropical Andes

(III) Europe and Central Asia

10. Caucasus
11. Irano-Anatolian
12. Mediterranean Basin
13. Mountains of Central Asia

(IV) Africa

14. Cape Floristic Region
15. Coastal Forests of Eastern Africa
16. Eastern Afromontane
17. Guinean Forests of West Africa
18. Horn of Africa
19. Madagascar and the Indian Ocean Islands
20. Maputal-Pondal and Albany
21. Succulent Karoo

(V) Asia–Pacific

22. East-Melanesian Islands
23. Himalaya
24. Indo-Burma
25. Japan
26. Mountains of South-West China

27. New Caledonia
28. New Zealand
29. Philippines
30. Polynesia–Micronesia
31. South-West Australia
32. Sundaland
33. Wallacea
34. Western Ghats and Sri Lanka

16.26 POPULATION GROWTH

If population having initial size–N_0, increases to size N_t after time t, then the change in population size is given as,

$$N_t = N_0 + B + I - D - E$$

where

N_0 = Initial population
B = Natality rate
I = Immigration
D = Mortality rate
E = Emigration

Various theories have been proposed to describe population growth. Two most popular theories are: Malthus theory and Logistic growth theory.

16.26.1 Malthus Theory

Background: Thomas Malthus an English economist, in 1798, published his theory as "An Essay on the Principle of Population" to explain human population growth.

Postulates:

(i) Food is essential for the existence of man.
(ii)· Passion between sexes is natural and necessary, and shall remain nearly in its present form.

Statement: "Population increases by Geometric Progression (2, 4, 8, 16, 32, 64, ...) whereas, the means of its existence increases by Arithmetic Progression (2, 4, 6, 8, 10, 12, ...)."

Explanation: This theory advocates that population growth overtakes food supplies in due course of time and is limited by poverty, starvation, famines, diseases, calamities, wars, etc. According to Malthus, moral restraint was the primary solution to human population growth. He suggested 'Preventive Checks' (postponement of marriage and restraint on reproduction) and 'Positive Checks' (diseases, epidemics, natural calamities, wars, etc.) to check human population growth.

16.26.2 Objections to Malthus Theory

(i) Marx objected that there could be no universal law of population. The overpopulation was not due to human biological power of reproduction, but due to capitalistic mode of production.

(ii) There are many historical cases to point that the means of subsistence of a population (food) has increased faster than the population itself, more so due to agricultural revolutions, improved technologies and greater control exercised over cropping techniques.

(iii) The huge growth of population in underdeveloped and developing countries is more accounted for by fall in death rates, and not increase in birth rate as Malthus stated. This may be attributed to access to better medical facilities.

(iv) Many scientists believe that Malthus may have over generalized a particular case study.

16.26.3 Logistic Growth Theory

Pierre Verhulst's population theory in 1830's, attempted to quantify the trends in population growth.

Sigmoid (S-shaped) growth curve

Theory: Population normally grows in an orderly manner. It has a slow initial growth rate (owing to acclimatization and struggle for survival), which increases exponentially until it reaches a maximum (due to favourable conditions and then progressively becomes less as the population approaches an upper limit of its growth. This slow down is a result of increasing environmental resistance. This resulting curve is a predictable S-shaped curve.

The S-shaped sigmoid population growth curve (Figure 16.21) represented by following equation:

$$\frac{dN}{dt} = \frac{rn(K-N)}{K} = rn\left(1-\frac{N}{K}\right)$$

where

$\dfrac{dN}{dt}$ = Rate of change in population size

r = Biotic potential

N = Population size

K = Carrying capacity

$\left(\dfrac{K-N}{K}\right)$ = Environmental resistance

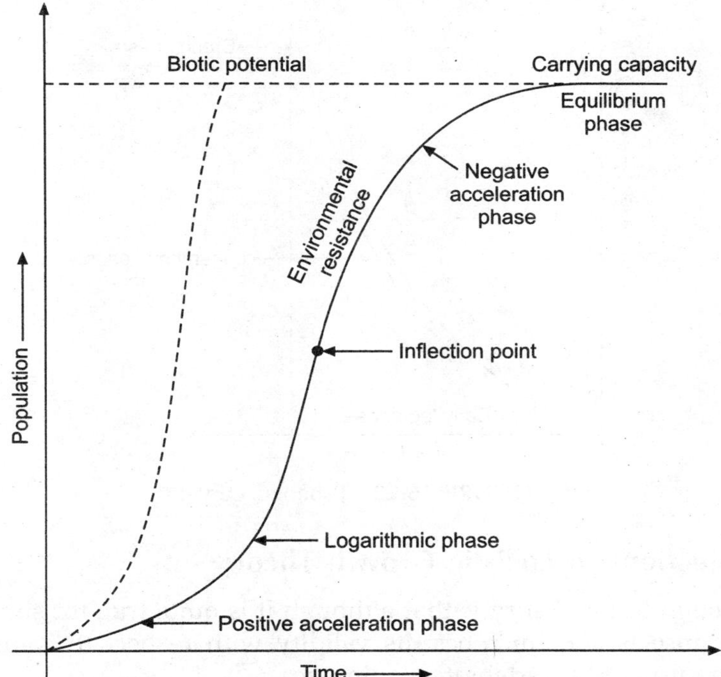

FIGURE 16.21 S-shaped population growth curve.

J-Shaped Growth Curve

In J-shaped growth, the population grows exponentially and after attaining a maximum/peak value, the population may abruptly crash. The ever changing environment and limitations of food and space, restrict the exponential growth.

For example, many insect populations show explosive increase in their numbers during the rainy season, and disappear at the end of the season. The following exponential equation represents the growth in J-shaped curve (Figure 16.22).

$$\frac{dN}{dt} = rN$$

where

$\dfrac{dN}{dt}$ = Rate of change in population size

N = Population size

r = Biotic potential

Figure 16.23 shows growth of Indian Population vis-a-vis world's population until 21st century.

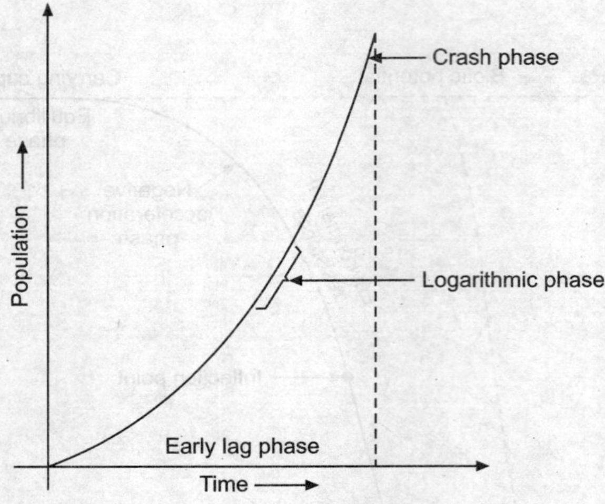

FIGURE 16.22 J-shaped curve.

16.26.4 Objections to Logistic Growth Theory

The main objection to the theory is that although it is quite true for simpler and lower organisms like insects and microbes, its validity with respect to complex life forms and higher organisms like vertebrates is limited.

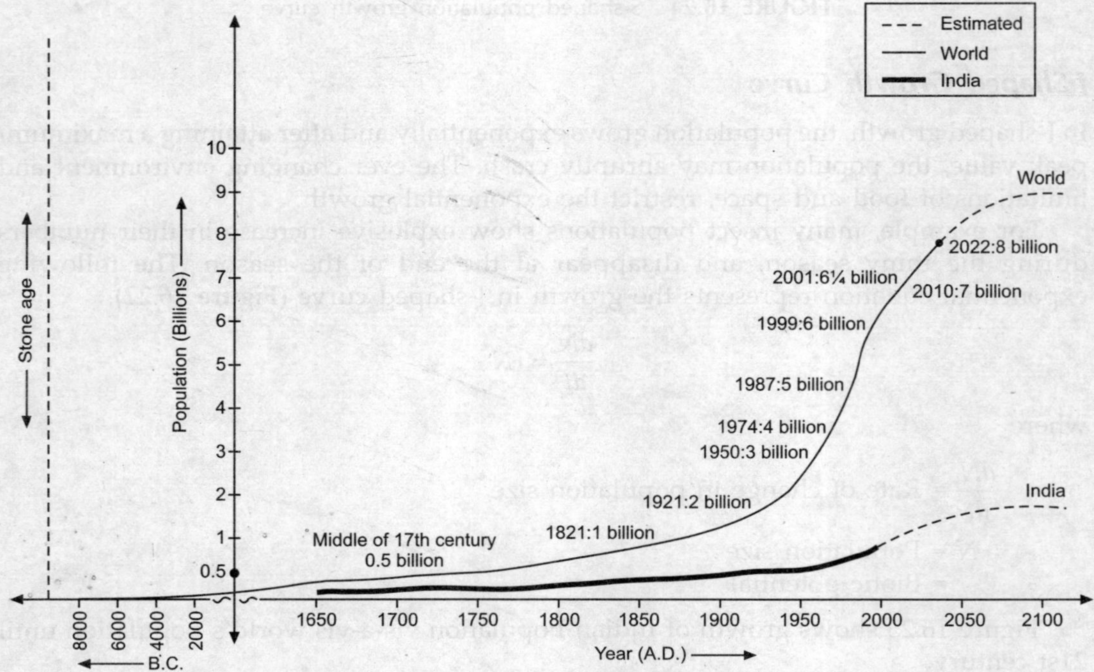

FIGURE 16.23 State of World and India's Population by 2100 AD.

16.27 IMPORTANT DAYS W.R.T. ENVIRONMENT

S. No.	Date	Significance
1.	2nd February	World wetland day
2.	28th February	National Science day
3.	21st March	World forest day
4.	7th April	World health day
5.	22nd April	Earth day
6.	31st May	Anti-tobacco day
7.	5th June	World environment day (UN conference as human environment, stockholm 5th–6th June, 1972)
8.	11th July	World population day
9.	16th September	World ozone day
10.	1st October	World wild life day (1st–7th Oct. wild life weak)
11.	3rd October	World nature day
12.	13th October	UN International day for lessening natural disaster
13.	16th October	World food day
14.	24th October	United nations day
15.	19th Novemeber-to- 18th December	National environmental awareness day (month)
16.	2nd December	National pollution prevention day
17.	3rd December	World conservation day
18.	10th December	Human rights day.

REVIEW QUESTIONS

1. Explain the multidisciplinary nature of environmental studies.
2. Briefly explain the components of environment.
3. Write short notes on:
   (a) Bio-geochemical cycles
   (b) Ten percent law
   (c) Ecological pyramids
   (d) Food web
   (e) Types of food chains.
4. Describe the characteristic features, structure and function of the grassland ecosystem.
5. Discuss the structure and function of a desert ecosystem.
6. Describe the characteristic features, structure and function of aquatic ecosystem.

7. Explain in detail:

    (a) Ecological succession
    (b) Hot spots of biodiversity
    (c) In-situ and Ex-situ conservation measures to conserve biodiversity.

8. Discuss the importance of glaciers.

9. Who governs the Antarctica? Explain in detail.

10. Define the following terms:

    (a) Extinct species
    (b) Endangered species
    (c) Vulnerable species
    (d) Rare species
    (e) Endemic species.

11. Explain: National park, wildlife sanctuary and biosphere reserve. Enumerate the differences between them.

12. Define pollution. Describe the causes, effects and control measures of air pollution.

13. Explain the effect of air pollutants on plants, materials and human health.

14. Write short notes on:

    (a) Greenhouse effect
    (b) Global warming
    (c) Effects of climate change.

15. Explain the following with well-labelled diagrams:

    (a) Bag house filters
    (b) Electrostatic precipitators
    (c) Cyclone separators
    (d) Spray towers
    (e) Gravitational settlers

16. Write an essay on ground water depletion in India.

SMART GRID

Smart Grids are necessary for existing grids to accept power injections from all distributed renewable sources to enhance economy and efficiency of the operating electric power networks.

GRID

An 'electricity grid' is a network of high voltage transmission lines interconnecting a number of electric power generating stations. A grid enables the generating plants to be pooled, thereby reduction of reserve capacity, allowing the cheapest generating plant to be used to meet the load demand.

STATE GRID

At the time of independence (1947), generating capacity in the country was only 1362 MW. To ensure faster development electricity (supply) Act 1948 came into being to allow formation of Electricity Board in each state. After formation of State Electricity Board in various states, first step was to interconnect various generating stations within state to form 'Stage Grid'. In initial stages, operating voltage of State Grid was 66 kV and 132 kV.

Indian power system grew as major hydro-electric projects were commissioned. Accordingly EHV (extra high voltage) transmission lines were commissioned (220 kV in 1959 and 400 kV is 1977).

REGIONAL GRIDS

With the formation of extensive transmission network, and increase in power generation capacity, it was realised that few states were surplus of electrical energy while others were deficient. To cope with this situation, concept of interconnecting grids of neighbouring states was promoted to form 'Regional Grids'. Today India has five large independent regional grids viz., Northern, Western, Southern, Eastern and North Eastern (see Figure A.1).

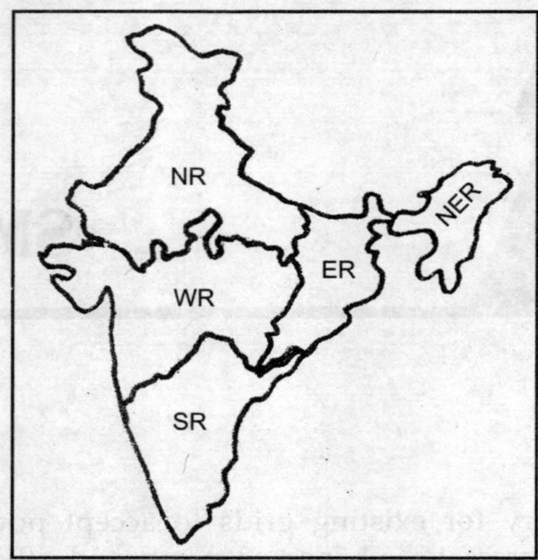

FIGURE A.1 Regional boundaries of Indian power system.

To cope with regional power shortages, inter-regional power transmission is designed. Present generating capacity of the country is 331117.58MW as on 31/10/2017. As generating capacity grew so, transmission system voltage had to be increased to cope with bulk power transmission over long distances to minimize line losses. These voltages with circuit kilo meter (ckm) as on 31/10/2017 are:

1. 765 kV = 33286 ckm
2. 400 kV = 166710 ckm
3. ±800/500 kV HVDC = 15556 ckm
4. 220 kV = 165419 ckm

Total transmission lines in the 5 regional grids = 380971 ckm
Addition during 13th plan (up to Oct. 2017) = 13120 ckm
Expected addition during 13th plan (2017–2022) = 130,000 ckm

EXTENSION TO SAARC GRID THROUGH HVDC SYSTEM

In order to harness capacities and resources of SAARC Nations and to address the growing energy demands a strong SAARC grid is envisioned by the South Asian countries. HVDC transmission provides necessary features that would provide system stability and assist in prevention of cascaded disturbances. SAARC countries are: India, Sri Lanka, Nepal, Bhutan, Bangladesh, Pakistan and Afghanistan.

TRADITIONAL POWER GRID

Present power grid constitute the following:

- Centralized bulk generation
- Heavy reliance on coal and oil
- Limited automation
- Consumers lack data to manage energy usage

Power grid is represented by Figure A.2.

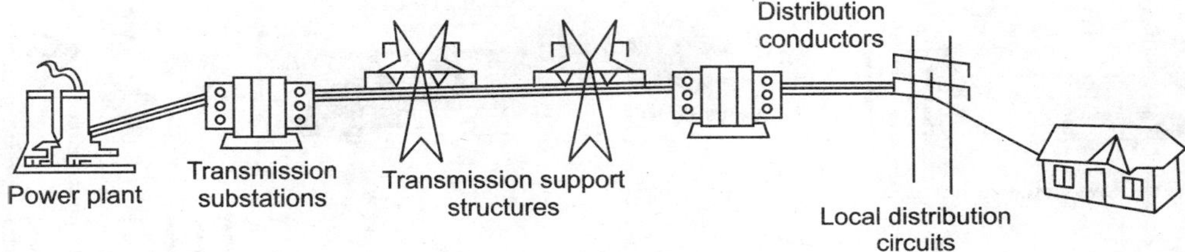

FIGURE A.2 Power Grid Representation.

What is Smart Grid?

According to US Department of Energy it is defined as:

"Smart grid" generally refers to a class of technologies that people are using to bring utility electricity delivery systems into the 21st century, using computer-based remote control and automation. These systems are made possible by two-way digital communication technologies and computer processing that has been used for decades in other industries. These are to be used in electricity networks, from the power plants and wind farms all the way to the consumers at homes and businesses. This offers many benefits to utilities in terms of energy efficiency and reliability. Smart Grid is represented in Figure A.3.

Smart Grid Functions are:

1. Improving Power Reliability and Quality
   - Better monitoring using sensor networks and communication
   - Better and faster balancing of supply and demand
2. Minimizing the need to construct Back-up (Peak Load) Power Plants
   - Better demand side management
   - By use of advanced metering infrastructures
3. Expanding deployment of Renewable and Distributed Energy Sources like solar, wind, biomass and small hydro turbines. Renewable energy generation can be interconnected at local level allowing residential and industrial consumers to self generate and sell excess power to the grid.
4. Automating maintenance and operation by:
   - Using sensor networks and communication
   - Distributed grid management and control

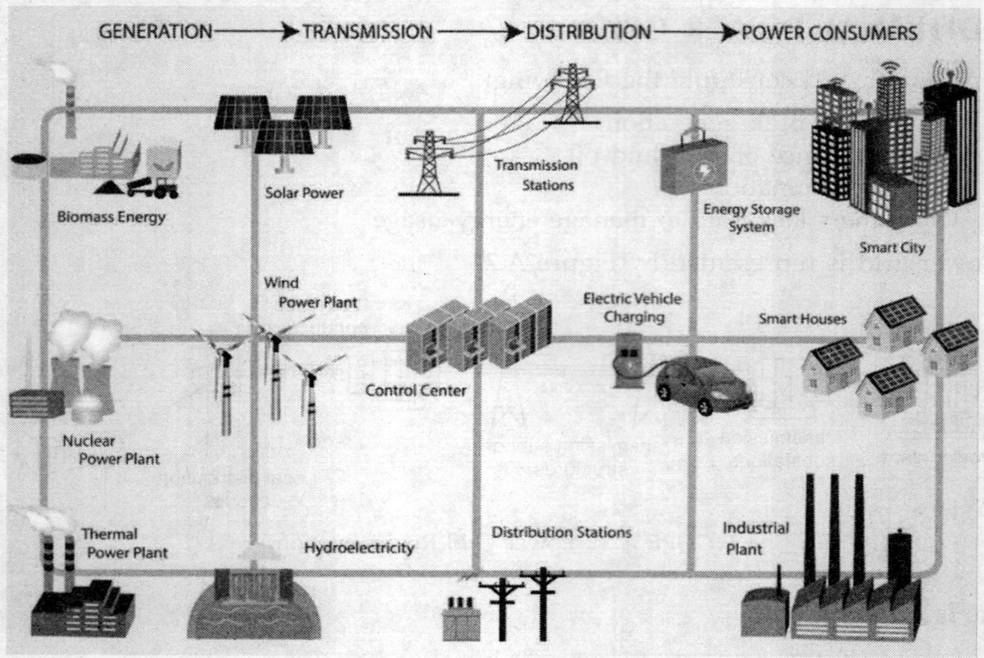

FIGURE A.3 Smart Grid Representation (Notional).

5. Reducing green house gas emissions by:
   - Encouraging the use of electric vehicles
   - Renewable power generation with low carbon footprint

6. Increasing consumer choice by:
   - The use of advanced metering infrastructure
   - Home automation
   - Energy smart appliances
   - Better demand side management

7. Energy theft prevention
   - A small grid replaces analog mechanical meters with digital meters and record usage in real time.

Conclusion: Smart grid would become common and distribution network will become bi-directional. There will be millions of energy producer (consumers) who would produce electricity using solar roof-tops and feed the excess to the distribution togrids. Vertical power companies with no asset of their own, would come into being. They would collect power generated from Individual roof-tops and renewable power plants and sell to other consumers. This will be the "Uber model" in the electricity sector. Companies would also emerge to install big utility scale electricity storage device and provide electricity to the grid for balancing, peaking and even base load requirement.

REMOTE VILLAGE ELECTRIFICATION BY RENEWABLE ENERGY

India constitutes 664,369 villages (2019) and Haryana is the first state to attain 100% electrification with 6764 villages. Programme of electrification in other states is in progress, so far approximately 598,000 villages were electrified until 2020, besides villages, electrification of hamlets is also in progress. Remote village electrification programme (RVEP) is for those unelectrified villages and hamlets which are not approachable by state grid power supply. Special attention is being given to the electrifications of 8 states in the North Eastern Region of the country, namely Arunachal Pradesh, Assam, Manipur, Meghalaya, Mizoram, Nagaland, Sikkim and Tripura. Few achievements are:

ARUNACHAL PRADESH

- 1306 villages (as per 2015 data) electrified by completing 114 small Hydro Power Projects
- 523 very remote border villages electrified by providing each household with one set of solar Home Lighting System.

OTHER STATES

- 100kW SPV power plant at Legislative Assembly Shillong installed in Meghalaya.
- SPV power plants having total capacity of 49 kWp were installed at various government buildings in Mizoram.
- In government institutions of Sikkim, SPV power plants with total capacity of 115 kWp installed and commissioned.

So far, 3494 remote villages and hamlets have been provided renewable energy based system and projects in another 119 remote villages and hamlets are under implementation in various North Eastern States.

Appendix C

INDIAN ELECTRICITY ACT 2003 SUPPORTS DEVELOPMENT OF RENEWABLE ENERGY

Electric power development in India began with the commissioning of a small hydroelectric plant of 200 kW in Darjeeling in 1897 to regulate generation of electricity, transmission, and distribution to consumers. The first Indian Electricity Act 1910 came into force with effect from 1st January, 1911. During preindependance period, power supply utilities were set up in private sector to meet load demand in urban areas.

After independence, Indian Electricity Supply Act 1948 was promulgated to form Electricity Boards in each state, to promote co-ordinated development of generation, transmission, and distribution of electric power to enhance industrial and agricultural production. Central Electricity Authority (CEA) was formed to evolve National Power Policy, formation of Regional and National Grid to rationalise power development in the country.

Electricity Supply Act 1948 was amended in 1976 for the formation of National Thermal Power Corporation (NTPC), National Hydro Power Corporation (NHPC), and Nuclear Power Corporation (NPC) under Govt. of India. To facilitate transfer of electric power within and across the regions, Power Grid Corporation of India started functioning.

Till the enactment of Electricity Act 2003, the renewable energy development was governed by the policies framed by State and Central Givernment. Enactment of Electricity Act, 2003 has brought out radical changes to legal and regulatory framework applicable to renewable sector, as it has provisions for promotion of renewable energy technologies including wind, biomass, small hydro, solar and ocean energy.

The Act provides for policy formulation by Govt. of India and mandates State Electricity Regulatory Commissions (SERC) to promote renewable sources of energy within their area of jurisdiction. Progressively the share of electricity from renewable and non-conventional sources of energy would be increased. Provisions of Electricity Act relevant to renewable energy are:

- To promote stand-alone and Renewable Energy (RE) for rural areas.
- Determination of tariff to consider promotion in electricity from RE sources.
- State electricity regulatory authorities to specify minimum percentage of Renewable Energy consumption.
- Central Electricity Regulatory Commission (CERC) set up to deal with licensing issue for power trading and inter state transmission and tariff issues.
- Efforts be made to reduce capital cost of projects based on renewable sources of energy.
- Generation and Distribution of electricity in rural areas has been de-licencesed.
- State Electricity Regulatory Commission (SERC) be constituted in each state to deal with tariff and licensing issues.

WAIVING OF TRANSMISSION CHARGES FOR SOLAR ENERGY

The electricity Act 2003 mandates the Central Electricity Regulatory Commission (CERC) to determine tariffs including transmission tariffs. The CERC has notified its regulations on 'Sharing of Inter State Transmission Charges and Losses' wherein no transmission charges for the use of Inter-State-Transmission System (ISTS) network shall be charged on solar based generation for the useful life. Also Cabinet Committee of Economic Affairs (CCEA) has approved the creation of ISTS in the states of Andhra Pradesh, Gujarat, Himachal Pradesh, Karnataka, Madhya Pradesh, Maharashtra and Rajasthan costing about ₹ 8548.68 crore with Govt. of India and contribution from the National Clean Energy Fund (NCEF) of ₹ 3419.47 crore.

BIBLIOGRAPHY

Abbasi, S.A., and N. Abbasi, *Renewable Energy Sources and Their Environmental Impact*, Prentice-Hall of India, 2001.

Annual Reports Ministry of New and Renewable Energy (MNRE), Govt. of India, 2002–03, 2003–04 and 2007–2008.

Bhattacharya, T., *Terrestrial Solar Photovoltaics*, Narosa Publishing House, 1998.

Bureau of Energy Efficiency, 2003, *Applications of Non-Conventional & Renewable Energy Sources*.

Chauhan, D.S., and S.K. Shrivastava, *Non-Conventional Energy Resources*, New Age International, New Delhi, 2004.

Chokroverty, A., *Biotechnology & Other Alternative Technologies*, Oxford & IBH Publishing, Kolkata, 1995.

Duffie, J.A., and W.A. Beckman, *Solar Engineering of Thermal Processes*, John Wiley, USA, 1980.

Garg, H.P., *Solar Energy*, TMH, New Delhi, 1997.

Integrated Energy Policy, Planning Commission, Govt. of India, August, 2006.

Kashkari, C., *Energy Resources, Demand & Conservation with Special References to India*, TMH, 1975.

Kothari, D.P., *Power System Engineering*, TMH, New Delhi, 2004.

Maheshwar Dayal, *Energy Today and Tomorrow*, Publication Division, Govt. of India, 1989.

Mangal, B.S., *Solar Power Engineering*, TMH, 1993 (Fourth Reprint).

Mathur, G.N., and D.K. Jain (Editors), *Renewable Energy International Conference 2004*, Central Board of Irrigation & Power, New Delhi, 2004.

Meinel, A., and M. Meinel, *Applied Solar Energy*, Addison-Wesley, London, 1977.

Ministry of New and Renewable Energy, Govt. of India, Annual Report 2005–2006 and 2007–2008.

Mittal, K.M., *Renewable Energy Systems*, Wheeler Publishing, New Delhi, 1997.

Mukherjee, D., *Renewable Energy Systems*, New Age International, New Delhi, 2004.

Rabel, A., "Comparison of Solar Concentrators", *Solar Energy*, 18(1976), 93.

Rao, S., and B.B. Parulekar, *Energy Technology: Non-conventional, Renewable and Conventional*, 3rd ed., Khanna Publishers, New Delhi, 1999.

Ravinder Nath, Usha Rao, Bhaskar Natranjan and P. Monga, *Renewable Energy & Environment: A Policy Analysis for India*, TMH Centre for Environmental Education, Ahmedabad, 2000.

Sukhatme, S.P., *Solar Energy*, 2nd ed., TMH, New Delhi, 1996.

Swinbank, W.C., "Longwave Radiation from Clear Skies," *Quart J. Roy. Metorol. Soc.*, 89, 1963, 539.

Tata Energy Research Institute (TERI), *Rural & Renewable Energy Perspective from Developing Countries*, New Delhi, 1997.

Tewari, G.N., *Renewable Energy Resources*, Narosa Publishing House, New Delhi, 2005.

The Institution of Engineers, *World Congress on Sustainable Development Vols. I & II*, Kolkata, 2000.

Twidell, J.W. and A.D. Weir, *Renewable Energy Resource*, ELBSUK, 1986.

Venikov, V.A., *Introduction to Energy Technology*, Mir Publisher, Moscow, 1984.

Verma, C.V.J. (Ed.), *International R&D Conference Water & Energy 2001*, CBI & P, New Delhi, 2001.

INDEX

Abiotic, 420
Adaptability, 439
Aerodynamic operation of wind turbines, 172
Air mass, 51
Air pollution, 425
Airfoil, 169
Albedo, 42, 52
Application of PV systems, 151
Aquatic, 421
Atmospheric pollution, 11, 30
Axial flow turbines, 226

Basic components, 236
Beam radiation, 60
Biochemical conversion, 328
Biodiesel, 322, 344
Biodiversity, 445, 450, 452, 453, 454
Bio-fouling, 315
Biofuels, 22, 325
Biogas, 325, 331
   plants, 324
   technology, 331
Biogas-solar thermal hybrid, 400
Biomass, 346, 347
   burner, 403
   classification, 322
   cogeneration, 339

conversion technologies, 326
energy, 26, 43, 351
gasification, 328
resources, 322
Biomass-diesel hybrid, 398
Biotic, 420
Bitz limit, 177
Black hydrogen, 385
Bulb turbine, 229

Carbon cycle, 412
Central electricity regulatory commission, 28
Circulating fluidized bed combustion (CFBC), 14
Classification of water turbines, 224
Clean coal technologies (CCTS), 14
Climate change, 435, 439
Coal, 11
Cogeneration, 20
Combined cycle gas-turbine (CCGT) plant, 7
Components of a tidal power, 284
Compound parabolic collector, 86
Compound parabolic concentrator, 96
Concentrating-type solar collector, 70
Conservation of energy, 1
Control system for wind farms, 218
Conventional source of energy, 3